FERROELECTRIC THIN FILMS: SYNTHESIS AND BASIC PROPERTIES

Ferroelectricity and Related Phenomena

Editor: **George W. Taylor**, Princeton Resources, Princeton, New Jersey, USA

Associate Editor: **Lev A. Shuvalov**, Institute of Crystallography, Russian Academy of Sciences, Moscow, Russia

Volume 1
An Introduction to the Physics of Ferroelectrics
T. Mitsui, I. Tatsuzaki and E. Nakamura

Volume 2
Sourcebook of Pyroelectricity
S. B. Lang

Volume 3
Ferroelectrics and Related Materials
G. A. Smolenskii, V. A. Bokov, V. A. Isupov, N. N. Krainik, R. E. Pasynkov and A. I. Sokolov

Volume 4
Piezoelectricity
Edited by G. W. Taylor, J. J. Gagnepain, T. R. Meeker, T. Nakamura and L. A. Shuvalov

Volume 5
Medical Applications of Piezoelectric Polymers
Edited by Pierre M. Galletti, Danilo E. De Rossi and Aime S. De Reggi

Volume 6
Defects and Structural Phase Transitions
A. P. Levanyuk and A. S. Sigov

Volume 7
Ferroelectric Liquid Crystals: Principles, Properties and Applications
J. W. Goodby, R. Blinc, N. A. Clark, S. T. Lagerwall, M. A. Osipov, S. A. Pikin, T. Sakurai, K. Yoshino and B. Zeks

Volume 8
The Photovoltaic and Photorefractive Effects in Noncentrosymmetric Materials
B. I. Sturman and V. M. Fridkin

Volume 9
Microwave Dielectric Spectroscopy of Ferroelectrics and Related Materials
J. Grigas

Volume 10
Ferroelectric Thin Films: Synthesis and Basic Properties
Edited by C. Paz de Araujo, J. F. Scott and G. W. Taylor

This book is part of a series. The publisher will accept continuation orders which may be cancelled at any time and which provide for automatic billing and shipping of each title in the series upon publication. Please write for details.

FERROELECTRIC THIN FILMS: SYNTHESIS AND BASIC PROPERTIES

Edited by

Carlos Paz de Araujo
University of Colorado at Colorado Springs, USA

James F. Scott
University of New South Wales Sydney, Australia

and

George W. Taylor
Princeton Resources, Inc.
Princeton, New Jersey, USA

Gordon and Breach Publishers

Australia Canada China France Germany India Japan Luxembourg Malaysia The Netherlands Russia Singapore Switzerland Thailand United Kingdom

 Printed in Singapore.

Emmaplein 5
1075 AW Amsterdam
The Netherlands

British Library Cataloguing in Publication Data

Ferroelectric thin films : synthesis and basic properties.
- (Ferroelectricity and related phenomena ; v. 10)
1. Ferroelectric thin films
I. Araujo, Carlos Paz de II. Scott, James F. III. Taylor, George W. (George William), 1934 -
537.2'448

ISBN 2-88449-189-9 (hardcover)
2-88449-197-X (softcover)

CONTENTS

Series Editor's Foreword vii

1 **Introduction** 1
Carlos Paz de Araujo, James F. Scott and George W. Taylor

I BASIC PROPERTIES

2 **Finite-Size Effects on Phase Transitions in Ferroelectrics** 11
D. R. Tilley

3 **Defect Chemistry, Conduction, and Breakdown Mechanism of Perovskite–Structure Titanates** 47
R. Waser and D. M. Smyth

4 **Ferroelectric-Electrode Interactions** 93
J. M. Bell, P. C. Knight and G. R. Johnston

5 **Polarization Reversals in Ferroelectrics** 135
Yoshihiro Ishibashi

6 **Fast Polarization Reversal Process: Evolution of Ferroelectric Domain Structure in Thin Films** 153
Vladimir Ya. Shur

7 **Electrode Materials for Ferroelectric Thin Film Capacitors and Their Effect on the Electrical Properties** 193
Husam N. Al-Shareef and Angus I. Kingon

II WET CHEMICAL SYNTHESIS AND DEPOSITION

8 **Metal Alkoxides – Precursors for Ferroelectric Materials** 233
N. Ya. Turova and M. I. Yanovskaya

9 **Sol-Gel Science and PE-MOCVD of Dielectric Perovskite Films** 329
Sandwip K. Dey

III PHYSICAL AND CHEMICAL VAPOR DEPOSITION

10 **Sputter-Deposition of Ferroelectric Films and Heterostructures: Analysis of Orientation–Microstructure–Property Relationships and *in situ*, Real-Time Characterization of Growth Processes** 393
Orlando Auciello, Alan R. Krauss and Ken D. Gifford

11 **Multi-Component Oxide Ferroelectric Thin Films by Multi-Ion-Beam Reactive Sputter (MIBERS) Deposition** 427
S. B. Krupanidhi, H. Hu and G. R. Fox

12 **Synthesis of Ferroelectric Thin Films by Organometallic Chemical Vapour Deposition** 485
M. de Keijser, G. J. M. Dormans and P. K. Larsen

13 **Pulsed Laser Ablation-Synthesis and Characterization of Ferroelectric Thin Films and Heterostructures** 525
Orlando Auciello, Rovindra Dat and Ramamoorthy Ramesh

Author Index 567

Subject Index 569

SERIES EDITOR'S FOREWORD

For more than forty years, scientists and engineers have been intrigued with the possibility of exploiting for computer memories the most intrinsic property of a ferroelectric material; namely, the electrical reversibility of its spontaneous polarization.

The initial attempts, in the period 1952 to 1965, to design ferroelectric memories were focused on direct matrix addressing of ferroelectric capacitors simply formed on bulk single crystals or ceramic ferroelectric materials. The lack of a true coercive field, needed for matrix addressing in a ferroelectric and limited endurance (i.e., significant fatigue) in the ferroelectric materials available at that time made this type of ferroelectric memory impractical.

In the period 1965 to 1975, several research groups tried to develop ferroelectric–semiconductor memories in which a thin film of a semiconductor material was deposited on a bulk single crystal or ceramic ferroelectric material. In such an integrated memory, the switching and matrix addressing functions are provided by the semiconductor and the ferroelectric provides the nonvolatile memory function (i.e., permanent retention of the stored information "1" or "0") without the need for any external sustaining power. Though the basic concept was valid, the instability of the semiconductor thin films of that time did not permit a viable memory to be built.

Since 1985 an ever-increasing research and development effort has occurred worldwide to combine ferroelectric thin films with standard CMOS silicon integrated circuit chips. This combination of a thin film ferroelectric with a bulk semiconductor chip has proved to be more practical than the earlier combination of a thin film semiconductor with a bulk ferroelectric. The present effort is concentrated on developing specialized ferroelectric–CMOS silicon i.c. memory chips which take advantage of the nonvolatile memory properties of the ferroelectric. In addition, the very high dielectric constant of the ferroelectric thin film is being developed as the critical component in manufacturing ultra high density dynamic random access CMOS silicon memories. Also, work is beginning on a variety of other types of integrated thin film ferroelectric –semiconductor devices which will utilize the unique piezoelectric, pyroelectric, and electro-optic properties of ferroelectric thin films.

The current market for all types of memories is about \$ 30 billion per year and it is the underpinning of the worldwide computer and communication revolutions. Presently, the two major commercial memory technologies are: (1) magnetic memories (tapes and discs) which are used for read only and nonvolatile memories. Magnetic memories have the important disadvantages of being slow and requiring mechanical movement—and (2) solid state CMOS i.c. memory chips that are used for random access memories.

The new integrated ferroelectric–semiconductor memories have the potential of eventually replacing a large portion of the magnetic memory market and providing unique and significant advantages to CMOS i.c. memories.

In view of the above, it is easy to understand why tremendous interest has grown in the science and technology of ferroelectric thin films and their integration with silicon chips for memory and other applications. This interest is manifest by an ever-increasing number of national and international conferences on the subject. The new journal INTEGRATED FERROELECTRICS was established in 1992 to serve as the publication medium for the rapidly expanding growth of papers in the field.

Given the amount of significant science that has and is being done on ferroelectric thin films and the engineering that is and will be applied to integrating these films into semiconductor memories and other devices with unique properties, it is most appropriate to publish a set of three related books with the general title *Integrated Ferroelectric Devices and Technologies*. Professors Carlos Paz de Araujo and James F. Scott have joined me in editing this volume and the future companion books that will comprise the set. The three of us have been fortunate to be pioneers and still to be actively involved in this very exciting field.

This book, the first of the set, focuses on the synthesis and basic properties of ferroelectric thin films. Following the introductory chapter are twelve chapters written by some of the leading researchers in the field, all of whom are internationally recognized for their many contributions to ferroelectric thin films.

I believe this book will be of great interest and value to physicists, chemists, ceramicists, and electronic engineers in industry and academia who are either currently or expect to be involved in this new emerging technology.

Two companion books are planned for publication in the near future. One will deal with the dielectric, switching, piezoelectric, pyroelectric, and electro-optic properties of ferroelectric thin films. The other will be devoted to integrated thin film ferroelectric–semiconductor devices that utilize these special properties of ferroelectric materials.

It is gratifying to many of us in the ferroelectrics community that after several premature starts, at last these fascinating materials are proving to be viable for commercial computer memories.

George W. Taylor

CHAPTER 1

INTRODUCTION

Carlos Paz de Araujo
Department of Electrical Engineering, University of Colorado at Colorado Springs, Colorado, USA

James F. Scott
Faculty of Science, University of New South Wales, Australia

George W. Taylor
Princeton Resources, Inc., Princeton, New Jersey, USA

In the past ten years there has occurred a paradigm shift in the science and engineering of ferroelectrics, away from dominance of single crystal and bulk ferroelectric devices into the realm of thin films.[1,2] Thin solid films and devices using ferroelectric materials are by no means new; what is new is the full integration of such films into the world of semiconductor integrated circuits.

As early as 1963, Moll and Tarui[3] began investigating the combining together of *bulk* ferroelectric and *thin film* semiconductors to form new types of electronic devices. In these devices, the electrically alterable state of the ferroelectric was used to modify the basic electrical properties of the semiconductor film. Their work and that of others during the period 1966 to 1975, (reviewed in reference 4) was focused on developing a variety of devices including individual and small arrays of adaptive transistors and small analog and non-volatile memory systems. The material combinations that were investigated at that time included semiconductor thin films of either Tellurium (Te), Cadmium Sulphide (CdS), Cadmium Selenate (CdSe) or Tin Oxide (SnO) and bulk ferroelectrics such as single crystal Barium Titanate ($BaTiO_3$) or Triglycine Sulphate (TGS) or ceramic Lead Zirconate Titanate (PZT). This work, while demonstrating the basic validity of the concept, was never able to be commercialized, primarily because of the inherent instabilities of the thin film semiconductor materials that were available at that time.

During the same approximate period (1960-1975), a number of researchers studied the properties of more than a dozen thin film (<10μm) ferroelectrics fabricated by a variety of techniques such as casting, hydrolysis, electrophoresis, melting and vacuum deposition (sputtering and evaporation). A review of this work is contained in Reference 4). In general, these films had poor ferroelectric, dielectric and mechanical properties and were unstable with time, temperature or humidity. An exception was the work of Francombe, Takei and Wu[5] who showed that excellent electrical and even electro-optic properties could be achieved in ferroelectric thin films of Bismuth Titanate ($Bi_4Ti_3O_{12}$) prepared by rf sputtering.

In a paper in 1978, Taylor[6] predicted that the integration of a stable ferroelectric thin film onto a CMOS silicon chip would make possible a high density, non-volatile integrated circuit memory chip. And, that furthermore such memories would find wide application and eventually capture a very large market.

Another seven years would elapse before these predictions would start to be fulfilled. Beginning about 1985, semiconductor memory companies, which drive the technology of the whole semiconductor industry, began looking for higher levels of integration (greater than 1 Gigabit devices) and new functions such as non-volatility with low power and infinite write endurance. With this need, ferroelectrics are, once again, entering the semiconductor world.

The first steps along this path were to select the ferroelectric materials that are the best suited and to develop the optimum ferroelectric thin film fabrication techniques that are compatible with the latest CMOS manufacturing technology.

The need to place exotic materials such as Strontium Titanate (STO), Lead Zirconate Titanate (PZT), and layered Pervoskites into semiconductor processing, has caused solid state physicists to work together with electrical engineers, ceramicists, electrochemists, and materials scientists in some new and demanding ways. New journals (for example, reference 7) and conferences have been spawned, and many established scientific societies have been drawn by or into this new field of science and technology. There has not been, however, a text available for researchers and newcomers which surveys the accomplishments of the last few years and focuses the diverse multi disciplinary areas of this field into a systematic body of knowledge.

In this first volume of a set of three complementary books with the general title, "Integrated Ferroelectrics Devices and Technologies," we have provided the basic introduction to the chemistry, materials science, processing, and physics - of the materials - from which integrated ferroelectrics are made. Other planned books in this series will address areas such as device engineering and applications.

Before we present the contents of this book in this introductory chapter, a quick digression is necessary in order to place the reader closer to the outstanding developments which have occurred in the last three years.

Since 1992, Integrated Ferroelectric Devices have progressively penetrated many areas of semiconductor integrated circuits. Although it is not possible to do justice to all contributions in such a short space, it is appropriate to point out the advances that led to the importance and attention this field is now receiving.

Firstly, the utilization of high dielectric constant Barium Strontium Titanate embedded in GaAs microwave circuits for use in PCS-band phones in Japan demonstrated a spectacular technological solution which only ferroelectric thin-films could have provided.[8] Today, over 90% of all PCS phones in Japan use three ferroelectric based integrated circuits. In the U.S., PCS phones will only start market testing in late 1997. But one can appreciate the value of PCS technology in knowing that the FCC auction for this band was well over U.S. $2 billion. If the Japanese trend continues in the U.S., ferroelectric based microwave monolithic circuits (Fe-MMICs) will grow into a large segment in the GaAs area.

As of the time of publication, the largest fully integrated ferroelectric non-volatile memory has already achieved 1 Mbit. This device was presented by NEC at the ISSCC in San Francisco.[9] When compared with the standard non-volatile semiconductor devices known as EEPROMs, this device shows the full strength of the advantages of using ferroelectrics.

In EEPROMs, the number of erase-write cycles can reach 10^5. Devices using the layered Perovskites (in fully integrated high densities of the order of 256 kbits) show better than 10^{13} erase-write cycles. Write cycles in EEPROMs are of the order of milliseconds per byte, at 12.5V. (This is the smallest programming voltage at the memory cell level.) Ferroelectric Memories (FeRAMs) such as this 1 Mb device shown in Figure 1, operate at 3V and write or read data in their storage cells at 60 nanoseconds. This is a six order of magnitude increase in speed at less than 25% of the voltage. Even the more sophisticated page-erased (not random access) FLASH EEPROMs cannot match these ferroelectric results in endurance, speed and power usage.

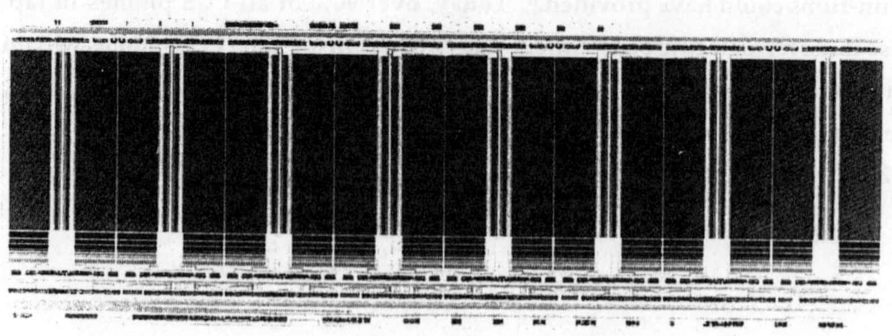

Figure 1. One megabit ferroelectric memory by NEC[9]
(See color plate at the back of the book)

With these two major applications in hand, ferroelectric devices have made a quantum leap from the frustrated attempts of the sixties and seventies, and the laboratory devices of the eighties. But, as if this was not enough, the current race of high density (greater than 1 Gbit) DRAMs are pushing ferroelectric deposition technology to the limit of state-of-the-art 0.2 micron design rules. It is possible to achieve a 300% improvement of the capacitance of each DRAM cell simply by replacing the standard silicon dioxide dielectric with Barium Strontium Titanate (BST). Efforts worldwide in this area have created a flurry of activity in areas such as MOCVD and metallorganic chemistry, which appear throughout this book.

Another major area of application of FeRAMs is in embedded micro controllers. A whole new generation of microcontrollers with FeRAM-stored flexible instruction sets promise to revolutionize the graphics intensive "multi-media appliances" and computers.[10] Also, the millimeter-wavelength device regime of extended antennas, phase-array radar, and related systems will also benefit from the new developments in ferroelectric-on-superconductor integrated thin film devices.[11]

Predictions for the total memory market at the turn of the century is over U.S. $100 billion. When one imagines the new paradigm for the personal computer as a portable "network appliance" linked to the Internet via PCS band digital phones, one can find several places for ferroelectrics. They may be in the microwave circuits, in the non-volatile portion of the memory, in the high density DRAMs and in the microcontrollers that support the multi-media aspects of this "network appliance." If this is so, FeRAMs may be a significant portion of the total memory market. It is, therefore, very timely that this book is being published. Ferroelectric thin-films are still hardly known by the typical semiconductor engineer. Graduate students and professionals in the field will find the chapters in this book self-contained and representative of the best research results up to now.

This book is divided into three parts: Part I - Basic Properties; Part II - Wet Chemical Synthesis and Deposition; and, Part III - Physical and Chemical Vapor Deposition. Each part is self-contained and represents the state-of-the-art, as well as the fundamental issues that affect materials and lead to well behaved device characteristics.

Chapter 2, by Professor D.R. Tilley, describes the effects of scaling ferroelectric materials in the thin-film regime. Professor Tilley makes detailed numerical predictions of the minimum thickness real ferroelectric films can be, without running into degradation of their polarization and Curie temperature.

In Chapter 3, Professors R. Waser and D.M. Smyth combine their well-known expertise to give a succinct yet comprehensive account of the defect chemistry of perovskite titanates, including leakage current and breakdown mechanisms. These two phenomena are the leading causes for concern in the present attempts to commercialize high density memories, and Waser and Smyth have probably been the greatest single contributors in the scientific community to their understanding.

Chapter 4 describes the important interfacial interactions between the ferroelectric films and the metal electrodes. Despite the high quality of ferroelectric films, per se, used in integrated ferroelectrics, it is the electrode interface that often dominates the system as a whole. In this chapter, Dr. J.M. Bell and his colleagues describe the effects that can arise from this interfacial structure.

Chapters 5 and 6 both address switching kinetics in the attempt to answer, among other related concerns, the important questions of what is the ultimate speed of a ferroelectric capacitor, and what are the rate-limiting parameters. In Chapter 6, Professor V. Ya. Shur provides fascinating stroboscopic data on the real-time evolution of domain walls during ferroelectric switching together with evidence for a newly discovered process of coherent nucleation near domain walls. In Chapter 5, Professor Y. Ishibashi offers a new theory, based on the Kolmorgorov-Avrami approach to nucleation and growth, of polarization reversal processes.

Part I concludes with an in-depth study by Dr. H.N. Al-Shareef and Professor A.I. Kingon, of suitable electrode materials for ferroelectric thin-films.

Part II leads off with a comprehensive review of metal alkoxides and their use as precursors for liquid-phase deposition of ferroelectric thin films (sol-gel, MOD, or liquid-phase CVD). This chapter, by Drs. N. Ya. Turova and M.I. Yanovskaya, two women of exceptional talent who pioneered such metal-organic chemistry in the 1970s, contains invaluable information for material synthesis.

Chapter 9, by Professor S.K. Dey, shows how the precursors described in Chapter 8 can be used to form sol-gels and MOCVD precursors. In sol-gel, a spin-on deposition technique is used. This provides a low-cost way of achieving excellent materials. Panasonic uses this technique for commercial GaAs MMICs.[8]

Advanced deposition techniques are required for high density memories and other devices with small capacitor areas. In Part III, three areas are covered in this highly specialized frontier: Sputter Deposition, Metallorganic Deposition (MOCVD), and Pulsed Laser Ablation (PLD).

Chapter 10, by Drs. O. Auciello, A.R. Krauss, and K.D. Gifford shows a careful analysis of the sputtering growth process. The authors are well-known in their achievements in sputter deposition and a thorough review of this process precedes these new results presented by in situ time of flight techniques.

Professor S.B. Krupanidhi and his co-workers, H. Hu and G.R. Fox, describe in Chapter 11 the important subject of Ion Beam Assisted Sputtering. This Penn State team is well-known for developing new techniques using Ion Beams. And this review of their work on multi-ion beam encompasses the state-of-the-art in this area.

In Chapter 12, Drs. M. de Keijser, G.J.M. Dormans, and P.K. Larsen describe the all important area of MOCVD. Their work at Phillips is among the best in the world in this area. MOCVD may be the most important technology to achieve high density devices.

The final chapter, 13, covers the new area of Pulsed Laser Ablation (PLD). Professors O. Auciello, R. Dat, and R. Ramesh present this subject with an in-depth analysis of the stoichiometeric control available via PLD.

Taken as a whole, the chapters in this volume take the reader from the requisite basic chemistry for making these films and the key deposition techniques for fabricating devices from them, through the cell design for integrating them into chips, and the physics and physical chemistry of their performance and failure. We, the Editors, are delighted that the most prominent contributors to this field have taken the time to provide chapters in their special areas of expertise.

REFERENCES

1 Scott, J.F., and Araujo, C.A., Science **246**, pp. 1400 (1989)

2 Paz de Araujo, C.A. and Taylor, G.W., Ferroelectrics, vol. **116**, p. 215 (1991)

3 Moll, J.L., and Tarui, Y., IEEE Trans. Electron Devices **ED10**, 328 (1963)

4 Burfoot, J.C., and Taylor, G.W., *Polar Dielectrics and their Applications,* MacMillan, London/University of California Press, Berkeley, 1979, MIR, Moscow 1981.

5 Francombe, M.H., Ferroelectrics **3**, 199 (1972)
Takei, W.J., Formigoni, N.P., and Francombe M.H., Appl Phys Letters **15**, 256 (1969)
Wu, S.Y., Takei, W.J., and Francombe, M.H., Ferroelectrics **10**, 209 (1974)

6 Taylor, G.W., Ferroelectrics, vol. **18**, 17 (1978)

7 Taylor, G.W., Integrated Ferroelectrics (editorial), vol. 1, number 1 (1992)

8 Nagata, S., Ueda, T., Noma, A., Koizumi, H., Kanazawa, K., Ishida, H., Ueda, D., Kazumura, M., Kano, G., and Paz de Araujo, C.A., Proceedings of the ISSCC, Feb. 10, 1994.

9 Koike, H., Otsuki, T., Kimura, T., Fukuma, M., Hayashi, Y., Maejima, Y., Amamuma, K., Tanabe, N., Matsuki, T., Salto, S., Takeuchi, T., Kobayashi S., Kunio, T., Hase, T., Miyasaka, Y., Shomata, N., and Takada, M., Proceedings of the ISSCC, February 8, 1996.

10 Sumi, T., Azuma, M., Otsuki, T., Gregory, J.W., and Paz de Araujo, C.A., Proceedings of the ISSCC, February 16, 1995.

11 Hermann, A.M., Scott, J.F. etal, Journal of Superconductivity **7**, 463 (1994)

PART I

BASIC PROPERTIES

CHAPTER 2

FINITE-SIZE EFFECTS ON PHASE TRANSITIONS IN FERROELECTRICS

D.R. Tilley
School of Physics, Universiti Sains Malaysia, 11800 USM Penang, Malaysia
and
Department of Physics, University of Essex, Colchester CO4 3SQ, UK.

1. INTRODUCTION

The presence of surfaces or interfaces can influence solid-state phase transitions and the nature of the ordering that occurs in one or other phase. The effect is well known in magnetic insulators, where ions in a surface layer may have a different number of nearest neighbours from those in the bulk, and exchange and anisotropy constants near a surface may differ from the bulk values. For both these reasons, the local magnetization may vary in the vicinity of a surface and phase transitions may occur in which only a surface layer orders. In the last ten years a great deal of attention has been devoted to magnetic superlattices, in which layers of two different materials, one at least being magnetic, alternate. In these, particularly superlattices involving rare earths, magnetization configurations can appear which are determined by the interfaces and are specific to the superlattice structure.[1]

The corresponding phenomenon in ferroelectric materials has been much less studied. However, with an increasing emphasis on ferroelectric thin films in applications, it is becoming more important to understand the physics of phase transitions in finite-size ferroelectrics.

The experimental situation has been reviewed recently[2] and a brief summary will be presented in section 2. Various investigations over some twenty years on a variety of systems using a variety of techniques have produced evidence for surface order, or sometimes disorder, distinct from what occurs in the bulk. However, the results do not lead to any easy generalizations.

Theoretically two rather different mechanisms have been considered. First, in a ferroelectric film with polarization normal to the surfaces the depolarization field depresses the transition to ferroelectricity. The magnitude of the effect depends on any electrodes that have been deposited on the film. If there are no electrodes the phase transition is completely suppressed. Conversely, for the idealised case of perfectly conducting electrodes the polarization change is fully compensated and the critical temperature is unaltered from the bulk value.

The second mechanism envisages an intrinsic modification of the order parameter in the vicinity of a surface or interface. This may be described within the two main models for the thermodynamics of ferroelectrics. The Ising model in a transverse field (IMTF) describes order-disorder ferroelectrics like potassium di-hydrogen phosphate (KPD);[3] the polarization is proportional to the thermal average $\langle S_z \rangle$ of the pseudo-spin component S_z. As happens for other magnetic Hamiltonians, $\langle S_z \rangle$ becomes a function of position near a surface or interface because of alteration in the nearest-neighbour number and perhaps also exchange constants. An alternative and more widely used approach is the Landau-Devonshire theory[4,5] in which the free energy F is expanded as a Taylor series in the order parameter, namely the polarization **P**. For a bulk material, the point-group symmetry determines which terms appear in the expansion. In the presence of a surface additional terms, typically of the form $|\mathbf{n} . \nabla \mathbf{P}|^2$, are allowed by symmetry; these lead ultimately to a boundary condition on **P**. This means that **P** is a function of position, so terms like $|\nabla \mathbf{P}|^2$ are added to the bulk free-energy density. The algebraic equations for **P** that suffice for the bulk are replaced by an Euler-Lagrange differential equation plus boundary condition. Because of its plausibility and tractability the implications of this approach have been studied in some detail, although direct experimental support is largely lacking.

In section 3 we discuss the theory without including depolarization effects; in practice this means we consider a film or surface with in-plane polarization. The emphasis will be on the Landau formalism, although the main results of IMTF calculations will be described. In section 4 depolarization is included within the Landau formalism to give a unified description. Section 5 is concerned with the case when only depolarization operates; the formalism is a special case of the general results of section 4.

2. EXPERIMENTAL RESULTS

The ultrasonic measurements of Höchli and Rohrer[6] on $SrTiO_3$ platelets with thicknesses in the range 100 to 600 μm showed the persistence of an ordered surface layer in a range of one or two degrees above the bulk critical temperature T_c. The upper critical temperature for the surface layer increased with decreasing platelet thickness.

A sustained study, mainly using far-infrared (FIR) transmission on tri-glycine sulphate (TGS) plates, has been carried out by Hadni, Gerbaux and co-workers.[7-13] They find that Beer's law is not obeyed at 4K in TGS plates of a range of thicknesses, that is, the FIR transmission does not decrease with increasing thickness in the way it would if the plates were optically uniform. The full data shows first that Beer's law is obeyed for plates of thickness up to about 13 μm and secondly that Beer's law is obeyed at 80K and above. Gerbaux and Hadni deduce that at some temperature between 80K and 4K the bulk of the plate undergoes a transition into a phase which is transparent in the FIR, but the surface remains in the high-temperature phase. Similar results have been obtained for guanidine aluminium sulphate (GASH)[12] and Rochelle salt,[13] but not for sodium ammonium tartrate.[13]

A much more dramatic difference between bulk and film phase transitions is found in KNO_3. In the bulk, the ferroelectric phase (phase III) is reentrant, observed only between about 113°C and 120°C at ambient pressure. Figure 1 shows the result of a Raman study[14] of the phase diagram of thin films; it is seen that the ferroelectric phase occupies a much greater temperature range than in the bulk.

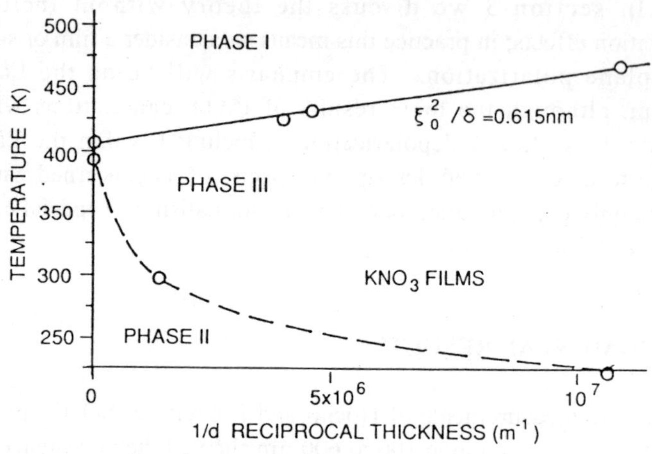

FIGURE 1. Thickness dependence of phase transitions in KNO_3 films. [After ref. 14.]

A number of studies have been carried out on small particles or microcrystals rather than films or platelets. Marquardt and Gleiter[15] found evidence of a lowering of the critical temperature by a few degrees in a DTA analysis of $NaNO_2$ microcrystals of size 5 to 20 nm suspended in PVA foam. Ishikawa et al.[16] carried out a Raman study of ultrafine $PbTiO_3$ particles; by varying the sintering temperature they were able to vary the average particle size. They found that the frequency of the soft optic phonon decreased with decreasing particle size. Figure 2 shows how the transition temperature, defined as the temperature at which the soft-phonon Raman line disappears, decreases with decreasing particle size. The empirical curve passing through the data points is given by a similar expression to that derived subsequently, eq. (65), for the depression of the critical temperature of a film due to depolarization effects.

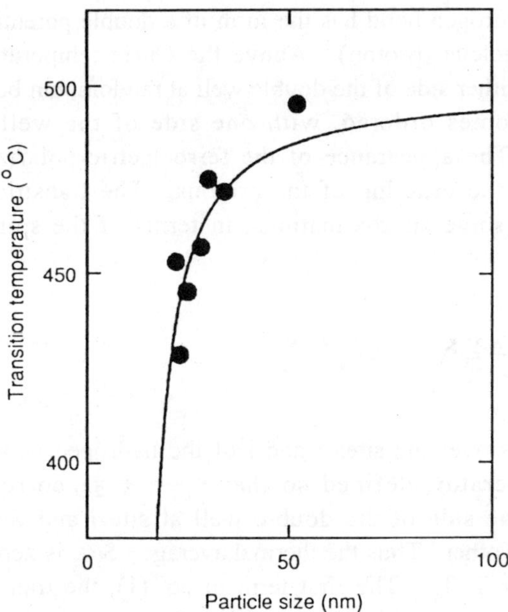

FIGURE 2. Transition temperature T_c versus average particle size D for sintered $PbTiO_3$ particles. The observed values are full circles. The curve is the empirical expression $T_c = 500 - 588.5/(D - 12.6)$ where D is the particle diameter in nm. [After ref. 16.]

3. THEORY: NO DEPOLARIZATION

3.1. Theoretical Models

As mentioned in section 1, the two main models that apply for bulk ferroelectrics are the IMTF for order-disorder ferroelectrics and the Landau form of mean field theory, applicable in principle to both order-disorder and displacive ferroelectrics near the Curie temperature. Here we review these two models and explain the extensions needed in order for them to apply for surfaces and films. More detailed accounts for the bulk materials are given in the standard texts.[3,4]

KDP and similar materials contain hydrogen bonds linking PO_4 groups. Each hydrogen bond has the form of a double potential well for the hydrogen nucleus (proton). Above the Curie temperature T_c the protons occupy either side of the double well at random, but below T_c the occupation becomes ordered, with one side of the well occupied preferentially. The appearance of the ferroelectric polarization **P** is associated with the ordering of the protons. The transition may be described, with some approximations, in terms of the spin $\frac{1}{2}$ IMTF Hamiltonian

$$H = -\Delta \sum_i S_i^x - \frac{1}{2} \sum_{i,j} J_{ij} S_i^z S_j^z \tag{1}$$

where the sum is over the sites i and j of the hydrogen bonds. **S** is a pseudo-spin operator, defined so that $S_i^z = +\frac{1}{2}$ corresponds to occupancy of one side of the double well at site i and $S_i^z = -\frac{1}{2}$ to occupancy of the other. Thus the thermal average $\langle S_i^z \rangle$ is zero for $T>T_c$ and non-zero for $T<T_c$. The first term in eq. (1), the transverse-field term, corresponds to tunnelling of the proton from one side of the double well to the other. The second term describes interaction between neighbouring sites in terms of the effective exchange constant J_{ij}. For the bulk material J_{ij} is usually taken non-zero only for nearest-neighbour i and j and to be independent of site position $J_{ij} = J$.

The mean-field treatment of eq. (1) gives expressions for T_c and the variation with temperature of $\langle S^2 \rangle$, to which **P** is proportional. The tunnelling term is essentially disordering and there is no transition to a ferroelectric phase for

$$\Delta > \frac{1}{2} \sum_j J_{ij} \ .$$

Near a surface or interface Δ and J may become a function of site position and the number of nearest neighbours changes from the bulk value. The delicate problem of calculating Δ and J from a microscopic model has not been tackled, but the consequences of some simple possible changes have been explored. An account will be given in later sections.

The simplest form of the Landau theory for a bulk ferroelectric starts from an expansion of the free energy in terms of the polarization.

$$F = \frac{1}{2}AP^2 + \frac{1}{4}BP^4 + \frac{1}{6}GP^6 \tag{2}$$

where

$$A = A_o(T - T_{co}) \tag{3}$$

and T_{co} is the critical temperature of the bulk material.

For a second-order phase transition $B > 0$ and the term in P^6 can be ignored; for a first-order phase transition $B < 0$ and $G > 0$. The implications of this expansion are discussed in some standard solid-state texts as well as in the more specialized monographs. The terms that may appear in a Landau expansion of the form of eq. (2) are the invariants of the symmetry group of the paraelectric phase.[5]

In the presence of surfaces **P**, like $<S_z>$ in the IMTF, may become a function of position. A term in $|\nabla \mathbf{P}|^2$ is then allowable in the free-energy expansion, together with a surface term, so eq. (2) is replaced by

$$\frac{F}{S} = \int d^3r \left\{ \frac{1}{2}AP^2 + \frac{1}{4}BP^4 + \frac{1}{6}GP^6 + \frac{1}{2}C|\nabla P|^2 \right\} + \frac{1}{2}C\delta^{-1}\int dS(\mathbf{n}.\mathbf{P})^2 \tag{4}$$

where S is the surface area. The integrals are over the bulk and surface respectively of the ferroelectric. The algebraic equation for a minimum in eq. (2) is replaced by an Euler-Lagrange differential equation for $\mathbf{P}(\mathbf{r})$ together with boundary conditions arising from the surface integral. Although eq. (4) is based here on the simplest possible invariants, it can be derived from the IMTF with the assumption of slow spatial variation of $<S^z>$.[17]

3.2. Semi-Infinite Medium

We discuss here a semi-infinite medium with a surface at z = 0, say. The detailed implications of the IMTF depend on the crystal structure, the orientation of the surface with respect to the crystal axes and the site dependence of the Hamiltonian parameters Δ_i and J_{ij}. As an example,[17] we consider a simple cubic array of spins with the coordinate axes parallel to the cube axes and $\Delta_i = \Delta_s$ if site i is in the surface layer and $\Delta_i = \Delta$ otherwise. J_{ij} is taken non-zero only for nearest-neighbour sites i and j, with $J_{ij} = J_s$ if both spins are in the surface layer and $J_{ij} = J$ otherwise. Mean field theory gives

$$R_1^z = \left[(4J_sR_1^z + JR_2^z)/2\gamma_1\right] \tanh(\gamma_1/2k_BT) \tag{5}$$

$$R_n^z = \left[J(4R_n^z + R_{n+1}^z + R_{n-1}^z)/2\gamma_n\right] \tanh(\gamma_n/2k_BT) \quad n\geq 2 \tag{6}$$

where

$$\gamma_n = \begin{cases} \left[\Delta_s^2 + (4J_sR_1^z + JR_2^z)^2\right]^{1/2} & n=1 \\ \left[\Delta^2 + J^2(4R_n^z + R_{n+1}^z + R_{n-1}^z)^2\right]^{1/2} & n\geq 2 \end{cases} \tag{7}$$

Here $R_n^z = \langle S_n^z \rangle$ denotes the spin average in layer n from the surface.

Equations (5) and (6) are coupled nonlinear equations for R_n^z. The critical temperature is found from the linearized equations, which can be solved exactly.[17] For $\Delta_s = \Delta$, it is found that for $J_s < 5J/4$ there is a single phase transition at the bulk critical temperature T_c given by eq. (2). For $J_s > 5J/4$, however, there is a transition at a higher critical temperature T_c^* to a surface-ordered state with $R_1^z \neq 0$ and $R_n^z \to 0$ as $n \to \infty$. This is followed by a second phase transition at T_c to the bulk-ordered state.

The temperature dependence of the layer-spin averages R_n^z is obtained by numerical solution of eqs. (5) and (6). An example is shown in figure 3 for two values of J_s/J. In figure 3(a), eq. (13) gives $T_c^* = 1.32\ T_c$. The figure shows how the phase transition at T_c^* is confined to the first few surface layers. In figure 3(b), $J_s = 0.25J$. It is seen that there is only one phase transition and the small value of J_s, together with the lower coordination number at the surface, leads to a depression of R_1^z at all temperatures.

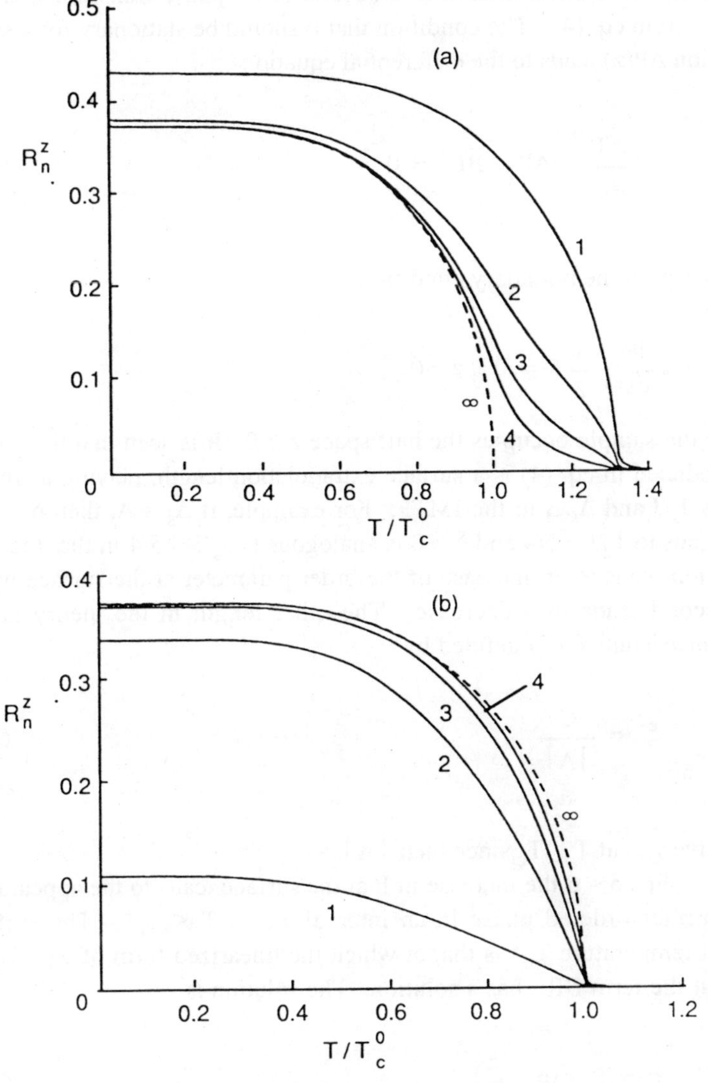

FIGURE 3. Spin averages $R_n{}^z$ plotted as a function of T/T_c for n = 1,2,3,4 (full curves) and n = ∞ (broken curves) for (a) $J_S/J = 1.75$ and (b) $J_S/J = 0.25$. It is assumed that $\Omega_S = \Omega = 2J$. [After ref. 17.]

We now turn to the analogous Landau theory results.[18,19] For simplicity we restrict attention to second-order phase transitions, B > 0 and G = 0 in eq. (4). The condition that F should be stationary for a small variation $\Delta P(z)$ leads to the differential equation

$$C\frac{d^2P}{dz^2} - AP - BP^3 = 0 \tag{8}$$

together with the boundary condition

$$\frac{dP}{dz} - \frac{1}{\delta}P = 0 \text{ at } z = 0 \tag{9}$$

where the sample occupies the half space z > 0. It is seen that the length δ introduced in eq. (4) is a surface extrapolation length, playing a similar role to J_s/J and Δ_s/Δ in the IMTF. For example, if $\Delta_s = \Delta$, then $\delta < 0$ is analogous to $J_s/J > 5/4$ and $\delta > 0$ is analogous to $J_s/J < 5/4$ in that the first condition leads to an increase of the order parameter at the surface while the second leads to a decrease. The other length in the theory is the coherence length ξ (T) defined by

$$\xi^2 = \frac{C}{|A|} \tag{10}$$

This diverges at $T = T_c$ since then $|A| = 0$.

For $\delta < 0$ the increase in P at the surface leads to the appearance of a surface-ordered phase in an interval $T_{co} < T < T_c^*$. The surface critical temperature T_c^* is that at which the linearized form of eq. (8), ie. without the term BP^3, has a solution. The solution is

$$P = P_0 \exp\left(-\frac{z}{\xi}\right) \tag{11}$$

with P_0 arbitrary, and the condition for this to satisfy eq. (9) is

$$\xi(T_c^*) = |\delta| \tag{12}$$

which is the equation for T_c^*.

Solutions of eqs. (8) and (9) can be found at all temperatures[17] and used to draw order-parameter profiles. Examples are shown in figure 4, in which the simple temperature dependences $A = A_0\,(T/T_{co}-1)$, B = constant, C = constant are assumed. Figure 4 may be compared with figure 3, which shows similar effects for a different model.

Wang[20] has pointed out that for $\delta < 0$ eq. (8) has a second solution with $P < 0$ at the surface and $P > 0$ as $z \rightarrow \infty$ with the sign reversal at $z = z_0$; it is identified by Wang as a simple form of domain structure.

3.3. Films

Both the models described in section 3.2 have been applied to films. In this case, since the thickness is finite, it is not possible to have separate surface and bulk phase transitions. It is a question of the critical temperature of the film as a whole, either elevated for $J_s > 5J/4$ or $\delta < 0$, or depressed otherwise.

Results for the IMTF within mean field theory are given by Wang et al.,[21] who solve the linearized form of a finite set of coupled equations like eqs. (5) and (6) to find the Curie temperature T_c. An example of the dependence of T_c on J_s is shown in figure 5. As for the semi-infinite medium, the value $J_s = 5J/4$ is singled out: whatever the thickness of the film, T_c is less than the bulk value for $J_s < 5J/4$ and greater for $J_s > 5J/4$. Sarmento and Tucker[22] have carried out a similar study using a more refined effective field theory.

Landau theory in the form of eq. (4) has also been applied to films.[23,24] It must be emphasized that eq. (4) does not include the effects of depolarization which in fact are crucial when **P** is normal to the film surfaces. The results to be presented therefore apply only when **P** is in plane. For a film with plane surfaces at $z \pm \frac{1}{2}\ell$, the condition for the free energy F of eq. (4) to be stationary is that P should satisfy eq. (8) with boundary conditions

$$\frac{dP}{dz} \pm \frac{1}{\delta} P = 0 \text{ at } z = \pm \frac{1}{2}\ell \tag{13}$$

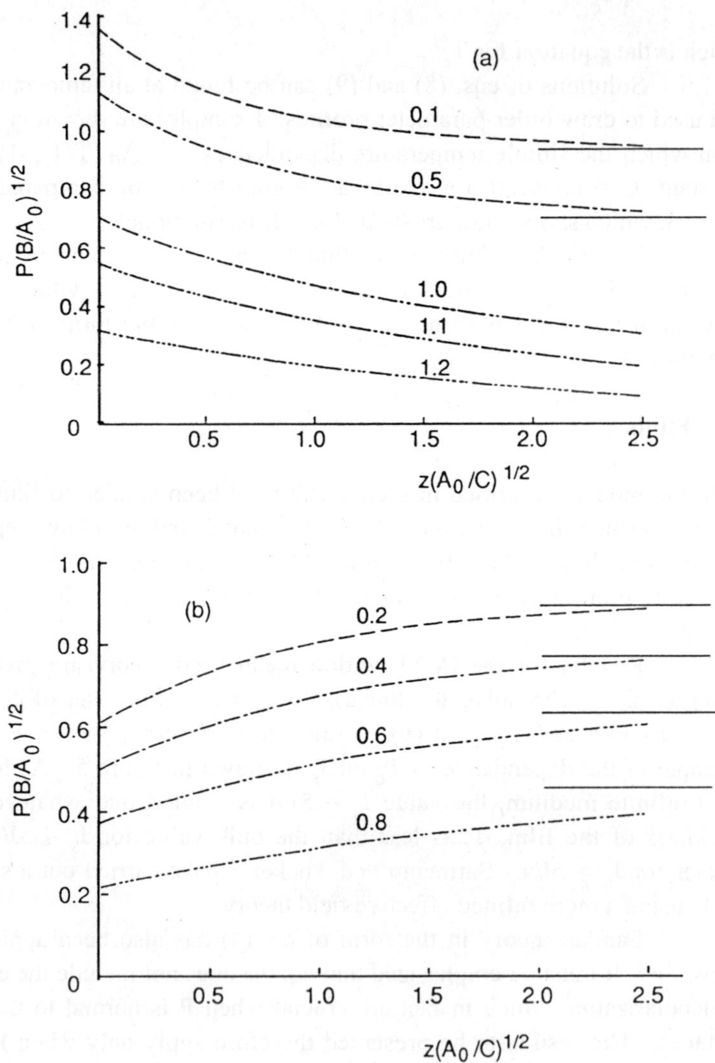

FIGURE 4. Equilibrium polarizations for (a) $\delta\,(A_0/C)^{1/2} = -2$, (b) $\delta\,(A_0/C)^{1/2} = 2$. Values of T/T_c are marked. The asymptotic forms for large z are $P(B/A_0)^{1/2} \sim (1 - t)^{1/2}$ for $t < 1$ and $P \sim 0$ for $t \geq 1$ in (a). [After ref. 17.]

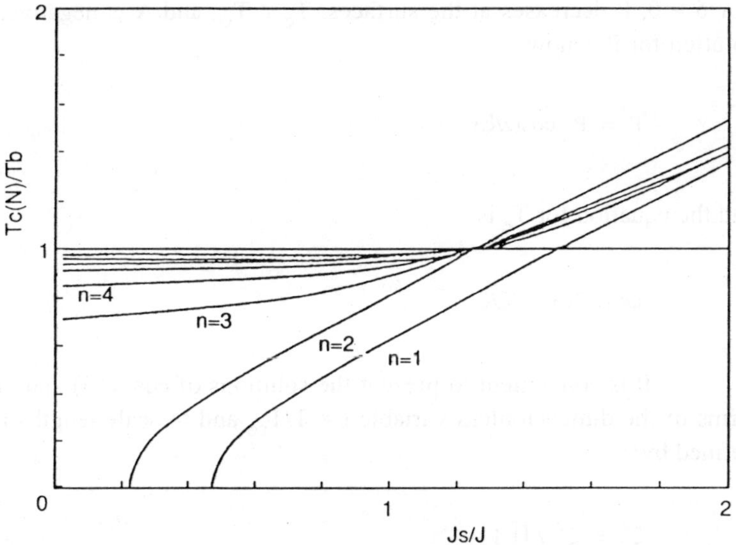

FIGURE 5. Variation of film Curie temperature with surface exchange constant J_S in the IMTF for a simple cubic film with surfaces normal to one of the crystal axes and nearest-neighbour exchange constants J_S for both spins in the surface layer, J otherwise. Tunneling integral $\Delta = J$. Parameter n is the number of layers. [After ref. 21.]

As before, the film critical temperature T_c is given by the condition that the linearized form of eq. (8) together with eq. (13) should have a solution. For $\delta < 0$, P increases at the surfaces, so $T_c > T_{co}$ and the coefficient A in eq. (8) is positive. The solution of the linearized equation is

$$P = P_o \cosh(z/\xi) \qquad (14)$$

where ξ is the coherence length given by eq. (18). Application of the boundary conditions then gives the equation for T_c, contained implicitly in ξ, in the form

$$\tanh(\ell/2\xi) = -\xi/\delta \qquad (15)$$

For $\delta > 0$, P decreases at the surfaces; $T_c < T_{co}$ and A is negative. The solution for P is now

$$P = P_o \cos(z/\xi) \tag{16}$$

and the equation for T_c is

$$\tan(\ell/2\xi) = \xi/\delta \tag{17}$$

It is convenient to present the solutions of eqs. (15) and (17) in terms of the dimensionless variable $t = T/T_{co}$ and to scale lengths by ξ_0, defined by

$$\xi^2 = \xi_o^2 / |t-1| \tag{18}$$

so that $\xi_0{}^2 = CT_{co}/A_o$ where A_o is defined in eq. (3). The asymptotic solutions of eq. (15) for the enhancement of T_c in the case of negative δ are[2]

$$t - 1 \sim 2\xi_0^2/\ell\,|\delta| \quad \text{as } \ell \to 0 \tag{19}$$

and

$$t - 1 \to \xi_0^2 / |\delta|^2 \quad \text{as } \ell \to \infty \tag{20}$$

The first of these shows that T_c increases indefinitely as ℓ decreases. The limit in eq. (20) is independent of ℓ; it is just the surface critical temperature T_c* of eq. (12). For positive δ and a reduction of T_c below T_{co} the asymptotic solutions of eq. (17) are

$$1 - t \sim 2\xi_0^2 / \ell\delta \quad \text{as } \ell \to 0 \tag{21}$$

and

$$1 - t \sim \pi^2 / \xi_0^2 \ell^2 \quad \text{as } \ell \to \infty \tag{22}$$

The first limit is purely formal since t cannot be negative; $t = 0$ already corresponds to complete suppression of the phase transition. It is easily seen from eq. (17) that $t = 0$ corresponds to

$$\ell = \ell_c = 2\xi_0 \tan^{-1}(\xi_0/\delta) \tag{23}$$

so that there is no phase transition for $\ell < \ell_c$.

Curves for T_c resulting from the numerical solution of eqs. (15) and (17) are shown in figure 6. The full form of eq. (8), including the nonlinear term, can be solved in terms of elliptic functions.[23] The solutions can be used to generate order-parameter profiles analogous to those shown in figure 4 for the semi-infinite medium. Examples of such profiles are shown in figure 7.

The above analysis is all concerned with the case that the bulk phase transition is second-order. In fact, many ferroelectric phase transitions are first-order so that the term P^6 must be included in eq. (4) and subsequent equations. The extension of the foregoing analysis to first-order phase transitions is due to Duiker.[24] For a bulk first-order transition P changes discontinuously at the critical temperature T_c from zero to some non-zero value P_o. In order to find T_c, which is not equal to T_{co} that appears in eq. (3), it is necessary to equate the ordered-state free energy $F(P_o)$ to the normal-state free energy, namely zero as eq. (2) is written. For the phase transition in a film, P(z) also changes discontinuously at T_c. This means that the differential equation for P(z) cannot be linearized. It is necessary to find P(z) numerically and then evaluate the free energy; the phase transition occurs at the value of T for which the ferroelectric and paraelectric ($P(z) = 0$) phase free energies are equal.

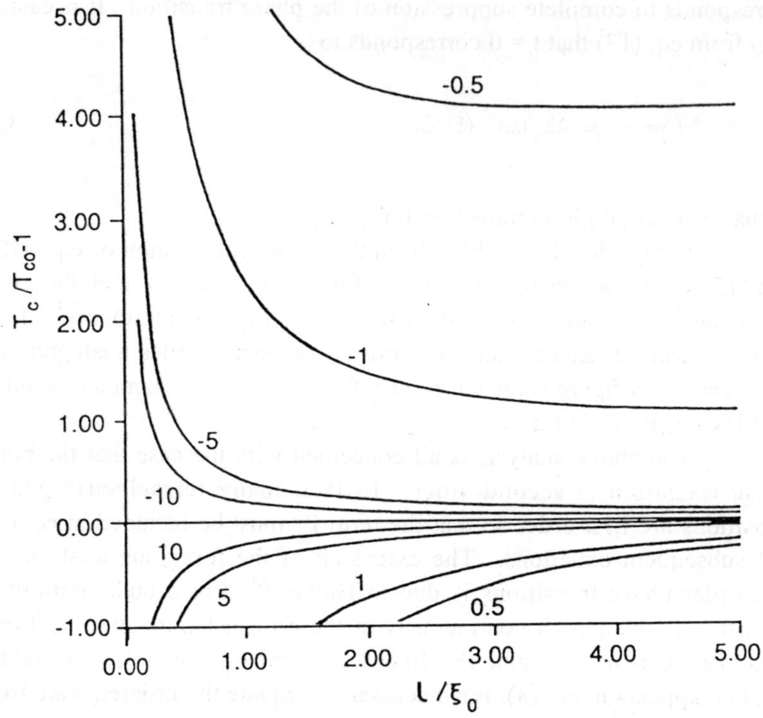

FIGURE 6. Solutions of eqs (15) and (17) for the critical temperature T_c of a film as a function of film thickness ℓ. The curves are labeled by values of δ/ξ_0. [After ref. 2.]

Duiker[24] writes the free energy in the form

$$\frac{F}{S} = \int_{-\ell/2}^{\ell/2} dz \left\{ \frac{1}{2}\xi_0^2 \left(\frac{dP}{dz}\right)^2 + \frac{1}{2}(1+t)\,P^2 - P^4 + \frac{1}{2}P^6 \right\}$$

$$+ \frac{\xi_0^2}{2\delta}(P_+^2 + P_-^2) \tag{24}$$

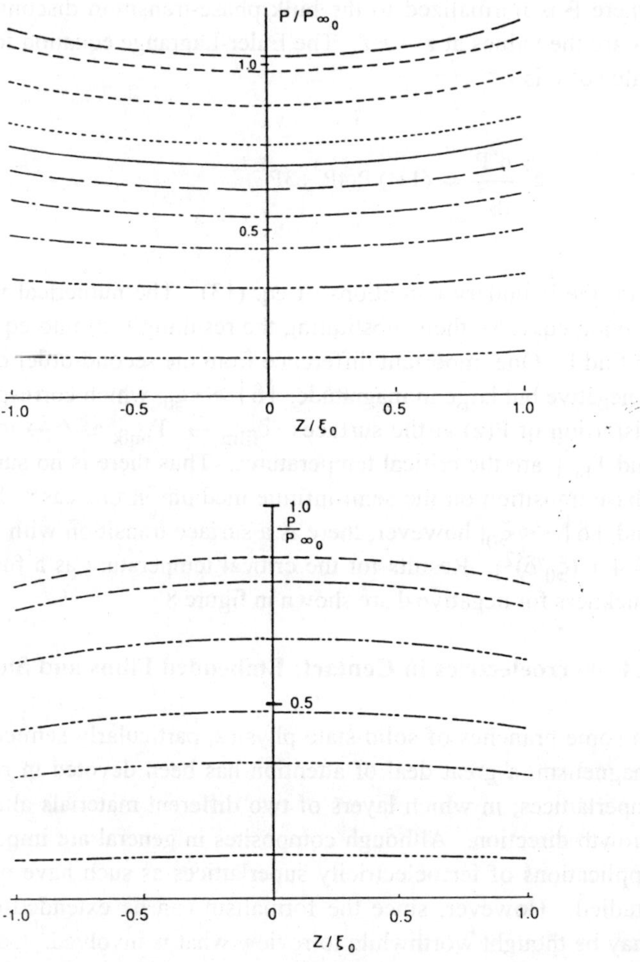

FIGURE 7. Order-parameter profiles, P vs z/ξ_0, for $\delta = \pm\, 5\xi_0$ and $\ell = 2\xi_0$, where $A = A_0\,(T/T_{co} - 1)$ and $\xi_0 = (CT_{co}/A_0)^{1/2}$. The P axis is scaled by $P_{\infty 0} = (A_0/B)^{1/2}$

(a) $\delta = -5\xi_0$, $T_c = 1.21\ T_{co}$. T/T_{co} values, from the top, are 0.1, 0.2, 0.4, 0.6, 0.7, 0.8, 0.9, 1.0, 1.1, 1.2.

(b) $\delta = 5\xi_0$, $T_c = 0.813\ T_{co}$. T/T_{co} values, from the top, are 0.1, 0.2, 0.4, 0.6, 0.7, 0.8.

[After ref. 23.]

where P is normalized to the bulk phase-transition discontinuity P_o and $P\pm$ are the values at $z = \pm \ell$. The Euler-Lagrange equation for a stationary value of F is

$$\xi^2 \frac{d^2P}{dz^2} = (1+t)\,P - 4P^3 + 3P^5 \qquad (25)$$

with the boundary conditions of eq. (13). The numerical work involves solving eq. (25), then substituting the resulting P(z) into eq. (24) in order to find F. One important difference from the second-order case is that for δ negative but large in magnitude, $|\delta| >> \xi_0$, which corresponds to weak distortion of P(z) at the surfaces, $T_{film} \rightarrow T_{bulk}$ as $\ell \rightarrow \infty$, where T_{film} and T_{bulk} are the critical temperatures. Thus there is no surface-ordering phase transition on the semi-infinite medium in this case. For δ negative and $|\delta| << \xi_0$, however, there is a surface transition with $T_{film} \sim T_{bulk} [3/4 + (\xi_0/\delta)^2]$. Results for the critical temperature as a function of film thickness for negative d are shown in figure 8.

3.4. Ferroelectrics in Contact: Embedded Films and Superlattices

In some branches of solid-state physics, particularly semiconductors and magnetism, a great deal of attention has been devoted in recent years to superlattices, in which layers of two different materials alternate along a growth direction. Although composites in general are important in some applications of ferroelectricity superlattices as such have not been much studied. However, since the formalism can be extended quite easily it may be thought worthwhile to review what is involved. Coupling of two magnetic materials across an interface results from an exchange term of the form

$$H_I = \sum_{i,j} I_{ij} S^z_{1i} S^z_{2j} \qquad (26)$$

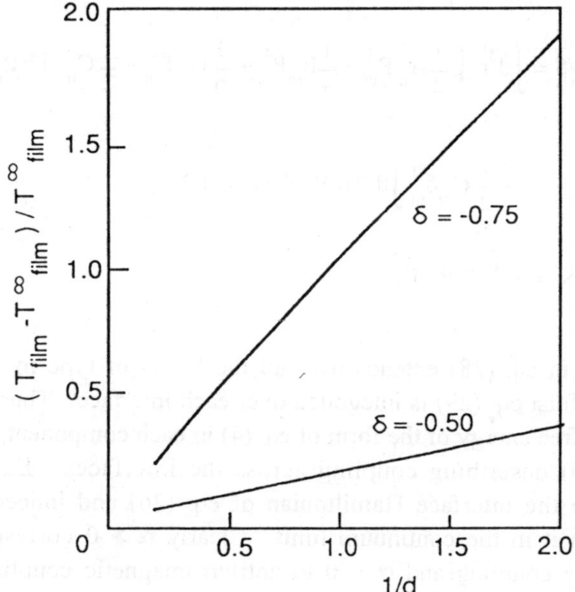

FIGURE 8. Film critical temperature $T_c(\ell)$ for negative δ in the case of first-order phase transitions. [After ref. 24.]

for an Ising system. The sum is over spins S_{1i} and S_{2j} on either side of the interface; often only the nearest-neighbour spins immediately adjacent to the interface would enter. The exchange I_{ij} might be positive or negative, corresponding to ferromagnetic or antiferromagnetic coupling. One might speculate by analogy that a superlattice of two different order-disorder ferroelectrics would be described by the IMTF Hamiltonian of eq. (1) in each layer together with terms of the form of eq. (26) at each interface. Mean-field equations analogous to eqs. (5) and (6) could be applied to investigate the phase diagram, but as far as we are aware this has not yet been done.

A Landau-theory formulation can be set up as follows.[25] The free energy is

$$F = F_1+F_2+F_I \tag{27}$$

$$F_m/S = \int d^3r \left\{ \frac{1}{2}A_m P_m^2 + \frac{1}{4}B_m P_m^4 + \frac{1}{6}G_m P_m^6 + \frac{1}{2}C_m \, |\nabla P_m|^2 \right\}$$

$$+ \frac{1}{2}C_m \delta_m^{-1} \int dS(\mathbf{n}.\mathbf{P}_m)^2 \quad m = 1,2 \tag{28}$$

$$F_I/S = \alpha \int dS \, \mathbf{P}_1.\mathbf{P}_2 \tag{29}$$

The integral in eq. (28) extends over all the layers of type m and their interfaces, whilst eq. (29) is integrated over each interface. Thus eq. (27) comprises a free energy of the form of eq. (4) in each component, together with eq. (29) describing coupling across the interfaces. Eq. (29) is analogous to the interface Hamiltonian of eq. (26) and indeed can be derived from it in the continuum limit. Clearly $\alpha > 0$ corresponds to ferromagnetic coupling and $\alpha < 0$ to antiferromagnetic coupling. The Euler-Lagrange equations for F to be stationary are

$$C_m \frac{d^2P_m}{dz^2} - A_m P_m - B_m P_m^3 - G_m P_m^5 = 0 \tag{30}$$

in each medium (m = 1,2), together with boundary conditions at each interface

$$C_1(\mathbf{n}_1.\nabla P_1 + \delta_1^{-1} P_1) + \alpha P_2 = 0 \tag{31}$$

$$C_1(\mathbf{n}_2.\nabla P_2 + \delta_2^{-1} P_2) + \alpha P_1 = 0 \tag{32}$$

where $\mathbf{n}_1$ and $\mathbf{n}_2$ are unit normals to the interface directed away from the medium in question.

We now review one or two implications for the case of second-order transitions, $B_m > 0$ and the terms in P^6 omitted. We assume that medium 2 has the higher bulk Curie temperature, $T_{20} > T_{10}$. The simplest example is two semi-infinite media in contact, and we ask how the phase transitions in 2 are affected by the juxtaposition of 1. The interesting case is $\delta_2 < 0$, since this corresponds to a surface-ordering transition in 2. In order to find the critical temperature it is only necessary to use the solutions of the linearized form of eq. (30), which are

$$P_1 = P_{10} \exp(z/\xi_1) \tag{33}$$

$$P_2 = P_{20} \exp(-z/\xi_2) \tag{34}$$

where it is assumed that medium 1 occupies the half space $z < 0$. Substitution of these expressions into eqs. (31) and (32) gives two linear, homogeneous equations for the amplitudes P_{10} and P_{20}; the solvability condition is

$$C_1 C_2 (\xi_1^{-1} + \delta_1^{-1})(\xi_2^{-1} + \delta_2^{-1}) - \alpha^2 = 0 \tag{35}$$

This may be compared with eq. (12) for the surface-ordering critical temperature of medium 2 on its own. Since α^2/C_1C_2 and $\xi_1^{-1} + \delta_1^{-1}$ are positive, $\xi_2^{-1} + \delta_2^{-1}$ must also be positive when eq. (35) is satisfied. This means that the surface-ordering temperature is increased by the juxtaposition of 1. Note that this conclusion is independent of the sign of α: it is the juxtaposition of a potentially ferroelectric medium, regardless of the sign of the coupling, that increases the critical temperature.

Some further implications of this model are discussed in ref. 25. If a film of medium 2 is embedded in medium 1, then again the critical temperature is greater than that of the isolated film, whether α is positive or negative. This applies whether δ_2 is positive or negative. Finally, ref. 25 contains an account of the critical temperature of a superlattice comprising alternating layers of thickness D_1 of material 1 and D_2 of material 2. If material 2 has the higher bulk critical temperature, the main result is that the critical temperature of the superlattice is greater than that of an isolated film of thickness D_2 of material 2 embedded in material 1. This is not very surprising since the volume fraction of material 2 is larger in the superlattice.

3.5. Cylinders

Since cylindrical rods may be used in composite materials, it is worth considering the extension of the previous formalism to cylinders;[26] for simplicity we deal only with a free-standing cylinder of radius R. An analysis similar to that in section 3.3. gives expressions to determine T_c, namely

$$\frac{J_1(R/\xi)}{J_0(R/\xi)} = \frac{\xi}{\delta} \tag{36}$$

for $\delta > 0$ and $T_c < T_{co}$, and

$$\frac{I_1(R/\xi)}{I_0(R/\xi)} = -\frac{\xi}{\delta} \tag{37}$$

for $\delta < 0$ and $T_c > T_{co}$.

Equations (36) and (37) may be compared with the analogous eqs. (17) and (15) for a film. The Bessel functions $J_0(x)$ and $J_1(x)$ may be regarded as the cylindrical equivalents of cos(x) and sin(x), and likewise $I_0(x)$ and $I_1(x)$ are the equivalents of cosh(x) and sinh(x). Thus the analogy is quite close.

4. THEORY: DEPOLARIZATION INCLUDED

4.1. General Information

As was mentioned, the results of section 3 do not include the effects of depolarization, so they apply only when the polarization vector **P** is parallel to the surface or surfaces on which the boundary conditions are applied. The more usual geometry in applications involves a film with **P** normal to the surfaces. In this case the effects of depolarization are important and must be included.

A sustained account of the influence of depolarization was given in refs. 27 to 32. It is shown there that the free energy of the electrodes must be included and in fact for semiconducting electrodes the electrode free energy term can change the order of the phase transition from second to first. These references do not include any intrinsic tendency for the polarization to change near a surface, as modeled by the boundary condition of eq. (9). This kind of effect, together with depolarization, is included by Kretschmer and Binder.[19,33] They make the approximation of perfectly conducting electrodes and deal explicitly mainly with a single surface of a thick ferroelectric film. Much of their work is concerned with the critical exponents of the surface-related phase transitions.

A synthesis of the approaches of refs. 27 to 32, together with the Landau-theory approach of the previous section, was presented in ref. 2 and will now be reviewed. The system to be analysed is a ferroelectric film of finite thickness placed between metal electrodes. The free energy of eq. (4) is augmented by terms accounting for the electrostatic energy of the electrodes and the film. The notation used, which is that of Mehta et al.,[29] is indicated in figure 9, together with sketches of the expected spatial dependence of the fields **D**, **E** and **P**, all of them normal to the interfaces (z-components only). Regions $-L/2 < z < -\ell/2$ and $\ell/2 < z < L/2$ are occupied by the electrodes and the ferroelectric film occupies $-\ell/2 < z < \ell/2$.

The free energy of the film including the electrostatic terms is

$$F = F_0 + F_E \tag{38}$$

where F_0 is given by eq. (4) and F_E is the electrostatic energy

$$F_E/S = -\int_{-\ell/2}^{\ell/2} dz \left\{\frac{1}{2} E_d(z)P(z) + E_{ext}P(z)\right\} \tag{39}$$

Here E_{ext} is the applied field

$$E_{ext} = -V_0/L \tag{40}$$

where V_0 is the applied voltage. $E_d(z)$ is the dipole field, related to P(z) by Poisson's equation

$$\varepsilon_0\varepsilon_F\, dE_d/dz = -dP/dz \tag{41}$$

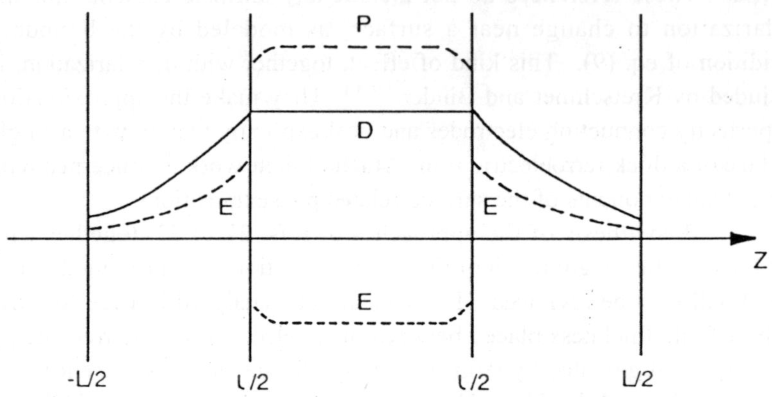

FIGURE 9. Notation for calculation of electrostatic energies and qualitative variation of electrostatic field quantities. [After ref. 2.]

where ε_F is the dielectric constant of the ferroelectric. Within the electrodes, it is assumed that the electric field is governed by Thomas-Fermi screening

$$\frac{d^2E}{dz^2} = \frac{1}{\ell_s^2} E \tag{42}$$

where the screening length ℓ_s is typically of order 1Å, for example ℓ_s= 0.83Å for Au and 0.58Å for Al.[34] The electrostatic energy is evaluated from the solution of eq. (42) together with the standard expression for electrostatic energy density. The final expression for the free energy of the whole system, film plus electrodes, is[2]

$$F/S = \int_{-\ell/2}^{\ell/2} dz \left\{ \frac{1}{2}AP^2 + \frac{1}{4}BP^4 + \frac{1}{2}C\left(\frac{dP}{dz}\right)^2 + \frac{V_0}{(2\alpha\ell_s+\ell)} P + \frac{1}{2\varepsilon_0\varepsilon_F} P^2 \right\}$$

$$- \frac{1}{2\varepsilon_0\varepsilon_F(2\alpha\ell_s+\ell)} \left[\int_{-\ell/2}^{\ell/2} Pdz \right]^2$$

$$+ \frac{\varepsilon_0\varepsilon_F^2\ell_s\beta}{4\varepsilon_e(2\alpha\ell_s+\ell)^2} \left\{ \frac{1}{\varepsilon_0^2\varepsilon_F^2} \left[\int_{-\ell/2}^{\ell/2} Pdz \right]^2 - \frac{2V_0}{\varepsilon_0\varepsilon_F} \int_{-\ell/2}^{\ell/2} Pdz + V_0^2 \right\}$$

$$+ \frac{1}{2} C\delta^{-1} \left[P^2\left(-\frac{\ell}{2}\right) + P^2\left(\frac{\ell}{2}\right) \right] \tag{43}$$

where ε_e is the dielectric constant of the electrode material and the quantities α and β are given by

$$\alpha = \{\cosh[(L-\ell)/2\ell_s] - 1\} / \sinh[(L-\ell)/2\ell_s] \tag{44}$$

and

$$\beta = \frac{\sinh[(L-\ell)/\ell_s] - (L-\ell)/\ell_s}{\sinh^2[(L-\ell)/\ell_s]} \tag{45}$$

In practice, one would generally have $L - \ell >> \ell_s$ so that $\alpha = \beta = 1$.

Equation (43) does not include a term in P^6, as would be required for materials with a first-order phase transition in the bulk. No analysis for this case has yet been carried out, and what follows applies for second-order transition materials.

4.2. Perfectly Conducting Electrodes

In the limit $\ell_s \to 0$, eq. (43) simplifies to

$$F/S = \int_{-\ell/2}^{\ell/2} dz \left\{ \frac{1}{2}AP^2 + \frac{1}{4}BP^4 + \frac{1}{2}C\left(\frac{dP}{dz}\right)^2 + \frac{V_0}{\ell}P + \frac{1}{2\varepsilon_0\varepsilon_F}P^2 \right\}$$

$$
-\frac{1}{2\varepsilon_0\varepsilon_F}\left[\int_{-\ell/2}^{\ell/2} P dz\right]^2 + \frac{1}{2}C\delta^{-1}\left[P^2(-\tfrac{\ell}{2}) + P^2(\tfrac{\ell}{2})\right] \tag{46}
$$

This is the expression for perfectly conducting electrodes that was applied to the semi-infinite medium by Kretschmer and Binder.[19,33] It is of interest to compare the implications for the phase transition in a film[2] with the results of section 3.3.

The Euler-Lagrange equation resulting from eq. (46) is

$$
C\frac{d^2P}{dz^2} - AP - BP^3 - \frac{P}{\varepsilon_0\varepsilon_F} + \frac{1}{\varepsilon_0\varepsilon_F\ell}\int_{-\ell/2}^{\ell/2} P dz = 0 \tag{47}
$$

with the boundary conditions of eq. (13). In order to find the critical temperature we omit the nonlinear term BP^3 and write eq. (47) as

$$
\frac{d^2P}{dz^2} - \frac{1}{\xi_0^2}(t - 1 + t_1) P + \frac{t_1}{\xi_0^2} R = 0 \tag{48}
$$

where $t = T_c/T_{co}$,

$$
t_1 = \xi_0^2 / \varepsilon_0\varepsilon_F C \tag{49}
$$

and ξ_0 is defined in section 3.3. R is the spatial average of the polarization:

$$
R = \frac{1}{\ell}\int_{-\ell/2}^{\ell/2} P dz \tag{50}
$$

For $\delta < 0$ (enhancement of T_c), the solution of eq. (48) is

$$P = \frac{t_1}{t - 1 + t_1} R + P_0 \cosh \frac{z}{\xi} \tag{51}$$

where now $\xi^2 = \xi_0{}^2 /(t\text{-}1\text{+ } t_1)$. Substitution of eq. (51) into the boundary conditions and the definition of R, eq. (50), yields two equations for R and P_0:

$$(s/\xi + c/\delta)\, P_0 + t_1 R/\delta\, (t - 1 + t_1) = 0 \tag{52}$$

$$2s\xi P_0/\ell + (1 - t)\, R/(t - 1 + t_1) = 0 \tag{53}$$

where $s = \sinh(\ell/2\xi)$ and $c = \cosh(\ell/2\xi)$. For $t_1 = 0$, eq. (52) reduces to eq. (15), so t_1 effectively includes the depolarization effects. The critical temperature is given by the vanishing of the determinant of coefficients in eqs. (52) and (53).

The corresponding analysis for $\delta > 0$ (reduction of T_c) leads to[2]

$$(s_1/\xi_a - c_1/\delta)\, P_0 + t_1 R/\delta(1 - t - t_1) = 0 \tag{54}$$

$$2s_1\xi_a P_0/\ell - (1 - t)\, R/(1 - t + t_1) = 0 \tag{55}$$

where $s_1 = \sin(\ell/2\xi_a)$, $c_1 = \cos(\ell/2\xi_a)$ and $\xi_a{}^2 = \xi_0{}^2/(1\text{-}t\text{-}t_1)$ provided $T_c/T_{co} > t_1$. For $T_c/T_{co} < t_1$, on the other hand, eqs. (52) and (53) still apply.

The general correspondence of these results is that T_c is reduced below the values that are found in the absence of depolarization. In the case $\delta < 0$, this may already be seen from the asymptotic expressions for large and small values of the film thickness ℓ. The most important concerns large ℓ. Without depolarization, eq. (15) shows that there is always enhancement of T_c above the bulk value T_{co}, corresponding to the transition at temperature $T_c{}^* = T_{co}(1+ \xi_0{}^2 / |\delta|^2)$ to a phase in which the surface is ordered but the bulk is not. Now, however, there is enhancement only if $t_1 < \xi_0{}^2 / |\delta|^2$, that is, if depolarization effects are not too strong. The various asymptotic expressions are given in Table 1.

TABLE 1. Limiting values of T_c for $\delta < 0$ with and without depolarization. The variables are $\tau = T_c/T_{co} - 1$, $d = |\delta|/\xi_0$ and $\lambda = \ell/\xi_0$.

	No depolarization	Depolarization	Depolarization
		$\frac{1}{d^2} > t_1$	$\frac{1}{d^2} < t_1$
$\lambda \to \infty$	$\tau = \frac{1}{d^2}$	$\tau = \frac{1}{d^2} - t_1$	$\tau \sim \frac{2t_1^{1/2}}{(t_1^{1/2} d - 1)} \frac{1}{\lambda}$
$\lambda \to 0$	$\tau \sim \frac{2}{d\lambda}$	$\tau \sim \frac{2}{d\lambda} - t_1$	$\tau \sim \frac{2}{d\lambda} - t_1$

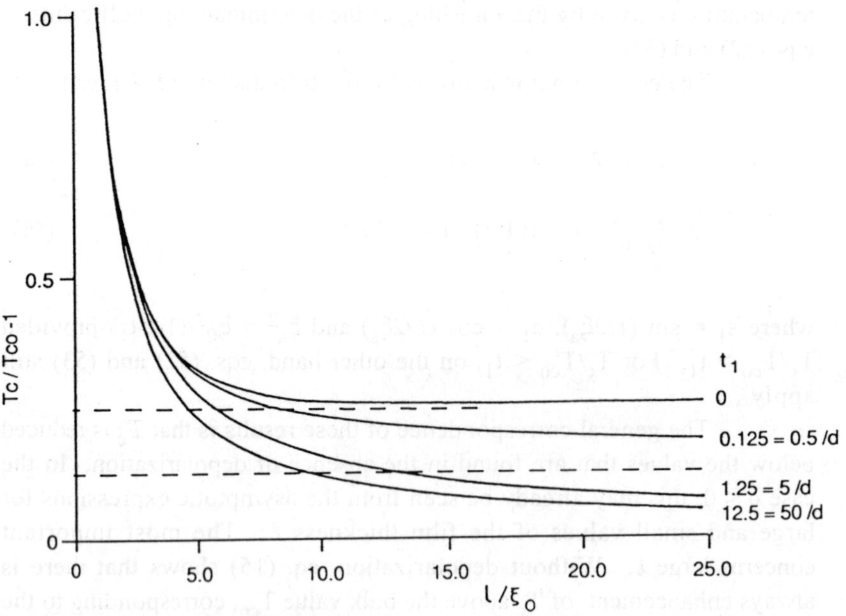

FIGURE 10. Solutions of eqs. (52) and (53) for the critical temperature of a film in the case $\delta < 0$ (enhancement of T_c) with depolarization included. The curves are drawn for fixed $|\delta|/\xi_0 = 2$ and various values of the depolarization parameter t_1 (marked). [After ref. 2.]

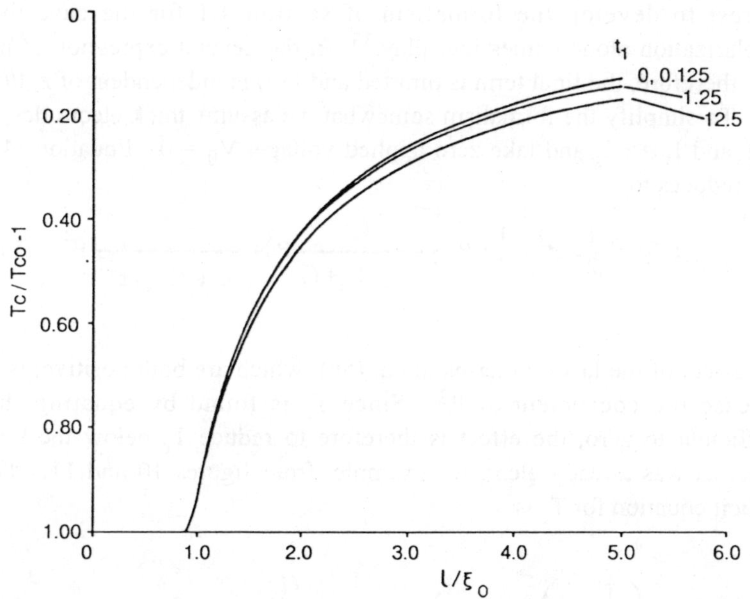

FIGURE 11. Solutions of eqs. (52) to (55) for the critical temperature of a film in the case $\delta > 0$ (depression of T_c) with depolarization included. The curves are for the same values $\delta/\xi_0 = 2$ and $t_1 = 0, 0.125, 1.25$ and 12.5 as in figure 10; the curves for $t_1 = 0$ and $t_1 = 0.125$ are not resolved. [After ref. 2.]

An illustration of the general reduction of T_c by depolarization effects is given in figure 10, which shows T_c/T_{co} -1 plotted versus ℓ/ξ_0 for fixed $|\delta|/\xi_0$ and various values of t_1. Similar graphical results for $\delta > 0$ are presented in figure 11. Here again, the inclusion of depolarization leads to a further depression of T_c, but the effects are less striking than those for negative δ.

4.3. Depolarization only

As was seen in section 2, there is as yet no unequivocal evidence that boundary conditions of the form of eq. (13) affect ferroelectric phase transitions, whereas the influence of depolarization is obvious on general grounds and is also experimentally established.[32] It is therefore of interest to develop the formalism of section 4.1 for the case that depolarization alone comes into play.[35] In the general expression of eq. (43), therefore, the final term is omitted and P(z) is independent of z, P(z) = P. To simplify the formalism somewhat we assume thick electrodes, L >> ℓ and L >> ℓ_s and take zero applied voltage, $V_0 = 0$. Equation (43) then reduces to

$$F/S\ell = \frac{1}{2}AP^2 + \frac{1}{4}BP^4 + \frac{\ell_s}{\varepsilon_0\varepsilon_F(2\ell_s+\ell)}P^2 + \frac{\ell\ell_s}{4\varepsilon_0\varepsilon_e(2\ell_s+\ell)^2}P^2 \qquad (56)$$

The effect of the last two terms in eq. (56), which are both positive, is to increase the coefficient of P^2. Since T_c is found by equating this coefficient to zero, the effect is therefore to reduce T_c below the bulk value, as was already clear, for example, from figures 10 and 11. The explicit equation for T_c is

$$A_0\left(\frac{T}{T_{co}} - 1\right) + \frac{2\ell_s}{\varepsilon_0\varepsilon_F(2\ell_s+\ell)} + \frac{\ell\ell_s}{(2\ell_s+\ell)^2 2\varepsilon_0\varepsilon_e} = 0 \qquad (57)$$

The two limits $\ell_s \to 0$ (infinite conductivity) and $\ell_s \to \infty$ (free standing film without electrodes) are apparent. For $\ell_s \to 0$ the second and third terms in eq. (57) vanish so that T_c is not shifted from T_{co}. For $\ell_s \to \infty$, the second term is retained:

$$A_0\left(\frac{T_c}{T_{co}} - 1\right) + \frac{1}{\varepsilon_0\varepsilon_F} = 0 \qquad (58)$$

This can be derived in a more elementary way.[36] For a free-standing film with polarization normal to the interfaces the free energy is

$$F = \frac{1}{2}AP^2 + \frac{1}{4}BP^4 - \frac{1}{2}EP \qquad (59)$$

and the boundary condition gives

$$P + \varepsilon_0\varepsilon_F E = 0 \tag{60}$$

Substituting for E from eq. (60) into eq. (59) and equating the coefficient of P^2 to zero one recovers eq. (58). It will be seen from the numerical estimates to be presented below that in fact eq. (58) corresponds to complete suppression of the phase transition.

One further result may be derived from eq. (57). It is clearly of interest to know the thickness ℓ_c for which T_c is reduced to zero. This follows from putting $T_c = 0$ in eq. (57):

$$\frac{2\ell_s}{(2\ell_s+\ell_c)\varepsilon_0\varepsilon_F} + \frac{\ell_c\ell_s}{(2\ell_s+\ell_c)^2 2\varepsilon_0\varepsilon_e} = A_0 \tag{61}$$

In order to make further progress it is necessary to know some numerical magnitudes. The bulk susceptibility for $T > T_c$ is $\chi = A_0^{-1}(T - T_c)^{-1}$, so A_0 is just the Curie constant; a typical value for a displacive ferroelectric[37] is 2×10^5 Jm K^{-1} C^{-2}. With the typical value $\varepsilon_F = 10^3$, eq. (61) for the critical value ℓ_c becomes

$$1.771 \times 10^{-3} = \frac{2}{2+x_c} + \frac{\varepsilon_F}{\varepsilon_e}\,\frac{x_c}{2(2+x_c)^2} \tag{62}$$

where $x_c = \ell_c/\ell_s$. Clearly the solution is $x_c >> 1$, so it is satisfactory to retain only leading terms in; this gives

$$x_c = \frac{4+\varepsilon_F/\varepsilon_e}{2A_0\varepsilon_0\varepsilon_F} \tag{63}$$

Here the main uncertainty is the value of ε_e. It is the long-wavelength limit of Thomas-Ferni static dielectric function, which is formally infinite.[34] It enters eq. (56) via the boundary condition on normal **D** at the electrode-ferroelectric interface. Since this could be modified, for example by trapped charge and interface roughness, it may be satisfactory to treat $\varepsilon_F/\varepsilon_e$ as a parameter. With $\varepsilon_F/\varepsilon_e = 0$ and the values used in eq. (62), eq. (63) gives

$$x_c = 1.13 \times 10^3 \tag{64}$$

For Au electrodes, for example, this gives $\ell_c = 940Å$.

The variation of T_c with ℓ is given by eq. (57). With the retention of only leading terms in $x = \ell/\ell_s >> 1$, eqs. (57) and (63) lead to

$$1 - t = \frac{x_c}{x} \tag{65}$$

where as before $t = T_c/T_{co}$. This simple variation, which was previously derived by Batra and Silverman,[27] is illustrated in figure 12. As remarked before, it has a similar form to the observed results for Pb Ti O_3 particles.[16]

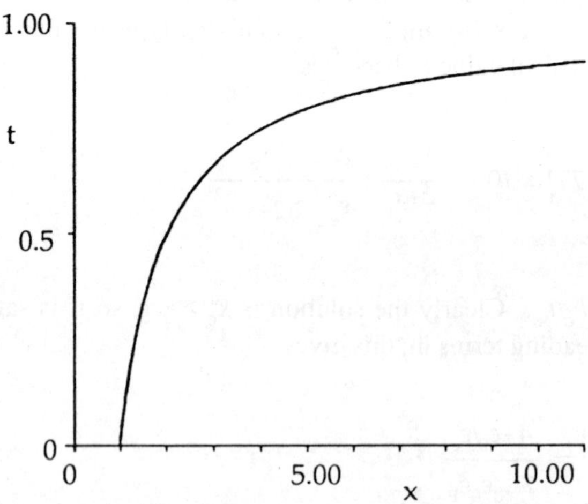

FIGURE 12. Variation of film critical temperature $t = T_c/T_{co}$ with thickness $x = \ell/\ell_c$ resulting from depolarization alone. Here ℓ_c is the thickness for which T_c = 0, given by eq. (63). [After ref. 35.]

Finally, it was remarked that in practice eq. (58) describes complete suppression of the phase transition by the depolarization field for the free standing film. In fact, with the numerical value of $A_0\varepsilon_0\varepsilon_F$ used on the left-hand side of eq. (62), eq. (58) becomes

$$T_c/T_{co} - 1 = -5.65 \times 10^2 \tag{66}$$

Clearly this has no solution for $T_c > 0$.

5. CONCLUSIONS

The theoretical framework that has been presented includes both depolarization effects and a possible intrinsic variation of **P** in the vicinity of a surface, modeled in the Landau theory by a boundary condition of the form of eq. (9). Depolarization effects alone have been quite extensively studied[28-32] and a certain amount of experimental work has been performed.[32] It would seem, however, that a more complete approach should include domain structures. The simple derivation of eq. (58) amounts to saying that the electrostatic energy outside a monodomain film is much larger than the free energy difference ΔF between the paraelectric and ferroelectric phases; this electrostatic energy is reduced by the appearance of a domain structure.

The alternative mechanism of an intrinsic variation in **P** is possible on general grounds, but is not yet clearly supported by experiment. It is to be hoped that current rapid improvements in techniques for preparation of high-quality films will lead to results that permit a full critical comparison.

REFERENCES

1. C.F. MAJKRZAK, J. KWO, M. HONG, Y. YAFET, D. GIBBS, C.L. CHEN, and J. BOHR, Advances in Physics, **40**, 99 (1991).
2. D.R. TILLEY, "Phase Transitions in Thin Films" in Ferroelectric Ceramics, edited by N. Setter and E.L. Colla (Birkhäuser Verlag, Basel, 1993), 163-83.
3. R. BLINC and B. ZEKS, Soft Modes in Ferroelectrics and Antiferroelectrics (Elsevier, Amsterdam, 1974).
4. M.E. LINES and A.M. GLASS, Principles and Applications of Ferroelectrics and Related Materials (Clarendon, Oxford, 1977).
5. J.C. TOLEDANO and P. TOLEDANO, Landau Theory of Phase Transitions (World Scientific, Singapore, 1987).
6. U.T. HÖCHLI and H. ROHRER, Phys. Rev. Lett., **48**, 188 (1982).
7. A. HADNI and X. GERBAUX, Optics Comm., **50**, 231 (1984).
8. X. GERBAUX, A. HADNI, and A. KITADE, Infrared Phys., **24**, 543 (1984).
9. X. GERBAUX, A. HADNI, and A. KITADE, Phys. Stat. Sol. (a), **88**, 165 (1985).
10. A. HADNI and X. GERBAUX, J. Phys., C **18**, 4805 (1985).
11. X. GERBAUX and A. HADNI, Ferroelectrics, **70**, 175 (1986).
12. X. GERBAUX and A. HADNI, Phase Transitions, **14**, 117 (1989).
13. X. GERBAUX, A. HADNI, and A. KITADE, Phys. Stat. Sol. (a), **115**, 587 (1989).
14. J.F. SCOTT, M.S. ZHENG, R. BRUCE GODFREY, C. ARAUJO, and L. McMILLAN, Phys. Rev. B, **35**, 4044 (1987).
15. P. MARQUARDT and H. GLEITER, Phys. Rev. Lett., **48**, 1423 (1982).
16. K. ISHIKAWA, K. YOSHIKAWA, and N. OKADA, Phys. Rev. B, **37**, 5852 (1988).
17. M.G. COTTAM, D.R. TILLEY, and B. ZEKS, J. Phys. C: Solid State Phys., **17**, 1793 (1984).
18. T.C. LUBENSKY and M.H. RUBIN, Phys. Rev. B, **12**, 3885 (1975).
19. R. KRETSCHMER and K. BINDER, Phys. Rev. B, **20**, 1065 (1979).
20. C.L. WANG, Solid State Commun., **82**, 743 (1992).
21. C.L. WANG, W.L. ZHONG, and P.L. ZHANG, J. Phys.: Condens. Matter, **3**, 4743 (1992).

22. E.F. SARMENTO and J.W. TUCKER, J. Magn. Magn. Mater., **118**, 133 (1993).
23. D.R. TILLEY and B. ZEKS, Solid State Commun., **49**, 823 (1984).
24. H.M. DUIKER, Ph.D. Thesis, University of Colorado.
25. D.R. TILLEY, Solid State Commun., **65**, 657 (1988).
26. N. NAPP, unpublished.
27. I.P. BATRA and B.D. SILVERMAN, Solid State Commun., **11**, 291 (1972).
28. I.P. BATRA, P. WURFEL, and B.D. SILVERMAN, Phys. Rev. Lett., **30**, 384 (1973).
29. R.R. MEHTA, B.D. SILVERMAN, and J.T. JACOB, J. Appl. Phys., **44**, 3379 (1973).
30. I.P. BATRA, P. WURFEL, and B.D. SILVERMAN, Phys. Rev. B, **8**, 3257 (1973).
31. I.P. BATRA, P. WURFEL and B.D. SILVERMAN, J. Vac. Sci. Technol., **10**, 687 (1973).
32. P. WURFEL and I.P. BATRA, Phys. Rev. B, **8**, 5126 (1973).
33. K. BINDER, Ferroelectrics, **35**, 99 (1981).
34. N.W. ASHCROFT and N.D. MERMIN, Solid State Physics, (Holt, Rinehart and Winston, New York, 1976).
35. D.R. TILLEY and B. ZEKS, Ferroelectrics, **134**, 313 (1992).
36. J. FOUSEK, private communication.
37. J.F. SCOTT, private communication.

CHAPTER 3

DEFECT CHEMISTRY, CONDUCTION, AND BREAKDOWN MECHANISM OF PEROVSKITE-STRUCTURE TITANATES

R. Waser[1] and D.M. Smyth[2]

1. Institut für Werkstoffe der Elektrotechnik, Lehrstuhl II, RWTH Aachen, 52074 Aachen, Germany

2. Lehigh University, Department of Materials Science & Engineering, Whitaker Laboratory, 5 East Packer Avenue, Bethlehem, PA 18015-3195, USA

1. Introduction

The majority of thin film materials investigated and employed for integrated ferroelectrics today are based on perovskite–structure oxides. The electrical characteristics of these ferroelectric thin films such as the switching behaviour, the fatigue, the retention, the leakage current, and the reliability under high fields is closely related to the defect chemistry which is determined by the types, the locations, and the mobilities of point defects in the bulk and interface regions of the films.

Ferroelectric thin films are extremely complex electrical systems. Firstly, they are *thin* films. Their dielectric thickness typically ranges from 0.1 to 1 μm. Consequently, the influence of electronic interface states (up to now virtually unknown) will be at least much more pronounced than in conventional compact ceramic samples, or perhaps will even completely dominate the behaviour of the systems. In addition, the microstructure of the films usually is very different from the bulk ceramics or bulk single crystals. Secondly, the films are *ferroelectric*. Effects caused by the ferroelectric polarization are superimposed or interact with defect–chemistry related effects such as charge transfer processes and space charge formation. Hence, the ferroelectric behaviour superimposes an additional degree of complexity. In order to provide a basic understanding of the nature and impact of defect chemistry, we will restrict ourselves in this report to non-ferroelectric perovskite–structure titanates such as $SrTiO_3$ and cubic $BaTiO_3$ (i.e. at temperatures above the Curie temperature). These materials show the same crystal structure and a similar band gap as, for example, the PZT (lead–zirconate–titanate) used for integrated ferroelectric devices and are assumed to be governed by a comparable defect structure.

In order to approach the thin films starting from a relatively simple situation and increasing in complexity, the treatise will proceed from the bulk behaviour (of an infinite crystal lattice) via the discussion of grain boundaries in ceramics, the electrode interfaces of single crystals, to thin film devices. In Section 2, the *bulk* defect chemistry of undoped and doped titanates will be comprehensively described. The high and low temperature behaviour will be evaluated and the consequences of the defect situation on the conduction mechanism will be explained. In Section 3, the discussion will be extended to external and internal *interfaces* and will emphasize the conduction through single crystals, ceramics, and thin films of the same composition. A slightly acceptor doped $SrTiO_3$ is considered for this comparison. In Section 4, the high field behaviour is described and electrical failure modes such as the instantaneous high field breakdown and the long-term resistance degradation are discussed.

2. Defect Chemistry of Perovskite–Structure Titanates

2.1. General Considerations

The defect chemistry of any compound depends primarily on its crystal structure and on the chemical properties of the constituent chemical species. The structure influences the types of lattice defects that may be formed in significant concentrations, and also influences the mobilities of the defects and hence of the chemical species. These mobilities determine whether or not defect equilibria can be achieved within pertinent times at various temperatures, and at what point during the cooling of the material from processing temperatures the population of lattice defects becomes effectively quenched. The sizes and charges of the atomic species affect the selection of the most favored defects, and their ability to be either oxidized or reduced determines the direction and amount of nonstoichiometry and the resulting enhanced electronic carrier concentrations. The volatility of a component can also affect equilibration and defect choice; this is of particular importance for Pb–based compounds, due to the volatility of PbO.

The perovskite structure is the structure of choice for compounds having the generic formula ABO_3, where one of the cations, traditionally the A ion, is substantially larger than the other. Oxides having the perovskite structure exist with all combinations of cation oxidation states that add up to +6, e.g. $KTaO_3$ (1–5), $SrTiO_3$ (2–4), and $LaCrO_3$ (3–3), and there are also perovskite halides such as $KNiF_3$. This structure seems to be a particularly favorable configuration, because it is found for an extremely wide range of compounds, and its basic pattern is frequently found in compounds that differ significantly from the ideal composition [1][2]. Moreover, it is almost certainly the most abundant naturally occurring crystal structure on our planet, since a major constituent of the earth's mantle, $MgSiO_3$, is squeezed into a distorted perovskite structure at the ambient pressures and temperatures [3].

From the above discussion, it is apparent that the important defects in the perovskite structure are vacancies on all three sublattices, electrons, holes, and substitutional impurities. Protons represent a possible exception to this rule, but they will not be dealt with in this review. The role of protons in perovskite–structure titanates has been investigated [4]. Interstitial defects and ions on the wrong lattice site play no significant role. Thus the only feasible type of intrinsic ionic disorder, i.e. an electrically neutral combination of lattice defects that result from spontaneous disorder, is Schottky disorder; this conclusion is supported by theoretical calculations of defect energetics [5]. Schottky disorder involves the formation of all three types of vacancies in stoichiometric ratio

$$nil \leftrightarrows V_M^{''} + V_{Ti}^{''''} + 3V_O^{\cdot\cdot} \qquad (2.1)$$

where nil represents the defect–free crystal, the thermodynamic standard state. M^{++} will frequently be used to represent any large, divalent cation, such as Sr^{++}, Ba^{++}, or Pb^{++}, in a generic perovskite titanate of formula $MTiO_3$. At equilibrium, the concentrations of these defects are related through a mass–action expression

$$[V_{Ba}''] \, [V_{Ti}''''] \, [V_O^{\cdot\cdot}]^3 = K_s(T) = K_s' \, e^{\Delta S_s/k} \, e^{-\Delta H_s/kT} \tag{2. 2}$$

where the square brackets denote the concentration of the enclosed species. The right-hand-side is the mass action constant where ΔH_s and ΔS_s are the standard enthalpy and entropy of the disorder reaction, Eq. (2. 1), respectively. The defect notation is the conventional one due to Kröger and Vink [6], where the main symbol represents the species located at the site in question, with V representing a vacancy, the subscript indicates the site, and the superscript is the difference in charge from that at the same site in the perfect crystal. A dot represents an extra positive charge (or missing negative charge), while a slash represents an extra negative charge (or a missing positive charge). If Eq. (2. 1) is the only significant source of vacancies, then their concentrations are related, and can be expressed by

$$[V_M''] \approx [V_{Ti}''''] \approx \frac{1}{3}[V_O^{\cdot\cdot}] \approx \left(\frac{K_s'}{27}\right)^{1/5} e^{\Delta S_s/5k} \, e^{-\Delta H_s/5kT} \tag{2. 3}$$

There has been no experimental determination of ΔH_s for a perovskite compound, but theoretical calculations for $BaTiO_3$ give the reasonable value of 2.3 eV for $\Delta H_s/5$, the enthalpy per defect [5], and this is consistent with experimental behavior. This gives values for the exponential enthalpy term in Eq. (2. 3) of $8 \cdot 10^{-10}$ (0.0008 ppm) at 1273 K and $1.2 \cdot 10^{-7}$ (0.12 ppm) at 1673 K, the usual maximum sintering temperature for $BaTiO_3$, for example. These values set the scale for the defect concentrations; the entropy term might increase these values by as much as two orders of magnitude to 0.08 ppm and 12 ppm, respectively. These values are still small compared with any realistic impurity content, and indicate that under most circumstances, intrinsic ionic disorder plays very little role in the defect chemistry of $BaTiO_3$. However, it is undoubtedly important in such processes as sintering where transport of all of the ionic species is necessary, even when the transport of some species may occur by a defect whose concentration is small compared with those of the more favored defects.

The perovskite titanates are typical of transition metal oxides in having band gaps of the order of 3–4 eV. This is sufficiently low to give carrier concentrations at high temperatures that support conductivity levels characteristic of semiconductors, while maintaining the possibility of highly insulating properties near room temperature.

2.2 The Incorporation of Aliovalent Impurities

From the above discussion, it is clear that impurities play an important role in the normal equilibrium defect chemistry of the perovskite titanates. Since there is no evidence for interstitial impurities, only substitutional impurities need to be considered. Isovalent substitutional impurities have no first order effect on the defect population, because they do not change the local charge. On the other hand, aliovalent substitutional impurities, those that have a charge that differs from that of the ions they replace, require the formation of compensating defects of opposite charge in order to maintain bulk charge neutrality.

Substitutional cations of lesser charge than the host ion they replace are commonly called acceptor impurities. Since they represent a deficiency of positive charge, and hence carry a negative defect charge, they require the presence of a positively charged compensating defect such as an oxygen vacancy or a hole. Typical acceptor impurity centers in $MTiO_3$ include Al'_{Ti}, Na'_M, and Mg''_{Ti}. Substitutional cations of higher charge are called donor impurities and can be compensated by electrons, or by either M–site or titanium vacancies. Typical examples include $La^{\bullet}_M$ and $Nb^{\bullet}_{Ti}$. Since chemical species with low oxidation states are much more abundant in nature than those with higher charges, the undoped perovskite titanates invariably contain an excess of naturally–occurring acceptor–type impurities in sufficient amount to dominate the defect chemistry under most circumstances [7][8][9].

There is usually a distinct site preference for cation impurities, based on size and charge. Large cations preferentially substitute for the larger M ion, e.g. K^+ for M^{++} in $MTiO_3$, while small cations replace the B ion, e.g. Nb^{+5} for Ti^{+4} in $MTiO_3$. These cases are particularly clearcut, because substitution for the other host cation would result in impurity centers with prohibitively high defect charges, e.g. K'''_{Ti} or $Nb^{\bullet\bullet\bullet}_M$. Such highly charged defects are very unlikely when a lesser charged alternative is available. For trivalent impurities in the alkaline earth titanates, the charge distinction is not apparent; the charge discrepancy is one unit no matter which host cation is replaced. In the case of $BaTiO_3$, it has been found that there is a gradual switch–over in site preference with the size of the trivalent impurity [10]. La^{+3} (0.1032 nm) appears to substitute almost entirely on Ba sites as a donor, $La^{\bullet}_{Ba}$, while Al^{+3} (0.0535 nm) resides primarily on Ti sites as an acceptor, Al'_{Ti}, Y^{+3} and Er^{+3} (0.090 and 0.089 nm, almost exactly midway between the sizes of Ba^{+2} and Ti^{+4}) can occupy either site to a significant degree, depending on whether the preparation is rich in BaO or TiO_2. It will be seen that the site selection for trivalent cations can have profound effects on the room temperature properties of these materials.

2.3 Undoped and Acceptor–Doped Perovskite Titanates

Undoped and acceptor–doped materials are discussed together, because they actually differ only in the amount of acceptor impurity present [7][8][9]. As stated above, even the undoped material invariably contains an excess of naturally–occurring acceptor impurities. As a typical example, we will describe $MTiO_3$ with a small amount of the TiO_2 content replaced by a generic trivalent acceptor oxide, A_2O_3. The stoichiometric composition of the doped material, where there are neither excess electrons nor excess holes, corresponds to the combination of stoichiometric binary constituents, MO + $(1-x)TiO_2 + (x/2)A_2O_3$, or $MTi_{1-x}A_xO_{3-x/2}$. The latter formula makes it clear that the negatively–charged acceptor centers are compensated by oxygen vacancies in the stoichiometric composition. The impurity incorporation reaction can then be written as

$$2MO + A_2O_3 \rightarrow 2M_M + 2A'_{Ti} + 5O_O + V^{\bullet\bullet}_O \qquad (2.4)$$

and charge neutrality is maintained by the relationship

$$[A'_{Ti}] \approx 2\,[V_O^{\cdot\cdot}] \tag{2.5}$$

The incorporation of the impurity cations into the perovskite lattice occurs during processing and grain growth, and is not an equilibrium process. Once the impurities are in the lattice, they are effectively frozen in place by extremely slow diffusion rates. In this model we have assumed that there is no significant association between the oppositely charged acceptor centers and oxygen vacancies.

The stoichiometric composition corresponds to the condition of a nominally filled valence band, and a nominally empty conduction band. In that case, electrons and holes result only from ionization across the band gap

$$nil \leftrightarrows e' + h^{\cdot} \tag{2.6}$$

and their concentrations are related by the mass–action expression

$$np = K_i\, e^{-E_g^o/kT} \tag{2.7}$$

where $n = [e']$ and $p = [h^{\cdot}]$, and $E_g{}^o$ is the band gap at 0 K or the enthalpy of the ionization reaction, Eq. (2. 6).

The equilibrium electrical conductivity of $BaTiO_3$ as a function of the oxygen partial pressure, $P(O_2)$, over the temperature range 873–1273 K is shown in Fig.2–1 as a typical example.

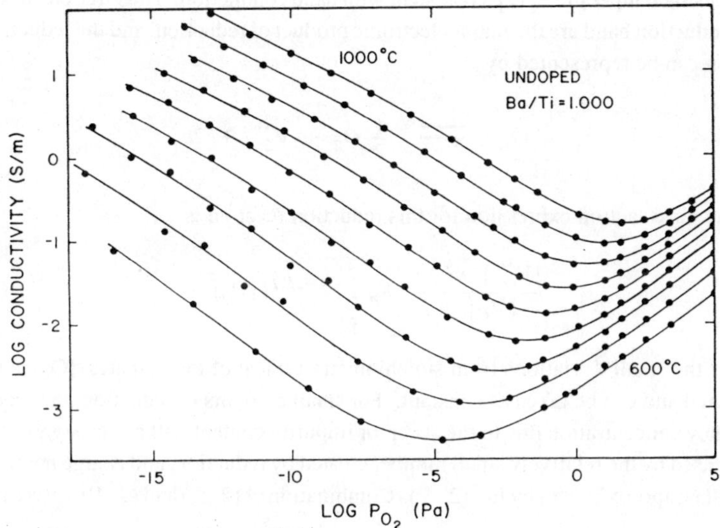

FIGURE 2–1: The equilibrium electrical conductivity of undoped, polycrystalline $BaTiO_3$. The isothermal lines are separated by 50°C intervals [7].

The reduction regime, where the conductivity is increasing with decreasing $P(O_2)$, is separated from the oxidation regime, where the conductivity is increasing with increasing $P(O_2)$, by a minimum in the conductivity. The minima occur at specific combinations of temperature and $P(O_2)$, and correspond to equal contributions to the conductivity by electrons and holes, such that

$$n\mu_n = p\mu_p \tag{2.8}$$

where μ_n and μ_p are the mobilities of the electrons and holes. This is displaced from the stoichiometric composition only by the mobility ratio of the two carriers, if there are no effective trapping centers in the band gap.

2.3.1 The Reduction Regime

For $P(O_2)$ values below that at the stoichiometric composition, and at temperatures where equilibration can occur in reasonable times, e.g. above 873 K, the oxide will lose oxygen to stay in thermodynamic equilibrium with the atmosphere. The departing neutral oxygen atoms will leave behind additional vacant oxygen sites as well as the electrons with which they had been combined in the form of oxygen ions. The positively-charged oxygen vacancies are donor-like centers that could possibly serve as traps for the electrons created by reduction. Since such reduced material is semiconducting near room temperature, with conductivities close to those measured at equilibration temperatures, it is clear that trapping of electrons is not significant, certainly not during equilibration. The donor levels associated with oxygen vacancies must then lie very close to the conduction band. The electron mobility decreases with increasing temperature [11], consistent with band conduction. Thus free electrons in the conduction band are the major electronic product of reduction, and the reduction reaction can be represented by

$$O_O \rightleftharpoons \frac{1}{2} O_2 + V_O^{\cdot\cdot} + 2e' \tag{2.9}$$

The mass-action expression for this reduction reaction is

$$\frac{[V_O^{\cdot\cdot}]\, n^2}{[O_O]} = K_n\, e^{-\Delta H_n/kT} P(O_2)^{-1/2} \tag{2.10}$$

For the small deviations from stoichiometry typical of the titanates, $[O_o]$ is little affected and can be taken as constant. For small amounts of reduction, the oxygen vacancy concentration due to the acceptor impurity content will not be significantly increased by the relatively small number created by reduction, and charge neutrality can still be approximated by Eq. (2.5). Combination of (2.5) and (2.10) gives (no trapping, no complexes)

$$n \approx \left(\frac{2K_n[O_O]}{[A']}\right)^{1/2} e^{-\Delta H_n/2kT} P(O_2)^{-1/4} \tag{2. 11}$$

This pressure dependence of the conductivity is seen on the low $P(O_2)$ side of the conductivity minima in Fig. 2–1. When the reduction reaction becomes the major source of oxygen vacancies, the approximate condition of charge neutrality changes to

$$n \approx 2\ [V_O^{\cdot\cdot}] \tag{2. 12}$$

which in combination with Eq. (2. 10) gives (no trapping, no complexes)

$$n \approx \left(2K_n\ [O_O]\right)^{1/3} e^{-\Delta H_n/3kT} P(O_2)^{-1/6} \tag{2. 13}$$

This dependence of the conductivity on $P(O_2)$ can be seen at the highest temperatures and lowest values of $P(O_2)$ in Fig. 2–1. Experimentally determined values for the enthalpy of reduction, ΔH_n, have been reported to be 5.90 eV per oxygen atom in $BaTiO_3$ [8], and 5.19 eV in $Ba_{0.03}Sr_{0.97}TiO_3$ [12].

Since the electrons that result from reduction remain free to conduct at room temperature, the reduced materials are dark–colored semiconductors. The oxygen vacancies are thus always doubly–charged for conditions of interest. Such low oxygen activities are obviously to be avoided during processing of devices that require that the titanates be insulating, as in ceramic capacitors.

There is experimental evidence at low temperatures for defect complexes formed by electrostatic attraction between the negatively–charged acceptors and the positively–charged oxygen vacancies [13][14]. The defect model just described assumes that there is no interaction between the acceptor centers and the oxygen vacancies under the equilibration conditions. It is easily shown that the behavior of the equilibrium conductivity is inconsistent with significant complex formation at these higher temperatures.

2.3.2 The Oxidation Regime

2.3.2.1 High Temperature Equilibrium Conditions Equilibrium at $P(O_2)$ values above that of the stoichiometric composition is achieved by partial filling of some of the extrinsic oxygen vacancies that are present because of the acceptor impurity content, either accidental or deliberately added. The availability of extrinsic oxygen vacancies gives a very favorable mechanism for oxidation, since the concentration of lattice defects is actually decreased by oxidation. In the absence of such a mechanism, oxidation would result in the formation of either oxygen interstitials or cation vacancies and would be much more costly in enthalpy, and the oxidation regime would occur only far above normal atmospheric pressure. The neutral oxygen atoms that enter the lattice require two electrons to become oxygen ions, and these electrons are taken from the valence band, leaving an equivalent number of holes. The oxidation reaction for the creation of free holes is given by

$$V_O^{\bullet\bullet} + \frac{1}{2} O_2 \leftrightarrows O_O + 2h^{\bullet} \qquad (2.14)$$

The mass-action expression for this oxidation reaction is

$$\frac{[O_O]\, p^2}{[V_O^{\bullet\bullet}]} = K_p\, e^{-\Delta H_p/kT} P(O_2)^{1/2} \qquad (2.15)$$

where again $[O_O]$ can be taken as approximately constant.

Since these oxygen-excess materials are insulators at room temperature, it is clear that the hole concentration has been reduced from the equilibrium value to some very small value in the cooled material. The obvious conclusion is that the holes have been trapped by the negatively-charged acceptor centers at low temperatures. The combination of a hole and a singly-charged acceptor gives the neutral species A^x. The ionization reaction for these trapped holes can be written as

$$A^x \leftrightarrows A' + h^{\bullet} \qquad (2.16)$$

with the mass-action expression

$$\frac{[A']}{[A^x]}\, p = K_A'\, e^{-E_A'/kT} \qquad (2.17)$$

where E'_A is the ionization energy, the height of the acceptor level above the valence band, and the superscript indicates that the acceptor is singly-charged. One must then consider whether or not trapping of the holes is significant under equilibration conditions. If it is not, and the holes are almost all free, then the defects have the following relative magnitudes

$$[A'] \approx 2[V_O^{\bullet\bullet}] \gg p \approx 2\delta \gg [A^x] \qquad (2.18)$$

where δ is the amount of excess oxygen. As long as only a small fraction of the extrinsic oxygen vacancies are filled by oxidation, charge neutrality can still be approximated by Eq. (2. 5), i.e. the first two terms in Eq. (2. 18). When this is combined with Eq. (2. 15), the hole concentration is given by (no trapping, no complexes)

$$p \approx \left(\frac{K_p[A']}{2[O_O]}\right)^{1/2} e^{-\Delta H_p/2kT}\, P(O_2)^{1/4} \qquad (2.19)$$

This dependence of the p-type conductivity on $P(O_2)$ is seen for the $P(O_2)$ region above the conductivity minima in Fig. 2-1. The dependence of the hole concentration on the square root of the acceptor content is also consistent with experimental observations [15]. The enthalpy of oxidation is 0.92 eV per oxygen atom for $BaTiO_3$ [8] and 1.30 eV for $Sr_{0.97}Ba_{0.03}TiO_3$ [12].

The algebraic sum of the reduction reaction, Eq. (2. 9), and the oxidation reaction, Eq. (2. 14), is equal to twice the intrinsic ionization reaction, Eq. (2. 6). Therefore, if the model is correct, the enthalpies of these reactions should be related as

$$\Delta H_n + \Delta H_p = 2E_g^0 \tag{2. 20}$$

For the case of $BaTiO_3$, (5.90 + 0.92)/2 = 3.41 eV, which is in excellent agreement with the value of E_g^o obtained from the temperature dependence of the conductivity minima [8]. The agreement is equally good for the data obtained from $Sr_{0.97}Ba_{0.03}TiO_3$ by Choi et al., (5.19 + 1.30)/2 = 3.24 eV which compares well with their experimental value for E_g^o of 3.26 eV [12]. The defect model that results from the preceding discussion, which neglects hole–trapping under the conditions of equilibration, is summarized in the form of a Kröger–Vink diagram in Fig. 2–2. This is a schematic log–log plot of the concentrations of the various defect species as a function of the oxygen activity at constant temperature.

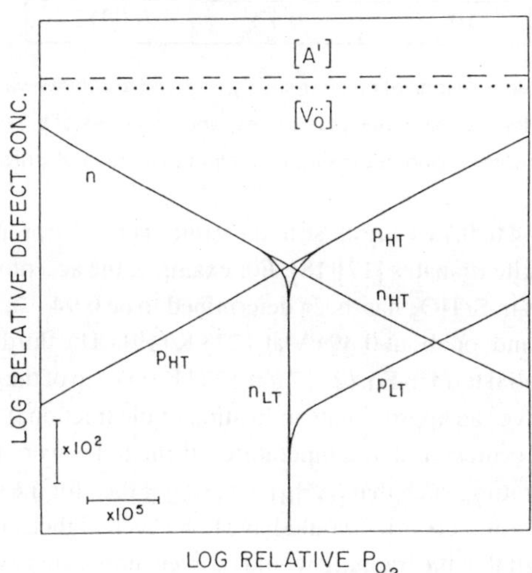

FIGURE 2–2: An idealized Kröger–Vink diagram for an acceptor–doped perovskite for the case of negligible hole–trapping at high temperatures (HT), and extensive hole–trapping at low temperatures (LT) [19].

An estimate of the ionization energies necessary to account for the insulating properties at room temperature can be obtained from the change in conductivity from equilibrium to quenched conditions by taking ratios of Eq. (2. 17) for the two conditions. If the equilibration (high) temperature is taken as 1273 K, designated as HT, and the (low) temperature of measurement is taken as 298 K, designated as LT, then E'_A is given by

$$E'_A \approx 0.077 \log \frac{(p/[A^x])_{HT}}{(p/[A^x])_{LT}} \; eV \tag{2. 21}$$

in which it is assumed that [A'] is the same at both temperatures. For the limit that hole-trapping is extensive at all temperatures, $[A^x]$ will not be significantly changed by quenching, and the log term in Eq. (2. 21) is merely the ratio of the hole concentrations, and can be approximated by the ratio of the conductivities. For the case of $SrTiO_3$, the equilibrium conductivity at 1273 K in one atmosphere of oxygen is $2 \cdot 10^{-3}$ S/cm [12][16], but the room temperature hole conductivity is not accurately known. A reasonable range of values is given in Table I along with the resulting values of the ionization energies obtained from Eq. (2. 21) for the case of a sample quenched from 1273 K to 298 K in one atmosphere of oxygen.

σ_{LT} (S/cm)	E_A (eV)	$e^{-E_A/1273K \cdot k}$
10^{-8}	0.41	0.024
10^{-10}	0.56	0.006
10^{-12}	0.72	0.0015

TABLE I: Calculated hole ionization energies E_A required to achieve various conductivities at room temperature, σ_{LT}, plus corresponding values of the exponential term at 1273 K representing approximate fractional degrees of ionization.

The values of 0.4 to 0.7 eV are close to the values generally proposed for acceptor levels in the perovskite titanates [17][18]. For example, the acceptor level associated with Fe^{+3} on Ti sites in $SrTiO_3$ has been determined to be $0.94 - 3.5 \cdot 10^{-4} \cdot (T/K)$ eV above the valence band, or about 0.49 eV at 1273 K [19]. The third column gives the value of the exponential term in Eq. (2. 17) at 1273 K for each of the calculated ionization energies; this gives an approximate indication of the fractional amount of ionization of the acceptor centers at that temperature. If the holes were mostly free at the equilibration temperature, such that $[A^x]_{HT} << [A^x]_{LT}$, then for the same ratio of conductivities, the ionization energies would have to be even higher. This would lead to the inconsistent result that the trapping would be even more extensive at the high temperature. From this analysis, one can conclude that if trapping is extensive enough near room temperature to give highly insulating materials, it is also likely to be very significant under the high temperature equilibration conditions. As a result, the defect model summarized in Fig. 2–2 needs to be modified. Since only a small fraction of the extrinsic oxygen vacancies are filled by the oxidation reaction, the relative magnitudes of the defect concentrations are then given by

$$[A'] \approx 2[V_O^{\cdot\cdot}] \gg [A^x] \approx 2\delta \gg p \tag{2. 22}$$

The concentration of free holes is still given by Eq. (2. 19), and the concentration of trapped holes can be obtained by combination of Eqs. (2. 15), (2. 17), and (2. 22)

(trapping by A', no complexes)

$$[A^x] \approx \frac{1}{K_A'}\left(\frac{K_p}{2[O_O]}\right)^{1/2}[A']^{3/2}\, e^{-\left(\frac{\Delta H_p}{2}-E_A'\right)/kT}\, P(O_2)^{1/4} \qquad (2.\ 23)$$

An alternative depiction of the oxidation reaction can be obtained by subtracting two times Eq. (2. 16) from Eq. (2. 14) to give the oxidation reaction for the formation of trapped holes

$$2A' + V_O^{\cdot\cdot} + \frac{1}{2}O_2 \leftrightharpoons O_O + 2A^x \qquad (2.\ 24)$$

with the mass-action expression

$$\frac{[O_O][A^x]^2}{[V_O^{\cdot\cdot}][A']^2} = K_{ox}'\, e^{-\Delta H_{ox}'/kT}\, P(O_2)^{1/2} \qquad (2.\ 25)$$

Under these equilibrium conditions, the trapped hole concentration is given by (trapping by A', no complexes)

$$[A^x] \approx \left(\frac{K_{ox}'}{2[O_O]}\right)^{1/2}[A']^{3/2}\, e^{-\Delta H_{ox}'/2kT}\, P(O_2)^{1/4} \qquad (2.\ 26)$$

Comparison of Eqs. (2.23) and (2.26) indicates that the enthalpies are related as

$$\Delta H_{ox}' = \Delta H_p - 2\,E_A' \qquad (2.\ 27)$$

The enthalpy of oxidation to form trapped holes is less than that to form free holes by twice the acceptor ionization energy.

A modified Kröger–Vink diagram that incorporates trapped holes at a level higher than that of free holes, and is shown in Fig. 2-3 [19].

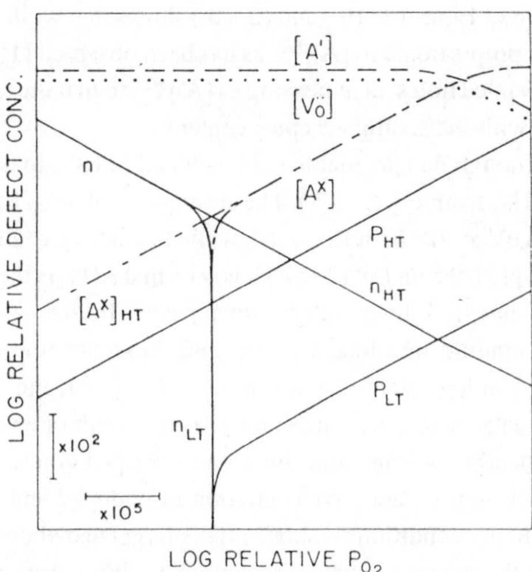

FIGURE 2-3: An idealized Kröger–Vink diagram for an acceptor-doped perovskite with predominant hole-trapping at both high and low temperatures. (Reproduced from [19] with permission from the American Ceramic Society.)

The trapped holes and the free holes have the same dependence on $P(O_2)$. However, it should be noted that the stoichiometric composition is displaced from the conductivity minima and corresponds to the condition

$$n \approx 2[A^x] \tag{2. 28}$$

The stoichiometric composition occurs where the sum of the concentrations of all electron species, trapped and untrapped, equals the sum of the concentrations of all hole species, trapped and untrapped. In the case of the perovskite titanates, this appears to occur when the concentration of free electrons equals the concentration of trapped holes. As a result, it should be possible to obtain insulating materials for equilibration with $P(O_2)$ values below the conductivity minima, down to that at the stoichiometric composition. This is of some practical interest in the processing of multilayer ceramic capacitors with base–metal electrodes, such as nickel. The sintering of such devices must occur at an oxygen activity that is low enough to avoid oxidation of the nickel electrodes, but high enough to avoid reduction of the ceramic dielectric to a semiconducting state. It has been thought that this required oxygen activities higher than that at the conductivity minima, but, if hole trapping is extensive, the stoichiometric composition will occur at an oxygen activity below that at the minimum, and the sintering conditions can be extended to more reducing conditions.

The same $P(O_2)$–dependence of the hole conductivity is predicted whether or not trapping is significant, and the relationship between the enthalpies of hole and electron formation and the band–gap given by Eq. (2. 20) is still valid in both cases. Thus measurement of the equilibrium conductivity does not distinguish between the two models.

Comparison of Eqns. (2. 19) and (2. 23) shows that while the equilibrium hole concentration is proportional to $[A']^{1/2}$, as has been observed [15][16], the concentration of A^x, the trapped holes, is increasing as $[A']^{3/2}$, so that the ratio of trapped to free holes increases with increasing acceptor content.

There are enough data available for Fe–doped $SrTiO_3$ to determine an approximate value of ΔH_{ox} from Eq. (2. 27). The acceptor level, E'_A, for this case was found to be about 0.60 eV at 973 K in Ref. [19], while the data of Choi et al. give a value of 1.29 eV for ΔH_p [12]. From Eq. (2. 27) it is seen that ΔH_{ox} is then 0.09 eV, indicating that Eq. (2. 23) has little temperature dependence. In that case, under equilibrium conditions, the amount of added oxygen, and the concentration of trapped holes, would be almost independent of temperature. However, the concentration of free holes increases with increasing temperature, as the result of an increasing amount of ionization of a nearly constant concentration of trapped holes.

Since the oxygen vacancy concentration in undoped and acceptor–doped titanates under oxidizing conditions is substantially larger than the concentrations of electronic carriers at the stoichiometric composition, it is not surprising that ionic conduction has been detected at high temperatures. It is particularly noticeable near the conductivity minima, where the electronic contributions to the total conductivity are the least. As is apparent from Fig. 2–3, at $P(O_2)$ values sufficiently far above that at the minimum so that the electron conductivity can be neglected, the total conductivity, σ_T, should be a combination of hole conduction varying as $P(O_2)^{1/4}$, and an oxygen vacancy contribution that is independent of $P(O_2)$

$$\sigma_T \approx \sigma_I + \sigma_p^o P(O_2)^{1/4} \tag{2.29}$$

Thus a plot of the total conductivity, σ_T, versus $P(O_2)^{1/4}$ should give a straight line whose slope is the hole conduction at one atmosphere, σ_p^o, and whose intercept is the vacancy contribution, σ_I. A similar argument can be made for the reduction side of the conductivity minima, where the electronic conductivity is n-type. Plots of this type are shown in Fig. 2-4 for undoped and acceptor-doped $BaTiO_3$ where it is seen that plots from both sides of the minima give the same intercept. Oxygen concentration cell measurements have quantitatively confirmed these values of ionic conductivity [20].

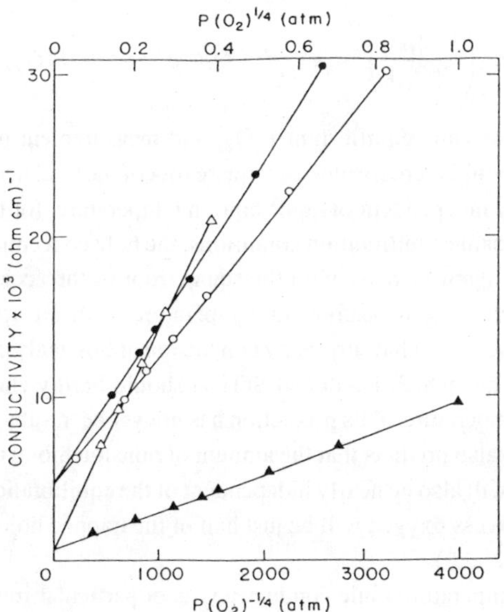

FIGURE 2-4: The total equilibrium conductivity of $BaTiO_3$ vs. $P(O_2)^{\pm 1/4}$. Triangles represent undoped $BaTiO_3$ while the circles represent an acceptor-doped composition, $BaTi_{0.99}Ca_{0.01}O_{2.99}$. Solid symbols refer to p-type conduction and the upper abscissa, while open symbols refer to n-type conduction and the lower abscissa. (Reproduced from [20] with permission from the authors and from the American Ceramic Society.)

2.3.2.2 Low Temperature Properties In terms of the room temperature properties, the charged acceptor centers are very effective hole traps, i.e. the acceptor levels lie quite far above the valence band. As a result, the holes are thoroughly trapped at ordinary use temperatures, and the materials are insulators. On the other hand, donor levels such as those associated with oxygen vacancies lie very close to the conduction band, and the reduced materials are semiconductors near room temperature. This difference in the depth of donor and acceptor levels is typical of transition metal oxides.

In compounds that contain a reducible cation, such as Ti^{+4}, donor levels are usually very shallow, while in the absence of an oxidizable cation, the acceptor levels are quite deep.

It appears that a model that involves extensive hole-trapping under equilibration conditions is the most realistic. In this case, since only a small fraction of the extrinsic oxygen vacancies are filled by the oxidation reaction, i.e. $[A'] >> \delta$, $[A^x]$ will be much less than $[A']$ under equilibrium conditions and this will still be true near room temperature. In the case of extensive trapping at the equilibration temperature, the trapped hole concentration near room temperature will be essentially the same as that at equilibrium, i.e. Eq. (2. 26), with the temperature being that at equilibrium. With these conditions, Eq. (2. 26) can be substituted into Eq. (2. 17) to give an expression for the hole concentration in quenched samples (for the case $p_{HT} << [A^x]_{HT}$)

$$p_{LT} \approx K'_A \left(\frac{K'_{ox}[A']}{2[O_O]} \right)^{1/2} e^{-\Delta H_{ox}/2kT_{HT}} \, e^{-E'_A/kT_{LT}} \, P(O_2)_{HT}^{1/4} \qquad (2.\ 30\ )$$

Thus for the same equilibrium $P(O_2)$ and measurement temperature, the hole concentration should increase with the square root of the total net acceptor excess, and should be nearly independent of equilibration temperature, for the case of Fe-doped $SrTiO_3$. For the same equilibration conditions, the hole concentration near room temperature should again increase with the square root of the acceptor content, and increase with increasing measurement temperature with an activation enthalpy of $E'_A \sim 0.6$ eV. The somewhat surprising conclusion of this analysis is that the bulk hole conductivity of quenched, Fe-doped $SrTiO_3$ should hardly change with increasing equilibration temperature. This prediction has not yet been subjected to experimental test. The model also predicts that the amount of nonstoichiometry, i.e. the amount of excess oxygen, will also be nearly independent of the equilibration temperature, since the amount of excess oxygen will be just half of the trapped hole concentration as expressed in Eq. (2. 22).

The low temperature ionic conductivity is of particular importance for the consideration of degradation mechanisms for devices subjected to voltage-temperature stress. It is generally accepted that oxygen vacancies are the only significantly mobile ionic defects. If complexes between acceptors and oxygen vacancies are in the minority even at low temperatures, the vacancy concentration will be given by $[V_O^{\cdot\cdot}] = 1/2[A']$ or $[V_O^{\cdot\cdot}] = [A'']$ and will thus be temperature independent. For either kind of acceptor, the oxygen vacancy dependence on the acceptor content will become increasingly sublinear with increasing importance of complex formation. For acceptor (Ni)-doped $SrTiO_3$, the increase of ionic conduction with increasing Ni concentration was indeed found to be much less than proportional [19]. This is consistent with a significant influence of complex formation.

2.4 Donor-Doped Perovskite Titanates

A cationic donor impurity has a higher formal oxidation state than the host cation it replaces. Thus its oxide carries more oxygen per cation than does the binary oxide of the host cation it replaces, e.g. La_2O_3 substituted for BaO or Nb_2O_5 substituted for TiO_2 in $BaTiO_3$. The electrical properties of the resulting material are crucially dependent on the fate of the extra oxygen. As in the case of the acceptor–doped compounds, the stoichiometric composition for donor–doped materials will correspond to the combination of the stoichiometric binary components, e.g. (1–x)BaO + (x/2)La_2O_3 + TiO_2 or $Ba_{1-x}La_xTiO_{3+x/2}$, or BaO + (1–x)TiO_2 + (x/2)Nb_2O_5 or $BaTi_{1-x}Nb_xO_{3+x/2}$. However, since the perovskite structure does not tolerate oxygen interstitials, these formulas cannot be literally correct. Under reducing conditions, the excess oxygen is lost to the atmosphere, leaving behind an equivalent charge in excess electrons to compensate the charged donor centers, e.g. $Ba_{1-x}La_xTie'_xO_3$ and $BaTi_{1-x}Nb_xe'_xO_3$. Because the donor levels are so close to the conduction band in the titanates, they are semiconductors at room temperature, and even far below. They are also nonstoichiometric by definition, because of the loss of some of the oxygen associated with the original binary components. When subsequently equilibrated under oxidizing conditions, and when x > 0.005, these materials quantitatively gain back the amount of oxygen necessary to restore the stoichiometric state and an insulating condition [21]. Thus if the extra oxygen associated with the donor oxide is retained in the lattice, the titanates are stoichiometric insulators, while if it is lost, the materials are nonstoichiometric, n–type semiconductors.

Since interstitial oxygen does not occur in any detectable amount in the perovskite structure, the positively–charged donor centers must be compensated by cation vacancies in the oxidized, stoichiometric state. Indeed, there is convincing evidence that donors are compensated by Ti vacancies in the stoichiometric composition of $BaTiO_3$ [22][23], while there is circumstantial evidence that Sr vacancies are preferred in $SrTiO_3$ [24]. Compositions in these two cases can be written $BaTi_{1-5x/4}Nb_xO_3$ and $Sr_{1-x/2}Ti_{1-x}Nb_xO_3$. Thus the incorporation of Nb_2O_5 into $BaTiO_3$ under reducing conditions can be written as

$$BaO + (1-x)TiO_2 + (x/2)Nb_2O_5 \rightarrow Ba_{Ba} + (1-x)Ti_{Ti} + xNb_{Ti}^{\cdot} + 3O_O + (x/4)O_2 + xe' \qquad (2.31)$$

The product is a single–phase, nonstoichiometric, semiconducting material. Upon oxidation, the lost oxygen is regained and a Ti–rich phase, $Ba_6Ti_{17}O_{40}$, is exsolved in order to leave behind the proper amount of Ti vacancies [23]. If the exsolved phase is simplified as TiO_2, the oxidation reaction can be written as

$$Ti_{Ti} + O_2 + 4e' \leftrightarrows TiO_2 + V_{Ti}'''' \qquad (2.32)$$

The reaction for forming single–phase, stoichiometric, insulating material directly from the binary components under oxidizing conditions is

$$BaO + (1 - 5x/4)TiO_2 + (x/2)Nb_2O_5 \rightarrow Ba_{Ba} + (1 - 5x/4)Ti_{Ti} + xNb_{Ti}^{\cdot} + (x/4)V_{Ti}'''' + 3O_O \qquad (2.33)$$

$BaTiO_3$ represents an unusual case, since for donor concentrations up to about 0.5 atom %, the donors are compensated by electrons according to Eq. (2. 31), even when the processing is carried out in air [21][25]. For larger donor concentrations, there is an abrupt switch to compensation by Ti vacancies, and a corresponding abrupt switch from semiconducting to insulating properties. The change is accompanied by a similarly abrupt reduction in grain size. While no satisfactory explanation has yet been proposed for the abrupt change in grain-size, the change in defect chemistry is elegantly explained by Wernicke on the base of the small diffusion coefficient of cation vacancies [26].

The presence of donor impurities can substantially suppress the concentration of oxygen vacancies, regardless of the compensating defect. If charge compensation is achieved by cation vacancies, then the oxygen vacancies are reduced through the Schottky equilibrium, Eq. (2. 2). If electrons are the compensating species, then the oxygen vacancies are suppressed through the reduction equilibrium, Eq. (2. 10), [25]. An example of a schematic defect diagram for the case of cation vacancy and electron compensation is shown in Fig. 2-5.

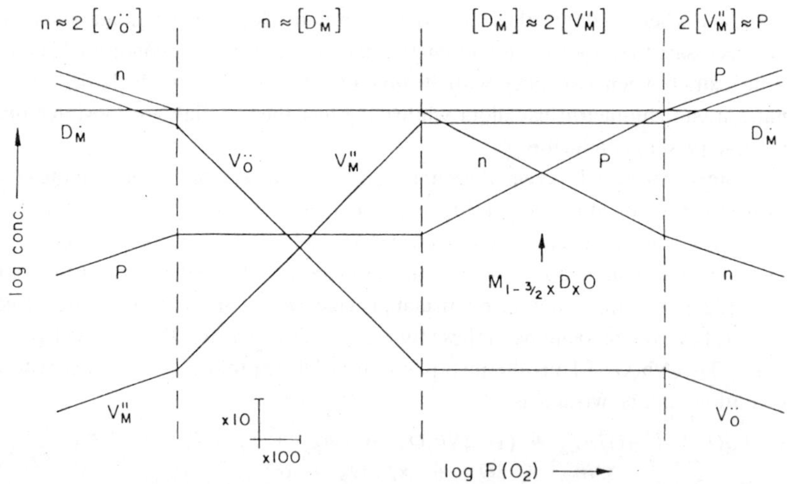

FIGURE 2-5: An idealized Kröger-Vink diagram for a hypothetical donor-doped oxide, MO

Since the concentrations of both cation vacancies and electrons are normally very low under oxidizing conditions, perhaps of the order of parts per million, and can reach the order of percent in donor-doped compositions, the suppression of oxygen vacancies can be very substantial. This can be observed as a drastic reduction in equilibration rates with changes in the oxygen content of the ambient [7] [25]. It also results in a reduction in the concentration of ionic mobile species that can drift in an electric field under practical use conditions. This can be a great advantage in reducing the rate of device degradation under dc electrical fields (see Section 4.3).

For compounds that are significantly deficient in one of the binary constituents, the effectiveness of donor impurities in suppressing the oxygen vacancy content may be greatly reduced. This is particularly pertinent for compounds in which one of the binary constituents is volatile, as in the case of $PbTiO_3$. Thus if the PbO–deficiency is greater than the donor content, the excess oxygen associated with the donor oxide can be accommodated by partial filling of the oxygen vacancies that result from the cation nonstoichiometry. In effect, oxygen vacancies are partially replaced by the donor centers as charge–compensating defects for the lead vacancies. In this type of situation, donor impurities will not greatly suppress the oxygen vacancy concentration unless the loss of PbO is avoided, or, in a more general sense, unless both sets of cation sites are kept almost ideally filled.

2.5 Summary of Defect Chemistry

* For applications that require insulating properties, the perovskite titanates must be in an oxygen–excess state of nonstoichiometry.
* The strong trapping of holes by acceptor centers in oxygen–excess material implies that there will be significant, and perhaps dominating, hole–trapping even at equilibration temperatures.
* When hole–trapping is important at equilibration temperatures, the stoichiometric composition will generally occur at oxygen activities significantly below those at the conductivity minima.
* Even though the p–type equilibrium conductivity increases with increasing equilibration temperature , it is likely that the amount of excess oxygen increases much less with increasing temperature, and may even decrease in some cases.
* The formation of defect complexes between acceptor centers and oxygen vacancies does not seem to play an important role at equilibration temperatures. They are known to occur at lower temperatures, but the extent of complex formation is not accurately known.
* Donor impurities can greatly suppress the concentration of oxygen vacancies, but only when the A/B cation ratio is maintained near unity with a precision that exceeds the donor content.

3. Charge Transport through Titanate Systems

3.1 Impedance Analysis in the Time Domain

The impedance of real samples is not only determined by the conductivity of the bulk crystal lattice discussed in Section 2. The electrode interfaces (EI) and – in the case of ceramics – the grain boundaries (GB) contribute to the total impedance as illustrated by the simplified equivalent network in Fig. 3–1. In principle, a further RC branch would have to be included due to the reorientation of defect complexes in the bulk lattice, i. e. associates between acceptors and oxygen vacancies (see Sec. 2.3.3). As described in Sec. 3.3, however, these complexes were found to play no dominant role.

In the present study, impedance analyses in the time domain are reported. The method is based on recording the time evolution of the current response after applying a dc voltage step. Details of this method and the transformation between the time and the frequency domain are described in Ref. [42]. It is well known that ionically conducting ceramics as, for instance, stabilized ZrO_2 [27] as well as mixed conducting ceramics such as titanates [28] [29] tend to build highly resistive layers at electrode and GB interfaces. These are commonly interpreted as depletion space charge layers similar to Schottky barriers at semiconductor interfaces [30]. Consequently, the average conductivities in these layers, σ_{EI} and σ_{GB}, are usually very small compared to the bulk conductivity σ_B because the corresponding layers are depleted by the charge carriers. In conventional ceramics, the depletion layer thicknesses d_{GB} and d_{EI} are much smaller than the grain size:

$$d_{gr} \gg d_{GB}, d_{EI} \quad . \tag{3. 1}$$

Hence, the capacities generated by the interfaces (i. e. branches 'EI' and 'GB' of the equivalent network in Fig. 3–1) are large compared to the capacity caused by the bulk of the grains (branch 'B'). In addition, the GB contributions perpendicular to the electrodes can be neglected because of relation (3. 1). Still, there is a certain mobility of oxygen ions along GBs. This becomes obvious in either donor doped titanates where the ionic conduction in the bulk of the grains is negligible (see Sec. 2.4) or in very fine grain acceptor doped ceramics where relation (3. 1) is not valid any more (see Sec. 3.4). The permittivity in the depletion layers, ε_{GB} and ε_{EI}, is approx. equal to the permittivity ε_B of the crystal lattice in paraelectric ceramics as has been found recently for $SrTiO_3$ [31],i.e

$$\varepsilon_{GB} \; , \varepsilon_{EI} \;\; \approx \;\; \varepsilon_B \tag{3. 2}$$

Usually, the large number of grain boundaries crossing the current path between the electrodes outweighs the influence of the just two electrode interfaces (EI) on the impedance of a ceramic component. Applying a dc voltage step to a network (Fig. 3–1) with two dominating branches leads to a Maxwell–Wagner polarization. In Fig. 3–2, the voltage stimulation as well as the current response is sketched for a charging period and a discharging period. At very short times, when the space charge polarization at the grain boundaries is still negligible, the current density J is determined by a short–term conductivity $\sigma_s := \sigma(t \rightarrow 0)$. If the relation (3. 1) is valid, σ_s is approx. equal to the bulk

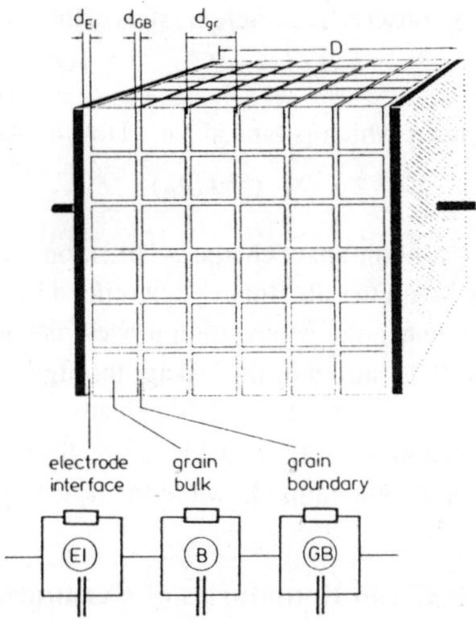

FIGURE 3-1: Simplified block model of a ceramic and simplified electrical equivalent network illustrating the conduction and polarization contributions. The contributions of both electrode interfaces are merged into the branch 'EI'. The contributions of all grains and all grain boundaries are merged into the branch 'B' and the branch 'GB', respectively.

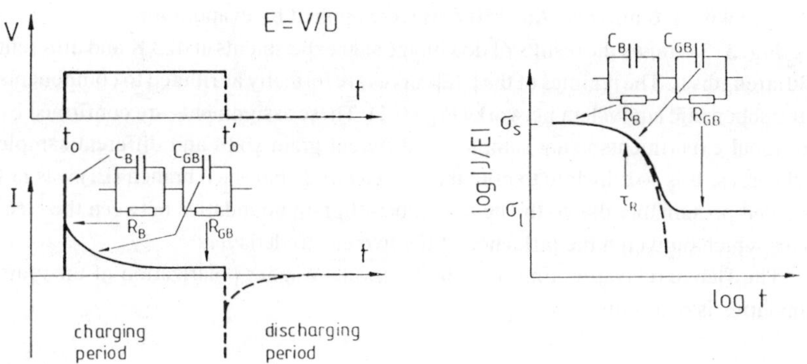

FIGURE 3-2: Sketch of a dc voltage step experiment. The voltage stimulation V(t) and the current density response J(t) are illustrated for a charging period starting at t_0 and a discharging period starting at t_0'. In the present paper, the response is shown on a log-log scale. Furthermore, the unsigned current density - field ratio $|J/E|$ is often used in order to obtain data in the conductivity unit and to compile the charging and discharging response in one diagram. For an explanation of the short term conductivity, σ_S, the Debye relaxation time, τ_R, and the long-term conductivity σ_l: see text.

conductivity σ_B which is also observed in single crystals of the same composition, i. e.

$$\sigma_B \approx \sigma_s \tag{3.3}$$

The current decays in a manner which is typical for a Debye relaxation:

$$J \propto \exp(-t/\tau_R) \tag{3.4}$$

where τ_R is the relaxation time of the space charge polarization at the grain boundaries (and should not be confused with the life time τ_L determined by the degradation, see Sec. 4.3). After a long time, when the polarization process is complete, the residual conductivity σ_l remains left. It is caused by the leakage through the GB depletion layers. Switching off the applied voltage leads to a depolarization current of reversed sign which proceeds until the capacitive interface elements are discharged. This discharging behaviour is included in the diagrams shown in the next three sections.

3.2 Conduction through Grain Boundaries of Ceramics

In this paper, Ni acceptor doped $SrTiO_3$ is employed for a comparison between ceramics, single crystals, and thin films [32]. A ceramic of 0.1 at% Ni–doped $SrTiO_3$ was prepared using the conventional mixed–oxide technique. For details see Ref. [19]. After sintering at 1613 K for 6 h in oxygen and subsequent isostatic hot–pressing at 1573 K for 1 h, a porefree ceramic with a uniform microstructure and a grain size of 2.5 ± 1.0 μm was obtained. Discs of 125 μm, 250 μm, and 500 μm thickness were cut, lapped, polished, and annealed at 973 K for 6 h in oxygen. After quenching, electrodes of NiCr alloy (thickness: 6 nm) and Au (150 nm) were applied by evaporation.

Fig. 3–3a shows the results of dc voltage step experiments at 423 K and different field strengths E. The regimes of the |J/E| curves are formally attributed to components of the schematic equivalent network (Fig. 3–1). These assignments are confirmed by additional experiments using samples of different grain sizes and different sample thicknesses. It is worthwhile to note that the electrode interface branch (EI) was not observed presumably due to the large number of grain boundaries between the electrodes which outweigh the influence of the two electrode layers.

The Debye relaxation time τ of the Maxwell–Wagner polarization of the grain boundaries is determined by

$$\tau_R = \frac{\varepsilon_{eff}\,\varepsilon_0}{\sigma_s} \tag{3.5}$$

where ε_{eff} denotes the effective permittivity of the material when the grains act as shorts i. e. in the long term behaviour. ε_0 is the vacuum permittivity (8.854 10^{-14} As/Vcm). Based on geometrical considerations for the simple block model of the ceramic and relations (3. 1) and (3. 2), ε_{eff} obviously is determined by the ratio of the grain size d_{gr} to the GB depletion layer width d_{GB}:

$$\varepsilon_{eff} / \varepsilon_{GB} = d_{gr} / d_{GB} \tag{3.6}$$

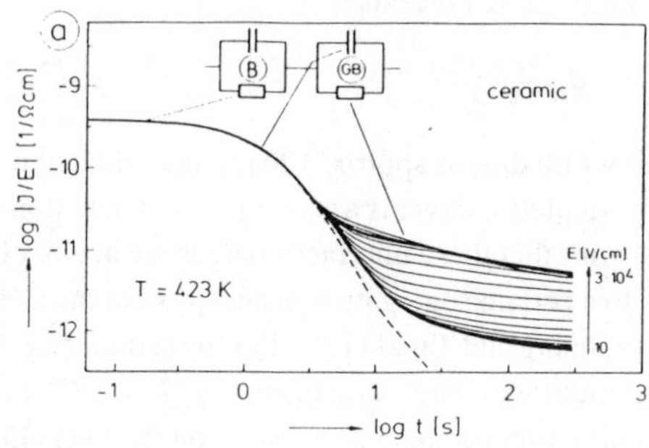

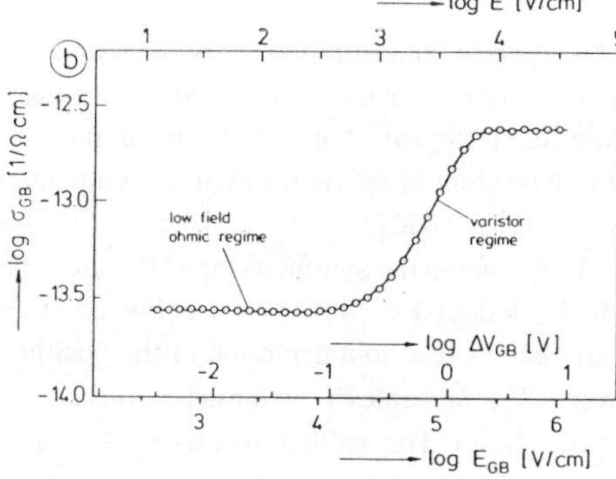

FIGURE 3-3 :
(a) Current density-field ratio $|J/E|$ vs. time t for a fine grained, 0.1 at% Ni-doped $SrTiO_3$ ceramic at T = 423K for different dc voltage steps corresponding to dc fields in the range E = 10 V/cm to $3 \cdot 10^4$ V/cm. The determining components of the equivalent network are sketched on top. The discharging curve (dashed line) is shown for E = 300 V/cm. (b) Field dependence of the conductivity at t = 300 s in diagram (a), interpreted as the conductivity in the GB space charge layer σ_{GB} [27]. Abszissa from top to bottom: externally applied field at the sample E = V/d, voltage drop at a grain boundary ΔV_{GB}, and field across a GB depletion layer E_{GB}. Thickness of the sample: D = 500 μm. Average grain size: d_{gr} = 2.5 μm.

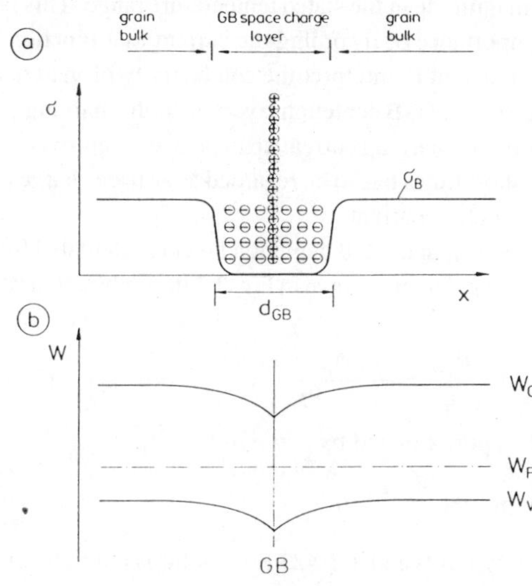

FIGURE 3-4:
Sketch of the conductivity profile (a) and the energy band diagram (b) across a GB in acceptor doped titanate ceramics.
In (a), (+) indicate the positively charged (i. e. donor type) GB interface states and (–) illustrate the uncompensated acceptor centers forming a negative space charge region. In (b), W_C, W_V, and W_F are the conduction and valence band edges and the Fermi level, respectively.

Inserting Eqs. (3. 3) and (3. 5) into (3. 6) reveals

$$d_{GB} = d_{gr} \frac{\varepsilon_{GB} \ \varepsilon_0}{\tau \ \sigma_B} \qquad (3.7)$$

With Eq. (3. 2), a GB depletion layer width d_{GB} of approx. 100 nm is evaluated from the data in Fig. 3–3a. The origin of the depletion layer is assumed to be due to donor–type interface states. There are indications that these interface charges are active even during the sintering process and lead to a certain segregation of acceptors at the GB as reported in a comprehensive study by Chiang and Tagaki [33]. For the ceramic used in the present study, a relatively high GB interface charge Q_{GB} of approx. $7 \cdot 10^{-5}$ C/cm^2 is determined using the Schottky approximation for depletion space charge layers [34]:

$$Q_{GB} \approx e_0 \ d_{GB} \ z_{Ni,av} \ [Ni]_{av} \qquad (3.8)$$

In this relation, $[Ni]_{av}$ is the average Ni acceptor concentration in the depletion layer taking into account the segregation. $z_{Ni,av}$ represents the average valency of the Ni centers related to the substituted Ti ions and is approx. 2. e_0 denotes the unit charge.

The magnitude of d_{GB} is almost independent of the field and of the temperature within a measured range from approx. 350 K to 750 K [31]. This is typical for a Schottky–type depletion space charge layer [30]. Due to the symmetry of a GB, the depletion layer can be regarded as a back–to–back double Schottky barrier (Fig. 3–4). According to this model, donor–type GB charges lead to a depletion of the positively charged mobile carriers, oxygen vacancies $V_O^{\cdot\cdot}$ and holes $h^{\cdot}$. In the depletion layer, the immobile acceptors form a negative space charge. The width d_{GB} of the depletion layer is established in such a way that the positive GB charges are just compensated by the negative space charge. Fig. 3–4b qualitatively illustrates the corresponding bending of the energy bands.

Since the bulk conductivity σ_B is thermally activated (see Sec. 2), the relaxation time τ_R varies by many orders of magnitude in the stated temperature range. This fact emphasizes the often neglected importance of recording the current as a function of time when it is intended to determine and to interpret the conductivity of materials. Another clear indication for the presence of GB depletion layers is the discharging current which is indistinguishable from the charging current up to (at least) approx. $t \approx \tau_R$ (Fig. 3–3a). Hence, the current at short times has to be regarded as a space charge polarization current and not as a conduction current.

The average local conductivity σ_{GB} in the GB depletion layer is calculated from the long–term conductivity σ_l and the geometry shown in Fig. 3.1 in combination with relation (3. 1):

$$\sigma_{GB} = \frac{d_{GB}}{d_{gr}} \cdot \frac{\sigma_l \ \sigma_B}{\sigma_B - \sigma_l} \qquad (3.9)$$

For $\sigma_l << \sigma_b$, this equation can be approximated by

$$\sigma_{GB} = \frac{d_{GB}}{d_{gr}} \cdot \sigma_l \qquad (3.10)$$

For 0.1 at% Ni–doped $SrTiO_3$ at low fields and T = 423 K, σ_{GB} turns out to be more than **four** magnitudes below the conductivity σ_b in the bulk of the grains (Fig. 3–3a,b).

In combination with the fact that the ionic conductivity in the crystalline bulk of e. g. Ni–doped $SrTiO_3$ is only slightly dominating over the hole conductivity (see Sec.2.3.2.2), this means that the GB depletion layers build significant **barriers for the cross transport of both ionic and electronic charges**. A simplified sketch of a conductivity profile cross section is shown in Fig. 3–7a.

In Fig. 3–3b, σ_{GB} is shown as a function of the externally applied field E = V/D. After the Maxwell–Wagner polarization is settled, the grain bulk regions are virtually field free. Based on this, additional abscissa are included in Fig. 3–3b showing the actual field across the GB depletion layer E_{GB} and the voltage drop $\Delta V_{GB} = E_{GB} \cdot d_{GB}$. At a low voltage drop, σ_{GB} exhibits an ohmic behaviour. Above approx. ΔV_{GB}= 0.5 V, σ_{GB} starts to enhance strongly with incremented voltages. This varistor–type regime is succeeded at very high fields by another ohmic regime. Experimentally, a further increase of the external field E is not possible although the ceramic does not yet suffer an instantaneous breakdown. Rather, the onset of the resistance degradation (see Sec. 4.3) is shifted to such short times at very high fields that the Maxwell–Wagner polarization of the grain boundaries remains incomplete.

The ohmic regime at low voltages and the varistor regime of the GB depletion space charge layer in n–type semiconducting ceramics has been consistently interpreted in terms of *charge transport over a Schottky barrier* (see Ref. [34] and references cited therein). In these ceramics, the energy distribution of the densities of interface states $N_{GB}(W)$ was found to affect the voltage ΔV_{GB} at which the enhancement of the conductivity sets in, the slope of the conductivity increases, and the fine structure of the J–E curve. We assume a similar process to be responsible for the charge transport across the GB depletion layers in the case of *acceptor* doped titanates. Further details and possible implications of the predominant ionic conduction superimposed on the p–type conduction in the titanates will be reported in Ref. [35].

3.3 Conduction through Electrode Interfaces at Single Crystals

In ceramics, the large number of grain boundaries masked the influence of electrode interfaces in the overall device impedance according to Fig. 3–1. In single crystals, however, the effects of the electrode interface branch 'EI' and the bulk branch 'B' of the equivalent network can be investigated.

Samples were cut from a single crystal boule of 0.1 at% Ni–doped $SrTiO_3$ (Commercial Crystals Inc.), lapped, and polished to optical quality. Any mechanical damage to the surface was found to be below the detectability limit of a SEM [36]. An example of voltage step measurements is given in Fig. 3–5. Based on an interpretation according to the network model of Fig. 3–1, the electrode interface give rise to depletion space charge layers similar to grain boundaries in ceramics. The same experiments were performed for different surface treatments (e. g. annealing at 973 K in oxygen before or after polishing) and for different electrode metals (e. g. Au, Al, Cu, Pt). Although in all cases a Maxwell–Wagner relaxation was observed, the surface treatments and the type of metal clearly showed an influence on the characteristics of the results. This indicates that the polarization is indeed – at least to large extent – caused by *space charge effect at the electrode interfaces.* Still, as an additional effect the *reorientation of defect complexes* (Sec. 2.3.3 and 2.3.4) in the bulk of the crystal may well be present. As mentioned in Sec. 3.1, such a reorientation would show electrically a Debye relaxation behaviour as well. Up to now, the contribution of this effect is not yet studied.

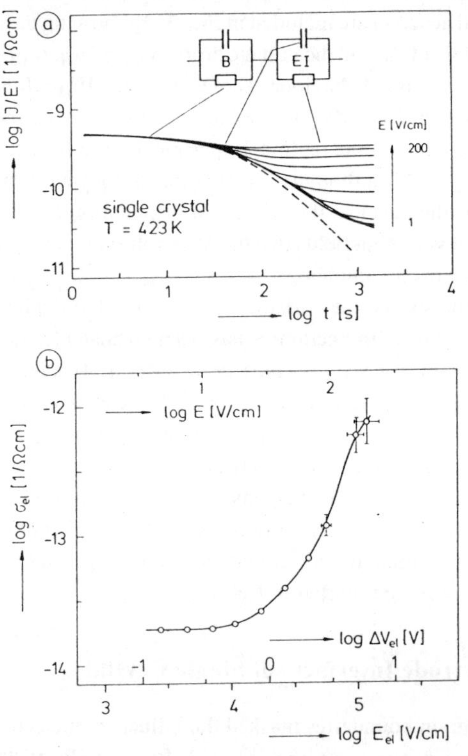

FIGURE 3-5:(a) Current density – field ratio |J/E| vs. time t for a 0.1 at% Ni-doped $SrTiO_3$ single crystal at T = 423 K for different dc voltage steps corresponding to dc fields in the range E = 1 V/cm to 200 V/cm. The determining components of the equivalent network are sketched on top. The discharging curve (dashed line) is shown for E = 50 V/cm.
(b) Field dependence of the conductivity at t = 1500 s in diagram (a), interpreted as the conductivity in the electrode interface space charge layers σ_{EI}. Abszissa from top to bottom: applied field at the sample E = V/D, voltage drop at an electrode interface ΔV_{EI} assuming equal distribution at both electrodes, and field across a depletion layer E_{EI}. For three data error bars are included indicating the strong decrease in the reliability when σ_I gets close to σ_S at high fields. Thickness of the sample: D = 1500 μm.

Based on the evaluation according to an electrode interface polarization, a total depletion layer width of approx. 1000 nm can be calculated from the data using Eq. (3. 7) appropriately. For a low voltage step stimulation, we assume that the total layer width of approx. 1000 nm in Fig. 3–1a is equally distributed between both electrodes, i. e. d_{el} = 500 nm. At higher voltages, the widths at the anode and cathode may be different. This aspect awaits further clarification using samples with asymmetric electrodes. Compared to the GB depletion layer d_{GB}, the depletion layer at the electrode d_{EI} is unexpectedly broad. This may either be attributed to a high surface charge or it may be enhanced by artefacts introduced during the polishing procedure. For instance, such an artefact could be a near-surface layer of a high dislocation density possibly establishing additional donor states or Ruddlesdon–Popper phases [37].

Fig. 3–5b illustrates the field dependence of the conductivity σ_{EI} in the depletion layer evaluated in the same way as σ_{GB} of the ceramics adapting Eq. (3. 9). A comparison of Fig. 3–3b and Fig. 3–5b shows that the conductivity in the depletion layers of grain boundaries and at electrodes are quite similar. In the low field ohmic regime, σ_{EI} is in the same order of magnitude as σ_{GB} in ceramics indicating the same situation concerning the degree of depletion of charge carriers. Above a voltage drop ΔV_{EI} in the depletion layer of approx. 0.5 V (as at grain boundaries in ceramics), the conductivity departs from the ohmic behaviour and exhibits a pronounced varistor–type behaviour. At high voltages, σ_I gets close to σ_s and, hence, the reliability of the data evaluated using Eq. (3. 10) drastically decreases (as indicated by the error bars in Fig. 3–5b). Based on the present data it is therefore not possible to judge whether the slope of the varistor regime of σ_{EI} decreases towards high fields before σ_{EI} comes close to σ_B.

As mentioned, both the width of the depletion layer d_{EI} as well as the conductivity σ_{EI} and its field dependence are influenced by the annealing before or after the polishing and by the electrode metal. Details will be given in a forthcoming paper [35].

Depletion space charge layers are present at the *free surfaces* of acceptor doped titanate crystals as well, caused by donor type surface states. In the transition region between the LT and the HT regime, these depletion layers decrease the rate of the uptake or release of oxygen according to Eqs. (2. 9) and (2. 14). This aspect is relevant for the use of acceptor doped $SrTiO_3$ as material in resistive oxygen sensors. Quantitative data are reported in comprehensive studies based on dynamic resistance measurements [38] and an optical absorption technique [39].

3.4 Conduction through Thin Films

A modified sol–gel process has been developed to deposit doped and undoped $SrTiO_3$ and $BaTiO_3$ films on Si(100) substrates with a Ti/Pt electrode layer [40]. Solutions containing Ba, Sr, Ti, Ni, Mn, and Ce as metal alkoxides, metal carboxylates, and metal acetylacetonates dissolved in butanol were deposited by spin coating and subsequent firing at 773 to 973 K in oxygen. Perovskite films with thicknesses of 55 nm and 110 nm per coating were obtained using 0.2 M and 0.4 M solutions, respectively. To increase the thickness up to the desired value, the deposition process was repeated several times. X–ray diffraction (XRD) and scanning electron microscopy (SEM) studies revealed that all the films reported here crystallize in a single phase perovskite structure with a random orientation of the crystals on the substrate plane. Due to the low reaction temperatures of the sol–gel process extremely fine grained ceramic films with grain sizes of 40–90 nm are obtained as indicated both by SEM as well as by evaluating the broadening of the XRD lines.

Fig. 3–6 shows the charging and discharging behaviour of a 0.1 at% Ni–doped $SrTiO_3$ thin film for a voltage V = 2 V (representing $E = 3.64 \cdot 10^4$ V/cm) for different temperatures. As in the ceramic and the single crystal sample, the polarization current predominates in the beginning while at long times the current is determined by the conduction through the film. The transition times τ_T between the polarization and the con–

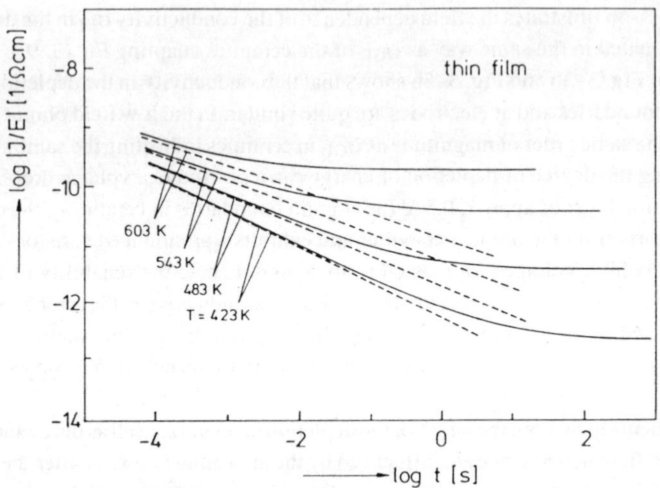

FIGURE 3-6: Current density – field ratio |J/E| vs. time t for a 0.1 at% Ni-doped $SrTiO_3$ thin film at T = 423 K, 483 K, 543 K, and 604 K. The charging (solid lines) and discharging (dashed lines) response for a dc voltage step of 2 V is shown. Thickness of the film: 0.55 μm. Area of an electrode dot: 0.011 cm^2.

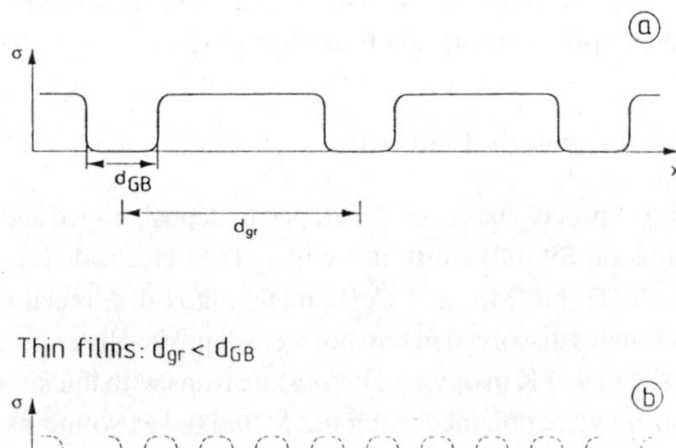

FIGURE 3-7: Schematic illustration of the conductivity profiles in a cross section of conventional ceramics (a) and ultrafine grained thin films (b) as proposed to explain the different dc voltage step responses. d_{gr} is the mean grain size.

duction regime are approx. 0.0015 s (T = 603 K), 0.02 s (543 K), 0.3 s (483 K), and 4 s (423 K). An extrapolation to 298 K reveals a τ_T of 10^4 s to 10^5 s. This means that any (low voltage) dc measurement at room temperature has to be carried out for up to one day to determine the true conductivity with a reasonable reliability. Since PZT films (prepared along two alternative sol/gel routes) exhibit qualitatively the same dc voltage step responses [41], we expect similar amounts of time to be needed for reliable conductivity measurements at room temperature.

The eminent difference between the space charge polarization behaviour of the ceramic and the single crystal on the one hand and the thin film on the other hand is given by the fact that the first shows a Debye relaxation, Eq. (3. 4), while the latter is described by a Curie - von Schweidler law (the so-called "universal dynamic response") [42]

$$J \propto t^{-n} \tag{3. 11}$$

For the discharging curves in Fig. 3-6 (dashed lines), n is decreasing from 0.63 to 0.44 when increasing the temperature from 423 K to 603 K.

The Curie - von Schweidler law is reported for a wide range of disordered systems such as partially or totally amorphous solids and different materials such as organic polymers and inorganic compounds. In contrast to a Debye relaxation which can be represented by an equivalent network consisting of a resistor and a capacitor in series, a Curie - von Schweidler relaxation corresponds to a distributed RC line (with unequal R's and C's). Different explanations have been offered to explain the behaviour expressed by Eq. (3. 11):

- **Injection model**
  For thin films of a good insulator such as SiO_2, electrons injected from the metal cathode may get trapped and generate a negative space charge region in front of the cathode [43]. The formation of this space charge subsequently lowers the electric field at the cathode and, hence, reduces the injection current according to Eq. (3. 11).

- **Many body interaction model**
  The hopping motion of a charged particle will always affect the motion of all neighbouring charges. As shown theoretically in the references cited in [44], the correlation of the charge transport is correlated due to the many-body interaction between the charges leading to Curie - von Schweidler type equations.

- **Distribution of Relaxation Times (DRT) model**
  A superposition of Debye-type relaxations, Eq. (3. 4), with a (large) distribution of relaxation times, τ_R, unequivocally leads to Curie - von Schweidler type equations. The exponent n is determined by the specific distribution of the individual Debye-type processes. This distribution may be caused by a variation in a charge transport barrier. Only a relatively small variation is required for the energy barrier to obtain a spectrum of relaxation times over many orders of magnitude [45].

For $SrTiO_3$ thin films, the *injection model* appears to be unlikely for two reasons. Firstly, the dielectric relaxation of the titanate films clearly is a bulk phenomenon which is independent of the electrode metal. As will be described below, the electron emission properties of the metal used in this study are extremely different. Since the injection model relies on electron emission from the cathode as the first step, a significant influence of the electrode metal would be expected - in contrast to our findings. Secondly, during the discharging process, the emptying of the traps would happen randomly towards *both* electrodes. Consequently, no - or at least only a very small - depolarization current would be expected - in contrast to our findings.

In principle, the hopping of the polaron–type electronic carriers as well as of the ionic carriers may be correlated to a certain extent. Therefore, the *correlation model* cannot be discarded at first glance. However, there are two indications which render the dominance of this mechanism less likely. First, the correlation of the charge motion would be possible in conventional bulk ceramics in a similar manner as in the thin films. Yet, for conventional acceptor–doped $SrTiO_3$ ceramics, a *Debye-type relaxation* was observed and fully explained by a grain - grain boundary Maxwell–Wagner polarization (Sec. 3.2 and 3.3). Second, the temperature dependence is different to what is expected from the correlation model ([42], Chap. 8.4).

Let us consider the *DRT model*. As described in Section 3.2, an energy band bending and the formation of a depletion space charge layer of width d_{GB} is observed at interfaces in acceptor–doped titanate ceramics due to donor–type interface states. In Ref. [40], we proposed a complete carrier depletion in nanocrystalline films and a distributed spectrum of local conductivities because the grain size, d_{gr}, is below d_{GB} (Fig. 3–7). To look into more details, this also means that the extent of the band bending will start to depend on the distance, d_{gr}. Hence, because of the natural variation of the grain sizes in a given film, the energy barrier W_B will show a scatter in its height across the thin film. This is exactly what is required by the DRT model. Based on this fact and the finding that the *relaxation current is a bulk phenomenon* of the films independent of the electrode metal, the authors considers the process according to the DRT model as the dominant mechanism. Details are revealed in Ref. [46].

Columnar epitaxial films show a Curie - von Schweidler behaviour despite their different microstructure. First, this may be due to the surface states which cause depletion space charge layers extending through the complete films if the thickness is not too large. Second, to our experience, even high quality films incorporate sufficient artefacts (e. g. micropores, dislocations, non–equilibrium point defects) to generate a relatively high density of interface states throughout the film. Both effects can easily lead to a situation of complete depletion compared to single crystals of the same composition. A similar DRT description is obtained when the ceramic film is considered as a disordered semiconductor in which the charge transport takes place as a hopping process between distributed trapping centers.

Concerning the long–term conductivity (i. e. the leakage time regime), the influence of the type of electrode materials has been studied for Ni–acceptor doped $SrTiO_3$ thin films [46][47]. Pt bottom electrodes were employed in all cases, while e. g. Pt, Au, Al, Ti were used as top electrodes. J–t tests for both polarities unequivocally showed that *only the cathode* determines the leakage behaviour of thin titanate films. The polarization and depolarization currents are found to be independent of the polarity and

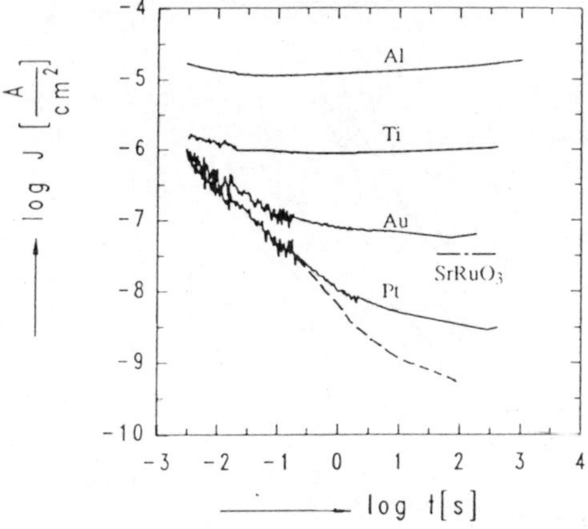

FIGURE 3-8: Current density J vs. time t for 0.1 at% Ni-doped $SrTiO_3$ films at 425 K and 0.2 V. Comparison of different top electrode materials as cathodes. All films have Pt bottom electrodes.
Film thickness:
230 nm (for Al, Ti, Au, Pt electrodes)
290 nm (for Pd electrode),
660 nm (for $SrRuO_3$ electrode).
Only one representative depolarization current curve is shown (dashed), because these curves are the same within the experimental scatter.

the electrode metals within the experimental scatter for all combinations as mentioned above.

In Fig. 3–8, the J–t curves for a variety of *cathode materials* are presented. The $SrRuO_3$ top electrode and the Pd top electrode were only available on films of different thickness. The corresponding measurements were recorded at the same voltage and the results are included in Fig. 3–8 for an – at least semi-quantitative – comparison with the other cathode materials. As an immediate consequence of the results, the leakage and, hence, the overall film resistance have to be regarded as primarily determined by the electrode–film interface contact and are found to vary over many orders of magnitude depending on the type of metal. For increasing film thicknesses, however, it is evident that the relative impact of the film bulk resistance on the overall film resistance must increase [47].

In order to obtain a first impression concerning the influence of the ceramic *thin film processing* on the leakage behaviour, films made along two different sol/gel routes, the hexanoate route and the acetate route (see Ref. [41] for corresponding PZT films), were compared. Although the microstructure and the average grain size are different, no significant effect of the preparation route on the leakage currents was found as long as the cathode metals were kept identical. This, again, emphasizes the dominant importance of the cathodic contact for the leakage behaviour.

4. Breakdown mechanism

4.1 Nature of the Field Enhanced Conductivity

In Fig. 4–1, the long-term conductivity σ_l of various titanate thin films at T = 423 K is displayed as a function of the applied field E. Besides Ni-doped $SrTiO_3$ which was used as a model material throughout this paper, Ce as a donor dope and Mn as an acceptor dope in $BaTiO_3$ were studied in addition. The field dependence of the conductivity of the thin films exhibits a similar characteristic as the conductivity of the depletion

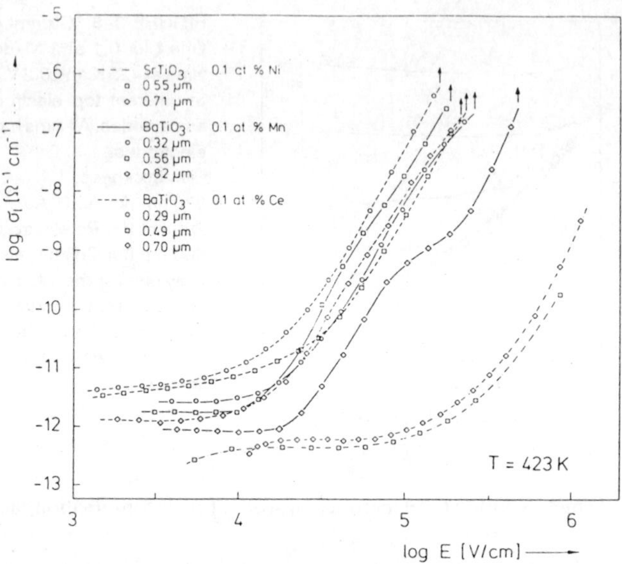

FIGURE 4-1: Field dependence of the long-term conductivity σ_l (recorded at t = 300 s after the dc voltage step) for 0.1 at% Ni-doped $SrTiO_3$, 0.1 at% Mn-doped $BaTiO_3$, and 0.1 at% Ce-doped $BaTiO_3$ thin films at T = 423 K. The curves must be considered as preliminary results because no statistical evaluation was performed. The arrows indicate breakdown incidences.

layers at GBs and electrode interfaces. At low fields, an ohmic regime is observed. Above a field of $1.5\text{–}3 \cdot 10^4$ V/cm for the $BaTiO_3$ films which corresponds to voltages in the range from 0.6 to 2 V, the conductivity starts to show an enhancement over many orders of magnitude. In the case of the Ni-doped $SrTiO_3$, the onset of the conductivity enhancement occurs at significantly higher fields ($E \approx 2 \cdot 10^5$ V/cm, corresponding to 10 - 15 V). PZT films were found to behave similar to $BaTiO_3$ films [41].

Reproducibility tests were performed for the Ce-doped $BaTiO_3$ films. In the ohmic regime, the scatter of σ_l is less than a factor of two for each film. In the varistor regime (e. g. at $E = 10^5$ V/cm), the scatter increases to a factor of up to five for the individual films. Due to this fact, the data reported in Fig. 4-1 should be regarded as showing trends only. Despite of this relatively large scatter, it can be stated that there is obviously a pronounced influence of the host material (i. e. $SrTiO_3$ and $BaTiO_3$) on the *onset field of the varistor regime* while the type of dopant in $BaTiO_3$ (i. e. Mn and Ce) at least in the concentrations used here does not lead to a significant difference. The latter may be indicative for the fact that the high density of interface states or impurity states suppresses the influence of small concentrations of additional donors or acceptors on the conduction process.

The thickness dependence of the $\sigma_l(E)$ was studied for acceptor doped $SrTiO_3$ films with thicknesses in the range between 0.3 and 2 μm [40]. Based on these data, the film thickness seems to impose no significant influence on the onset field of the varistor regime and on the characteristics of the varistor regime. Final conclusions have to await further detailed studies.

It can be estimated that the conductivity in the varistor regime must be predominantly of electronic nature. The estimation is based on the mass transfer in the case of an ionic current and the resulting time of material degradation (see Sec. 4.3). In the ohmic regime the dominant conduction type is not known yet.

In general, the leakage current through a MIM system is determined by the bulk of the film and by the electrode interfaces. Often, one of the contributions is highly dominating. Several limiting cases are known under which simplified mechanistic models are applicable:

❍ **Tunneling**

Electrons may tunnel through barriers which are sufficiently thin (usually < 10 nm) and energetically high [30]. Typical configurations are very thin films of high band gap insulators such as Al_2O_3 and SiO_2 between metal or semiconductor electrodes. Injection into a thick film or bulk insulator may take place through the interface barrier by tunneling under very high fields (Fowler–Nordheim emission).

❍ **Space Charge Limiting Currents (SCLC)**

In the case of flat–band or *enrichment* space charge contacts, SCLCs according to

$$J_L(V) \propto V^2/d^3 \tag{4. 1}$$

are observed above a critical onset voltage, where d denotes the film thickness. Many modifications of this relationship are discussed in the literature depending on the specific MIM system. The onset voltage strongly depends on the trap density in the film.

❍ **Poole–Frenkel Effect**

The conductivity of films which show hopping conduction of electronic carriers may be determined by an equilibrium between an absorption and an emssion of free carriers by traps. High electric fields in the films disturb this equilibrium and shift it towards larger free carrier densities. Obviously, the Poole–Frenkel effect is a typical bulk effect.

❍ **Thermoionic Emission**

Whenever there is a (not too thin) energy barrier W_B for an electron transfer, obviously the charge carrier needs sufficient thermal energy to circumvent the barrier. The temperature dependence of this thermoionic emission is given by

$$J_L = A^* T^2 \exp\left(-\frac{W_B}{kT}\right) \tag{4. 2}$$

A* denotes the effective Richardson constant for the emission from the metal into the insulator. The field dependence for a charge transport from a metal into or across an insulator is determined by the Schottky effect. This is based on the fact that charges in the insulator close to the interface induce mirror charges in the metal and, hence, lead to a field dependent decrease of the barrier height, ΔW. Consequently, the zero–field barrier height W_B in Eq. (4. 2) has to be replaced by $W_B - \Delta W$ with

$$\Delta W = e_0 \sqrt{\frac{e_0}{4\pi\varepsilon_r\varepsilon_0} E} \qquad (4.3)$$

ε_r denotes the permittivity of the titanate film.

Using this (simplified) terminology for the charge transfer mechanism, the experimental results and additional information from the literature will be used to obtain a deeper insight into the processes. Emphasis will be put on the influence of the electrode metal, the temperature and film thickness effect, and especially on the non-linear voltage dependence, J_L–V. A comprehensive comparison of the above mentioned effects is given by O'Dwyer [48].

Tunneling of carriers through the entire titanate film can certainly be excluded as the dominant charge transfer process in our metal–insulator–metal (MIM) systems considered here due to the large film thickness. In addition, the authors regard tunneling through parts of the film e. g. energy barriers at grain boundaries as very unlikely due to the insufficient energetic height of the possible barriers.

In Section 4.1, the very pronounced dependence of the leakage current density J on the type of *cathode* metal was shown. This strong influence of the cathodic metal/titanate interface renders those mechanisms as less likely which assume a dominant contribution of the film bulk to the overall resistance of the MIM system, i. e. the SCLC model and the Poole–Frenkel effect. As stated, the *anode* metal does not exhibit an effect on the leakage. Hence, the hole injection at the anode can be disregarded as a significant process.

The *temperature dependence*, in addition, can be used to discriminate between models. For checking the Schottky emission, the results for different metals are compared in a $\ln(J/T^2)$ vs. $1/T$ plot according to Eq. (4. 2) Straight lines are obtained in all cases indicating the applicability of this model [47]. The calculated barrier heights W_B show reasonable correlation with the work functions of the electrode metals. The *field dependence* of the Schottky emission reveals a linear dependence between $\ln(J_L)$ and $\sqrt{E}$ according to Eqs. (4. 2) and (4. 3) [47]. Obviously, the data satisfyingly fit the Schottky emission model with respect to both, the temperature and the field dependence. This supports the judgement that the leakage current across $SrTiO_3$ thin films is caused by an electron emission over a distributed Schottky barrier [32]. In this Ref., the conclusions are based on a comprehensive comparison of single crystals, ceramics, and thin film devices of the same titanate composition. However, the authors emphasize their opinion that the field dependence of the leakage current alone is insufficient to qualify the model mechanism. This is due to the simplified assumptions of all model mentioned above. For example, the specific (and usually unknown) density of state distributions for interface states or traps in the films are completely ignored.

For PZT films, different models for the charge transport across MIM devices are discussed. They include the SCLC model [49]][50][51] with different modifications as well as the Schottky emission model [52]][53][54].

A comparable correlation between the *breakdown* field and the metal work function was found by Scott et al. [55]. Obviously, this finding has the same origin, since impact thermal breakdown was confirmed as the dominant mechanism. Hence, the high–field leakage current density is determining the breakdown. Therefore, the temperature and field dependence of the leakage current density on the one hand and the breakdown field on the other hand , both are supposed to show the same correlation to the work function of the metals.

4.2 Instantaneous High–Field Breakdown

In a simple classification, the instantaneous breakdown in a dielectric under very high fields may be either a purely electrically or a thermally triggered process. In the first case, an impact ionization in the material causes an avalanche of electronic carriers and/or a chemical deterioration [48] [56]. A thermal breakdown occurs when the generated Joule heating in the system determined by the product of the current and the applied voltage leads to an excessive temperature rise. Because of the positive temperature coefficient of the conductivity, this rise causes an instability and a subsequent thermal runaway.

Using the data at the upper end of the $\sigma_l(E)$ curves in Fig.4–1 , a calculation of the generated Joule powers and the dissipated heat in combination with the estimated thermal transfer coefficient of the films and the thermal conductivity of the substrate reveals a situation which is typical for a thermal runaway. This finding is supported by ramp voltage breakdown tests at room temperature. In contrast to Al_2O_3 thin films, the current evolution for our $BaTiO_3$ films always exhibits a field enhanced conductivity before the final breakdown sets in [57]. According to O'Dwyer, the instantaneous high field breakdown in solids is usually of a mixed non–thermal and thermal nature. The essential question is whether the breakdown process is *started* by a thermal runaway. Compared to bulk solids and free–standing films, the thermal runaway is significantly impeded in *thin* films on substrates due to the heat sink behaviour of the latter [58]. Besides the (rare) avalanche breakdown process reviewed by O'Dywer, more recent studies by Wolter revealed a further, purely electrical breakdown mechanism using SiO_2 as a model system [56]. It can be assumed, however, that the breakdown of titanate thin films is *initiated thermally* due to the strongly enhanced high field conductivity. The impact thermal breakdown character was confirmed by Scott et. al. for $(Ba,Sr)TiO_3$ films [59]. This thermal breakdown process is possibly supported by an extremely accelerated resistance degradation mechanism. This mechanism which also takes place at fields much below the onset fields for the instantaneous breakdown is described in the next section.

4.3 Long–term Resistance Degradation

The *resistance degradation* of dielectric ceramics is characterized by a slow increase of the leakage currents under simultaneous temperature and dc field stress [60]. The process may take place at temperatures and fields much below the critical onset values for the failure of the dieletric and/or thermal breakdown. Degradation represents a

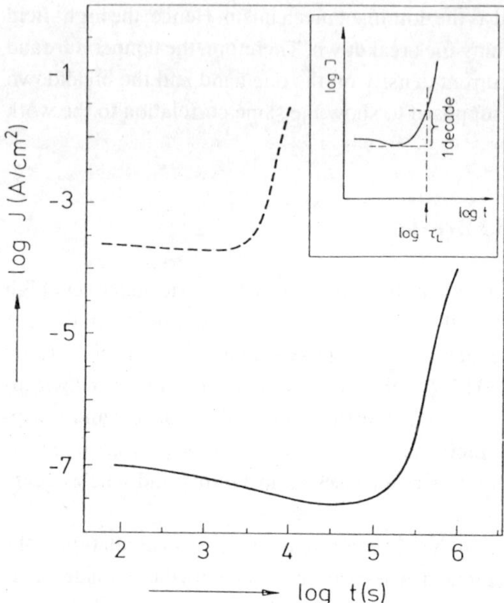

FIGURE 4-2: Examples for the degradation of titanate dielectrics: current density J vs. time t at T = 483 K and a dc electric field E = 8.0 kV/cm. The life time τ_L is defined as the time at which J increased one order of magnitude above its minimum value.
(-------) Ceramic multilayer capacitor (Z5U-type) made from technical grade $(BaCa)(TiZr)O_3$ ceramic. Thickness of the dielectric layers D = 34 μm.
(- - - - -)Ceramic disc made from 0.2 at% Al-doped $SrTiO_3$ ceramic, sintered at 1633 K for 6 h and subsequently at 1753 K for 3 h in oxygen. Thickness D = 500 μm.

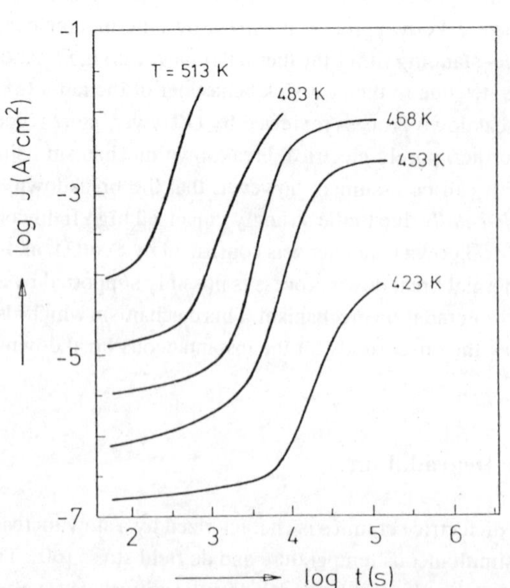

FIGURE 4-3:
Resistance degradation of 0.06 at% Ni-doped $SrTiO_3$ single crystals: current density J vs. time t at E = 1.2 kV/cm and different temperatures. Dielectric thickness of the samples D = 2000 μm.

typical wear–out process and, in general, it is the limiting factor for the lifetime of ceramic multilayer devices. It is not a fatal breakdown process by itself but it often leads into a thermal run–away situation.

As an example of the resistance degradation, the time evolution of the leakage current is shown in Fig. 4–2 for two ceramic components under dc voltage load at an elevated temperature. After a certain time of normal operation, the leakage current starts to increase by several orders of magnitude. Another example is given in Fig. 4–3 which shows the degradation behaviour of Ni–doped $SrTiO_3$ single crystals at different temperatures.

In order to evaluate the process quantitatively, a characteristic life time τ_L is introduced as a reciprocal measure of the degradation rate, $\beta = 1/\tau_L$. The arbitrary definition of τ_L is given in the inset in Fig. 4–2.

The temperature dependence of the degradation is that of a typical thermally activated process:

$$\tau_L \propto \exp(W_A/kT) \qquad (4.4)$$

where T denotes the absolute temperature, k represents the Boltzmann constant, and W_A is the activation energy of the degradation process. W_A is found to be in the range from approx. 1.1 to 1.4 eV for various titanate ceramics [61] [62] and from 0.6 to 1.0 eV for differently acceptor–doped $SrTiO_3$ single crystals [63].

The dependence of the life time on the electrical field strength E, the voltage V, and the dielectric thickness D can be described within good approximation by power laws [64]. One of several equal–ranking expressions is:

$$\tau_L \propto V^{n_1}\Big|_{D=const} \cdot D^{n_2}\Big|_{E=const} \cdot E^{n_3}\Big|_{V=const} \qquad (4.5)$$

Due to the constraint established by the relation E = V/D, the exponents in Eq. (4.5) are not mutually independent but they must satisfy the condition $n_1 - n_2 - n_3 = 0$. The exponent at constant voltage n_1 is usually in the range from approx. −1.5 to −2.5 for different titanate ceramics. For Ni–doped $SrTiO_3$ single crystals, a n_1 value of approx. −1 is found [63]. In the case of Al–doped $SrTiO_3$ ceramics, an exponent at constant field n_2 of approx. 1 is observed [62]. This may be regarded as a typical value which is independent of the microstructure within first approximation.

The type of dopant shows a striking influence on the degradation characteristics as already reported by Lehovec and Shirn [60]. While undoped and acceptor–doped alkaline earth titanates are prone to degradation, donor–doped materials (in as far as they are prepared electrically insulating [64]) are found to exhibit life times which are by orders of magnitude higher. This is attributed to the lack of a significant oxygen vacancy concentration, $[V_O^{\cdot\cdot}]$, in donor–doped titanates (see Sec. 2.4).

Besides the type of dopant, the microstructure of the titanate samples strongly affects the life time since grain boundaries act as barriers for cross transport of charges as explained in Sec. 3.2. Their impact on the degradation process will be described in Sec. 4.3.3.

In order to study the degradation mechanism, various methods to look **inside** the dielectrics between the electrodes have been employed. For this objective, samples under degradation are favourably considered as solid–electrolyte cells composed of the titanate dielectrics (s), two attached metal electrodes (m+, m–), and the ambient gas phase (g) as sketched in Fig. 4–4 .

Recently, the electro–optical Kerr effect [65] as well as an electron beam deflection technique [66] were employed as non–contact gauching methods to determine the time evolution of the conductivity distribution between the anode (m+) and the cathode (m–). It was found that at the beginning the conductivity is relatively homogeneous. As the degradation proceeds, the conductivity increases in the areas near both electrodes while it remains relatively low in a center part of the sample.

The specific optical absorption bands of transition–metal ions used as dopants in $SrTiO_3$ can be employed to reveal information about the local oxidation state of the dopant ions [67] [68]. This feature was used by Blanc and Staebler to interpret their electrocolouration results obtained for acceptor–doped $SrTiO_3$ [69]. The color changes after applying a dc voltage clearly indicate an oxidation process emerging from the anode (m+) into the interior of the dielectric and a reduction process starting at the cathode (m–). The suggested formation of a p–conducting anodic region and a n–conducting cathodic region was confirmed by a numerical simulation which led to a quantitative description of the degradation process [70] (see Sec. 4.3.2).

Electrocolouration studies under oxidizing and reducing ambient atmospheres (g) did not reveal a significant influence of the oxygen partial pressure on the degradation process up to temperatures of approx. 600 K [63]. Firstly, this indicates that the oxygen exchange via the free surface (s/g) can be neglected within good approximation at lower temperatures. Secondly, this shows that the electrochemical O_2–incorporation at the cathode as well as the O_2–release at the anode are both almost completely

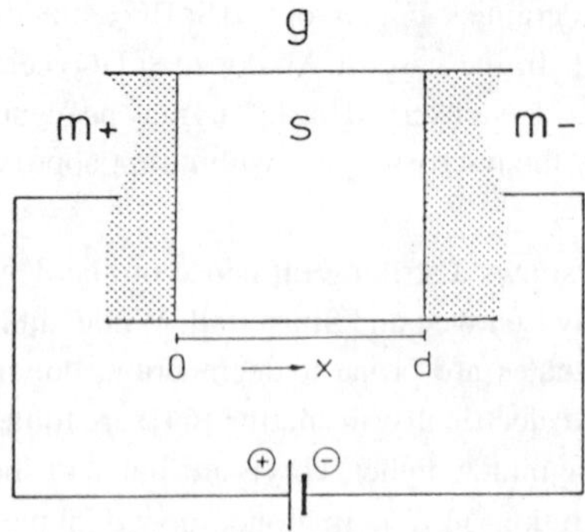

FIGURE 4–4: Schematic illustration of a solid–electrolyte cell as a model of a dielectric component subjected to a dc voltage stress. Phases: s = titanate dielectrics; m+, m– = anode and cathode metal; g = ambient gas phase. The interface transfer processes are discussed in the text.

suppressed i. e. the ionic transfer via the electrode interfaces is blocked. Both effects are probably due to the depletion space charge layer at the free surface and at the electrode interfaces EI (Sec. 3.3). As we describe later, this oxygen transfer blockage is an essential prerequisit for the degradation. Furthermore, the results show why samples with embedded electrodes such as multilayer devices can be modelled by samples with externally applied electrodes as far as the degradation behaviour is concerned.

4.3.1 Linear Transport Theory

Based on the defect chemistry of acceptor-doped titanates and a numerical simulation of the charge transport, a quantitative model for the dc voltage-induced resistance degradation was developed. While it is comprehensively described in Ref. [70], a brief outline is given Section 4.3.2. As an introduction, the effect of voltages which are sufficiently small to keep the process in the validity range of the linear transport theory will be discussed.

In general, the total conductivity in a solid is determined by

$$\sigma_{tot} = e_0 \sum \mu_k \; |z_k| \; [k] \qquad (4.\,6)$$

where e_0 represents the unity charge. z_k denotes the charge number, μ_k the mobility, and [k] the concentration of the mobile species k.

The current density J in a mixed-conducting solid-electrolyte phase is composed of an ionic and an electronic part, J_{ion} and J_{el},

$$J = J_{ion} + J_{el} \qquad (4.\,7)$$

Each of the partial current densities obeys the general transport law,

$$J_k = J_k^{Drift} + J_k^{Diff}$$

$$= \sigma_k E - z_k e_0 D \frac{\partial [k]}{\partial x} \qquad (\text{k = ion, el}) \qquad (4.\,8)$$

where D denotes the ambipolar diffusion coefficient. Eq. (4. 8) is only valid for conservative ensembles, i.e. for ensembles of species which are defined in a way that the source terms vanish [71]. In the case discussed here, this is true for [ion] = $[V_O^{\cdot\cdot}]$. The concentration of the electronic ensemble [el] depends on the difference n−p between the electron and hole concentration as well as on the concentrations of other charged defects [70].

The electrocoloration studies mentioned above clearly indicate that the oxygen transfer through the electrodes is blocked to a much larger degree than the electronic charge transfer. The sample must be regarded as de-coupled from the ambient atmosphere and the total amount of $[V_O^{\cdot\cdot}]$ is conserved in the solid-electrolyte. Application of a dc voltage leads to a **concentration polarization** between the electrodes. The transport problem, Eq. (4. 8), can easily be solved analytically when the field as well as the conductivities remain homogeneous and steady, i. e. $E \neq E(x,t)$ and $\sigma_k \neq \sigma_k(x, t)$. The latter condition is satisfied as long as the concentration changes $\Delta[k]$ are negligible compared to the total concentrations [k], i.e. $\Delta[k] << [k]$. For sufficiently low dc voltages V applied to the electrodes, these conditions are fulfilled.

Due to the ionically blocking electrodes, the ionic partial current must vanish, $J_{Vo} = 0$, at steady state, $t \rightarrow \infty$. Hence, from Eq. (4. 7), the concentration gradient, $\partial[V_O^{\cdot\cdot}]/\partial x$, can be calculated:

$$\frac{\partial[V_O^{\cdot\cdot}]}{\partial x} = \frac{\sigma_{Vo}E}{2e_0 D} \qquad (4.\ 9)$$

Integration of Eq. (4. 9) shows that the concentration difference between the cathode and the anode $\Delta[V_O^{\cdot\cdot}]$ is proportional to the applied voltage V:

$$V = \frac{2e_0 D}{\sigma_{Vo}} \Delta[V_O^{\cdot\cdot}] \qquad (4.\ 10)$$

Because of the conservation of the total amount of $V_O^{\cdot\cdot}$, the $[V_O^{\cdot\cdot}](x)$ profile is determined by:

$$[V_O^{\cdot\cdot}](x) = [V_O^{\cdot\cdot}]_0 + \left(\frac{x}{d} - \frac{1}{2}\right) \Delta[V_O^{\cdot\cdot}] \qquad (4.\ 11)$$

where $[V_O^{\cdot\cdot}]_0$ represents the initial homogeneous oxygen vacancy concentration. An appropriate profile of the electronic defect concentrations [el] is formed to maintain the local electroneutrality. The calculation of the time evolution of a linear $[V_O^{\cdot\cdot}](x)$ profile according to Eq. (4. 11) which develops under a sufficiently small voltage, V = 3 mV, is shown in Fig. 4–5a.

4.3.2 Degradation as a Non–Linear Process

A concentration polarization of the oxygen vacancies under the conditions mentioned in Sec. 4.2 does not lead to a resistance degradation but, instead, gives rise to a decrease of the total current density J due to the decay of the ionic contribution, J_{ion}. Inserting the oxygen vacancy mobility and concentration for acceptor–doped titanates (Sec. 2) into Eq. (4. 10) shows that the validity limit for the condition $\Delta[k] << [k]$ is in the few mV range. If this range is exceeded, the change $\Delta[V_O^{\cdot\cdot}]$ imposed by the demixing process soon gets into the range of the absolute value of $[V_O^{\cdot\cdot}]$ and concentration differences of several orders of magnitude can build up between the anodic and the cathodic region (Fig. 4–5b). As derived in detail in Ref. [70], this process strongly affects the local defect situation and leads to an increase of the p–type conductivity in the anodic region and the formation of a n–type conductivity in the cathodic region (Fig. 4–6). Therefore, despite the decay of the ionic conductivity, the total current density increases with time due to the pronounced rise in the electronic carrier concentrations (Fig. 4–7). For the parameters used in Fig. 4–7, the *onset voltage* of the degradation is between 0.1 and 1 V. The shape of the degradation evolution in Fig. 4–7 including the stabilization of the current density after the degradation period reproduces all aspects of the experimental curves (Fig. 4–3).

The degradation process represents an inhomogeneous and extremely non–linear transport problem because E in Eq. (4. 8) depends strongly on σ_k. Hence, a quantitative solution can hardly be achieved by analytical means. Therefore, the results shown in Figs. 4–5b, 4–6, and 4–7 have been generated by a numerical simulation based on a finite–element solution of the transport problem. Impedance spectroscopical results confirm the formation of a forward–biased pn–junction during degradation [72].

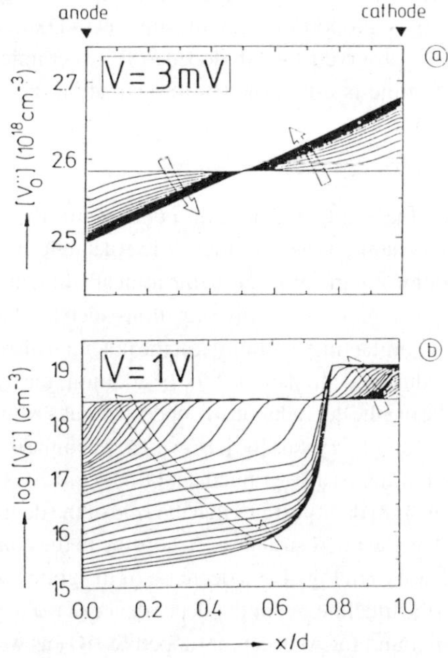

FIGURE 4–5:
Calculated time evolution of the oxygen vacancy profiles after applying a dc voltage V to 0.1 at% acceptor-doped $SrTiO_3$ crystals of the thickness D = 0.1 cm at T = 500 K for two different voltages V:

(a) low voltage stress, V = 3 mV, within the validity range of the linear transport theory;

(b) high voltage stress, V = 1 V, leading to resistance degradation.

Please note the very different scales of the ordinate: (a) a small interval on the linear scale; (b) several orders of magnitude on the logarithmic scale.

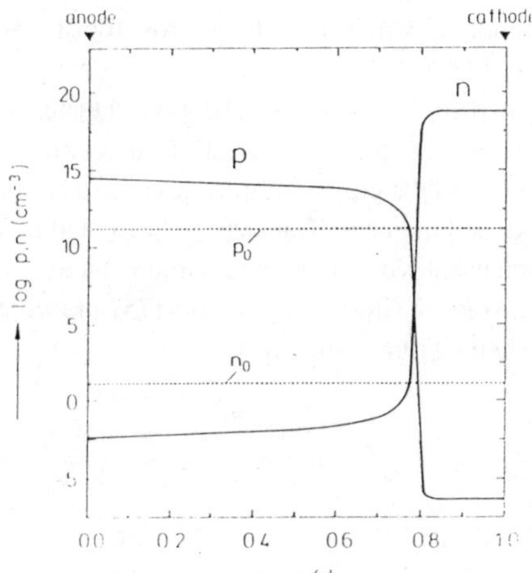

FIGURE 4–6:
Calculated p(x) and n(x) profiles after the degradation of a 0.1 at% acceptor-doped $SrTiO_3$ single crystal of the thickness D = 0.1 cm at T = 500 K and V = 1 V. p_0 and n_0 denote the initial profiles. Please note the extremely extended log scale used for the carrier concentrations. The resistance degradation is caused by an over–all increase of the electronic carrier concentrations by some orders of magnitude. This process is accompanied by the formation of a pn junction.

The results of the simulation concerning the influence of the field show that the degradation rate, $\beta = 1/\tau_L$, is determined in good approximation by the average drift velocity of the oxygen vacancies,

$$v_{V_O^{\cdot\cdot}} = \mu_{V_O^{\cdot\cdot}} E \qquad (4.12)$$

This suggests an exponent $n_1 = -1$ in Eq. (4.5) for single crystalline samples of constant dielectric thickness D in reasonable agreement with the experimental findings. Simulations for constant E reveal that τ_L is proportional to the thickness D of the sample, i.e. $n_2 = 1$. Such a dependence is observed for Al–doped $SrTiO_3$ ceramics. The temperature dependence of the life time is established by the activation energy of the oxygen vacancy mobility, $\mu_{V_O^{\cdot\cdot}}$ [63].

4.3.3 Impact of Grain Boundaries

Macroscopically (i. e. by averaging over the microstructure), the *character* of the resistance degradation of acceptor–doped ceramics is very similar to that of single crystals. It is determined by a $V_O^{\cdot\cdot}$ concentration polarization due to the ionically blocking electrodes. This statement is based on position resolving investigations such as electron beam deflection studies which show similar time evolutions of the potential distribution for ceramics and single crystals during degradation [66]. In addition, macroscopic coloration effects are observed during the degradation of doped ceramics which are comparable to the electrocoloration of single crystals [60]. However, a comparison of the degradation curves of single crystalline, coarse–grained, and fine–grained Ni–doped $SrTiO_3$ (see long–term part of Fig. 4–8) demonstrates that the *rate* of the degradation process is very strongly affected by the microstructure. According to the comparison of coarse– and fine–grained ceramics in Fig. 4–8 a decrease of the grain size leads to a over–proportional increase in the life time. A similar dependence of the degradation on the microstructure has been found for Al acceptor–doped $SrTiO_3$ as well [72].

The reason for the strong influence of the microstructure on the degradation rate obviously must be attributed to the barrier character of the GB depletion space charge layers concerning the cross transport of ionic (and electronic) charges described above. This barrier character of grain boundaries with respect to the cross transport of $V_O^{\cdot\cdot}$ is also known from diffusion studies [73] [74].

As mentioned in the introductory part of Sec. 4.3, donor–doped titanates are known to be highly degradation stable. However, there are indications that – in contrast to acceptor–doped titanates – a decrease of the grain sizes increases the degradation rate in ceramics made from donor–doped titanates. This may be attributed to an alternative degradation mechanism which is known from n–type semiconducting ceramics such as $BaTiO_3$ based PTC thermistors or ZnO based varistors [75] and which is based on an oxygen ion transport *along* the grain boundaries.

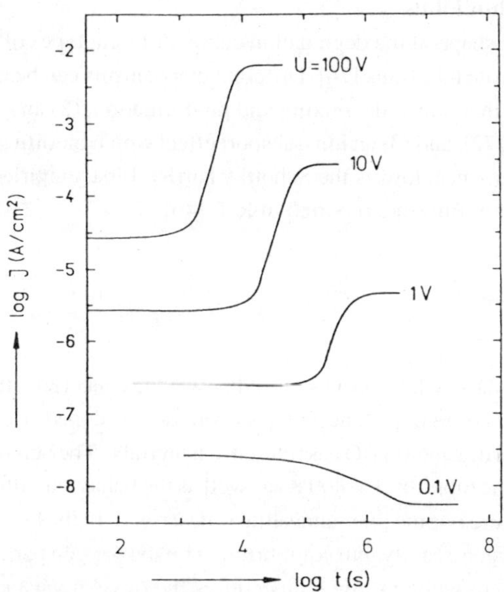

FIGURE 4–7: Calculated current density J vs. time t after applying a dc voltage V to 0.1 at% acceptor–doped $SrTiO_3$ crystals of thickness D = 0.1 cm at T = 500 K. The simulation results for different voltages V are shown.

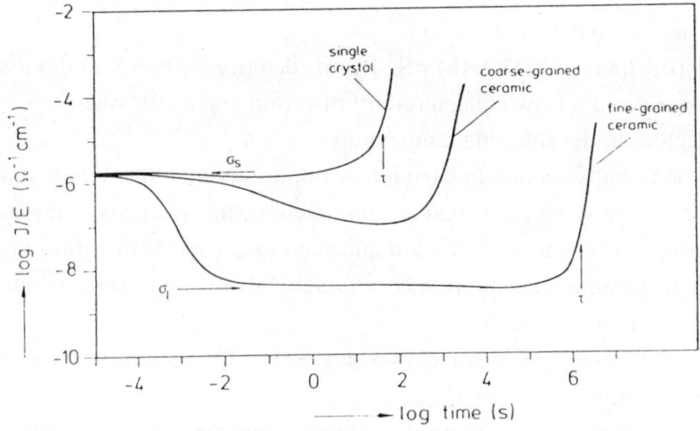

FIGURE 4–8: Maxwell–Wagner polarization and degradation of 0.1 at% Ni–doped $SrTiO_3$ at T = 600 K and E = 0.625 kV/cm. Comparison of different microstructures: single crystal, coarse–grained ceramic ($d_{gr} \approx 50\,\mu m$), fine–grained ceramic ($d_{gr} \approx 2.5\,\mu m$). Thickness of the samples D = 500 μm.

4.3.4 Ceramic Thin Films

Depending on the shape of the degradation curve, different types of degradations are discussed for titanate thin films [76]. Different mechanisms can be imagined, including (1) the above mentioned de–mixing and pn–formation, (2) an oxygen loss along grain boundaries [72], and (3) an ion transport effect which modifies the electrode interface states and, hence, lowers the Schottky barrier. First, empirical data of the life time for $SrTiO_3$ thin films are reported in Ref. [46].

5. Conclusions

This review summarizes the present state of knowledge concerning the defect chemistry and conduction of undoped, acceptor–doped, donor–doped perovskite–structure titanates using $SrTiO_3$ and $BaTiO_3$ as reference materials. The behaviour in the high–temperature regime (approx. $T > 800$ K) as well as the behaviour after quenching into the low–temperature regime after annealing is described. In the high–temperature regime the material gets readily into equilibrium with the oxygen partial pressure of the ambient atmosphere while at low temperatures the oxygen vacancy content of the samples can be regarded as virtually frozen–in. The depletion space charge layers at interfaces are discussed by comparing the impedance behaviour of single crystals, ceramics, and thin films of the same nominal composition. Finally, the instantaneous and long–term breakdown behaviour is outlined. The following conclusions can be drawn.

Conclusions on the defect chemistry:

* Undoped titanates $MTiO_3$ (M = Sr, Ba) are determined by a *Schottky disorder*, i. e. a formation of a (low concentration) of cation and anion vacancies at temperatures close to the sintering temperatures.
* Substitutionally accommodated *donor* impurities are either compensated by excess electrons (when annealed under reducing atmospheres) or they are compensated by cation vacancies (when annealed under oxidizing atmospheres). The oxygen vacancy concentration is extremely low in donor–doped titanates due to the Schottky equilibrium.
* Donor centers are shallow i. e. their energy level lies slightly below the conduction band edge.
* Substitutionally accommodated *acceptor* impurities are usually compensated by oxygen vacancies.
* Acceptor dopants are always deep traps i. e. their energy levels are relatively deep in the band gap.
* Defect complex formation between (negatively charged) acceptors and (positively charged) oxygen vacancies is not relevant at high temperature but may be significant at low temperatures.

Conclusions on the conduction mechanism in the bulk crystal lattice:

* The electronic charge carriers, electrons and holes, move by a polaron hopping in the titanate lattice. Their mobility at 700 K is approx. 0.1 cm^2/Vs for holes and 0.2 cm^2/Vs for electrons, showing a slight negative temperature dependence.
* Cation vacancies are relatively immobile and can be regarded as frozen–in at temperatures below approx. 1400 K.
* Oxygen vacancies are significantly more mobile and give rise to a detectable conductivity even at room temperature. Their migration enthalpy is approx. 1 eV.
* The *high–temperature conductivity* of acceptor–doped titanates is predominantly electronic with a significant ionic contribution in the vicinity of the intrinsic conductivity minimum.
* The *low–temperature conductivity* of acceptor–doped titanates annealed under oxidizing or moderately reducing conditions is predominantly ionic with a significant electronic contribution by holes.

Conclusions on the space charge layers at interfaces in acceptor–doped titanates:

* At internal and external interfaces (grain boundaries, electrode interfaces, free surfaces) depletion space charge layers are formed due to donor–type interface states. They can be regarded as back–to–back Schottky diodes.
* The *width* of the depletion space charge layer (approx. 100 nm) is virtually independent of the field, the temperature and the annealing oxygen partial pressure. It is slightly decreasing with increasing acceptor concentration.
* The depletion space charge layers act as barriers for the cross transport of positively charged carriers (here: oxygen vacancies and holes).
* The impact of the depletion space charge layer on the total impedance of the system increases with decreasing temperature. This is due to the activation energy of the residual conductivity of approx. 1.2 eV which is higher than for the bulk conductivity mainly determined by the oxygen vacancy migration enthalpy. At low temperatures (approx. $T < 500$ K) the averaged *residual conductivity* in the space charge layer can be more than four orders of magnitude below the bulk conductivity.
* The conduction through a depletion space charge layer shows a low–field *ohmic regime* and a conductivity enhancement at high fields (*varistor–regime*). The characteristics can be explained by a charge transport over a Schottky barrier.
* Ultrafine–grained *thin films* are completely depleted (compared to conventional bulk material) presumably due to the high interface density, i. e. films resemble distributed Schottky barriers.
* The conduction through thin films can be regarded as a thermionic emission over a distributed Schottky barrier.

Conclusions on the electrical breakdown mechanisms:

* In thin films, the varistor behaviour leads to a thermal instability at very high fields. Therefore, in contrast to e. g. SiO_2 films on substrates, the breakdown in titanate thin films seems to be *initiated thermally.*

* In general, acceptor doped titanates suffer from a dc voltage–induced long–term *resistance degradation* caused by a piling–up of the oxygen vacancies near the cathode. Due to the specific defect chemistry, this leads to an increase of the p–type conductivity in the extended anodic region, in which the oxygen vacancies are depleted. In the cathodic region a n–type conductivity is developed. Thus, during degradation a p–n junction biased in forward direction is generated.
* Donor–doped titanates are relatively degradation stable due to the lack of mobile ions.
* The rate of the resistance degradation in acceptor doped titanates strongly decreases with decreasing grain size of the ceramic. This is due to the barrier character of the depletion space charge layers at grain boundaries.

ACKNOWLEDGEMENT: The support of the Division of Materials Research of the National Science Foundation for an extended study of the basic defect chemistry of the perovskite titanates is gratefully acknowledged.

REFERENCES

[1] D. M. Smyth, in Ann. Rev. of Mat. Sci., edited by R. A. Huggins et al., Annual Reviews, Inc., Palo Alto, (1985), pp. 329-357.

[2] D. M. Smyth, Cryst. Latt. Defects and Amorph. Mat., 18, 355 (1989).

[3] A. Navrotsky and D. J. Weidner, in Perovskite: A Structure of Great Interest to Geophysic and Materials Science, edited by A. Navrotsky and D. J. Weidner, American Geophysical Union, Washington, D.C., (1989), p. xi.

[4] R. Waser, J. Am. Ceram. Soc., 71(1), 58 (1988).

[5] G. V. Lewis and C. R. A. Catlow, Rad. Eff., 73(1-4), 307-14 (1983).

[6] F. A. Kröger and H. J. Vink, in Solid State Physics, edited by F. Seitz and D. Turnbull, Academic Press, New York, (1956), p. 307.

[7] N.-H. Chan and D. M. Smyth, J. Electrochem. Soc., 123(10), 1584-85 (1976).

[8] N.-H. Chan, R. K. Sharma, and D. M. Smyth, J. Am. Ceram. Soc., 64(9), 556-62 (1981).

[9] D. M. Smyth, Prog. Solid St. Chem., 15, 145-71 (1984).

[10] K. Takada, E. Chang, and D. M. Smyth, Advances in Ceramics, 19, 147 (1987).

[11] A. M. J. H. Seuter, Philips Res. Rept. Suppl., 3 (1974).

[12] G. M. Choi, H. L. Tuller, and D. Goldschmidt, Phys. Rev. B, 34(10), 6972 (1986).

[13] E. S. Kirkpatrick, K. A. Müller, and R. S. Rubins, Phys. Rev., 135, A86 (1964).

[14] D. J. A. Gainon, J. Appl. Phys., 36, 2325 (1965).

[15] N.-H. Chan, R. K. Sharma, and D. M. Smyth, J. Am. Ceram. Soc., 65(3), 167-70 (1982).

[16] N.-H. Chan, R. K. Sharma, and D. M. Smyth, J. Electrochem. Soc., 128(8), 1762-69 (1981).

[17] H.-J. Hagemann and D. Hennings, J. Am. Ceram. Soc., 64(10), 590 (1981).

[18] R. Waser, T. Bieger, and J. Maier, Solid State Commum., 76(8), 1077 (1990).

[19] R. Waser, J. Am. Ceram. Soc., 74, (8) 1934-40 (1991).

[20] E. K. Chang, A. Mehta, and D. M. Smyth, in Proc. of Symp. on Electro-Ceramics and Solid-State Ionics, edited by H. L. Tuller and D. M. Smyth, The Electrochem. Soc., Inc., Pennington, NJ (1988), p. 35.

[21] N. G. Eror and D. M. Smyth, in: The Chemistry of Extended Defects in Non-Metallic Solids, L. Eyring and M.O. Keeffe, Eds., North Holland, Amsterdam (1970), pp. 62-74. J. Solid State Chem., 24(3-4), 235-44 (1978).

[22] G. H. Jonker and E. E. Havinga, Mat. Res. Bull., 17, 345 (1982).

[23] H. M. Chan, M. P. Harmer, and D. M. Smyth, J. Am. Ceram. Soc., 69, 507 (1986).

[24] I. Burn and S. Neirman, J. Mat. Sci., 17, 3510-24 (1982).

[25] N.-H. Chan and D. M. Smyth, J. Am. Ceram. Soc., 67(4), 285 (1984).

[26] R. Wernicke, phys. stat. sol.(a), 47, 139-144 (1978)

[27] T. G. Stratton, D. Reed, and H. L. Tuller, *Adv. Ceram.*, 1, 114 (1983).

[28] H.-Y. Lee, S. S. Villamil, and L. C. Burton, *IEEE Proc. of the 6th ISAF*, Bethlehem, PA, June 8-11, 1986.

[29] H. Neumann and G. Arlt, *Ferroelectrics*, 69, 179 (1986).

[30] S. M. Sze, "Physics of Semiconductor Devices", John Wiley & Sons, New York 1981.

[31] M. Vollmann and R. Waser, J. Am. Ceram. Soc., 77 [1] 235-43 (1994)

[32] R. Waser and M. Klee, (Proceed. of the 3rd ISIF, Colorado Springs, 1991), *Integrated Ferroelectrics*, 2, 23-40 (1992)

[33] Y.-M. Chiang and T. Tagaki, *J. Am. Ceram. Soc.*, 73, 3278 (1990).

[34] G. E. Pike and C. H. Seager, *Adv. Ceram.*, 1, 53 (1983).

[35] R. Hagenbeck, M. Vollmann and R. Waser, to be published.

[36] H. Oepen, private communication.

[37] K. Szot et al. , to be published

[38] Ch. Tragut and K. H. Härdtl, *Sensors and Actuators B*, 4, 425 (1991).

[39] T. Bieger, Joachim Maier, and R. Waser, *Sensors and Actuators B*, 7, 763-768 (1992)

[40] M. Klee and R. Waser, to be published.

[41] M. Klee, R. Eusemann, R. Waser, W. Brand, and H. van Hal (1992), "Processing and electri cal properties of $Pb(Zr_xTi_{1-x})O_3$ (x = 0.2 - 0.75) films: Comparison of metallo-organic de composition and sol-gel processes", J. Appl. Phys. **72**, 1566-76.

[42] A. K. Jonscher, "Dielectric Relaxation in Solids", Chelsea Dielectrics Press Ltd., London 1983.

[43] D. R. Wolters and A. T. A. Zegers–van Duynhoven (1988), Electron Charge Transport in Thin SiO_2 Films, in: R. A. B. Devine, The Physics and Technology of Amorphous SiO_2, 391–409, Plenum Press.

[44] D. R. Wolters and J. J. van der Schoot, *J. Appl. Phys.* 58, 831 (1985).

[45] H. Kliem (1989), Dielectric Small–signal Response by Protons in Amorphous Insulators, *IEEE Transactions on Electrical Insulation* **24**, 185–197.

[46] R. Waser, Proceedings of the NATO ARW, Italy, June 20–25, 1994 (in print).

[47] W. Antpöhler, G. Dietz, M. Klee, and R. Waser, to be published;

[48] J. J. O'Dwyer, "The Theory of Electrical Conduction and Breakdown in Solid Dielectrics", Clarendon Press, Oxford 1973.

[49] J. F. Scott, C. A. Paz de Araujo, B. M. Melnick, L. D. McMillan, and R. Zuleeg (1991), *J. Appl. Phys.* **70**, 382 .

[50] B. M. Melnick, J. F. Scott, C. A. Paz de Araujo, and L. D. McMillan, *Ferroelectrics* **135**, 163 (1992).

[51] H. Hu and S. B. Krupanidhi (1994), Current–Voltage Characteristics of Ultrafine–Grained Ferroelectric $Pb(Zr,Ti)O_3$ Thin Films, *J. Mater. Res.* **9**, 1484–1498.

[52] X. Chen, A. I. Kingon, L. Mantese, O. Auciello, and K. Y. Hsieh, *Integr. Ferroelectrics* **3**, 259 (1993).

[53] X. Chen, A. I. Kingon, H. Al–Shreef, and K. R. Bellur, Proc. IMF 8, Aug. 1993, accepted by *Integr. Ferroelectrics.*

[54] S. Desu, Proc. of the 8th IMF, Gaithersburg, 1993; submitted to Ferroelectrics.

[55] Scott et al., Integrated ferroelectrics, 4, 61–84 (1994)

[56] D. R. Wolters and A. T. A. Zegers–van Duijnhoven, *J. Vac. Sci. Technol.*, A5, 1563 (1987).

[57] G. Greuel, unpublished results.

[58] R. Coelho, "Physics of Dielectrics", Elsevier, Lausanne, 1979.

[59] J. F. Scott, B. M. Melnick, L. D. McMillan, and C. A. Paz de Araujo, *Integr. Ferroelectrics* **3**, 129 (1993) .

[60] K. Lehovec and G. A. Shirn, *J. Apply. Phys.*, 33, 2036 (1962).

[61] W. J. Minford, *IEEE Trans. Components, Hybrids, Manuf. Technol.*, CHMT–5, 297–300 (1982).

[62] R. Waser, T. Baiatu, and K. H. Härdtl, *J. Am. Ceram. Soc.*, 73, 1645 (1990).

[63] R. Waser, T. Baiatu, and K. H. Härdtl, *J. Am. Ceram. Soc.*, 73, 1654 (1990).

[64] R. Wernicke, *phys. stat. sol. (a)*, 47, 139–44 (1978).

[65] T. Baiatu, K. H. Härdtl, and R. Waser, *Brit. Ceram. Proceed.*, 41, 149 (1989).

[66] A. Raith, P. Reijnen, and R. Waser, *Ber. Bunsenges. Phys. Chem.*, 92, 1516 (1988).

[67] B. W. Faughnan, *Phys. Rev. B: Condens. Matter.* 4, 3623 (1971).

[68] O. F. Schirmer, W. Berlinger, and K. A. Müller, *Solid State Comm.*, 16, 1289 (1975).

[69] J. Blanc and D. L. Staebler, *Phys. Rev. B: Condens. Matter.* 4, 3548 (1971).

[70] T. Baiatu, K. H. Härdtl, and R. Waser, *J. Am. Ceram. Soc.*, 73, 1663 (1990).

[71] J. Maier and G. Schwitzgebel, *phys. stat. sol. (b)*, 113, 535 (1982).

[72] R. Waser, *J. Am. Ceram. Soc.*, 72, 2234 (1989).

[73] C. Schaffrin, *phys. stat. sol. (a)*, 35, 79 (1976).

[74] A. Müller, *VDI Fortschrittsberichte Umwelttechnik*, 54, VDI–Verlag, Dusseldorf, 1988

[75] K. Okazaki, *Adv. Ceram.*, 1, 23 (1979).

[76] X. Chen, A. I. Kingon, K. Bellur, and O. Auciello, DC Leakage and Failure of PZT Thin Film Capacitors for Non–Volatile Ferroelectric Memory and DRAM Applications, *Integr. Ferroelectrics* (accepted for publication).

CHAPTER 4

FERROELECTRIC-ELECTRODE INTERACTIONS

J.M. Bell, P.C. Knight and G.R. Johnston†
Department of Applied Physics, University of Technology, Sydney, Broadway, NSW, Australia
† Materials Research Laboratory, Defence Science and Technology Organisation, Ascot Vale, Australia

1. INTRODUCTION

The interaction between a thin film of a ferroelectric and its substrate can be a very important factor in determining the electrical properties of the film. When such a film is integrated into a device, such as a surface acoustic wave device or a non-volatile memory device, there is an interface between the film and its substrate, and between the film and one or more electrodes deposited over the ferroelectric material. The device is usually subjected to further processing, involving heating the system to several hundred degrees, so that interactions between the ferroelectric and the electrodes become even more pronounced. Understanding when electrode-substrate interactions are important, such as when they affect device performance, and how they can be controlled by judicious choices of materials and processing parameters, is now an important part of ferroelectric device design.

As an example, it is becoming increasingly accepted that interactions between electrodes and lead zirconate titanate (PZT) films in ferroelectric random access memory (FRAM) devices play an important role in the fatigue behaviour of these films[1,2]. Fatigue of these devices[3] is one of the major problems facing the successful development of a commercially viable FRAM technology. The ability to minimise electrode interactions, or the effects of these interactions, is therefore important.

The basic function of the electrode in any device is to provide electrical contact to the device. This contact should generally be an ideal contact, and not in itself alter the electrical characteristics of the device. An ohmic, or non-rectifying contact[4,5] is usually required in most devices. The basic nature of the interface between two materials, and in particular between a metallic conductor and a dielectric or semiconductor, will often result in a non-ohmic contact, causing asymmetric device characteristics due to a self-biased heterojunction effect at the ferroelectric-electrode interface[6,7]. A discussion of ohmic contacts is given in Section 1.2 below.

This chapter will consider the problem of ferroelectric-electrode interactions in thin film ferroelectric devices from two perspective's: the physical and chemical nature of the interactions, and the mechanisms (principally diffusion) which cause these interactions; and the consequences of these interfacial interactions on the electrical and electronic properties of the ferroelectric devices. An additional area, the effect of the substrate-film interaction on growth and mechanical properties of the ferroelectric material, will be discussed briefly. The subject is very broad, particularly as there are many different combinations of substrates, ferroelectric materials and overlayers which are used in the production of devices based on ferroelectric materials. A brief summary of some of the principal applications of ferroelectric thin films, and the types of interfaces encountered in these materials, is given in Section 1.1. A more detailed study of PZT-electrode interfaces is given in Section 4.1. Many of the properties of interfaces between conductors (or semiconductors) and dielectrics have been studied extensively, and in many cases understood, as a result of developments in the semiconductor device fabrication industry. However, there has been no substantial research devoted specifically to ferroelectric-electrode interactions. Section 1.2 discusses the lessons which can be learned from the semiconductor industry about electrode interactions in thin films.

Section 2 of this chapter deals with the fundamental processes which lead to ferroelectric-electrode interactions, and ways of minimising the interactions, both by an appropriate choice of materials, or by careful control of the processing used in device fabrication. The processes include diffusion and chemical interaction at the film-electrode interface. The effects of substrates on the mechanical properties and growth of ferroelectric films are also discussed. The consequences of electrode-dielectric interactions on the electrical properties of a ferroelectric device are described in Section 3. Section 4 describes briefly experimental techniques which can be used to probe the interfaces between layers in a multilayer thin film system, and presents some typical results for thin PZT films on metallic and semiconducting substrates. Finally, Section 5 discusses some of the factors identified as important in controlling the extent of interactions between electrodes and ferroelectrics, and discusses the future directions which may help overcome the problems which are associated with ferroelectric-electrode interactions.

Throughout this chapter, emphasis will be placed on interactions between PZT thin films and metallic electrodes. A substantial amount of research on ferroelectrics in the past few years has concentrated in this area, owing to the emergence of FRAM technology and the dominance of PZT in this field of research. However, other systems and applications of ferroelectric materials are also important, and most of the discussion is sufficiently general to apply to other systems.

1.1 Applications of Ferroelectric Thin Films

A substantial review of thin film ferroelectrics has been given by Swartz[8]. This review covers deposition technologies and applications. There are four basic properties of ferroelectric materials which are utilised in device applications: dielectric properties; polarisation properties; piezoelectric properties; and electro-optic properties. In almost all applications of thin film ferroelectrics, the material must be coupled to external circuitry through electrode termination. Some examples of ferroelectric devices are: non-volatile memory devices based on ferroelectric capacitors (ferroelectric random access memory - FRAM)[2,3,9], electro-optic (EO) devices based on thin film ferroelectric materials ($LiNbO_3$[10], $LiTaO_3$[11] and more recently, PZT/PLZT[12]), including optical waveguides, and optical information storage systems[13], and mechanical applications based on piezoelectric properties, such as Surface Acoustic Wave (SAW) devices (where the mechanical stresses caused by the electrodes can result in frequency shifts in the device response[14]) and other mechanical applications[8] including micromotors, and micromechanical actuators.

The deposition of a ferroelectric thin film usually involves processing at high temperatures, either during sputter deposition[15,16], or during post-annealing when metal-organic decomposition[17,18] or sol-gel processing[19,20,21] is used. The subsequent deposition of the top electrode and passivation layers can also produce interactions between the electrode and the ferroelectric film. Platinum metal[22,23,24] is most commonly used as the electrode material in ferroelectric devices, although a range of other metals (eg electroless nickel[25], gold[26], stainless steel[27] and titanium[28,29]), semiconductors[6,21] and alloys, such as transition metal silicides[17], have been used. More recently there has been a significant amount of interest in oxide electrodes for ferroelectrics, such as SnO_2[30], $Sn{:}In_2O_3$[31], RuO_2[32,33] and the high temperature superconductor $YBa_2Cu_3O_7$[34,35].

The performance of FRAM devices, whether based on PZT or other materials[36], is currently limited by aging (or change in small signal response)[2] and fatigue (or change in hysteresis loop characteristics)[37,38,39]. Most FRAM devices based on PZT exhibit fatigue after approximately 10^{12} cycles, while 10^{15} cycles are required for commercially viable devices. Significant evidence is accumulating that the interface between the electrodes and the ferroelectric is responsible for the fatigue phenomenon[1,35,40].

1.2 Electrodes in Thin Film Semiconductor Processing

A principal requirement in semiconductor device processing is to achieve ohmic contacts to the semiconductors. Since dielectrics and ferroelectrics can be regarded as wide-band gap semiconductors, many of the principles followed in the semiconductor fabrication industry can be used in ferroelectric device processing. This section outlines the properties of metal-semiconductor and semiconductor-semiconductor junctions. This provides useful insights into the types of junctions that can arise both in semiconductor processing and processing of integrated ferroelectric devices. A more complete discussion of the electronic properties of metal-semiconductor junctions and semiconductor junctions can be found in many references and texts on solid-state physics and semiconductor processing[4,41,42,43].

When a metal and a semiconductor are brought together charge will flow to equalise the Fermi levels of the two materials[44]. The layer of space-charge at the interface supports a potential drop $V_{bi} = \phi_m - \phi_s$ called the *built-in* or *diffusion voltage*. The energy difference or barrier for carrier injection from the metal to the semiconductor is termed a Schottky barrier. In the simplest model this barrier height, denoted ϕ_{Bn} for an n-type semiconductor (ϕ_{Bp} for a p-type material), is the difference between the semiconductor electron affinity (χ_s), which is the energy difference between the conduction band edge and the vacuum level (see figure 1) and the metallic work function:

n-type $$\phi_{Bn} = \phi_m - \chi_s \quad (1)$$

p-type $$\phi_{Bp} = E_g - (\phi_m - \chi_s) \quad (2)$$

where E_g is the energy gap of the semiconductor. This Schottky barrier height is independent of the applied voltage (V), whereas the semiconductor to metal potential depends on the junction bias V, and is given by $V_{bi} - V$. Figure 1 illustrates one possible energy diagram for a metal with $\phi_m > \phi_s$ contacting an n-type semiconductor. This figure illustrates the formation of a potential barrier, but the magnitude of the barrier is different for each combination of work function, electron affinity and carrier type.

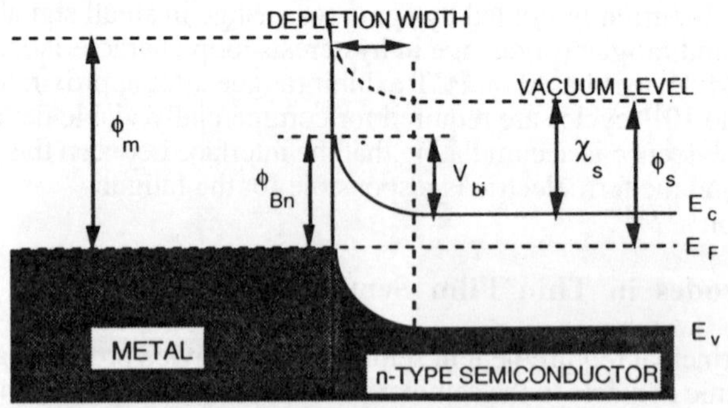

FIGURE 1. The energy diagram for a metal-n-type semiconductor interface with $\phi_m > \phi_s$

In general, conduction across a Schottky barrier is rectifying and can involve several different conduction mechanisms, including tunnelling and thermionic emission. One of the conduction mechanisms is usually dominant in any given range of field (*i.e.* applied potential) and temperature. In moderately doped semiconductors at room temperature, the dominant conduction process is usually thermionic emission. At high doping levels the depletion with becomes significantly smaller, and tunnelling can become a significant transport process.

One of the most important aspects of semiconductor device processing is to produce ohmic contacts from these metal-semiconductor junctions. An ohmic contact does not significantly change the characteristics of a device. Generally it should have small parasitic impedance and nonrectifying behavior[44]. For this to occur the barrier height must be small (compared to thermal energies) so that carriers can flow freely between metal and semiconductor in either direction.

Metallic work functions are typically larger than semiconductor electron affinities so that metal to p-type semiconductor contacts are generally ohmic, and metal to n-type semiconductors have a significant barrier height, and show non-ohmic behaviour. There are processing techniques, however, that can modify the potential barrier and result in a quasi-ohmic contact with a low contact resistance, and little rectifying behaviour over a small range of potential. One of these methods is heavily doping the semiconductor surface[4] as illustrated in figure 2. This is typically achieved by controlled diffusion of the electrode metal into the semiconductor. The increased majority carrier concentration near the interface causes a narrowing of the depletion layer, (*i.e.* the barrier width), therefore promoting tunnelling conduction.

The work functions and heats of formation of the oxide of several metals commonly used for electroding are listed in Table 1. From the

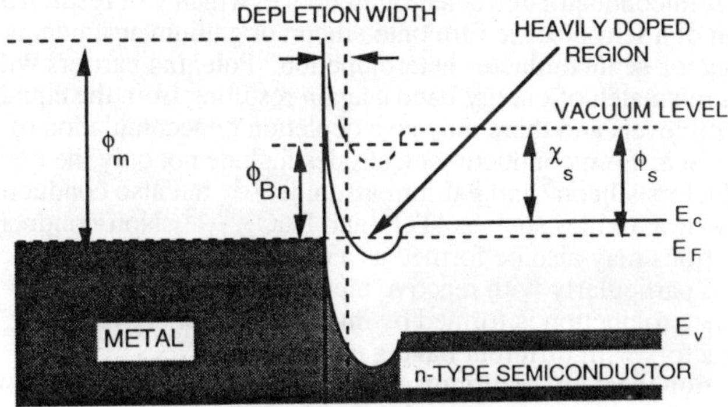

FIGURE 2. Energy diagram of a metal - n-type semiconductor ($\phi_m > \chi_s$) where doping has reduced the barrier width.

TABLE 1. Work function and heat of formation of the oxide for various metals commonly used as electrodes.

Atomic No.	Element	ϕ_m (eV) (Work Function)[45]	$-\Delta H$ 10^{-6} ($J.mol^{-1}$) (heat of formation of oxide at 298K)[46]	
78	Pt	5.65	56 522	(PtO)
			69 082	(PtO_2)
79	Au	5.1	125 604	(Au_2O_3)
22	Ti	4.33	518 744	(TiO)
			941 611	(TiO_2)
28	Ni	5.15	244 509	(NiO)
13	Al	4.28	1675 557	(Al_2O_3)
14	Si	4.85	437 939	(SiO)
			858 294	(SiO_2)

discussion above it and the numbers in table 1 it is apparent that platinum and gold are the preferred electroding metals for p-type semiconductors, as these metals will result in the smallest barrier height. However, almost all metals are suitable for this purpose, and of more interest for p-type semiconductors is the heat of formation of the oxide. A high heat of formation indicates a stable oxide and therefore also a greater tendency to form the oxide. This depends on the processing used, but the formation of an oxide will give rise to heterojunction effects at the electrodes.

A semiconductor-ferroelectric interface, which will result from direct integration of a ferroelectric film onto silicon or gallium arsenide, is a semiconductor-semiconductor heterojunction. Potential barriers will form due to the mismatch of energy bands, again resulting from the equalisation of the Fermi levels and there may be a depletion or accumulation of carriers at the interface. Semiconducting electrodes include not only the typical semiconductors silicon[6] and gallium arsenide[21,47] but also conducting and non-conducting oxides such as ITO[31] and RuO_2[32,33]. Non-conducting oxide junctions may also be formed as a result of oxidation during processing, particularly with reactive metals such as titanium.

A heterojunction is formed by intimate contact between two semiconductors with different band gaps or carrier types. For any heterojunction there are rectifying effects and space charge regions which have an associated capacitance and built-in potential. Current transport across a heterojunction is similar to either the p-n homojunction or a metal-semiconductor barrier[43].

From the above discussion it is clear that in order to form an ohmic contact to an n-type ferroelectric there are two considerations: the choice of electrode material, including its work function, chemical reactivity, and diffusivity; and the choice of an annealing protocol (time and temperature) which allows sufficient diffusion of metal into the ferroelectric to adequately dope the material. Yoon *et al*[25] have investigated the current characteristics of electroless nickel contacts on n-type semiconducting $BaTiO_3$. The electrodes are rectifying when the contacts are deposited but become ohmic after heat treatment at 200°C. This illustrates the importance of annealing to dope the surface layer of the ferroelectric in appropriate cases.

2. FERROELECTRIC-ELECTRODE INTERACTION MECHANISMS

There are many ways in which an electrode can influence the behaviour of a ferroelectric material, from changing the chemical nature of the material by diffusion into and chemical reaction with the ferroelectric, to the introduction of strain into the material, causing spontaneous polarisation of the ferroelectric. The principal influence we will consider throughout this chapter is the chemical interaction caused by diffusion between the electrode and ferroelectric at the interface. This includes the diffusion of cations from the electrode into the ferroelectric, causing doping of the material, and the diffusion of oxygen through the electrode or gettering of oxygen from the ferroelectric, causing oxidation of the electrode at the interface. The basic theory of diffusion in solids is covered in many elementary and advanced texts[4,41,48]. In Section 2.1 we discuss reaction and diffusion, and the important subject of oxygen diffusion and oxidation effects at ferroelectric-electrode interfaces, and in Section 2.2 we describe the effect grain boundary diffusion can have on electrode interactions. Finally we discuss the use of oxide electrodes in Section 2.3, as these could prove to be a useful means of improving ferroelectric crystal growth habits and eliminating some of the deleterious effects of oxygen diffusion. The influence of the substrate and film during film growth (*i.e.* epitaxy etc) is often very important in determining final film properties, but will not be discussed further in this chapter.

2.1 Diffusion and Reactions

The general case of chemical reaction of a diffusing atom reacting with the host matrix is not adequately described by the standard diffusion equation, also known as Fick's Law of diffusion[48]. This does not take into account chemical reactions that may occur between diffusing species, since the newly created species will have its own diffusion activity[49] and may also affect the diffusion of other species. Diffusion is also driven by a concentration gradient and if a chemical reaction of one species occurs and reduces the concentration at that point the reaction creates a sink for that species. The simplest way of treating the effect of reactions on diffusion is to add a sink (or source) term to the standard diffusion equation[48]:

$$\frac{\partial C}{\partial t} = D\frac{\partial^2 C}{\partial x^2} - kC \qquad (3)$$

In this modified diffusion equation, D is the diffusion coefficient, C is the species concentration at position x and time t, and kC is a sink or source term. The sink or source term is written for a first order chemical reaction with a rate constant k. The reversibility of any reactions, and whether they can be driven by any other external influences such as an applied electric field are also important considerations, and this is discussed briefly below. An important aspect of diffusion is that the diffusion coefficient depends strongly on temperature since diffusion is a thermally activated process.

A simple model which illustrates the importance of reactions is the Deal and Grove[50] model for thermal oxidation of silicon. This model also highlights one of the important features of diffusion in thin films: films are of finite thickness, and the film interfaces can act as sources or sinks for diffusing species. The ideas in the Deal and Grove model are applicable to deposition and processing of many ferroelectrics on silicon substrates as the ferroelectrics, typically with a perovskite structure, tend to be reasonably good transporters of oxygen, and also contain substantial levels of oxygen. However, it is also instructive for other substrate and electrode systems, such as titanium metal, with the major differences being the reactivity of the material with oxygen and the diffusivity of oxygen in the oxide coatings.

The Deal and Grove model for thermal oxidation of silicon assumes that oxygen molecules diffuse through the SiO_2 surface layer to the interface where they react with the silicon substrate. The oxygen concentration at the surface of the SiO_2 layer is given by $C_0 = Kp$, where K is the solubility of oxygen in SiO_2, and p is the oxygen partial pressure, and it is assumed that silicon does not migrate significantly into the SiO_2 layer. Provided the diffusion process is a steady-state process (not strictly true) then we have that $\frac{\partial^2 C_0}{\partial x^2}=0$, which leads to $C_0(x,t) \approx a + bx$. An elegant solution of this model is given in Ghez[51]. The thickness of the of the oxide layer as a function of time ($L(t)$) is given by

$$\frac{L(t)}{k_L} + \frac{L(t)^2}{k_P} = t \qquad (4)$$

where $k_L = V_{SiO_2}kKp$ is the linear rate constant and $k_P = 2\,V_{SiO_2}DKp$, is the parabolic rate constant. D is the oxygen diffusion coefficient in SiO_2, and V_{SiO_2} is the molecular volume of a silica molecule. The diffusion

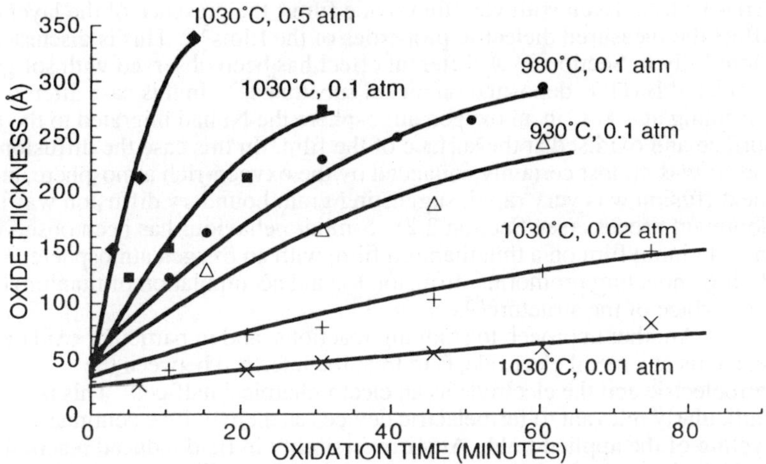

FIGURE 3. The thermal oxidation of silicon. Experimental results indicating the parabolic-linear relationship for the oxide thickness for different oxidation conditions (from Ng *et al*[52]).

profile predicted by this model, and experimental measurements of oxide thickness as a function of time are shown in figure 3.[52]

This model predicts quite different behaviour from the analytic behaviour for simple diffusion from a source into a material, and is applicable, at least in principle to the problem of diffusion between a reactive substrate (such as titanium, silicon or nickel or many of their alloys) and a ferroelectric material. In this situation the substrate electrode oxidises during processing, and growth of the oxide layer (which can have significant effects on the measured electrical properties of the film - see Section 3) depends on the rate of reaction, the concentration of oxygen at the surface of the growing oxide layer, and the rate of diffusion of oxygen molecules and oxygen atoms through the oxide layer. Most titanate ferroelectrics require annealing in an oxygen rich atmosphere, so there is always a source of oxygen which can diffuse through the ferroelectric film and cause electrode oxidation. The formation of an oxide interface layer has been observed for PZT deposited on titanium substrates[28,29] and for $LiNbO_3$ deposited on Si[53] and $SrTiO_3$ deposited on silicon[54]. An example of oxidation at the substrate interface between a PZT film and a titanium substrate is discussed in Section 4.

An important consideration in the growth of the oxide layer is its thickness, as once the thickness is greater than a threshold value, electron tunnelling through this layer can no longer occur, and the films forms a semiconducting or insulating barrier between the electrode and the

ferroelectric. Even with very thin oxide films, the presence of the layer can affect the measured dielectric properties of the films[53]. This is discussed more fully in Section 3. A different effect has been observed with sol-gel deposited $BaTiO_3$ deposited on nickel substrates[25]. In this case after annealing at 700°C in an oxygen atmosphere the Ni had migrated to the film surface and oxidised at the surface of the film. In this case the diffusion of the Ni was almost certainly enhanced by the oxygen-rich atmosphere, and the diffusion was very rapid, suggesting grain boundary diffusion was the dominant process (see Section 2.2). Similar behaviour has been observed in a platinum film on a thin titanium film, with an oxygen atmosphere during annealing promoting diffusion to, and accumulation of titanium on, the surface of the structure[55].

Another approach to studying reactions, and in particular oxidation reactions at a metal electrode, is to treat the interface between the ferroelectric and the electrode as an electrochemical half-cell. This is particularly relevant to ferroelectric devices as there will be continual cycling of the applied field. Any irreversibility in field induced reactions, such as electrode oxidation or vacancy formation, will change the device properties slowly over time. In this picture the electrical double layer potential is nothing more than the potential barrier at the junction (see Section 1.2). Oxidation reactions at the electrode then result from the transfer of electrons, which contribute an electron current. In appropriate circumstances, this can lead to oxidation of the electrode and the formation of a stable oxide layer.

This effect is well known with platinum[56] in electrochemical processes. A monolayer of oxygen is readily adsorbed onto platinum metal and can be readily stripped electrochemically. There is some controversy as to whether the layer which forms is PtO or Pt-O. However, platinum may also absorb oxygen, by a dissolution process, and this oxygen can not be readily stripped electrochemically from the platinum. There have been reports that the adsorbed oxygen on platinum becomes more stable and tightly bound over time[56], and this indicates an irreversibility in the reaction process. This will lead to changes in the electrode properties and behaviour with time.

2.2 Diffusion in Polycrystalline Systems

The actual situation in most ferroelectric films is complicated still further by the nature of most ferroelectric thin films. Generally these are polycrystalline, and vacuum deposited metal films used as electrodes (both substrate and top electrodes) are also generally polycrystalline. Diffusion in polycrystalline materials can occur by grain boundary diffusion and bulk diffusion. Grain boundary diffusion is usually described by a quite different diffusion constant (denoted D_b) from bulk diffusion (D), and usually $D_b \gg D$. The activation energy for grain boundary diffusion is also often different from that for bulk diffusion. The type of diffusion behaviour observed in a specific experiment or system depends on many factors - the grain size, the relative magnitudes of the diffusion coefficients, the grain boundary width, the temperature, and the time scale of the observation of the diffusion process.

A relatively simple classification of the diffusion behaviour, introduced by Harrison[57], divides the behaviour into three categories: one for long diffusion processes or high bulk diffusivity, one for an intermediate regime where some bulk diffusion occurs, but grain boundary diffusion is dominant, and a third category for short diffusion processes or low bulk diffusivities. These are denoted type A, type B and type C kinetics respectively.

When individual "bulk" grains are very small, or the diffusion times are very long, there is ample time for diffusing atoms to diffuse into and through grains from grain boundaries, leading to a relatively uniform distribution of diffusant throughout the polycrystalline host. This is type A kinetics, and can be described by an average diffusion constant representative of the bulk and grain boundaries. In this regime the system appears to obey Fick's law, and the diffusion length increases as $t^{1/2}$.

At short times, and when $D_b \gg D$, the diffusion length in the grain boundary (which is $(D_b t)^{1/2}$) is much greater than the diffusion length in the grain $(Dt)^{1/2}$, and all diffusion appears to take place in the grains. This is type C kinetics. There is an additional constraint that the grain boundary width δ is much greater than the grain boundary diffusion length. In some experiments, particularly looking at concentration profiles averaged over an area containing many grains, the diffusion process again appears to obey Fick's law and the diffusion penetration length increases as $t^{1/2}$, although the diffusant is confined to the grain boundaries.

Between these extremes, there is some diffusion of material into grains, accompanied by substantial grain boundary diffusion, resulting in a very non-uniform concentration profile. In this case[58] the penetration of diffusant into the material increases only as $t^{1/4}$ (as observed in measurements of averaged concentration profiles)

The types of concentration profiles, and the conditions for the different types of diffusion are schematically illustrated in figure 4. Figure 5 is a map showing the type of diffusion kinetics found in a polycrystalline thin film. This map is drawn for FCC metals, using empirical relations for the diffusion coefficients in this system[59]. The principal changes for other systems are in the position of the diagonal lines on the left of the diagram, which represent different annealing times. The temperature increases along these lines to the right. The boundaries between the regions of different kinetics are determined by the grain boundary width, δ, and the grain size, d. For $(Dt)^{1/2} \lesssim 0.1\delta$ the kinetics are purely type C. The map in figure 5 has been drawn for δ=5Å, and type C kinetics can only be found for low temperatures ($T<0.35T_m$) or at slightly higher temperatures (up to $0.5T_m$, T_m is the melting temperature of the host material) for short diffusion annealing times (t<300s). The type B kinetics region occurs for $(Dt)^{1/2}$

>100δ, but is also restricted by the grain size, and only exists up to $(Dt)^{1/2} < d/20$[58]. For a grain boundary width of 5Å, this implies that for grain sizes less than 1μm, only type A and type C kinetics can occur, with type C kinetics occurring for short, low temperature annealing processes, and larger grain sizes, and type A kinetics occurring for small grain sizes (d< 100δ) and higher temperature and longer time annealing processes.

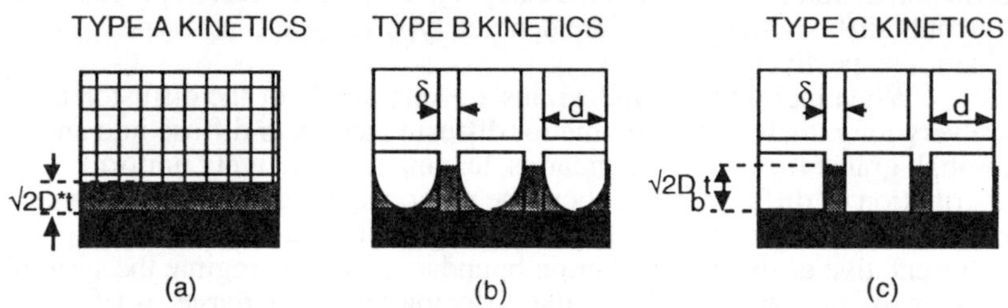

FIGURE 4. Schematic diagram showing the types of diffusion behaviour in an ideal polycrystalline system, characterised as (a) type A kinetics, (b) the intermediate type B kinetics, (c) and type C kinetics. D* is the effective diffusion coefficient for type A kinetics, δ is the grain boundary width and d the grain size.

Although figure 5 can only be applied strictly for FCC metals, changes for other materials only result in the adjustment of the reduced temperature scale, depending on the activation energies for bulk and grain boundary diffusion, and in the lateral position of the lines for constant annealing times. Data available for diffusion in oxides indicates significantly higher activation energies for both bulk and grain boundary diffusion (eg $D \sim \exp(-23.5 T_m/T)$ and $D_b \sim \exp(-18.9 T_m/T)$ for Ti^{4+} diffusion in $SrTiO_3$[58]) than are used in this diagram. This will result in a change in the slope of the lines, and they also shift further to the left in the diagram. However the conditions for finding type B kinetics are not changed significantly as the limits of this regime depend on grain size and grain boundary width.

In the case of ferroelectric thin films, grain sizes are generally quite small, in the range 50nm to 0.5μm[60,61], and the application of these principles implies that for low temperature processing ($T \ll T_m$) and for short annealing times, Type C kinetics will be found, and, under all other conditions, Type A kinetics will occur. This simplifies the situation to a certain extent, but there will always be some degree of overlap between the different type of diffusion kinetics. Two situations are most likely to arise, and these are the situations illustrated in figures 4(a) and 4(b). In most cases, unless the bulk diffusion constant is large, and the crystal also quite

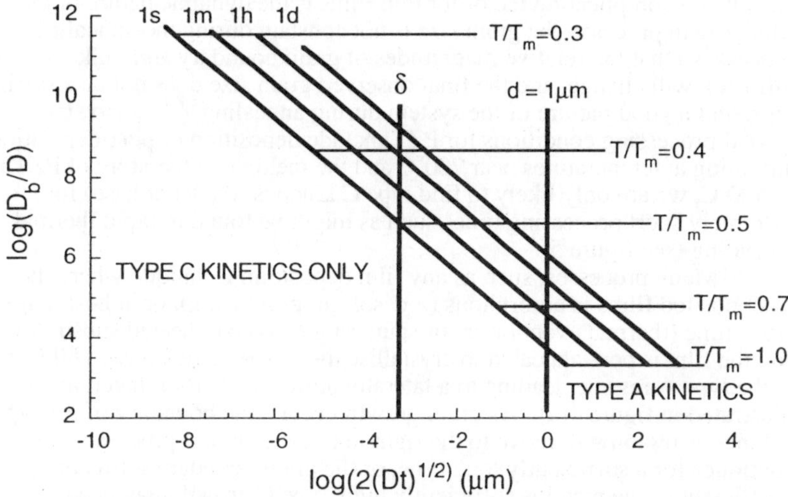

FIGURE 5. A map illustrating the occurrence of different diffusion regimes, as a function of bulk diffusion length (x-axis - $2\sqrt{(Dt)}$), ration of grain boundary to bulk diffusivity (D_b/D) and temperature, measured along the lines illustrated as a fraction of melting temperature for the host matrix. The boundaries between the different kinetics regimes are determined by the vertical lines in the diagram. This map uses the values $D=5\times10^{-5}\exp(-17.1 T_m/T)$, and $D_b=3\times10^{-5}\exp(-8.96 T_m/T)$, and $\delta=5$Å.

large, we can consider the ferroelectric grains to be either completely free of impurities from electrodes, or completely impregnated with diffusant.

In order to determine which of these situations will occur we first need to know the annealing time and temperature and the species which are likely to diffuse, and in which direction. For diffusion from the ferroelectric film into the electrode, as is likely to occur with oxygen and possibly lead (see Section 4), we require the diffusion coefficients and grain structure of the electrode, while for diffusion of electrode material into the ferroelectric we need the diffusion coefficients and grain size and grain boundary width for the ferroelectric. Unfortunately owing to the paucity of reliable data on diffusion in ferroelectrics, in particular grain boundary diffusion, detailed predictions can not be made. Grain boundary widths in bulk polycrystalline barium titanate have been observed to vary between 1 and 10μm for grain sizes ranging between 15.6μm and 300nm respectively[62]. These are quite large values, and suggest that type C kinetics will occur over a wide range of temperature and anneal times.

However a complicating factor for thin films is the dynamic nature of the film growth process - the grain size is not constant during the annealing process, so that the relative magnitudes of grain boundary and bulk diffusion will change, and the final observed grain size does not necessarily represent a good picture of the system during annealing[4,63]. Also since typical processing conditions for PZT include deposition or post-deposition annealing at temperatures near 700°C and the melting temperature of PZT is ~1500°C, we are only likely to find type C kinetics (the latter case) for extremely short processing times such as might be found in rapid thermal annealing (see figure 5).

Many processes, such as any film deposition technique where the as-deposited films is amorphous (*e.g.* sol-gel processing), or at best nano-crystalline (thermal evaporation or sputtering onto an unheated substrate), and the film is post-annealed to crystallise the ferroelectric phase, will start in the type A regime, leading to a laterally uniform diffusion layer, as illustrated in figure 4(a). As grain growth occurs, the boundary of the type A kinetics region will move to the right in figure 5 and, if processing continues for a sufficiently long time, or the grain boundary diffusion coefficient of the metal is sufficiently high, type C growth may occur, leading to diffusion of electrode material along grain boundaries (figure 4(b)). This may result in dendritic structures, or at the very least, sites for nucleation of dendritic structures. In extreme cases, if the diffusion is sufficiently rapid or the film thin enough, electrode material could diffuse right through the thin film. Depending on the composition of the grain boundary region, this could lead to short-circuiting the ferroelectric material, resulting in catastrophic failure of a device. Epitaxial growth, and sputter deposition at higher substrate temperatures, tends to produce larger grain sizes. Therefore we would expect less diffusion by type A kinetics, and more grain boundary diffusion. However grain sizes can still be still very small for thin epitaxial films, and crystallite sizes of 30nm for 300nm epitaxial $PbTiO_3$ on MgO have been observed[64], so the differences between epitaxial films and randomly oriented materials will not be significant.

2.3 Oxide Electrodes - Minimising Oxygen Diffusion Effects?

As noted in Section 1, the use of oxide electrodes has attracted significant interest recently, partly because there are some new electrode materials, such as the high temperature superconductor $YBa_2Cu_3O_{7-\delta}$, which have interesting properties and partly owing to the problems of fatigue which occur in ferroelectric devices, even when otherwise excellent material is prepared. It has been suggested[38,40] that the formation of oxygen defects in the material and the accumulation of oxygen defects at electrodes (via diffusion) is a mechanism for fatigue. The use of oxygen electrodes provides a possible way of overcoming this problem. Indium-tin oxide ($Sn:In_2O_3$ - ITO) electrodes are also extremely useful for optical applications owing to the high transparency of good quality ITO.

The advantage of a conducting oxide electrode is that oxygen diffusion and reaction are reduced owing to a less abrupt oxygen gradient at the interface. The conductivity of the oxide electrode is less sensitive to further oxidation during ferroelectric annealing than metallic electrodes.

Three particular oxide electrode materials are of great interest: ITO, RuO_2, and $YBa_2Cu_3O_{7-\delta}$. The latter material is very attractive as a substrate for ferroelectric thin film deposition as it has a tetragonal/ orthorhombic perovskite structure with a lattice parameters a≈b≈3.8Å, and c=11.68Å. Epitaxial growth of PZT on $YBa_2Cu_3O_{7-y}$ has been reported by some groups[34,35], and it has a high conductivity if well prepared (up to $10^6\Omega^{-1}m^{-1}$ at 300K[65]), which of course increases substantially at low temperatures. ITO is also an attractive electrode material, particularly for opto-electronic applications, as it is transparent (T>90% from 400nm to approximately 1.2μm)[66], and can achieve quite high conductivities (up to $10^4\Omega^{-1}m^{-1}$ at 300K[67]). RuO_2 has a very high conductivity ($\sigma = 2.84\times10^6\Omega^{-1}m^{-1}$ at 300K[46]), and has the rutile crystal structure with a=4.52Å and c=3.12Å.

The principal advantage of an electrode such as $YBa_2Cu_3O_{7-\delta}$ (123) lies in its ability to epitaxially match ferroelectric films such as PZT, $BaTiO_3$ and $PbTiO_3$. However, from the perspective of oxygen diffusion, there are other significant advantages. Firstly, 123 maintains its electrical conductivity over a relatively wide range of oxygen stoichiometry[68]. This means that during processing of an integrated device, some oxygen diffusion through the ferroelectric layer and into the 123 electrode will not destroy the electrode conductivity. Also the crystal structure does not change significantly during oxygen uptake or depletion, provided $|\delta|<0.4$. There is only a small change in lattice parameters as δ increases (ie oxygen is lost from the superconductor), from a=b=3.82Å in the tetragonal state to a=3.81Å, b=3.88Å in the orthorhombic state[68]. The situation is quite different from that in a metal such as titanium or aluminium. Oxidation of Ti changes the crystal structure significantly, and the molecular volume of TiO_2 is approximately twice that of Ti[46]. Similarly the coverage of a platinum electrode with as little as a monolayer of oxygen can change the lattice parameters significantly. Any oxidation or reduction occurring at a ferroelectric-metal interface is liable to produce large stresses in the films, affecting electrical properties and possibly also bonding between the film and the electrode. These problems are largely absent when using the high temperature superconductor. However other problems can arise with the 123 material. It is not as stable as most metals or metal oxides, and its properties are readily destroyed by cation diffusion into the structure.

There has been some work reported using ITO electrodes for PZT films[30,31,69]. However significant cation diffusion (particularly Pb) has been observed in some of this work[31], and strong thickness dependence of the dielectric properties of the PZT in the other work[30,66], which can also be accounted for by substantial interdiffusion (see Section 3). Another potential problem with ITO is that to achieve high conductivities, annealing in a reducing atmosphere is necessary[67]. Therefore subsequent processing in an oxidising atmosphere, as is required for most ferroelectrics, can degrade the conductivity of the ITO and reduce its effectiveness as an electrode.

3. EFFECTS OF ELECTRODE INTERACTIONS ON FERROELECTRIC FILM PROPERTIES

Layers of differing phase or differing stoichiometry formed by diffusion and reaction of species at an electrode-ferroelectric interface affect the overall dielectric properties of the entire device. In this section the effects of a ferroelectric electrode interface on the measurable dielectric and ferroelectric properties shall be considered. There are many empirical electrical circuit models of a ferroelectric device[70,71] . These models essentially fit an equivalent circuit to measured data to distinguish the electrical contributions of the measuring circuit, electrode influences and intrinsic parameters of a ferroelectric layer. We will consider in detail only a simple two-layer model for an electrode termination to a ferroelectric material. This model is described in Section 3.1, where a brief discussion of the film thickness dependence and frequency dispersion of dielectric properties of the system is also given. The effect on coercive field and switching times are considered, and limitations of the simple two-layer model discussed.

The simple model assumes there is ohmic contact between the electrode and ferroelectric material. Any space-charge associated with the electronic junction or interface region will polarise with an applied field. Therefore the choice of electrical contact to the ferroelectric is important when measuring the electrical properties of a film. Any contact resistance or accumulation of space charge will usually be a source of low frequency dispersion of the dielectric constant. Rectifying effects caused by a single electrode or asymmetric combination of electrodes can cause significant asymmetry in the measured conduction, ferroelectric hysteresis and switching characteristics of ferroelectric films. These effects are discussed in Sections 3.2 and 3.3.

3.1 The Two-Layer Capacitor Dielectric Model

A ferroelectric capacitor with an interface layer can be modelled by two capacitors in series[23,70,72]. This simple model is illustrated in figure 6. This model is clearly an idealised model of the real situation. Since the interface layer is caused by diffusion of material between the electrode and the ferroelectric, the concentration profile of elements in the interface layer will not be constant across the layer, but will vary according to the diffusion profiles predicted by the diffusion equation (see Section 2), and will not change discontinuously at the boundary between the different regions. The use of a two-layer model implies that some form of "average" dielectric properties can be used. An exact treatment of a similar problem, that of a graded AlGaAs layer, has been solved[73]. The results show only

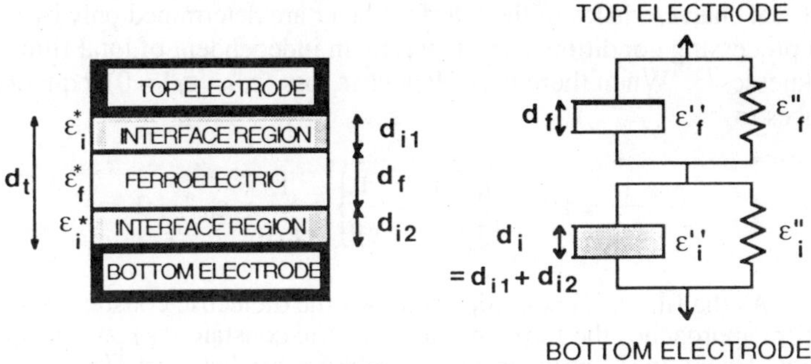

FIGURE 6. The basis of the two-layer capacitor model, and the electrical representation of a ferroelectric capacitor with an interface layer between the electrode and the ferroelectric on which the model is based.

negligible differences between the field distributions predicted by the exact and approximate solutions.

For a simplified discussion, the interface regions at both electrodes can be combined and represented electrically as a single layer with a thickness equal to the sum of the interface layer thicknesses.

The equations of this equivalent circuit will be expressed in two ways. The first form highlights the thickness dependence of the dielectric properties, while the second form highlights the effects of the interface on frequency dependent dielectric behaviour.

In the model the ferroelectric layer has a total thickness d_f and a complex permittivity $\varepsilon_f^* = \varepsilon_f' - i\varepsilon_f''$, while the interface layer has a thickness d_i and permittivity $\varepsilon_i^* = \varepsilon_i' - i\varepsilon_i''$. The total film thickness is $d_t = d_i + d_f$, and the measured permittivity of the device is $\varepsilon_m^* = \varepsilon_m' - i\varepsilon_m''$. The real and imaginary parts of the measured complex permittivity of this simplified device are

$$\varepsilon_m' = \frac{R}{R^2 + I^2}; \quad \varepsilon_m'' = \frac{I}{R^2 + I^2} \tag{5}$$

where
$$R = \frac{d_i}{d_t}\left\{\frac{\varepsilon'_i}{\varepsilon'^2_i + \varepsilon''^2_i} - \frac{\varepsilon'_f}{\varepsilon'^2_f + \varepsilon''^2_f}\right\} + \frac{\varepsilon'_f}{\varepsilon'^2_f + \varepsilon''^2_f} \tag{6}$$

and
$$I = \frac{d_i}{d_t}\left\{\frac{\varepsilon''_i}{\varepsilon'^2_i + \varepsilon''^2_i} - \frac{\varepsilon''_f}{\varepsilon'^2_f + \varepsilon''^2_f}\right\} + \frac{\varepsilon''_f}{\varepsilon'^2_f + \varepsilon''^2_f}. \tag{7}$$

3.1.1. Thickness dependence of dielectric properties. Eliminating d_f (= d_t - d_i) in equations (5)-(7), the permittivity can be expressed as a function of the ratio $^{d_i}/_{d_t}$. We assume that the thickness and electrical characteristics of the interface layer are determined only by the film processing conditions, *i.e.* they remain independent of total film thicknesses[74]. When there is no dielectric loss ($\varepsilon_i'' = \varepsilon_f'' = 0$), equation (5) reduces to

$$\frac{1}{\varepsilon_m'} = R = \frac{d_i}{d_t}\left\{\frac{1}{\varepsilon_i'} - \frac{1}{\varepsilon_f'}\right\} + \frac{1}{\varepsilon_f'} \; ; \; \varepsilon_m'' = 0 \qquad (8)$$

As the film thickness (d_t) increases the dielectric constant (ε_m') of the film approaches the ferroelectric dielectric constant (ε_f'), as one would expect. Equation (8) predicts a linear relationship between $^1/_{\varepsilon_m'}$ vs $^1/_{d_t}$ in the case of a lossless film, and by extrapolation ($d_t \to \infty$) a value for the ferroelectric dielectric constant ($1/\varepsilon'_f$) can be extracted.

Scaling of dielectric constant with total film thickness has been observed in many experiments[75,76,77], and with different substrates, in particular ITO[30], silicon[7], platinum and gold[26]. The variation of measured dielectric constant with total film thickness described by equation (5) is shown in figure 7 for three values of the loss tangent (tan δ) for the

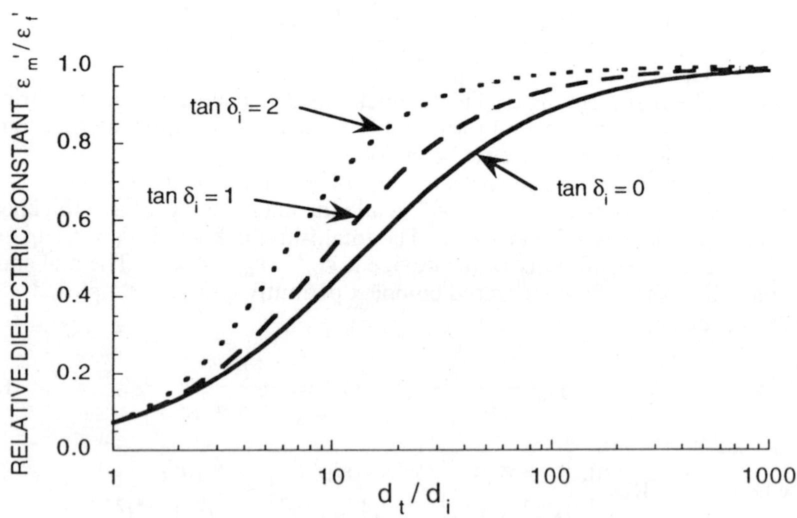

FIGURE 7 The ratio of the measured dielectric constant to the bulk ferroelectric value for a two-layer capacitor device as a function of total thickness to interface thickness. The dielectric properties ε_i'=50, ε_f'=700, and ε_f''=0.

interface layer. As we would expect, the more conducting the interface layer the more rapidly the dielectric constant approaches the intrinsic ferroelectric value. This illustrates that even a very thin non-conducting or semiconducting interface layer can have a pronounced effect on the measured dielectric response of a film, reducing the apparent dielectric constant of the ferroelectric, and reducing its usefulness in many applications.

*3.1.2 **Dispersion effects.*** The dispersion of the complex dielectric constant (ε_m^*) caused by time dependence of the space-charge polarisation is termed the Maxwell-Wagner effect[78], and in general the dielectric loss of each layer is frequency dependent, $\varepsilon''(\omega) = {}^{1}/_{\rho\omega\varepsilon_o}$ where ρ is the layer resistivity, ω is the angular frequency and ε_0 is the permittivity of free space. Equations (5)-(7) can then be written in the form of Debye relaxation formulae for the dielectric constant and the film resistivity:

$$\varepsilon'_m(\omega) = \varepsilon_\infty + \frac{\varepsilon_s - \varepsilon_\infty}{1 + \omega^2 \tau_\varepsilon^2} \tag{9}$$

$$\rho_m(\omega) = \rho_\infty + \frac{\rho_s - \rho_\infty}{1 + \omega^2 \tau_\rho^2} \tag{10}$$

where ε_∞ and ρ_∞ are the high frequency dielectric constant and resistivity of the film, ε_s and ρ_s are the static (d.c) dielectric constant and resistivity of the film, and τ_ε and τ_ρ the relaxation times for the dielectric constant and resistivity. The Debye factors can be found by fitting the relaxation formulae to the measured frequency dependence of the dielectric constant. The Debye factors can also be expressed in terms of the layer permittivities, $\varepsilon_f^* = \varepsilon_f' - i\varepsilon_f''$, $\varepsilon_i^* = \varepsilon_i' - i\varepsilon_i''$ and layer thicknesses $d_t = d_i + d_f$.

If either of the dielectric layers is a lossy dielectric then there will be an accumulation of space-charge between the ferroelectric and interface layers. A non-ohmic contact to the ferroelectric will have high resistance in the interface layer and a high space-charge capacitance. Under these conditions dispersion becomes important and for a range of measuring frequencies it is possible for the measured dielectric constant of the device to exceed the intrinsic dielectric constant of the ferroelectric layer for a small total film thickness. In this case there will be an associated large dielectric loss, and as the film thickness increases the measured dielectric constant decreases asymptotically towards the ferroelectric value. Figure 8 illustrates that when the capacitance and resistance of the interface layer are large compared to the ferroelectric layer then a stationary point forms in equations (9) and (10). This effect has been noticed by Sreenivas *et al*[79] with a SnO_2/a-$BaTiO_3$/Al capacitor device. Losses within the device therefore produce low frequency dispersion which increases the measured dielectric constant and dielectric loss at low frequencies.

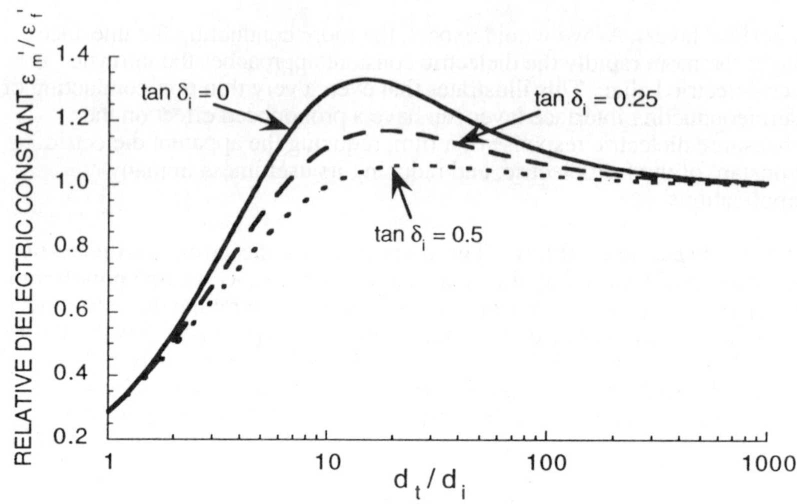

FIGURE 8. The measured dielectric constant of a two-layer capacitor device relative to the ferroelectric layer value illustrating the effect of a large dielectric loss in the interface layer. The values used in this simulation were $\varepsilon'_f = 700$, $\varepsilon''_f = 1400$, $\varepsilon'_i = 200$, $\varepsilon''_i = 0, 50, 100$ (corresponding ot $\tan\delta_i = 0, 0.7, 1.4$)

By fitting the frequency and thickness dependencies of the measured complex dielectric constant to the above equations it is possible that values of the dielectric constants and resistivities of both ferroelectric and interface layers can be extracted. However deviations from this behaviour are expected owing to effects such as non-uniform interface layers (*i.e.* with graded variation in properties), and the real significance of the two-layer model for a capacitive device lies not in the ability to extract intrinsic information about the ferroelectric material or any interface layers present, but in the implications for device behaviour. Therefore we will now discuss some of the elementary effects on device performance which are associated with the electric field distribution in devices resulting from the existence of an interface layer.

***3.1.3. Electric Field Distribution*.** One of the most important aspects of the presence of an interface layer at the electrode-ferroelectric boundary is the effect on the distribution of the electric field in the device. When an electric field is applied to series capacitors the field will be distributed so that the majority of the voltage drop occurs over the lower capacitance, or higher impedance, component. In the case of the two-layer model we are discussing, this is the low dielectric constant layer. For lossless films, the field across the ferroelectric is reduced by the ratio of the dielectric constants of the two layers, but is proportional to the layer thickness: $E_f = (\varepsilon'_i/\varepsilon'_f)\, E_i$. Even a very thin interface layer can have a large effect on the electric field distribution since the voltage drop is larger across a low dielectric region than across a high dielectric phase of similar thickness.

This effect can be illustrated by considering the coercive field of the ferroelectric in the two layer model. The coercive field is known to vary with film thickness[60,80], between $d_t^{-1/3}$ for thick films and $d_t^{-4/3}$ for thin films[2]. In the two-layer model the applied field E_c^m required to switch a ferroelectric layer with intrinsic coercive field E_{c0} is

$$E_c^m = E_{c0}\sqrt{\varepsilon'^2_f + \varepsilon''^2_f}\;\sqrt{R^2 + I^2}. \tag{11}$$

In the absence of dielectric loss, equation (11) reduces to

$$E_c^m = E_{c0}\,\varepsilon'_f\left\{\frac{d_i}{d_t}\left\{\frac{1}{\varepsilon'_i} - \frac{1}{\varepsilon'_f}\right\} + \frac{1}{\varepsilon'_f}\right\} \tag{12}$$

From equation (12) it is clear that the measured coercive field varies inversely with total film thickness and that as d_t increases the measured coercive field approaches the intrinsic coercive field of the ferroelectric ($E_c^m = E_{c0}$). This relation has been found for $(Ba,Sr)TiO_3$ films on steel[27], platinum[7] and $BaTiO_3$ crystals[81]. The variation of measured coercive field with total film depth calculated from equation (11) is illustrated in figure 9 for the same values of material properties as in figure 8. A large field is required at small film thicknesses to overcome the voltage drop across the interface layer.

The general case of equation (11) (*i.e.* including the effects of dielectric loss) is more complicated because the polarisation of space charge in the film produces fields associated with the charge separation. In figure 9 the thickness dependence changes from d_t^{-1} to d_t^0 as the thickness increases when there is dielectric loss in the interface.

Other important parameters for ferroelectric films, particularly for memory applications, are the switching time, t_s (*i.e.* for polarisation reversal), and peak switching current, i_{max}. Both the switching time and the peak switching current are known to be dependent on the field pulse applied to switch the ferroelectric[2,81]. The quantities are known to empirically vary according to

$$i_{max} = i_\infty\, e^{-\alpha/E_f}\,; \quad \text{and } t_s = t_\infty\, e^{+\alpha/E_f} \tag{13}$$

where i_∞ and t_∞ are the current and switching time at infinite field strength ($E_f \rightarrow \infty$), and α is a constant. E_f is the field across the ferroelectric. The

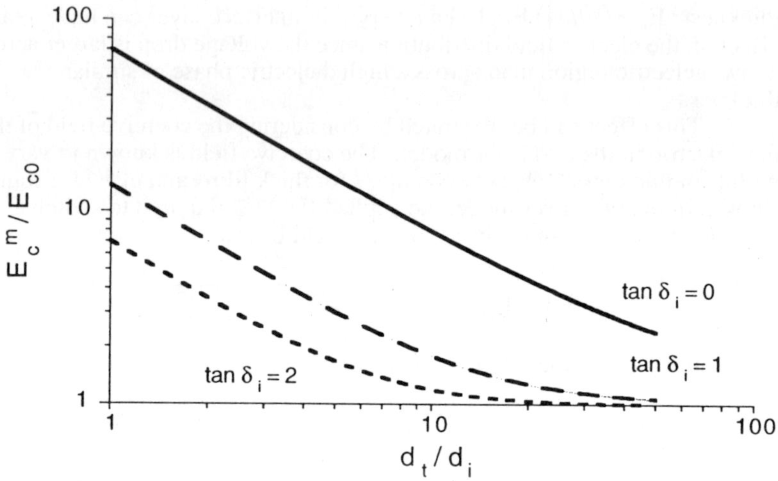

Figure 9. Theoretical prediction of log E_c^m/E_{c0} vs log d_t/d_i based on the two-layer capacitor model showing the power law variation of E_c^m on d_t. The material dielectric values used were $\varepsilon'_f = 700$, $\varepsilon''_f = 0$, $\varepsilon'_i = 50$, and $\varepsilon''_i = 0, 50, 100$.

presence of an interface layer in the device means the field switching the ferroelectric layer is dependent on the film thickness, and is related to the applied field according to equation (11). This can lead to dependence of the switching time and peak switching current on film thickness. This is an example of how an interface layer can have an indirect effect on electrical measurements, and in particular on important parameters for device operation. In this specific case, we are also considering a transient effect and the effects of space charge in the interface layer should also be considered. These will be discussed in the Section 3.2, but this simple example illustrates that care must be taken interpreting results when the presence of an interface layer is suspected.

3.1.4 . Other Interface Effects There are some obvious deficiencies of the model discussed in the previous two sections, some of which have been mentioned. The most obvious deficiency is that the model assumes that the dielectric constant is a single discrete value throughout the interface layer. However since the interface layer is most likely to be a result of diffusion processes, the dielectric properties are more likely to vary with a diffusion gradient as illustrated in Section 2.

Sections 3.1.1-3.1.3 describe the consequences of the existence of a well-defined interface layer on measured dielectric properties of a ferroelectric film . It should be noted that these results do not imply that all thickness effects are the result of an interface layer of the type described here. Thickness dependence can also be the result of modifications to grain shape and size in a ferroelectric at different film thicknesses, and even changes in crystal growth habit in thin films. It has been suggested that the crystallite size and the thickness may not be independent[79] and therefore ferroelectric properties that scale with grain size cannot be discounted. The grain boundary around grains is electrically equivalent to a low dielectric layer and consequently this model is also used to study the effects of a connecting grain boundary phase[62]. Further complications can arise because the volume fraction of the grain boundary phase is known to increase at smaller film thicknesses, corresponding to an increase in the interface depth d_i as d_t decreases.

Another approach which can be used to consider the effect of interactions between electrodes and a ferroelectric, and which is particularly applicable in thin films where grain boundary diffusion dominates, is to consider low dielectric constant regions growing through the film (*i.e.* forming along the grain boundaries). This will result in a parallel capacitor arrangement as illustrated in figure 10.

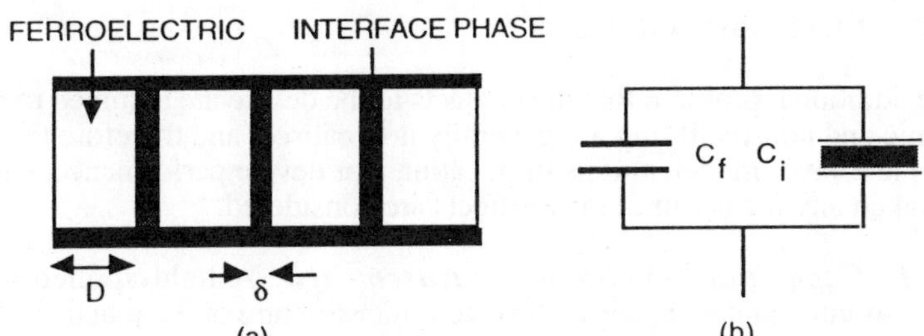

FIGURE 10. (a) Schematic representation of the parallel capacitor arrangement, where an interface phase forms between grains in a film, but perpendicular to the electrode, giving rise to the equivalent circuit shown in (b).

This effect has been extensively analysed[23,62,82], but the interface regions in parallel with a ferroelectric material are usually not considered to play a major role as compared to any low-dielectric constant series interface layer. This is principally owing to the different electric field distributions found in the two arrangements. In the parallel arrangement, the electric field across the ferroelectric is essentially the same as that applied to the entire device, so that field effects such as described in Section 3.1.3 are not seen. However, there are some situations where a small film thickness dependence of dielectric properties could be observed. If the grain size, D, in the thin film scales with film thickness, d, as is known to occur in many thin film systems[63] and has been observed in ferroelectrics[64], then a simple calculation for lossless materials (both the ferroelectric and the low dielectric grain boundary phases), with 2-dimensional grains of characteristic lateral size D shows that

$$\varepsilon_m' = \varepsilon_f' - \frac{4\delta}{\alpha d^{\beta}}(\varepsilon_f' - \varepsilon_i') \qquad (14)$$

where δ is the grain boundary width, and $D=\alpha d^{\beta}$. Experimental evidence suggests that $\beta \approx 1/2$ (for $PbTiO_3$[77]), and this leads to a very weak thickness dependence of dielectric properties. This type of electrode-ferroelectric interface structure is much more likely to play an important role when the grain boundary is quite conductive, and 'spikes' emerge from the electrodes as a results of diffusion. As noted in Section 2.2, these spikes could act as nucleation centers for dendritic growth of defects, since electric fields at the tips of the spikes will be significantly higher than in-the-plane electrodes[83], and this will act to shield the ferroelectric beneath the tips of the spikes from the electric field.

3.2 Electronic effects

The additional problem that the contacts to the device are assumed to be ohmic and non-rectifying are generally not realised, and therefore the simple 2-layer model, and its implications for device performance, can not be taken any further until these effects are considered.

3.2.1. Capacitance-Voltage Measurements. A field applied across a non-ohmic contact changes the electronic structure of the junction. Space-charge will polarise and, under conditions of reverse bias, a depletion layer forms. In terms of the 2-layer model, these effects change the apparent dielectric response of each layer, and the magnitude of the effects vary with field strength (*i.e.* applied potential).

The capacitance at different voltages (capacitance-voltage characteristic, or C-V) represents the changing polarisation in the film. In ferroelectric films there are contributions from hysteretic domain switching[7,83] and the accumulation or depletion of space-charge. C-V measurements are often used as a technique to determine carrier density profiles in semiconducting devices. Spierings *et al*[84] has used C-V characteristics to distinguish between two types of platinum electrode preparation and the effects of fatigue.

A ferroelectric is more than a simple dielectric. The polarisation and capacitance vary non-linearly with an applied field due to the domain structure. A typical C-V characteristic for a ferroelectric with ohmic contacts is shown in figure 11. The C-V profile is actually a measure of $\partial P/\partial E$, so the peak in the C-V characteristic corresponds to a large polarisation change, or the ferroelectric domains switching from one orientation to another. The "butterfly" curve shown in figure 11 has peaks at voltages corresponding to the coercive field strength of the film. At higher voltages the dielectric constant deceases. The curve in figure 11 is offset slightly in the horizontal direction (towards negative voltages).

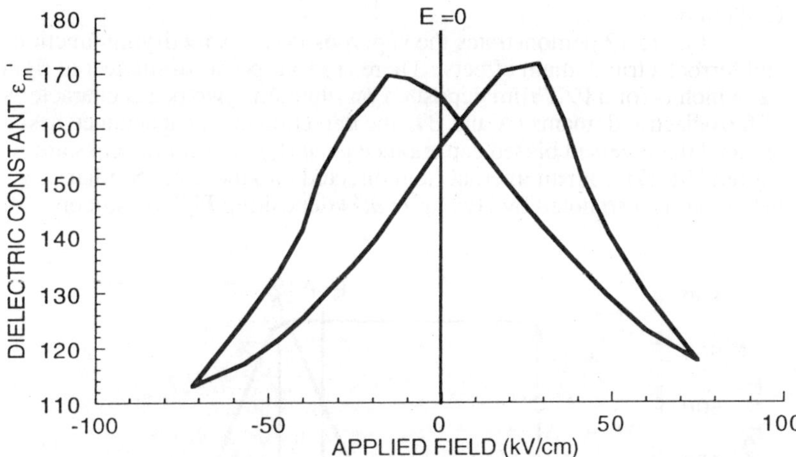

FIGURE 11. The CV characteristic of a ferroelectric PZT deposited by sol-gel processing on platinised silicon. The top electrode is gold. Note that the two peaks are of equal height, but there is a small offset towards positive bias.

The depletion capacitance of both a semiconductor-semiconductor p^+-n junction and a metal-semiconductor Schottky barrier have a similar characteristic form[4]:

$$C \propto \frac{1}{\sqrt{V_{bi} - V}} \qquad (15)$$

where V_{bi} is the built in voltage at the junction ($V_{bi} = \phi_m - \phi_s$ - see Section 1.2) and V is the bias voltage. At large reverse bias (V<0) the total capacitance is reduced by the series space charge capacitance at the electrode interface. This results in a suppression of the ferroelectric characteristic on the reverse bias side of the curve. The effects of the depletion will be very similar to the much studied metal-insulator-semiconductor (MIS) device in semiconductor applications. In a reverse biased MIS device the depletion width adds to the depth of the insulator reducing the overall capacitance of the structure. This reduction in capacitance has been observed by Roy *et al*[85] for $SrTiO_3$ on silicon. Domain effects are not prominent as there is a the low charge storage density.

It has been noted that even two ohmic electrodes with similar work functions and processing exhibit small shifts and a little asymmetry in the C-V curves[26,30]. An imbalance of space-charge at the two different electrode terminations may create an internal bias in the film that shifts the C-V curve.

Figure 12 demonstrates the superposition of a rectifying junction and ferroelectric domain effects. There are four points of interest in figure 12, which is for a PZT film deposited on titanium: two peaks characteristic of ferroelectric domains (A and B); the depletion layer capacitance has reduced the reversed biased capacitance peak (C); the curve peaks are shifted left (D) from an internal field directed into the film. Similar behaviour is also noted by Hwang *et al*[86] depositing PZT on silicon.

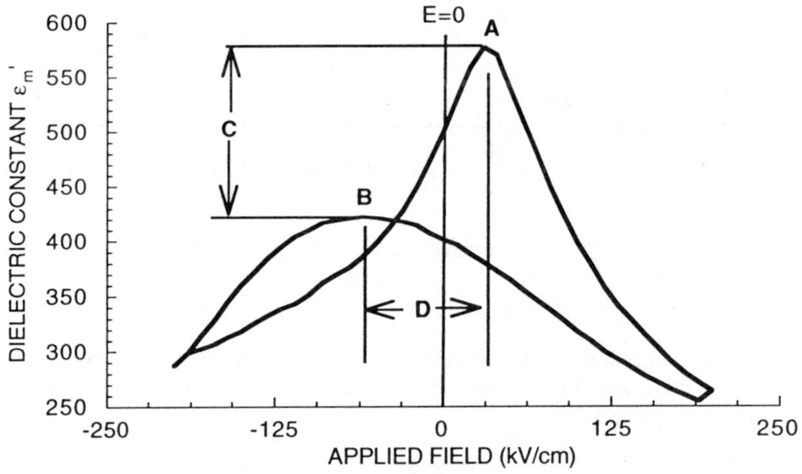

FIGURE 12. C-V of PZT on titanium. The film was deposited by sol-gel dip-coating and annealed at between 640˚C and 680˚C. The features marked are discussed in the text.

3.2.2. Hysteresis. Other measurements which can be affected by the electrode interface include hysteresis measurements, using for example a Sawyer-Tower circuit, or leakage current measurements. At frequencies typically used with the Sawyer-Tower circuit[87] (50-60Hz) dispersion from the space charge associated with an interface layer may be pronounced, particularly if the interface layer has a high dielectric loss. Any conduction through the film will also produce a phase shift in the polarisation relative to the applied cyclic field. A modified Sawyer-Tower circuit[88] adds a compensating phase shift to remove this effect. A time based ferroelectric hysteresis model has been developed by Miller *et al*[71, 89]. This model includes the effect of a non-switching dielectric layer and a sheet of space charge within the film. Results from this model indicate that a non-switching layer (such as an electrode-ferroelectric interaction region) changes the hysteresis loop considerably, both thinning the loop (*i.e.* reducing the coercive field) and adding a capacitive component to the loop. This type of behaviour has been observed by Sanchez *et al*[26], who found that gold readily diffused into PZT during alloying and that this resulted in a thinner hysteresis curve with a dramatic drop in coercive and saturation field.

3.2.3. Conduction. The majority carrier type in ferroelectric films will determine the electronic structure of the electrode interface. We have already seen in Section 1.2 the importance of n-type or p-type conduction in determining the metal-semiconductor contact type. PZT and $BaTiO_3$ are known p-type conductors, unless they are oxygen deficient when they become n-type. For thin films conduction may be modified by the presence of traps[90] and double injection of carriers may have to be considered[91]. Leakage current measurements made on thin films are dominated by the carrier injection/ejection from the metal/dielectric interfaces rather than by the bulk conduction properties of the film material.

Distinguishing between the various conduction mechanisms which are possible across a ferroelectric-electrode interface is not straightforward, and a discussion of this is beyond the scope of this chapter. However it has been noticed that the hysteresis loop and conductivity of PZT thin films are significantly different for different top contacts materials[40,91] implying that the contacts are not simply ohmic and that a metal-insulator-metal (MIM) model is necessary to fully describe conduction behaviour from both electrodes.

4. EXPERIMENTAL RESULTS

There is a wide range of different combinations of electrodes and ferroelectric materials, particularly as different deposition techniques for the same material can lead to different film growth characteristics and, as illustrated in Section 2.2, different interface interactions, especially at the film-substrate interface. In Section 4.1 we review some of the combinations of electrode materials which have been used for PZT films, and then in Section 4.2 discuss experimental techniques which can be used to probe the structure of the interface between an electrode and a ferroelectric. In Section 4.3 we present a case study of PZT deposited on titanium substrates. This illustrates some of the diffusion effects described in Section 2 and the consequences of the substrate-film interactions on the electrical properties of devices made using these films which are described in Section 3.

4.1 PZT-Electrode Combinations

For ferroelectric device work, the most commonly used substrate and base electrode is platinum. Platinum does not readily oxidise (although as noted in Section 2.2 it does develop a monolayer coverage of oxygen and has a reasonably high solubility for oxygen[56]) and is a reasonably effective diffusion barrier at PZT annealing and deposition temperatures. However, the situation is not as straightforward as might be hoped and there are still undesirable electrode effects with platinum electrodes. One of the major problems perceived with the performance of platinum electrodes is the fatigue in ferroelectric devices which use platinum electrodes. For integration into silicon devices, platinum has the added disadvantage of poor adhesion to SiO_2[56] and frequently a Pt/Ti /SiO_2/Si substrate is used to overcome this problem[26,55,92]. Other substrate/electrode materials have been used extensively tested in order to find a reliable substrate material which can allow easy integration into devices, and which produces fatigue-free films.

Table 2 is a summary of substrates which have been used for growth of PZT, using a variety of deposition and growth techniques. The final column in this table is a summary comment of the type of measurements and/or results described in the work. Again this list is not exhaustive, but it provides a good survey of the range of materials which have been used and the type of results obtained.

The most commonly used substrate for most of the ferroelectric memory research has been the Si/SiO_2/Ti/Pt structure, and most tests have made a device structure using a gold top electrode. The importance of careful study of the interface region, and it possible influence on the properties of the entire film are highlighted by the work of Abe *et al*[55] and Dana *et al*[92]. In both cases the substrate under the platinum film used as the substrate for ferroelectric film deposition was shown to play an important role in determining the final properties of the ferroelectric layer. X-ray diffraction patterns obtained by Abe *et al*[55]show that for the same rf sputter deposition conditions the perovskite PZT phase only forms if the platinum film has a titanium underlayer.

Another aspect of the ferroelectric-electrode interaction which is very important is the diffusing species. For example, in the devices with Au top electrodes and Pt bottom electrodes, Auger Electron Spectroscopy has shown that oxygen diffuses into both the platinum and gold electrodes, and that gold diffuses into the PZT to a much more significant extent than Pt[26,40], resulting in quite different electrical characteristics at the two interfaces.

TABLE 2. Substrate-PZT combinations

Material	Deposited on	Deposition method	Reference
PZT 50/50	TiO_x / SiO_2 TiN_x / SiO_2 ZrO_x / SiO_2 ZrN_x / SiO_2	Spin on pyrolysis	Parikh *et al*[93]
PZT 58/42	ITO	DC magnetron sputtering + rapid thermal processing	Sreenivas *et al*[31] Vasant Kumar *et al*[60]
PZT	Pt/SiO_2	rf sputtering	Abe *et al* [55]
PZT 52/48, 40/60, 42/58	Pt/MgO Pt/SiO_2 Pt/CaF_2	rf sputtering laser ablation	Etzold *et al* [23]
PZT	Pt/Ti/SiO_2/Si	Spin cast rf magnetron sputtering	Dana *et al*[92] Abe *et al*[55]
PZT	Pt/Ti/SiO_2/ SiN_x/GaAs Pt/Ti/SiO_2/Si	Spin cast Spin cast	Sanchez *et al*[26] Scott *et al*[40]
PZT	Pt-Ti on SiO_2	rf sputtered	Abe *et al*[55]
PZT	TiO_2/Ti	Hydrothermal	Shimomura *et al*[94]
PZT	Ti metal	dip-coating CVD	Bell *et al*[28] Yoon *et al*[29]
PZT (Nb doped)	p-Si (111), n-Si (100)	Spin cast	Xu *et al*[6]
PZT	p-Si	rf magnetron sputtering	Kalkur *et al*[7]
PT, PZ & PZT	Glass, Al_2O_3, TiO_2, MgO, YSZ & $SrTiO_3$ single crystals, Pt plates	Dip coat	Takahashi *et al*[95, 96]
PZT	Stainless Steel	Dip coating	Toghe *et al*[27]
PZT	Au (top electrode only)	Spin casting	Sanchez *et al*[26] Scott *et al* [40]
PZT	$YBa_2Cu_3O_7$ on $LaAlO_3$	Pulsed Laser Deposition	Ramesh *et al*[34,97] Lee *et al* [35]

4.2 Analytical Techniques for Compositional Profiling

A number of different modern analytical techniques can be used to determine the chemical composition of ferroelectric thin films and interfaces. Apart from studies of physical cross sections, through-film techniques include scanning Auger microscopy (SAM), Rutherford backscattering spectrometry (RBS), quantitative microprobe analysis, X-ray photoelectron spectroscopy (XPS), secondary ion mass spectrometry (SIMS), scanning electron microscopy (SEM) and transmission electron microscopy (TEM). No single technique, however, is able to fully characterise these systems, particularly the compositional variations that are present at the substrate/ferroelectric and electrode/ferroelectric interfaces. As discussed earlier, these variations in compositions arise from interactions at the interfaces as a result of either chemical reaction or diffusion.

One particularly powerful combination of complementary analytical techniques, and the combination that will be mainly discussed in this section, is RBS and SAM, although other combinations are also capable of producing meaningful analyses. Examples will be given (principally from our current research) of their use on PZT films on Ti substrates.

4.2.1. Rutherford Backscattering Spectrometry (RBS). RBS analysis involves the impingement of an energetic ion beam, typically ^{4}He ions with energies from 1 to 4 MeV, onto a target surface to be analysed. Ions with these high energies undergo elastic coulomb scattering with the target nuclei yielding information about the elemental composition of the surface to a depth of up to 10μm

The main advantages of RBS are that the technique is non-destructive; relatively rapid in producing the spectra; and can be calibrated to give quantitative concentration profiles for almost all the elements present in the near-surface region (≈10 μm). The identification of the heavier atoms is poor because of limitations in the detector resolution and the sensitivity for the detection of light elements is also poor, particularly when they are present in a heavier matrix. However, the reverse of these two limitations also holds, *i.e.* the technique is ideally suited to the analysis of heavy atoms in a light matrix, particularly if the identity of the heavy atom is known. Also relevant to the present application is that the technique is very well suited to the analysis of systems involving a thin film (or films) on a different substrate. The analysis of ferroelectric thin films such as PZT and $BaTiO_3$, containing heavy atoms, on substrates such as Si, Ti and glass, containing predominantly lighter atoms, is therefore ideally suited to RBS analysis.

Elemental concentration profiles are generated from an RBS spectrum by developing a model of the surface and, through the use of computer program based on well established physical relationships, matching iteratively the predicted spectrum to the experimentally-obtained spectrum[98]. Depth determinations from RBS are quantitative but are given in terms of atoms/cm^2, so that the generation of a linear depth scale requires a knowledge, or estimation, of the density of the surface volume analysed. For homogeneous thin films, the density is usually assumed to be the same as for the bulk material.

4.2.2. Scanning auger microscopy (SAM). High resolution SAM complements RBS extremely well in that its strengths compensate in areas where RBS is weak. All elements except hydrogen are detected with Auger spectroscopy, with the important lighter elements, C, N and O, being among the most sensitive. Depth resolution is a strength of SAM because it samples the first 4 to 10 atomic layers of the surface. However, depth information through a structure can only be obtained by milling the surface with ion beams.

Ion beam milling of multi-phase, multi-element surfaces to produce concentration profiles is at best semi-quantitative, because of preferential sputtering of some elements; different milling rates for different phases; and re-deposition of sputtered material on the area under analysis. SAM generated concentration profiles are therefore usually shown as relative thicknesses in terms of the ion milling time. However this still provides good comparative thicknesses for different films of the same (or similar) composition.

4.3 A Case Study: PZT on Ti

In this example PZT was deposited onto polished titanium substrates using the sol-gel synthesis route, and dip-coating. Final pyrolysis was between 600-700°C. At these temperatures the diffusion of the atoms is significant, and oxidation of the Ti substrate will occur, especially in the processing of very thin films of PZT. This case emphasises the effects of diffusion and substrate interaction.

In one experiment, a set of identical films consisting of 10 layers of dip-coated sol-gel PZT were prepared. A single substrate was used, and this was then cut up prior to final processing, and individual segments heat treated to the same temperature (≈680°C) for different lengths of time, ranging from 5 minutes to 40 minutes. These samples are referred to as A-*n*, where *n* represents the time (in minutes) of the annealing process. In another experiment, a single substrate was dipped in a manner that left successive layers exposed after each dipping. This film was then fired and characterisation of the successive layers, corresponding to different film thicknesses, was carried out. These films are referred to as B-*n*, where in this case *n* refers to the number of layers in the sample under examination.

4.3.1.Compositional Profiling. Auger depth profiles for samples A-5 and A-30 (*i.e.* heated for 5 minutes and 30 minutes) are given in figures 13a and 13c respectively. A depth profile for a single dipped coating B-1 is given in figure 13b for comparison. A number of observations can be made from these results:

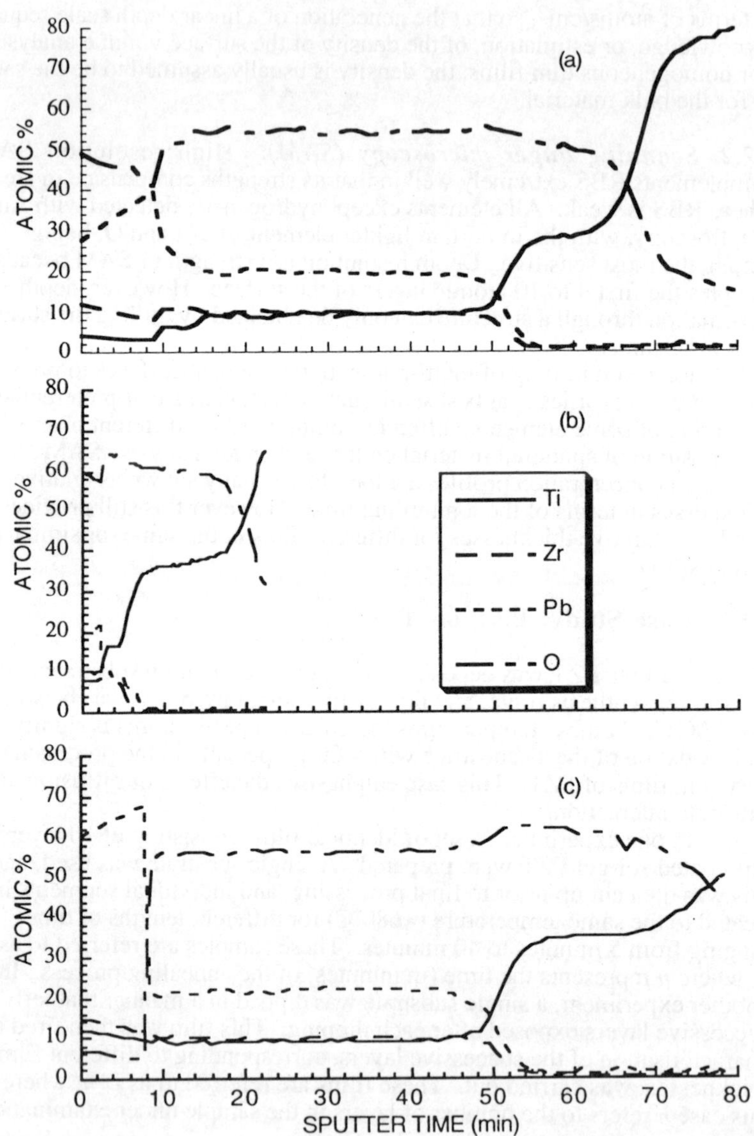

FIGURE 13. Compositional profiles through PZT on titanium sample derived using SAM. (a) sample A-5. (b) sample B-1. (c) sample A-30. Note the relative thickness of the 1-coat and 10-coat samples, and the uniformity of composition of the PZT film of a large depth in the 10-coat samples.

(i) Well-defined layers of PZT are present on a relatively thick TiO_x layer between the PZT and the Ti substrate. Similar results have been observed for CVD of PZT on Ti substrates[25].

(ii) The thickness of the sample B-1 (1 coat), expressed in relative sputter times, is approximately 4 minutes whereas the sputter times for A-5 and A-30 (10 coats) are about 42 minutes, or approximately ten times the thickness of the single coat. No loss in thickness, therefore, is observed as a result of the repeated coating process.

(iii) The concentration profiles for Pb, Zr and Ti remain reasonably constant throughout the thickness of the PZT films for all three samples, except in the near-surface region where Pb increases and Ti decreases, particularly for the sample B-1 and the sample A-5. A high level of Pb and a low level of O was present as a very thin layer on the surface of sample A-30, indicating an enrichment of Pb, possibly as elemental Pb, on the surface after the extended heat treatment. The relative concentrations of Pb, Zr and Ti remained constant at 2:1:1 throughout the 10-coat samples, in contrast to the 1-coat sample where the relative concentrations of Pb:Zr:Ti were approximately 1.0:0.9:1.6.

(iv) Interdiffusion between the PZT and TiO_x layers was not pronounced. In the sample B-1 Pb diffused an appreciable depth into the oxide, whereas Zr had a much sharper interface. While the diffusion depth of Pb was about the same as the thickness of the PZT film, it was still less than 10% of the thickness of the oxide interlayer. For the 10-coat samples the interdiffusion distance was an even smaller percentage of the oxide layer thickness. Interdiffusion of Pb and Zr in this system can, therefore, not be expected to have any significant influence on the electrical properties of the PZT films.

(v) The thickness of the oxide interface layer increased markedly for the sample A-30, heated for 30 minutes, in comparison with sample A-5, heated for 5 minutes. This indicates that oxygen diffuses readily through the PZT film allowing substantial oxidation of the Ti substrate to occur. It is this thickening of the oxide interlayer that potentially has the greatest effect on the electrical properties of these samples.

RBS results confirm the observations made from the SAM results. The lower Pb content and the increased diffusion of Pb into the TiO_x layer in the 1-coat film in comparison with the 10-coat film is observed, as well as the 10 fold difference in thickness between the 1-coat and the 10-coat systems.

The accumulation of Pb at the surface has been noted in other work[26,92], however, is not observed even though RBS is very sensitive to heavy elements such as Pb. Further work is being done to resolve this apparent contradiction between the results of the two analytical techniques.

Electrical characterisation of these samples confirms the effects of the diffusion which is apparent in the SAM and RBS results, and also confirms the behaviour described in Section 3 which results from an interfacial reaction. Figure 14 shows the dielectric constant for samples A-5 to A-40 as a function of annealing time. The initial increase in dielectric constant can be attributed to improved phase formation and growth of the perovskite phase on heat treatment for 10 minutes relative to a 5 minute anneal. The subsequent decrease in dielectric constant almost certainly reflects the increasing width of the TiO_x interface layer, causing a steady decrease in the measured value of ε' as the interface layer thickness increases. A similar observation for the effect of annealing temperature on dielectric constant has been made for PZT deposited on Si[7], with a maximum in ε' occurring when the temperature is sufficiently high to promote perovskite phase formation (650°C<T<750°C) overlaid on a steadily decreasing value of dielectric constant as the temperature increases.

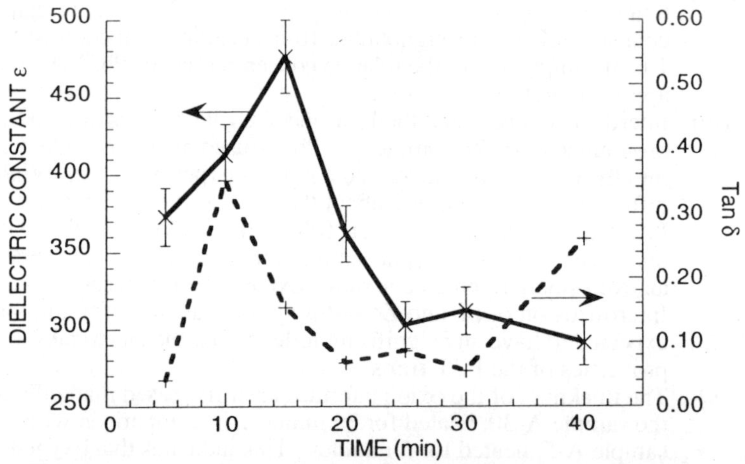

FIGURE 14. Dielectric constant measured at 1kHz as a function of anneal time for the A-5 to A-40 series of samples. The top electrode was a 25×25μm Au-Pd square. The increase in the dielectric constant up to 10 minutes is probably due to increasing formation of the perovskite phase and grain growth, while the decrease thereafter reflects the growth of an oxide interface layer.

Figure 15 shows the effect of increasing film thickness for the same processing conditions. Under the same processing conditions, we anticipate similar oxidation of the titanium substrate and therefore a similar interface layer thickness. For very thin films (eg the single coat film), there may be additional deviations from stoichiometry which affect the result. A similar effect of increasing dielectric constant with thickness, but with much more rapid increase in dielectric constant as thickness increases has been observed for many substrates, including platinum[26].

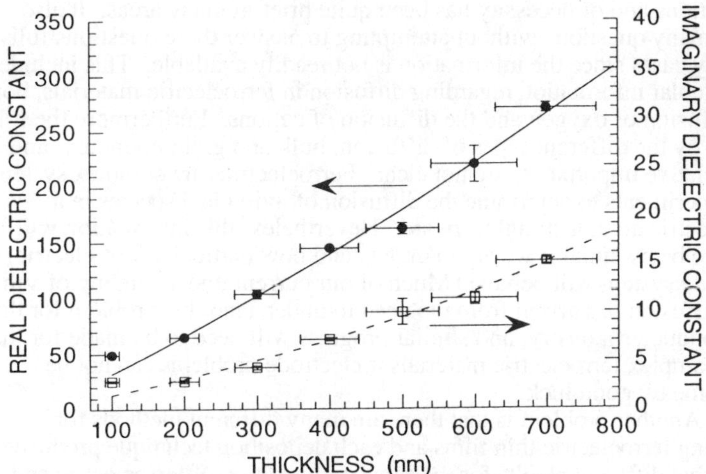

FIGURE 15. Dielectric behaviour measured at 10kHz of the B-*n* film series. The data has been fitted to equation (5). The thickness dependence of the dielectric constant is consistent with Figure 8. The ferroelectric layer dielectric constant is 1200 and Tan δ is 0.02.

The effect of the TiO_x interface layer on the C-V characteristics of the films on titanium are illustrated in figure 11. This shows rectifying behaviour at the electrode-ferroelectric interface and is another indication of the effect which diffusion and reaction at the interface may have on electrical properties of ferroelectric thin films.

From the RBS and SAM work there is no evidence of oxide "spikes" into the PZT, although these would be very difficult to detect owing to the similarities in elements in the ferroelectric film and the oxide interface layer. Scanning electron microscopy on cross-sections through these films has also shown no evidence for dendritic structures growing into the ferroelectric film from the substrate region. This suggests that the interface layer is uniform, following diffusion in the type A regime. This is not surprising given the amorphous, or at most nano-crystalline, nature of sol-gel deposited films prior to crystallisation.

5. FUTURE DIRECTIONS

This chapter has surveyed the whole area of ferroelectric-electrode interactions and of necessity has been quite brief in many areas. It also raises many questions without attempting to answer these questions fully, in many cases since the information is not readily available. This includes in particular information regarding diffusion in ferroelectric materials, both the diffusion of oxygen and the diffusion of cations. Furthermore the roles played by the different types of diffusion, bulk and grain boundary, and their relative importance, are not clear. Ferroelectrics are complex systems and experiments to determine the diffusion of individual species in a ferroelectric are not straightforward. Nevertheless this information will prove to be essential if we are to understand how particular ferroelectric-electrode system will behave. Much of our current understanding of solid-state diffusion has arisen from the need to understand this problem for the semiconductor industry, and similar progress will need to be made for the more complex ferroelectric materials if electrode problems can not be overcome by good luck.

Another problem is that there are many different methods for preparing ferroelectric thin films and each deposition technique produces somewhat different results for the same combination of ferroelectric and substrate. Again, basic research in the area of diffusion, and the relation between the growth mechanisms found using different deposition techniques and the diffusion processes which occur between the films and their substrates will be an important tool in overcoming problems with film-electrode interactions.

The experimental techniques commonly employed to study physical properties of interfaces are RBS and SAM. These are extremely useful techniques but frequently have insufficient sensitivity to be able to determine the diffusion coefficients, particularly if there is any material inhomogeneity in the diffusion process. Extraction of diffusion coefficients must be done by fitting diffusion profiles to interpreted atomic concentrations calculated from the experimental data. Techniques such as SIMS or radioactive tracer diffusion may provide more information particularly for important oxygen diffusion problems.

Throughout this chapter we have used examples from semiconductor industry to illustrate what can occur when a metallic electrode makes contact with a semiconductor or insulator. Learning the lessons from the semiconductor industry will be extremely important for successful integration of ferroelectrics into devices. Also since most ferroelectric materials are non-stoichiometric oxides there will always be a possibility for diffusion of oxygen vacancies, particularly under the influence of an applied electric field. It is also important to consider the electrode-ferroelectric system as an electrochemical half-cell and therefore to consider entire devices as solid state electrochemical cells. This suggests that battery technology and electrochemistry as other well-established areas, could provide useful insights into the behaviour of ferroelectric devices, particularly if electrode problems are suspected. This has already lead to some predictions of mechanism of fatigue in ferroelectrics.

Finally, in order to understand the influence of any of these physical processes occuring at the electrodes, we must know how they affect device performance. The elementary model of an interface layer forming at an electrode described in Section 3 shows that a number of different effects can be described in very simple terms by the presence of an interface layer. In order to fully understand the electrical characteristics of a ferroelectric device, more understanding of dielectric properties of very thin ferroelectric films, and the influences of small grain size on ferroelectric properties, is required.

REFERENCES

1. C.A. PAZ DE ARAUJO, L.D. MCMILLAN, B.M. MELNICK, J.D. CUCHIARO AND J.F. SCOTT, Ferroelectrics, **104,** 241-256 (1990).
2. J.F. SCOTT, C.A. ARAUJO AND L.D. MACMILLAN, "Ferroelectric Memory Applications", Proceedings of Ultrasonics Symposium, (IEEE, New York, 1989) 299-307.
3. D.E. FISCH, N.E. ABT, F.N. BENS, W.D. MILLER, T. PRAMANIK, W. SAIKI AND W.H. SHEPHERD, "Analysis of Thin Film Ferroelectric Aging", Proceedings on Reliability Physics, (IEEE, Ann Arbor, 1990) 237-242.
4. S.M. SZE, Physics of Semiconductor Devices (2nd Edition), (John Wiley and Sons, New York, 1981).
5. B. SCHWARTZ, Ohmic Contacts to Semiconductors, (Electronics Division of the Electrochemical Society, New York, 1969).
6. YUHUAN XU, CHING JIH CHEN, REN XU AND JOHN D. MACKENZIE, J. Appl. Phys., **67**, 2985-91 (1990).
7. T.S. KALKUR, GEORGE ARGOS AND LEE KAMMERDINER, Proc. Mat. Res. Soc., **200**, 313-318 (1990).
8. S.L. SWARTZ, IEEE Transactions on Electrical Insulation, **25**, 935-988 (1990).
9. NORMAN ABT, Proc. Mat. Res. Soc., **200**, 303-312 (1990).
10. G. GRIFFEL, S. RUSCHIN AND N. CROITORU, Appl. Phys. Lett., **54**, 1385-1387 (1989).
11. D.P. GIA RUSSO AND S.C. KUMAR, Appl. Phys. Lett., **23**, 229-231 (1974).
12. T. KAWAGUCHI, H. ADACHE, K. SETSUNE, O. YAMAZAKE, AND K. WASA, Appl. Opt., **23**, 2187-2191 (1984).
13. C.E. LAND, M.A. BULTLER, AND S.J. MARTIN, "New Devices Using Ferroelectric Thin Films", IEDM , 252-258, 1989.
14. B.K. SINHA AND STANLEY LOCKE, IEEE Transaction on Ultrasonics, Ferroelectrics and Frequency Control, **36**, 231-241 (1989).

15. R.A. ROY, K.F. ETZOLD AND J.J. CUOMO, Proc. Mat. Res. Soc., **200**, 77-82 (1990).
16. M.S. AMEEN, T.M. GRAETTINGER, S.H. ROU, H.N. AL-SHAREEF, K.D. GIFFORD, O. AUCIELLO AND A.I. KINGON, Proc. Mat. Res. Soc., **200**, 65-76 (1990).
17. F.W. AINGER, C.J. BRIERLEY, M.D. HUDSON, C. TRUNDLE AND R.W. WHATMORE, Proc. Mat. Res. Soc., **200**, 37-47 (1990).
18. A.I. KINGON, K.Y. HSIEH, L.L.H. KING, S.H. ROU, K.J. BACHMANN AND R.F. DAVIS, Proc. Mat. Res. Soc., **200**, 49-56 (1990).
19. J.M. BELL, B.BEN-NISSAN, M. ANAST, B.O. WEST, L. SPICCIA, J.D. CULLEN, I.D. WATKINS, D. DE VILLIERS AND G.R. JOHNSTON, Proc. Mat. Res. Soc., **180**, 453-456 (1990).
20. S. HIRANO AND K. KATO, Adv. Ceram. Mater., **3**, 503-506 (1988).
21. S.K. DEY AND R. ZULEEG, Ferroelectrics, **112**, 309-319 (1990).
22. K. IIJIMA, T. TERASHIMA, K. YAMAMOTO, K. HIRATA, AND Y. BANDO, Appl. Phys. Lett., **56**, 527-529 (1990).
23. K.F. ETZOLD, R.A. ROY, K.L. SAENGER AND J.J. CUOMO, Proc. Mat. Res. Soc., **200**, 297-302 (1990).
24. C. SUDHAMA, J.C. CARRANO, L.H. PARKER, V. CHIKARMANE, J.C. LEE, A.F. TASCH, W. MILLER, N. ABT AND W.H. SHEPHERD, Proc. Mat. Res. Soc., **200**, 331-336 (1990).
25. K. H. YOON, H. S. PARK, S.O. YOON AND HYO ILL SONG, J. Mat. Sci. Lett., **8**, 1442-1444 (1989).
26. L.E. SANCHEZ, D.T. DION, S.Y. WU AND I.K. NAIK, Ferroelectrics, **16**, 1-17 (1991).
27. N. TOHGE, S. TAKAHASHI AND T. MINAMI, J.Amer. Ceram. Soc., **74**, 67-71 (1991).
28. J.M. BELL AND P.C. KNIGHT, Integrated Ferroelectrics, accepted for publication (1994).
29. S.G. YOON, H.G. KIM, IEEE Transactions on Ultrasonics, Ferroelectrics and Frequency Control , **37**, 332-228 (1990).
30. G.YI, Z. WU, AND M. SAYER, J. Appl. Phys, **64**, 2717-2724 (1988).
31. K. SREENIVAS, M. SAYER, T LAURSEN, J.L. WHITTON, R. PASCUAL, D.J. JOHNSON, D.T. AMM, G. I. SPROULE, D.F. MITCHELL, M.J. GRAHAM, S.C. GUJRATHI AND K.OXORN, Proc. Mat. Res. Soc., **200**, 255-260 (1990).
32. S.D. BERNSTEIN, T.Y. WONG, YANINA KISLER AND R.W. TUSTISON, "Reactively sputtered RuO_x electrodes for ferroelectric PZT capacitors", paper presenterd at 4th International Conference on Integrated Ferroelectrics, 9-11 May, 1992, Monterey, CA.
33. C.K. KWOK, D.P. VIJAY, S.B. DESU, N.P. PARIKH AND E.A. HILL, "Conducting Oxide Electrodes for ferroelectric films", in

Proceedings 4th International Synposium on Integrated Ferroelectrics, edited by J.F. Scott, Monterey, CA.
34. R. RAMESH, A. INAM, W.K. CHAN, B. WILKENS, K. MYERS, K. RENSCHNIG, D.L. HART, AND J. M. TARASCON, Science, **252**, 944-6 (1990).
35. J.LEE AND A.SAFARI, "PZT/Y-Ba-Cu-O Heterostructures by pulsed laser deposition", paper presenterd at 4th International Conference on Integrated Ferroelectrics, 9-11 May, 1992, Monterey, CA.
36. A.K. KULKARNI AND G.A. ROEHRER, "Development of Ferroelectric Materials for Memory Applications", Proceedings 8th Biennial University/Governemnt/Industry Symposium, (IEEE, 1989), 150-155.
37. PAUL D. BEAL AND H.MATT DUIKER, Ferroelectrics, **116**, 111-116 (1991).
38. A.K. KULKARNI, G.A. ROHRER, LD. MACMILLAN AND S.E. ADAMS, "Fatigue Mechanisms in Thin Film Potsassium Nitrate Memory Devices", Proceedings on Reliability Physics, (IEEE, Ann Arbor,1989) 171-177.
39. D.J. JOHNSON, D.T. AMM, E. GRISWOLD, K. SREENIVAS, G. YI, AND M. SAYER, Proc. Mat. Res. Soc., **200**, 289-295 (1990).
40. J.F. SCOTT, C.A. ARAUJO, B.M. MELNICK, L.D. MACMILLAN, R. ZULEEG, J. Appl Phys, **70**, 382-388 (1991).
41. J. SEYMOUR, Electronic Devices and Components, (Longman Scientific and Technical, Harlow, 1986).
42. D.A. FRASER, The Physics of Semiconductor Devices, (Clarendon Press, Oxford, 1979).
43. DAVID H. NAVON, Semiconductor Microdevices and Materials, (CBS Publishing, Japan, 1986).
44. EDWARD YANG, Fundamentals of Semiconductor Devices, (McGraw-Hill, New York, 1978) .
45. DAVID R. LIDE, CRC Handbook Of Chemistry And Physics 71st Ed, (CRC Press, Boca Raton, 1991)
46. G.V.SAMSONOV, The Oxide Handbook, (IFI/Plenum, New York, 1973).
47. R. W. SCHWARTZ, Z. XU, D.A. PAYNE, T.A. DETEMPLE AND M.A. BRADLEY, Proc. Mat. Res. Soc., **200**, 167-172 (1990).
48. W. JOST, Diffusion, (Academic Press, New York, 1960).
49. S.-L ZHANG AND F.M. D'HEURLE, Phil. Mag. A, **66**, 415-424 (1992).
50. B.E. DEAL AND A.S. GROVE, J. Appl. Phys. , **36**, 3770-3778 (1965).
51. RICHARD GHEZ, A Primer on Diffusion Problems, (John Wiley and Sons, New York, 1988), 46-54.
52. K.K. NG, W.J. POLITO AND J.R. LIGENZA, J. Appl. Phys., **44**, 626 (1984).
53. R.C. BAUMANN, T.A. ROST AND T.A. RABSON, Proc. Mat. Res. Soc., **200**, 19-24 (1990).

54. S. MATSUBARA, T. SAKUMA, S. YAMAMICHI, H. YAMAGUCHI AND Y. MIYASAKA, Proc. Mat. Res. Soc., **200**, 243-253 (1990).
55. K. ABE, H. TOMITA, H. TOYODA, M. IMAI AND Y. YOKOTE, Japanese J. Appl. Phys., **30**, 2152-2154 (1991).
56. J.P. HOARE, The Electrochemistry of Oxygen, (Interscience Publishers, New York, 1968).
57. L.G. HARRISON, Trans. Faraday Soc., **57**, 1191 (1957).
58. INDERJEET KAUR, WOLFGANG GUST AND LASZLO KOZMA, Handbook of Grain and Interphase Boundary Diffusion Data (volume 2), (Ziegler Press, Stuttgart, 1989).
59. N.A. GJOSTEIN, in Diffusion, (Am Soc. for Metals, Ohio, 1973), 241-274.
60. C.V. VASANT KUMAR, M. SAYER, R. PASCUAL, D.T. AMM, Z. WU, AND D.M. SWANSTON, Appl. Phys. Lett., **58**, 1161-1163 (1991).
61. L.F. FRANCIS AND D.A. PAYNE, Proc. Mat. Res. Soc., **200**, 173-178 (1990).
62. AZIZ S. SHAIKH, ROBERT W. VEST AND GERALDINE M. VEST, IEEE Transactions on Ultrasonmics, Ferroelectrics and Frequency Control, **36**, 407-412 (1989).
63. K.L. CHOPRA, Thin Film Phenomena, (McGraw-Hill, New York, 1969).
64. M. DE KEIJSER, G.J.M. DORMANS, P.J. VAN VELDHOVEN AND D.M. DE LEEUW, Appl. Phys. Lett., **59**, 3556-3558 (1991).
65. T.T.M. PALSTRA, B.BATLOGG, R.B. VAN DOVER, L.F. SCHNEEMEYER AND J.V. WASZCZAK, Phys. Rev. B, **41**, 6621-6632 (1990).
66. C.H.L WEITJEINS AND P.A.C. VAN LOON, Thin Solid Films, **196**, 1-10 (1991).
67. N.J. ARFSTEN, J. Non-Crystalline Solids, **63**, 243-249 (1984).
68. J.R. LAGRAFF, P.D. HAN AND D.A. PAYNE, Phys. Rev. B, **43**, 441-447, (1991).
69. K. SREENIVAS, M. SAYER, D.J. BAAR AND M. NISHIOKA, Appl. Phys. Lett., **52**, 709-711 (1988).
70. P. GONNARD, R. BRIOT, Ferroelectrics, **93**, 117-126 (1989).
71. S.L.MILLER, R.D. NASBY, J.R. SCHWANK, M.S. RODGERS, P.V. DRESSENDORFER, J. Appl. Phys., **68**, 6463-6471 (1990).
72. L.E. SANCHEZ, S. WU, I.K. NAIK, Appl. Phys. Lett., **56**, 2399-2401 (1990).
73. S.C. KWOK, Electronic Letters, **26**, 894-896 (1990).
74. This is certainly a reasonable assumption for film deposition techniques using post-annealing (*e.g.* sol-gel deposition, sputtering or thermal evaporation onto unheated substrates etc), where the interaction will principally form during the annealing cycle rather than during the processing of deposition.

75. Y. MIYASAKA, T. SAKUMA, S. MATSUBARA, H. WATANABE AND K. KOYAMA, "High Dielectric Constant (Ba, Sr)TiO_3 thin films for DRAM Capacitor Application", paper presenterd at 4th International Conference on Integrated Ferroelectrics, 9-11 May, 1992, Monterey, CA.
76. A.A. GITEL'SON , A.M. LERER, V.S. MIKHALEVSKII, V.M. MUKHORTOV AND S.V. ORLOV, Sov. Phys. Solid State, **19**, 1121-24 (1977).
77. M. OKUYAMA AND Y. HAMAKAWA, Ferroelectrics, **63**, 243-53 (1985).
78. J. B. WHITEHEAD, Lectures on Dielectric Theory and Insulation, (McGraw-Hill, New York, 1927).
79. K.SREENIVAS, A.MANSINGH AND M.SAYER, J.Appl.Phys, **62**, 4475-4481 (1987).
80. S.B. DESU, C.H. PENG, L.KAMMERDINER, P.J. SCHELE, Proc. Mat. Res. Soc., **200**, 319-324 (1990).
81. W.J. MERZ, J. Appl. Phys., **27**, 938-943 (1956).
82. D.A. PAYNE, PhD Thesis, (Pennsylvania State University, Pennsylvania, 1973).
83. H.M. DUIKER AND P.D. BEALE, Phys. Rev., **41**, 490-495 (1990).
84. G.A.C.M.SPIERINGS, M.J.E. ULENAERS, G.L.M. KAMPSHOER, H.A.M. VAN HAL, P.K. LARSEN", J. Appl. Phys., **70**, 2290-2298 (1991).
85. D.ROY, C.J.PENG, S.B.KRUPANIDHI, Appl. Phys. Lett. , **60**, 2478 (1992).
86. C. HWANG, T.P. MA, Y. DI, J.W. GOLZ, B.L. HALPERN AND J.J SCHMITT, "Jet Vapor Deposited PZT on Si and MIS Capacitors", Proceedings 4th International Symposium on Integrated Ferroelectrics, edited by J.F. Scott, (1992).
87. C. B. SAWYER, C. H. TOWER, Phys. Rev, **35**, 269-273 (1930).
88. H. DIAMANT, K. DRENCK, R. PEPINSKY, Rev. Sci. Instr., **28**, 30-33 (1957).
89. S.L.MILLER, J.R. SCHWANK, R.D. NASBY, M.S. RODGERS, J. Appl. Phys., **70**, 2849-2860 (1991).
90. NORMAN M. TALLAN, Electrical Conductivity in Ceramics and Glass, (Marcel Dekker, New York, 1974)
91. C. SUDHAMA, A.C. CAMPBELL, P.D. MANIAR, R.E. JONES, R. MOAZZAMI, C.J. MOGAB, J.C. LEE, J. Appl. Phys., **75**, 1014-1022 (1994).
92. S.S. DANA, K.F. ETZOLD, AND J. CLABES, J. Appl. Phys., **69**, 4398-4403 (1991).
93. N.R.PARIKH, J. TODD STEPHEN M. L. SWANSON, AND E.R. MYERS, Proc. Mat. Res. Soc., **200**, 193-198 (1990).
94. K. SHIMOMURA, T.TSURUMI AND Y. OHBA, Jap.J.Appl.Phys., **30**, 2174-2177 (1991).

95. Y. TAKAHASHI, AND K. YAMAGUCHI, J. Mat. Sci., **24**, 3950-3955 (1990)
96. Y TAKAHASI, Y. MATSUOKA, K. YAMAGUCHI, M. MATSUKI, AND K. KOBAYASHI, J. Mat. Sci, **24**, 3960-3964 (1990).
97. R. RAMESH, A. INAM, W.K. CHAN, F. TILLEROT B. WILKENS, C.C. CHANG, T. SANDS, J.M. TARASCON AND V.G. KERAMIDAS, Appl. Phys. Lett., **59**, 3542-3544 (1991).
98. J. SAULITIS, G.R. JOHNSTON AND J.L. COCKING, Thin Solid Films, **166**, 201-211 (1988).

CHAPTER 5

POLARIZATION REVERSALS IN FERROELECTRICS

Yoshihiro ISHIBASHI
Synthetic Crystal Research Laboratory
School of Engineering, Nagoya University
Nagoya 464-01, Japan

1. INTRODUCTION

It is known that the polarization reversal in ferroelectrics does not proceed homogeneously through specimens, but inhomogeneously by the nucleation-growth mechanism, i.e., the nucleation of reversed nuclei followed by the forward and sidewise growth of the reversed areas. The growth in the forward direction takes place in a short time. To this kind of phenomena the Kolmogorov-Avrami (KA) theory is applicable,[1, 2] and very useful for analyzing the reversal processes. In fact, Scott et al.[3] have shown that the detailed current transients I (t) observed in the course of polarization reversal in several materials fit the theory of Ishibashi and Takagi,[4] who have expressed the essence of the KA theory convenience for its use in applications to ferroelectric polarization reversals. In sections 2 and 3, the KA theory will be briefly described.

On the other hand, in spite of great success of the KA theory, one should not overlook that it is constructed upon several simplified assumptions, some of which may not apply to actual polarization reversal processes in ferroelectrics. Therefore, it may be necessary to construct new models of polarization reversal, which must hopefully be easy to handle. In section 4, a lattice model, one of such models recently proposed, will be mentioned.[5] By using such a model, where the situation that in ferroelectrics nuclei for polarization reversals exist at predetermined sites is taken into account, one would be able to discuss the shape of hysteresis loops under various applied fields.

In section 5, concluding remarks will be given.

2. THE KOLMOGOROV-AVRAMI THEORY[1,2]

The purpose of arguments in this section is to clarify the relation between the reversed area and the elapsed time on the basis of the KA theory. We consider a ferroelectric thin plate where the polarization reversal takes place.

Let us assume that a plane is infinite, take a point P on it, and calculate the probability $q(t)$ that the point is not included at time t (after the start of polarization reversal) in the reversed domains. (Hereafter sometimes we use the word "domain" instead of "area"). If a nucleus appears at the point P' at the time τ ($\tau < t$), the reversed domain with the center at P' will expand, during the time interval $t - \tau$, by the distance of $v(t - \tau)$ (see Figure 1), where v is the velocity of the expansion of the reversed domains or of the motion of the domain walls (boundaries).* On the contrary, therefore, if a nucleus appears at time τ in the area of the radius $v(t - \tau)$ with P at the center, P will be included in the reversed domain at time t.

Let us denote such an area as $S(t, \tau)$, since S depends not only on t but also on the time of birth of the nucleus, τ. If the reversed domains are one-dimensional, $S(t, \tau)$ is given as

* That the velocity v is constant is an approximation. The velocity of the boundary between two states usually depends on the local curvature of the boundary, the difference of free energies of the states separated by the boundary and also on the interaction between boundaries.

$$S_1(t, \tau) = 2v(t-\tau), \tag{1}$$

and if they are two-dimensional, S (t, τ) is given as

$$S_2(t, \tau) = \pi v^2 (t-\tau)^2 \tag{2}$$

(see Figure 2).

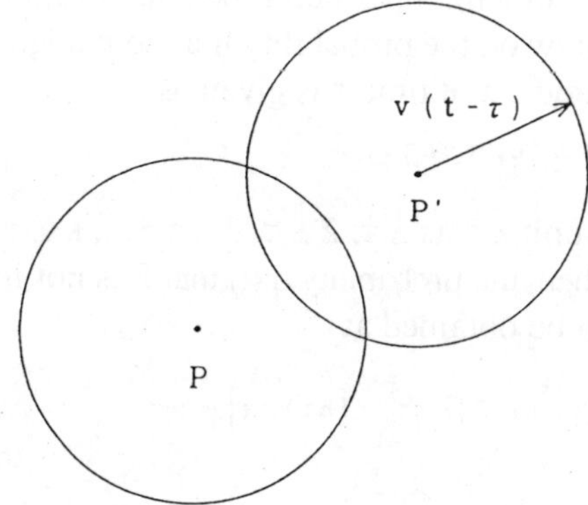

FIGURE 1. The area covered by the domain having started at the point P'.

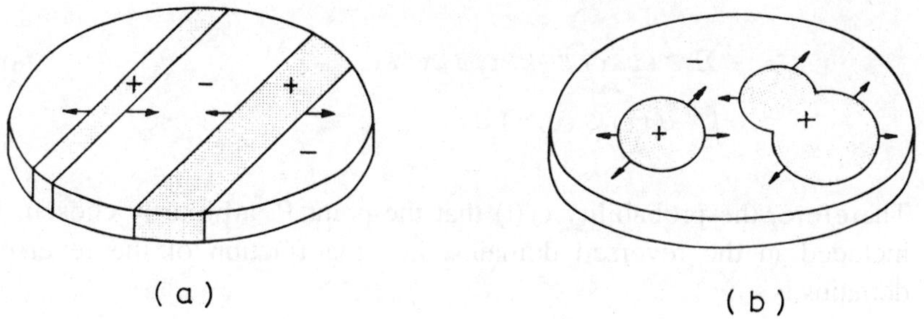

FIGURE 2. (a) One- and (b) two-dimensional domains.

In general, we can write S (t, τ) as

$$S_d(t,\tau) = C_d[v(t-\tau)]^d \tag{3}$$

where d means the dimensionality of the shape of reversed domains.

Let us denote the nucleation probability at time τ as J (τ), where we assume that it is spatially homogeneous. For one-dimensional and two-dimensional shapes of reversed domains, J (τ) means the numbers of appearing nuclei per unit length per unit time and per unit area per unit time, respectively. Anyway, the probability that no nucleus appears in S (t, τ) during a short time $\Delta\tau$ at time τ is given as

$$1-J(\tau)\,S(t,\tau)\,\Delta\tau. \tag{4}$$

We divide time t by a unit $\Delta\tau$ as $\Delta\tau$, $2\Delta\tau$, $3\Delta\tau$, ..., $k\Delta\tau$ and write $t = k\Delta\tau$ and $\tau = i\Delta\tau$. Then, the probability q (t) that P is not included in the reversed domains can be obtained as

$$q(t) = \prod_{i=0}^{k}\left[1-J(i\Delta\tau)\,S(k\Delta\tau,\ i\Delta\tau)\,\Delta\tau\right]. \tag{5}$$

Reducing $\Delta\tau \to 0$ and taking a logarithm of q (t), it is seen that

$$\begin{aligned} \ln q(t) &= \sum \ln\left[1-J(i\Delta\tau)\,S(k\Delta\tau,i\Delta\tau)\,\Delta\tau\right] \\ &\approx -\sum J(i\Delta\tau)\,S(k\Delta\tau,i\Delta\tau)\,\Delta\tau \\ &= -\int_0^t J(\tau)\,S(t,\tau)\,d\tau. \end{aligned} \tag{6}$$

Therefore, the probability c (t) that the point P, arbitrarily chosen, is included in the reversed domains, i.e., the fraction of the reversed domains is

$$c(t) = 1 - q(t)$$
$$= 1 - e^{-\int_0^t J(\tau)\, S(t,\tau)\, d\tau} . \tag{7}$$

In the case of polarization reversals in ferroelectrics, the reversed polarization P (t) and the switching current I (t) are given as

$$P(t) = 2P_s c(t) , \tag{8}$$

$$I(t) = 2P_s S_o \frac{dc(t)}{dt} , \tag{9}$$

where P_s and S_0 are the spontaneous polarization and the electrode area, respectively. It may be worthwhile to mention that equation (9) is approximate, because c (t) is obtained under the assumption that the system is infinite. If the system is finite as is meant by S_0, one may have to derive c (t) as a function of S_0.[6]

In applying the equation (7) to polarization reversals, let us consider two quite conceivable cases for J (τ). One is the case of

$$J(\tau) = R, \tag{10}$$

which means that the nucleation takes place with a constant rate, i.e., R nuclei are born per unit time unit length (d = 1) or unit area (d = 2). This case is classified as the category I,[4] or is called the homogeneous case.[7] The fraction of the reversed area in this case is obtained as

$$c(t) = 1 - \exp\left[-(t/t_I)^{d+1}\right], \tag{11}$$

where

$$\left(\frac{1}{t_I}\right)^{d+1} = R\frac{C_d v^d}{d+1} . \tag{12}$$

Another is the case of

$$J(\tau) = N\,\delta(\tau) , \tag{13}$$

which means that the nucleation takes place only at t = 0, or in other words, N latent nuclei per unit length (d = 1) or per unit area (d = 2) get active for polarization reversals. This case is classified as the category II, or is called the inhomogeneous case.[4, 7] The fraction of the reversed domains in this case is expressed as

$$c(t) = 1-\exp[-(t/t_{II})^d] \tag{14}$$

where

$$\left(\frac{1}{t_{II}}\right)^d = C_d N v^d . \tag{15}$$

In actual polarization reversals, both categories may work simultaneously, i.e., some reversed domains come from the latent nuclei and some others from the nuclei thermally activated. Then, c (t) may be written as

$$c(t) = 1-\exp[-(t/t_I)^{d+1} - (t/t_{II})^d] . \tag{16}$$

Since this expression is a little complicated, we would rather write c (t) as

$$c(t) = 1 - \exp[-(t/t_o)^n], \tag{17}$$

allowing n to take a non-integer value. It should be noted that then there appears only one characteristic time t_0.**

In Figure 3, the polarization reversal process in $(NH_2CH_2COOH)_3 \cdot H_2SO_4$ (TGS), which is ferroelectric below 49° C, is shown for various applied fields. The c (t) curves are well fitted by the index n ≈ 1.4. It has not been clarified, however, why such a non-integer value is obtained.

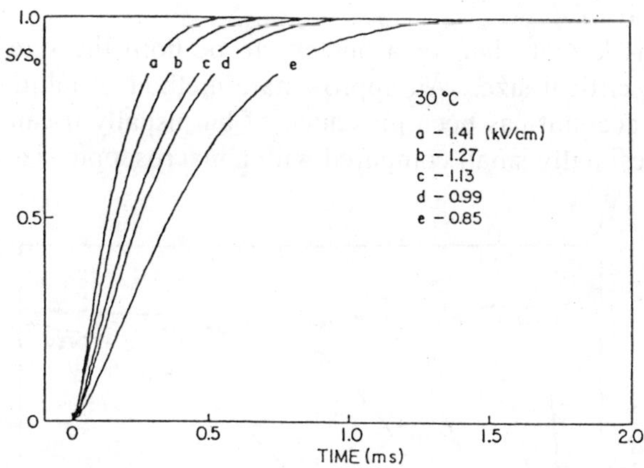

FIGURE 3. The observed c(t) curves in TGS (In the ordinate, c(t) is shown as S/S_0).

** As for the spatial scale it is found by dimension analyses that there is only one characteristic length ℓ for the category I in the case of the d dimensional shape of domains given as $\ell = (\frac{V}{R})^{1/(d+1)}$. For the category II, the characteristic length is given as $\ell = (1/N)^{1/d}$.

In Figure 4, the c(t) curves for a surface stabilized ferroelectric liquid crystal DOBAMBC are presented for various applied fields.[8] It is seen that the index n = 2 applies. There are two possibilities for n = 2, i.e., the d = 1 case for the category I and the d = 2 case for the category II. This has to be distinguished by some other experiments. Fortunately, the polarization reversal process in DOBAMBC can be observed in real time and in real space, using a polarizing microscope (Figure 5). It has been found by such observations that the latent nuclei get active at the instant of the application of the field, but almost no other nuclei are born afterwards. That is to say, the polarization reversal in DOBAMBC is found to be the case of the category II of d = 2.***

It is known that for a nucleus to be born the size of it has to exceed a critical size. An approximate method of taking the critical size into account has been presented,[4, 9] but usually it can be regarded as infinitesimally small compared with a macroscopic size of domains.

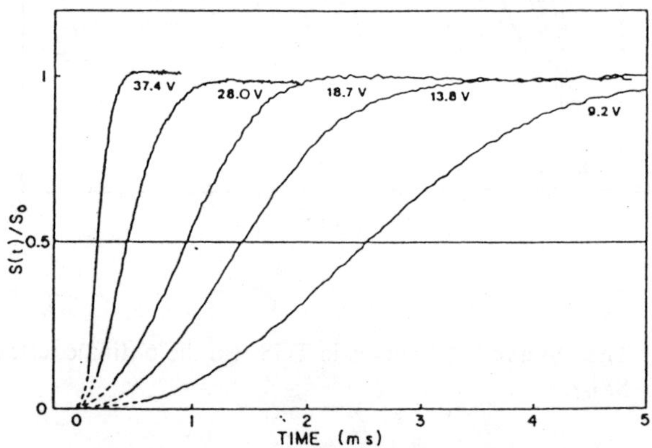

FIGURE 4. The observed c(t) curves in a ferroelectric liquid crystal DOBAMBC (c (t) is shown as S (t)/S_0).

*** In the case of DOBAMBC, the reversed domains are hexagons, but not circles. Therefore, the numerical factor π in (2) has to be modified accordingly.

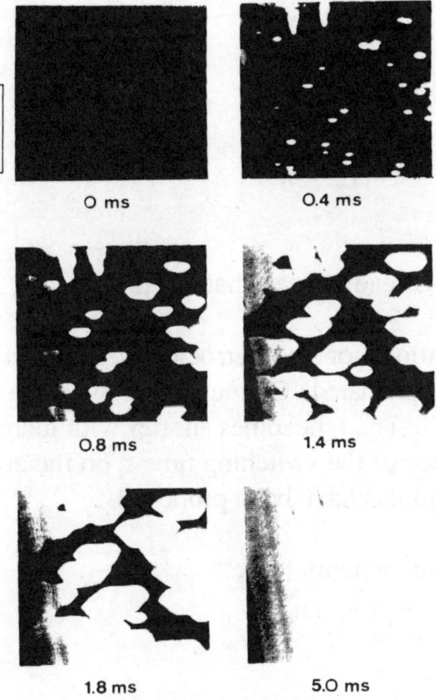

FIGURE 5. Domains observed in DOBAMBC with a polarizing microscope.

3. SWITCHING CURRENT AND SWITCHING TIME

To investigate polarization reversal processes the switching current is often observed. The switching current transient is given as

$$I(t) = 2P_s S_0 \frac{dc(t)}{dt}, \quad (18)$$

where P_s is the spontaneous polarization and S_0 the electrode area. On utilizing equation (14), I (t) can be expressed as

$$I(t) = 2P_s S_0 \frac{n}{t_o} \left(\frac{t}{t_o}\right)^{n-1} \exp\left[-(t/t_o)^n\right]. \quad (19)$$

The switching current maximum I_{max} and its time t_{max} are given as

$$I_{max} = 2P_sS_o\frac{n}{t_o}\left(\frac{n-1}{n}\right)^{\frac{n-1}{n}}\exp\left[-\left(\frac{n-1}{n}\right)\right], \tag{20}$$

$$t_{max} = t_0\left(\frac{n-1}{n}\right)^{1/n}, \tag{21}$$

respectively. It is readily seen that the product $I_{max} \cdot t_{max} / P_sS_0$ depends only on n.

In applications of the ferroelectric polarization reversal, fast reversal is quite appreciated. Of course, the switching time t_s is a function of the applied field, i.e., t_s becomes shorter with increasing the field. As for the dependence of the switching time t_s on the applied field, various experimental formulas have been proposed.

For example, the exponential law[10]

$$t_s \sim t_\infty \exp(E_0/E), \tag{22}$$

where E_0 is the activation field, was found for $BaTiO_3$ and a power law

$$t_s \sim E^{-3/2} \tag{23}$$

was reported for the high-field polarization reversal in $BaTiO_3$.[11] **** Unfortunately, however, no microscopic information has been made available at present, especially about fast reversals, and little is known about the microscopic mechanism governing the applied-field dependence of the switching time.

**** It is of interest to note that in the d = 2 case of the category II, the characteristic time is proportional to $E^{-3/2}$ if N and v are proportional to E. See equation (15).

4. LATTICE MODEL OF POLARIZATION REVERSAL[5]

It should be noted that the KA theory is only a very crude simplification of an actual polarization reversal process. For example, the KA theory is not easily applicable to interpreting the ferroelectric hysteresis loops through polarization reversals taking place under the changing applied field. This is because, in the KA theory only two states, i.e., the initial state destabilized by the application of a constant electric field and the final state now stabilized by it, are considered. In the KA theory presented in the previous section, a constant velocity v of sidewise domain growth is assumed a priori. In reality, however, the wall velocity is governed by the environment of the walls. The wall motion will be accelerated or decelerated due to interaction between neighbors. Under this situation, more microscopic models will be required.

Before describing a lattice model of polarization reversals, it may be worthwhile to mention the one-dimensional p^4 system, where the energy density is given as

$$f = \frac{1}{4}(1-p^2)^2 + \frac{k}{2}\left(\frac{dp}{dx}\right)^2, \quad k > 0\,, \tag{24}$$

and where the first and the second terms represent the local energy and the energy increase due to spatial inhomogeneity of polarization p, respectively, and the polarization corresponding to the equilibrium under the local potential (without the interaction with neighbors) is scaled to ± 1. Then, the domain structure is obtained by solving the Euler-Lagrange equation derived from (24),

$$k\frac{d^2p}{dx^2} + p - p^3 = 0 \tag{25}$$

under the boundary conditions that at p = ± 1 and $\frac{dp}{dx} = 0$ at x = ± ∞. Namely, we obtain

$$p = \tanh \frac{x}{\sqrt{2k}} . \tag{26}$$

The center of the wall is located at x = 0, and the width (thickness) of the domain wall is estimated as $\sqrt{2k}$. This implies that the domain wall becomes thicker as the interaction becomes stronger, i.e., as k increases.

To evaluate the velocity υ of the motion of the above wall under a small applied field e, let us equate the rate of the energy lost by viscosity and the electric energy supplied by the electric field.

The total energy of the system is given as

$$F = \int f(x)\, dx \tag{27}$$

(see equation (24)), and the time evolution can then be written as

$$\gamma \frac{dp}{dt} = - \frac{\delta F}{\delta p} . \tag{28}$$

Let us calculate dF/dt;

$$\frac{dF}{dt} = \int \frac{\delta F}{\delta p} \frac{dp}{dt}\, dx = -\gamma \int \left(\frac{dp}{dt}\right)^2 dx . \tag{29}$$

Writing the moving domain wall as

$$p(x, t) = p(x-vt) = \tanh \frac{x-vt}{\sqrt{2k}} , \tag{30}$$

we obtain

$$v = \frac{3}{2} \frac{\sqrt{2k}}{\gamma} e , \tag{31}$$

implying that the velocity is proportional to the wall thickness and the (small) applied field. The numerical factor 3/2 is characteristic of the p^4- model.

Let us now describe the lattice model of polarization reversals, and write the total energy of the system with the dipole moment at the (m, n) site, $p_{m,n}$, under the applied field e as

$$
\begin{aligned}
F = \Sigma \{ & \frac{1}{4} (p_0^2 - p_{m,n}2)^2 \\
& + \frac{k_1}{2} (p_{m+1,n} - p_{m,n})^2 + \frac{k_2}{2} (p_{m,n+1} - p_{m,n})^2 \\
& + \frac{k_3}{2} [(p_{m+1,n+1} - p_{m,n})^2 + (p_{m+1,n-1} - p_{m,n})^2] \\
& - p_{m,n} e \} \quad .
\end{aligned} \tag{32}
$$

This system is depicted in Figure 6.

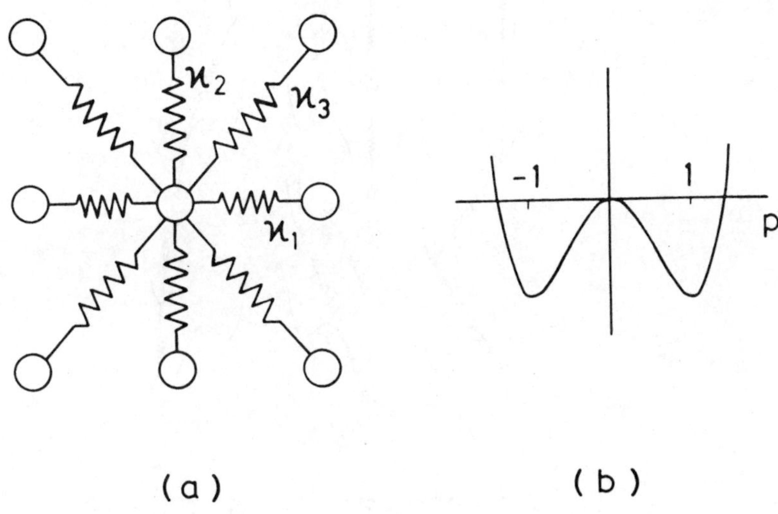

FIGURE 6. A two-dimensional lattice model. (a) Interactions with neighbors, and (b) The local potential where each atom is located.

As in the case of the polarization reversals in SSFLC,[8] nucleation events are not random in time, but rather the latent nuclei, which are distributed at pre-determined positions from the beginning, govern the process. It is not easy to model such latent nuclei. One of several possible ways is to place positive and negative nuclei randomly, and to assume that $p_{m,n}$ at the nucleus sites take the limited values, i.e.,

$$\begin{aligned} p_{m,n} &\geq p_0 , && \text{for positive nuclei,} \\ p_{m,n} &\leq -p_0 , && \text{for negative nuclei,} \end{aligned} \tag{33}$$

This situation is shown in Figure 7.

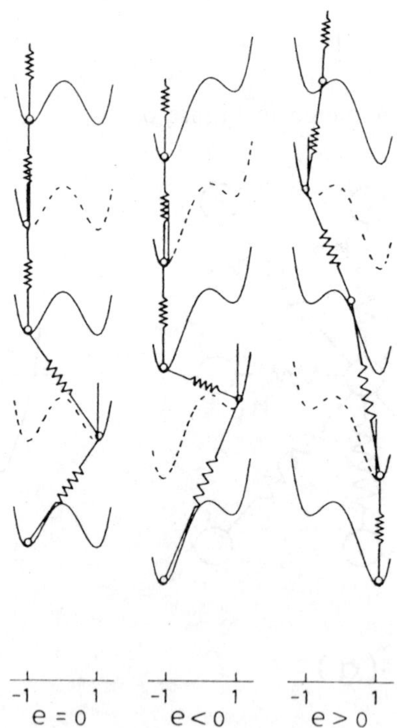

FIGURE 7. The local potentials at nucleus sites.

It is known that the number of the latent nuclei becoming active depends upon strength of the applied field e. It may be important to incorporate such features for discussing the field dependence of the switching time.

The kinetics of each dipole moment is governed by the rate equation;

$$\gamma \frac{dp_{m,n}}{dt} = -\frac{\partial F}{\partial p_{m,n}}, \tag{34}$$

where γ is the viscosity coefficient. The equation (34) forms a set of simultaneous nonlinear equations, and we have to resort to numerical solutions.

The results of the numerical analyses on the spatial profile of dipole moments during the polarization reversal caused by a step-like applied field are shown in Figure 8 for a one-dimensional system, where one can see how the domain walls move and coalesce.

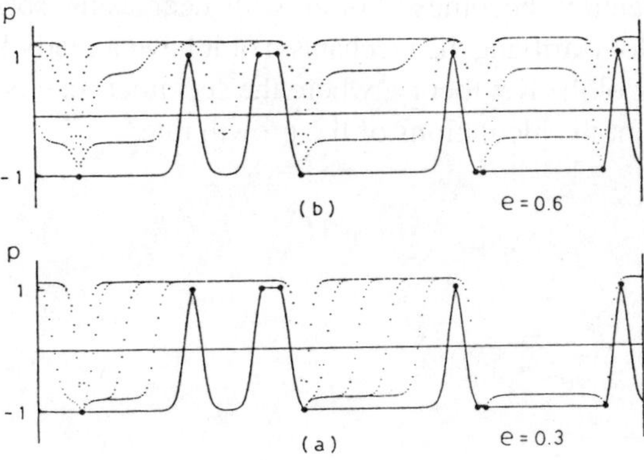

FIGURE 8. Temporal evolution (plotted at a constant time interval) of the spatial dipole distribution in a polarization reversal process. The solid line indicates the negatively polarized state. The domain wall motions can be seen.

By adopting a changing applied field like

$$e = e_0 \cdot \sin 2\pi ft\ , \tag{35}$$

f being the frequency and e_0 the amplitude, the present model can be used to reproduce the p-e hysteresis curves and to see their dependences upon e_0 and f. As for the details of outcomes of the model, readers are referred to a paper by Ohmura et al.[12]

5. REMARKS

In this paper the KA theory and its application for analyzing the polarization reversals were reviewed. A lattice model recently proposed for simulating polarization reversals was described. The physical natures of this lattice model have not yet been adequately clarified and are now under investigations.

One of the most interesting features, which has been discovered recently with respect to ferroelectric switching, is that the switching time in ferroelectrics becomes shorter with decreasing specimen size.[13, 14] Attempts for clarifying the mechanism of it have already been done in the framework of the KA theory, where the specimen size is explicitly taken into account in calculations of the growth rate.[6, 15, 16]

REFERENCES

1) A.N. KOLMOGOROV: Izv. Akad. Nauk USSR, Ser. Math. 3, 355 (1937).
2) M. AVRAMI: J. Chem. Phys. 7, 1103 (1939); 8, 212 (1939); 9, 77 (1941).
3) J.F. SCOTT and C.A. ARAUJO, H.B. MEADOWS, L.D. MCMILLAN and A. SHAWABKEH: J. Appl. Phys. 66, 1444 (1989).
4) Y. ISHIBASHI and Y. TAKAGI: J. Phys. Soc. Jpn. 31, 506 (1971).
5) Y. ISHIBASHI: J. Phys. Soc. Jpn. 59, 4148 (1990).
6) H. ORIHARA and Y. ISHIBASHI: J. Phys. Soc. Jpn. 61, 1919 (1992).
7) R.M. BRADLEY and P.N. STRENSKI: Phys. Rev. B40, 8967 (1989).
8) H. ORIHARA, Y. ISHIBASHI and Y. YAMADA: J. Phys. Soc. Jpn. 57, 4101 (1988).
9) T. OHTA, Y. ENOMOTO and R. KATO: Phys. Rev. B43 13262 (1991).
10) W.J. MERZ: J. Appl. Phys. 27, 938 (1956).
11) H.L. STADLER: J. Appl. Phys. 29, 1485 (1958).
12) M. OHMURA, H. ADACHI and Y. ISHIBASHI: Jpn. J. Appl. Phys. 30, No. 9B, 2384 (1991).
13) B.P. MODERIC, L.E. SANCHEZ and S.Y. WU: Ferroelectrics 116, 65 (1991).
14) T. HASE and T. SHIOSAKI: Jpn. J. Appl. Phys. 30, 2159 (1991).
15) Y. ISHIBASHI: Integrated Ferroelectrics 3, 351 (1993).
16) Y. ISHIBASHI and H. ORIHARA: J. Phys. Soc. Jpn. 61, 4650 (1992).

CHAPTER 6

FAST POLARIZATION REVERSAL PROCESS: EVOLUTION OF FERROELECTRIC DOMAIN STRUCTURE IN THIN FILMS

Vladimir Ya. Shur
Institute of Physics & Applied Mathematics
Ural State University,
Lenin Ave. 51, 620083 Ekaterinburg, Russia

1. INTRODUCTION

In recent years an interest in the investigation of the fast switching phenomena (with reversal time about 10 -100 ns) has grown rapidly in view of some immediate applications. Last advances in thin film technology have made it possible to create nonvolatile ferroelectric memory devices (such as FERAM) which are presently being marketed.[1] The most popular objects used for such applications are polycrystal ferroelectric thin films.

It is known that the switching process in ferroelectrics is a very complicated evolution of domain structure. The only way of obtaining direct and complete information about domain kinetics is real-time registration of domain patterns during switching by using optical or electron microscopy.[2,3] The observation of domain structure in ferroelectric thin films during fast switching requires the use of rather sophisticated apparatus. As a result there are almost no publications on this topic.

The investigation of fast polarization reversal in thin films can be provided by integral experimental methods only. The most popular method is the measuring of transient current.[4,5] In principle, the current pulse data contains a lot of information about domain kinetics, but it is impossible to extract it by traditional mathematical treatment based on the classical theory of Kolmogorov-Avrami.[6]

The way to solve this problem is as follows. At first, we summarize the results of experimental studying of domain kinetics in single crystals as a model objects. Secondly, we shall try to explain it from the general point of view. If we succeed, we can propose the new unified approach on the domain kinetics suitable for the treatment of integral data. As a result we obtain the essential information about domain kinetic in ferroelectric thin films using integral data e.g. transient current.

The above mentioned remarks determine the structure of the paper. The outline of the paper is as follows. In Sec.2 we introduce the general approach to the problem considering it as an example of phase transformation. In Sec.3 we present a brief review of experimental investigations of polarization reversal. It consist of the survey of several experimental techniques suitable for real-time investigation of domain kinetics with analyse of their merits and demerits. In Sec.4 we summarize our experimental results concerning domain kinetics in model single crystals. In Sec.5 we demonstrate fruitfulness of general approach based on nucleation theory for the description of domain evolution. In Sec.6 we discuss the problem of local electric field as a driving force of the nucleation. It includes the problems of depolarization field and internal and external screening. In Sec.7 we propose general description of transient current data. We applied the formalism of geometrical transformations/ catastrophes for the description of experimental data. We show that the proposed method of mathematical treatment allows to describe the switching process in terms of several kinetic parameters having clear physical meaning. The results are confirmed by computer simulation. Finally we demonstrate how the well-known results of integral measurements in thin films being interpreted by proposed approach can yield the essential information about domain kinetics even in any integrated device.

2. GENERAL CONSIDERATION

The possibility to change the direction of spontaneous polarization under the action of electric field is the main feature which singles out ferroelectrics among other materials. It is quite natural that the investigation of polarization reversal is always in fashion. Recently the interest in the exploration of fast switching process has grown rapidly in view of some promising applications of thin films mainly in memory devices.[1] The obtained experimental data in various ferroelectrics are enormous. Mainly they are different integral characteristics such as switching time or dielectric hysteresis loop parameters.

It must be mentioned that there exists local and integral levels of polarization reversal research. The first is connected with the direct observations of the individual domains behavior during switching. Thus the velocities of the domains arising per area and its sideways/forward growth are measured. The nucleation is the only mechanism for describing these phenomena. The next level is the measurement of integral characteristics of switching process. The field and temperature dependencies of switching time and shape of current pulse were measured in different ferroelectrics. It must be mentioned that statistical properties of domain evolution during switching must be taken into account for explanation of the behavior of any integral transient characteristic.

The only way to obtain complete and direct information about domain kinetics can be realized through real-time registration of domain structure momentary patterns. The direct observation of the fast and especially superfast reversal processes are not so popular because it require the application of rather complex technique.

The actual evolution of domain structure is the multiform unreproducible process not only for different materials but even for different samples. This hinders the understanding of the common laws governing the dynamics of ferroelectric domains. We shall try to reconsider all these complicated phenomena from the following point of view. Let us take into consideration that domains with different orientation of polarization coexisting in one and the same sample are none, but different phases divided by phase boundaries/ domain walls. Then the domain structure evolution during switching is none but an example of phase transformation - the first-order phase transition. This approach is rather fruitful. It is well-known, that dynamics of the first-order phase transition is achieved through the nucleation. So all different displays of domains kinetics can be attributed to the elementary processes of arising of the thermally activated nuclei with the preferred orientation of spontaneous polarization. It is worth to mention that the nucleation probability depends upon the strength of the local electric field, which is space averaged over the volume of the order of nucleus sizes.

Nucleation process. The switching process (similarly to any first-order phase transition) is achieved through arising of nuclei of different dimensionalities. The appearance of new domains is due to arising of three-dimensional **(3-D)** nuclei. The growth of domains/the motion of domain walls is the result of arising of **1-D** and **2-D** nuclei at the already existing domain walls. The nucleation probability during phase transformation in any media essentially depends upon its perfection. Usually the nucleation occurs at the sample surface and in the vicinity of defects.

Fast reversal processes become also understandable and predictable within the proposed approach. In this case one must take into account the time interval between the moment of applying of electric field and the arising of first nucleus, the so-called delay time, and «Zeldovitch time», which is necessary for the forming of steady equilibrium flow of arising nuclei.[7,8] It must be stressed, that it is impossible to create perfect single domain state even in single crystals of high quality, not to mention ceramics and thin films. Thus the switching process starts, usually, not through the arising of **3-D** nuclei, but by growing of already existing residual/remnant domains.

Local electric field. During switching the internal electric field is essentially spatially inhomogeneous. It is determined not only by the applied potential difference, but by bound charges of dipoles, screening charges on the sample surface and free charges in the bulk. Its behavior can be described in terms of depolarization field changing by several screening processes.

Depolarization field is produced by bound charges of dipoles arising at the polar surfaces of the samples during the switching of polarization and at the charged encounting (head to head or tail to tail) domain walls. For typical ferroelectrics the depolarization field strength in single domain plate cut perpendicular to the polar axis can reach the value of 10^8 V/m. The multidomainization reduces it sufficiently due to the strong dependence on the domain sizes and shapes. This field hampers the polarizing by external electric field and leads to spontaneous backswitching into multidomain state after the end of field pulse. The effect of depolarization field is more complicated as compare with ferromagnetic case, because of the existence of various screening processes due to the redistribution of charges. One can divide them in two groups: external screening and the bulk one.

External screening in ferroelectric capacitor (a ferroelectric platelet with polar facets completely coated by electrodes) is a result of external current and it depends upon the characteristic time of external circuit. In multidomain state the external screening of depolarization fields near the neutral domain walls even in steady state is incomplete due to features of surface layers. During switching the retardation effects lead to existence of partly compensated charge in a narrow strip/trail exactly behind the moving wall. Considered effects are most strikingly displayed in strong electric fields. During superfast switching one can observe the sideways motion of the walls of already existing domains through cooperative effect of correlated nucleation due to the peculiarities of field distribution in the vicinity of domain walls.

Bulk screening. There is the residual depolarization field in the bulk of the ferroelectric capacitor in single domain state even with short-circuited electrodes due to the existence of surface layers/dielectric gaps in which spontaneous polarization is absent.[9] The only way to compensate this field is the redistribution of the bulk charges.

3. EXPERIMENTAL STUDY OF POLARIZATION REVERSAL

There exist a lot of methods developed for the investigation of the polarization reversal. They can be cast in two main groups: integral and local. By using the integral technique one can obtain information about the reversal process in whole without any details. Such experiments provide us with effective averaged parameters only. The most popular integral technique is the measuring of the current flowing in the external circuit connecting the electrodes during polarization reversal The local technique provide us detail information about the evolution of the reversal process through direct observation of momentary domain patterns during switching.

3.1. Electrical Investigations of Switching Process in Ferroelectrics

Electrical technique can be used to study the fast switching process in almost all ferroelectric materials. This method being comparatively simple is rather complicated for the interpretation. The measurement of the current in the external circuit after application of opposite polarity voltage to ferroelectric sample is the basis of this technique. In the most informative variant of such technique proposed by Merz [4] the square voltage pulses have been applied. One can obtain from the transient current pulse the switching time $\mathbf{t_s}$, which is the most significant parameter describing the polarization reversal for memory application.

We provide only a brief review of classical experimental results. More complete information can be found in several reviews.[10-13] Numerous ferroelectrics were examined by measuring the field dependence of the switching time. There are some peculiarities in different crystals but usually it is possible to distinguish three main types of field dependencies.

In most ferroelectrics in some field range one obtains exponential law: $\mathbf{t_s(E) = t_{so}exp(\alpha/E)}$, where α is the so-called activation field. It is important to note that there is no threshold field for the switching. In other words the coercive field $\mathbf{E_c}$ determined from dielectric hysteresis loop is not a threshold one because even the fields much smaller than $\mathbf{E_c}$ can switch the sample if the voltage pulse is long enough. So one can say that such ferroelectrics does not have a true coercive field.[10] This is one of the main limitations for the application of the ferroelectrics as a memory elements.

At higher fields a power law of switching time field dependence is usually observed: $\mathbf{t_s(E) = kE^{-n}}$, where $\mathbf{n}$ is varying for different crystals from 1 (in sodium nitrite) to 7 (in tetramethylammonium trichloro mercurate).

In some materials, such as rare-earth molybdates, the linear field dependence of reciprocal switching time and domains' sideways velocity is obtained: **v(E)** = μ **(E - E$_{th}$)**, where μ is mobility and **E$_{th}$** is the threshold field.

3.2. Experimental Techniques of Domain Dynamics Investigation

At present more then twenty different experimental techniques are used to observe ferroelectric domains and its number constantly increases.[10-13] In principle it is possible to use all these methods for registration of domain dynamics, but in reality in the most part they are useful only for the investigation of the very slow switching. For the investigation of fast polarization reversal one must use partial switching by short field pulses and study the domain patterns during the long off-time between pulses. The drawback of these methods is that they do not provide any information about domain backswitching which may occur after the end of each pulse. Therefore, it is not entirely justified to assume that the domain patterns registered during partial switching are the same as obtained by applying one long pulse.[10] One can obtain detail and reliable information only by a few methods of the real-time direct observation of domain dynamics with high resolution: pyroelectric probe technique, electron microscopy, nematic liquid crystal and optical methods.

The pyroelectric probe technique is based upon the definition of the sign of pyroelectric signal measured by dynamic method of Chynoweth[14] obtained through local heating of ferroelectric platelet or film by scanning laser beam[15] or electron zond.[16] This method was essentially improved by Hadni *et al.*[17,18] They used laser scanning microscope with argon laser with a 200 kHz chopping frequency and phase detection of the pyroelectric signal sign.[18] This method was capable of inspecting a 100×100 μm^2 area of triglycine sulphate (**TGS**) sample at the display within few seconds with 2 μm limit of space resolution. It was possible to investigate by this method the domain structure in thin superficial layer with thickness of the order of 1 μm. This technique was used for the studying of domain wall motion in **TGS** and **NaNO$_2$**. It must be mentioned that dynamic possibilities of pyroelectric probe technique are restricted. The image of domain structure distributed over the area 0.1×2 mm^2 can be displayed within 0.02 s.[19] So this method is suitable for the real-time observation of slow domain dynamics and can be used only for the thin samples of ferroelectrics with large pyroelectric coefficient.

Electron microscopy is also used for the investigation of domain structure evolution during switching. By using TEM in stroboscope mode it was possible to observe the domain wall movement in thin films of **BaTiO$_3$** during switching in alternating fields of high frequency.[3] This method is capable of extremely high space resolution. The method of observing domain kinetics by SEM with the voltage contrast mode is more universal one.[20] The voltage contrast is based on the brightness modulation of secondary electron pictures, caused by the existence of different electric potentials on the sample surface.[21] As a result, this method can be used for real-time observation of dynamical processes in external electric field with TV-mode.[22] The achieved spatial resolution is higher then 1 μm. Commonly the electron zond changes the domain structure and so this method is a destructive one. At the same time it is possible to realize switching under specific conditions of the absence of an electrode at one of the surfaces using the electron beam for the applying of voltage.[23] It must be noted that it is impossible to use this method for the investigation of domain dynamics in uniform electric field and in the whole range of switching times.

Nematic liquid crystals (**NLC**) is the most simple and universal method. It was invented by Furuhata and Toriyama[24] and then widely used by Tikhomirova, Dontsova and Shuvalov for the investigation of domain dynamics in uniaxial ferroelectrics.[25-27] They distributed the thin layer of **NLC** (the thickness about 1-10 μm) with high specific resistance (up to 100 Ohm/m) between transparent electrodes and investigated sample. As a result sufficiently high space resolution 1-2 μm (of the order of the thickness of **NLC** layer) can be achieved. Nevertheless some limitations of this method must be mentioned. Firstly, by using **NLC** technique the voltage is applied to the sandwich structure consisting of investigated ferroelectric sample and **NLC** layers with nonlinear electrical characteristics. Due to this feature it is difficult to estimate the real magnitude of the voltage applied to the sample especially during the switching. Secondly, relatively slow change of optical properties of **NLC** (delay time about 0.01 s) restricts the possibility of the investigation of domain dynamics to the switching times exceeding seconds.

Optical methods of studying domain dynamics (not only ferroelectric one) by using of polarizing microscope are the most informative and suitable.[12] These methods are nondestructive, because the illumination with the proper chosen intensity and wavelength does not affect the domain kinetics. Even superfast switching can be investigated by using pulse lighting and high-speed photography.[28,29] The time resolution depends upon the duration of light pulse and can be shorter then 10 ns. The domains separated by 90° walls can be easily distinguished by polarizing microscope due to the different value of spontaneous birefringence.[30,31] This method had been used in investigation of a wide range of ferroelectrics. But usually it is impossible to use it for the 180° domains because optical indicatrixes in antiparallel domains are identical.

Let us consider more closely the crystals in which 180° domains can be distinguished by optical methods. In $\mathbf{BaTiO_3}$ antiparallel domains become optically distinguished when electric field[32,33] or corresponding mechanical forces[34] are applied perpendicular to polar axis. The induced strains are not strong enough to change domain structure, however the rotation of optical axis in opposite directions occurs in different domains. It leads to small difference between extinction positions of antiparallel domains allowing its visualization.

The investigation carried out by Miller and Savage[2,35] showed that during the polarization reversal in longitudinal field the antiparallel domains in **$BaTiO_3$** can be observed in transmitted light if the sample with transparent electrodes had been placed between crossed polarizers. Kobayashi *et al* [36] showed that in this case domains are distinguishable only under the illumination directed at small angles to polar axis. They explained this effect by the existence of some strain-induced birefringence in the vicinity of domain walls.

There are only 180° domains in Rochelle salt which are optically distinguished due to the different orientation of optical indicatrixes.[37,38] It must be noted that the same effect is observed in wide range of ferroelectric materials in which domains are also ferroelastic such as potassium dihydrogen phosphate,[38] bismuth titanate[39] and rare-earth molybdates.[40,41] $Pb_5Ge_3O_{11}$ is optically active crystal. In the domains with opposite orientation of polarization the directions of rotation are different.[42,43] Due to this effect the kinetics of 180° domains can be investigated using polarizing microscope with pulsed lighting.

It must be noted that optical methods usually can not be applied for the investigation of some crystals with 180° domains only, e.g. **GASH** and **TGS**. Nevertheless the optical methods if they can be utilized are the most suitable one for the real-time observation of the domain kinetics during fast switching

3.3. Evolution of Domain Structure During Switching

Most of the classical investigations of the domain dynamics during switching have been made on **$BaTiO_3$** by optical methods.[32,33,44,45] From this investigations a complete picture of the polarization reversal can be established. When a switching field is applied to the single domain sample, new domains start to form at first, mainly at the surface, and then grow through the sample thickness. After having grown through the sample they begin to expand sideways. Formation of new domains continues to take place as the first ones progress in their sideways expansion. Later they begin to join together (or to coalesce) until all of the unswitched region is completely overrun by them.[10] Thus the mechanism of polarization reversal involves the following steps: (a) nucleation of new domains; (b) forward growth of the domains through the thickness of the sample; (c) sideways expansion of the domains; (d) coalescence of residual domains (Fig.1).

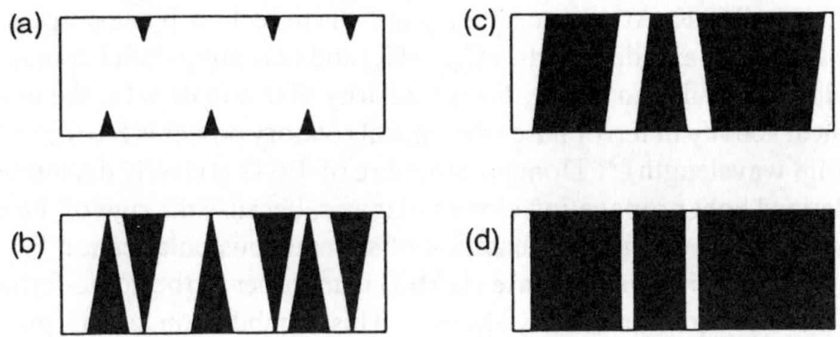

FIGURE 1. The main steps of polarization reversal in ferroelectrics.

4. KINETICS OF DOMAIN STRUCTURE IN MODEL CRYSTALS

The detail investigations of local behavior of domain kinetics carried out in real ferroelectrics by various authors differed sufficiently from sample to sample and did not present any unified picture of the phenomena. For clarifying the situation one have to use a model object.

4.1. The Model Crystals

Let us list the main requirements for those ferroelectrics which can be used as a model objects for the investigation of fast domain structure kinetics. It must be a proper chosen ferroelectric single crystal with following characteristics: 1) simple domain structure (uniaxial ferroelectric); 2) optically distinguished domains; 3) high quality; 4) easily reversible in electric field domain structure without damage of crystals; 5) experimentally convenient temperature region of ferroelectric phase existence. The use of such crystals permits to apply optical methods of real-time registration.

According to formulated requirements we have chosen model crystals clearly demonstrating the limiting geometric types of domain shape existing in uniaxial crystals. The first is the collinear ferroelectric-semiconductor lead germanate $\mathbf{Pb_5Ge_3O_{11}}$ with almost isotropic sideways growth of the domain walls. The other is improper ferroelectric-ferroelastic gadolinium molybdate $\mathbf{Gd_2(MoO_4)_3}$ with strong anisotropy of the domain walls' orientation.

Lead germanate (PGO) was synthesized for the first time by Speranskaya.[46] Its ferroelectric properties were confirmed independently by two Japanese research groups.[47,48] **PGO** undergoes ferroelectric phase transition at 450 K. At the transition point the reflection plane perpendicular to the symmetry axis disappears ($C_{3h} \rightarrow C_3$) and only antiparallel domains with 180° domain walls can occur. The peculiarity of this material is the existence of optical activity in ferrophase (the optical rotatory power is 5.6 °/mm for the 0.63 mm wavelength).[49] Domain structure of **PGO** is clearly distinguishable in polarized light propagating along polar axis, because the sign of the optical activity depends upon the orientation of spontaneous polarization.[42]

Gadolinium molybdate (GMO) is improper ferroelectric-ferroelastic so as several rare-earth molybdates.[50] This combination of the properties makes it possible to switch the domain structure by application of electric field or mechanical stress.[51] **GMO** undergoes phase transition of displacing type at 436 K from paraelectric tetragonal phase to ferroelectric orthorhombic one ($D_{2d}^3 \rightarrow C_{2v}^8$). There are four types of domains in **GMO**, but only two directions of spontaneous polarization (180° domains). In domains with opposite directions of polarization the orientation of crystallographic axes are different. The orientation of coherent domain walls in **GMO** coincides with twinning planes,[52] due to strong anisotropy of surface energy. Polarization reversal in weak fields is accompanied by sideways motion of planar domain walls. However nonplanar walls of other orientation (so-called incoherent domain walls) can arise during switching and then even become quasistable (long-lived metastable).[53] These walls bounded the lens-like and spike-like domains.[54,55] It was shown that polarization reversal in strong fields is accompanied not only by motion of existing planar and incoherent walls, but also by their mutual transformations.[55-58]

4.2. Experimental Technique

The investigated samples were platelets of the 0.15-2 mm thickness cut perpendicular to polar axis. The polar surfaces were electroded with **Sn** and **In** oxides. The transparent electrodes were sputtered by ionplasma reactive method from metal target in oxygen atmosphere and possessed resistance about 10-80 Ohm per square. The samples of **GMO** had rectangular shape with sides parallel to domain walls. It was necessary to apply a fine mechanical working of the side faces of the samples for diminishing of a number of mechanical defects which can serve as the nucleation sites for spike domains. The samples need special fastening to minimize the influence of uncontrolled mechanical stresses on spatial homogeneity of switching.

Linearly polarized light was transmitted along polar axis in order to observe domains (Fig. 2). Accordingly orientated analyzer made it possible to distinguish domains. In crossed polarizer and analyzer opposite domains

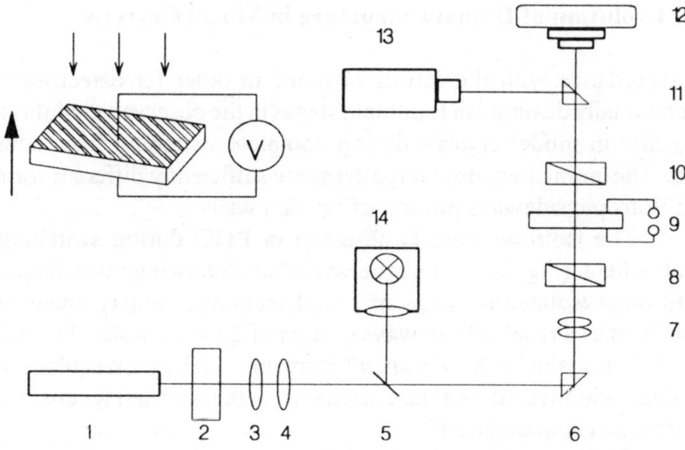

FIGURE 2. Experimental technique for real-time optical investigation of domain kinetics: 1 - pulse nitrogen laser; 2 - dye cell, 3,4 - lenses; 5 - semitransparent mirror; 6, 11 - prisms; 7 - objective; 8 - polarizer; 9 - sample; 10 - analyzer; 12 - camera; 13 - video-camera; 14 - permanent light source.

become indistinguishable and only dark lines were observed along domain boundaries. The width of «observed domain walls» was about 2-5 μm and it depends upon divergence of transmitted light beam. Naturally this value is not connected with the real width of domain walls. Its existence is due to the singular transmission of weakly divergent light in the vicinity of the boundary dividing regions with different optical properties, namely the sign of optical activity (in **PGO**) or the orientation of optical indicatrixes (in **GMO**).

In order to achieve real-time observation of domain structure kinetics during polarization reversal flashing illumination from dye laser pumped by nitrogen laser with light pulse width less than 10 ns was used (Fig. 2).[59] The measurement of switching currents revealed that illumination did not influence the polarization reversal process. Momentary domain patterns were photographed and microphotometered or registered by video system with subsequent processing of video signal.

The polarization was switched under the action of alternating-sign square voltage pulses. It was found that under these conditions domain structure changed repeatedly with detail reproducibility of momentary domain configurations. So we had the opportunity to use the stroboscopic method. Besides real-time registration of momentary domain patterns we measured the transient current pulses.

4.3. Evolution of Domain Structure in Model Crystals

In accordance with the results obtained in other ferroelectrics[10,13] one can conditionally distinguish four main stages in the observed evolution of domain structure in model crystals during complete switching from single domain state. The momentary domain patterns are sufficiently different for anisotropic and isotropic sideways motion of domain walls.

The isotropic case is observed in **PGO** during switching in strong electric field (Fig. 3a). The process include following main stages: 1) arising of through isolated domains of round section; 2) enlargement of domains' diameters as a result of sideways motion of domain walls; 3) coalescence of isolated domains with subsequent formation of large irregular-shaped ones; 4) disappearance of remnant domains with disadvantageous orientation of spontaneous polarization.

There are the qualitative difference of domain shape in **GMO** due to the high anisotropy of walls' energy. One can divide the domain structure evolution during switching in anisotropic case (Fig. 3b) into several stages: 1) arising of spike-like through domains with incoherent walls; 2) growth of spikes and formation of strip/lamellar domains; 3) expansion of strip domains as a result of sideways motion of planar walls; 4) domains' coalescence and disappearance of residual narrow strip domains.

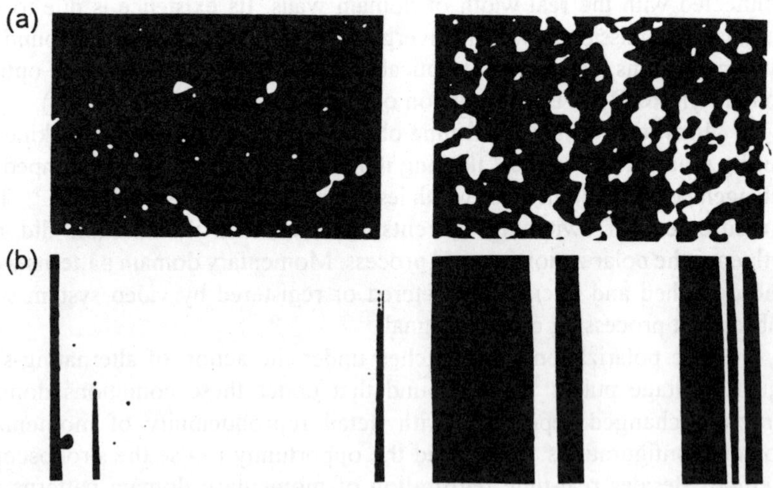

FIGURE 3. The main observed stages of domain structure evolution during switching (a) in PGO and (b) in GMO.

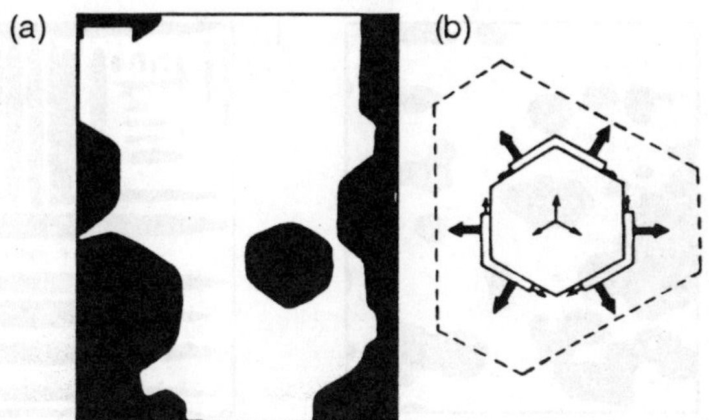

FIGURE 4. The switching in PGO in weak field: (a) - the pattern of the domain structure; (b) - the mechanism of hexagonal domain growing. The dashed line - the facets of single crystal.

4.4. Field Dependence of Domain Kinetics

It is well-known that domain kinetics is qualitatively changed with the increasing of field strength. We can divide the whole used field range into three regions, so-called weak, strong and superstrong fields.[61,62]

Weak field. In field less than 2×10^5 V/m (Fig 4a) growing domains in **PGO** become the regular hexagons. The domain walls are parallel to the facets of the crystal pulled along the polar axis. The field dependence of sideways motion velocity of these walls is approximately linear.[61,63]

In **GMO** such dependence is observed in wider field range.[64] Polarization is reversed by the sideways motion of plane domain walls which are parallel to the crystal facets (Fig. 5).

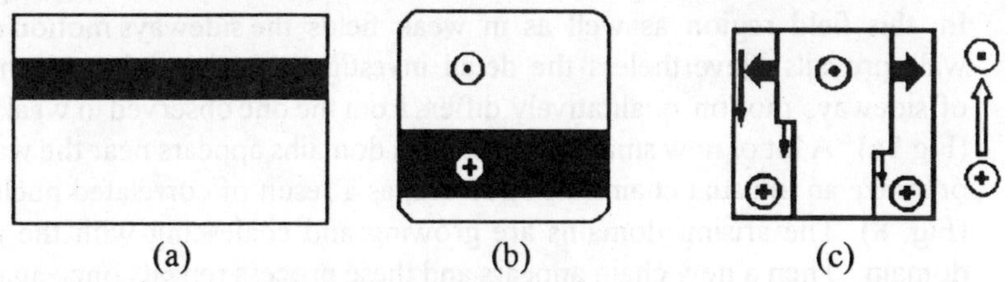

FIGURE 5. Switching in GMO in weak field: (a) - the pattern of the domain structure: (b) - the orientation correlation between the domain walls and the facets of crystals; (c) - the mechanism of plane domain walls motion during the switching from single domain state.

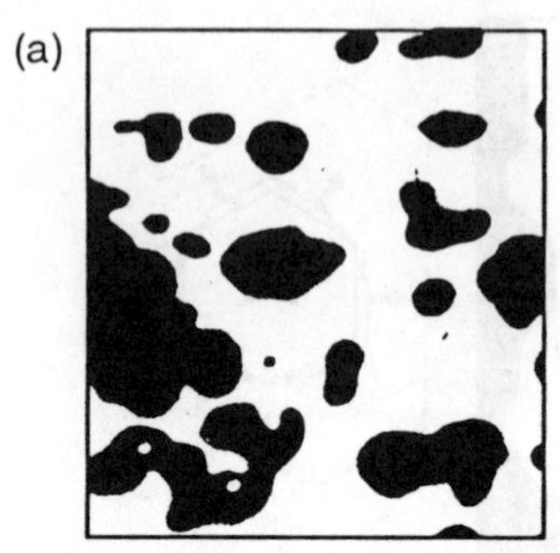

FIGURE 6. Switching in strong field. The momentary patterns of the domain structure: (a) - in PGO and (b) - in GMO.

Strong field. In field more than 3×10^5 V/m exponential field dependence of wall velocity has been observed in **PGO**.[63] The switching is accomplished through the arising of great number of new round section domains which become shapeless after coalescence (Fig.7a). Any preference in walls' orientation is not observed in this case. In **GMO** in strong field spike-like and lens-like domains oriented in two orthogonal directions are arising (Fig.7b). Considerable deviations of their walls from coherent directions are observed. In this case only «switching velocity» (reciprocal switching time) can be measured due to chaotic and irreproducible character of domain kinetics. It was shown that its value also depends exponentially on the electric field.[65]

Superstrong field. One would think that further increasing of the applied field must lead only to the increasing of the concentration of new domains, but in fact, in **PGO** in superstrong field (more than 1.5×10^6 V/m at RT) we have observed qualitatively different character of switching (Fig. 7).[59] In this field region as well as in weak fields the sideways motion of the walls prevails. Nevertheless the detail investigations show that mechanism of sideways motion qualitatively differs from the one observed in weak fields (Fig.3a). A lot of new small round section domains appears near the wall and organize an unusual chain-like structure, as a result of correlated nucleation (Fig. 8). The arising domains are growing and coalescing with the parent domain. Then a new chain appears and these process repeats once again.[60,66]

It is necessary to draw attention to the additional peculiarities of **PGO** and **GMO** domain structure evolution during switching. The behavior of domain structure essentially depends on the quantity and space distribution of nucleation sites and residual nonthrough domains, because they determine the number of newly arising domains. The switching always starts at the

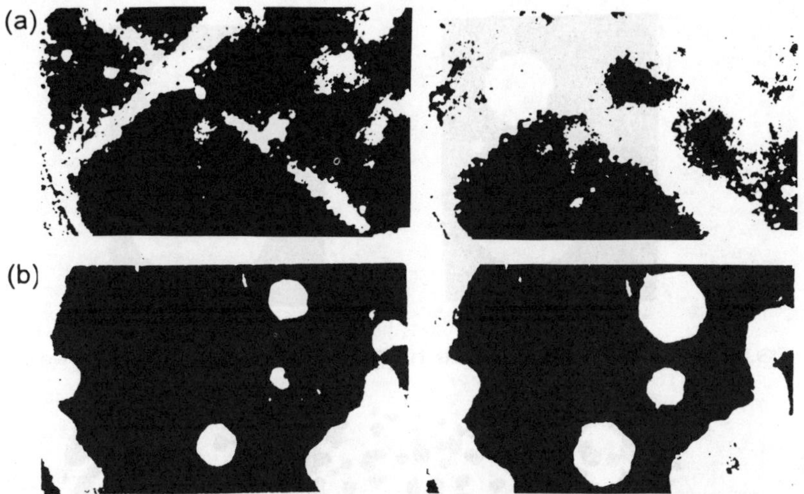

FIGURE 7. Switching in PGO in superstrong field. Comparison of momentary domain patterns arising in different fields: (a) - superstrong field, (b) - weak field.

electrode edges. This boundary effect is very important in weak or superstrong field, when the switching is achieved through the sideways wall motion, especially for samples with small electrodes area. The interaction of the walls of growing domains, such as wall-wall repulsion is observed. In weak field this interaction reveals itself through the significant slowing of wall motion before the coalescence (Fig. 9). In strong field in **PGO** it leads to the formation of raster-type structure of newly arising domains (Fig. 10).[66]

FIGURE 8. Chain-like momentary domain structure arising in PGO as a result of correlated nucleation in front of the moving domain wall in superstrong field.

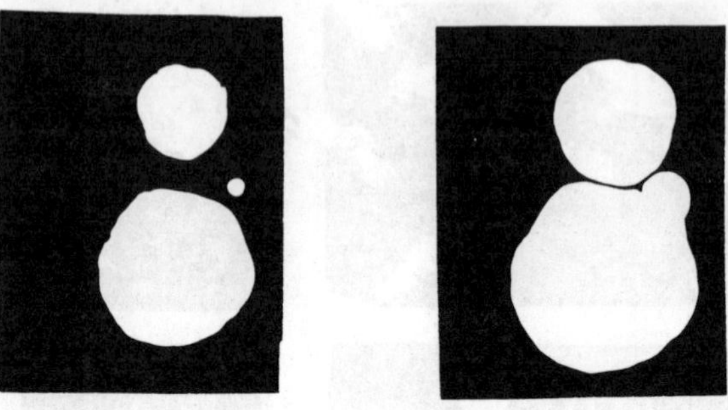

FIGURE 9. Interaction of domain walls in PGO during polarization reversal in weak field.

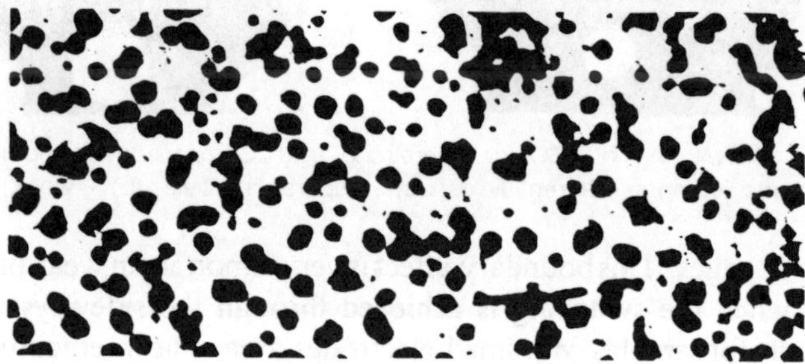

FIGURE 10. Raster-type domain structure arising in PGO during polarization reversal in strong field.

5. NUCLEATION PROCESS

The kinetics of the domain structure during switching is nothing but an example of phase transformation - the first-order phase transition. Now we are going to show the fruitfulness of this approach for the explanation of the details of ferroelectric domain kinetics under different conditions. The consideration will be carried out in the model crystals, as an example.

5.1. Nuclei Dimensionalities

The kinetics of the first-order phase transitions is well studied (both experimentally and theoretically) for the crystal growth from the melt.[67-69] In this case the driving force is the oversaturation of the melt which leads to the formation of solid phase through the nucleation process. All different nuclei arising during crystallization are commonly divided in three main types as regards to their dimentionalities and positions. They are: (a) **3-D** nuclei arising far from the phase boundary, (b) **2-D** one, which are forming on the plane phase boundary, and (c) **1-D** nuclei arising at the steps of the boundary (Fig.11). It is clear that the probabilities of occurrence of given type of nuclei depend in different manner upon the oversaturation degree and various external conditions. This phenomenon results in variety of observed types of crystal growth evolution.

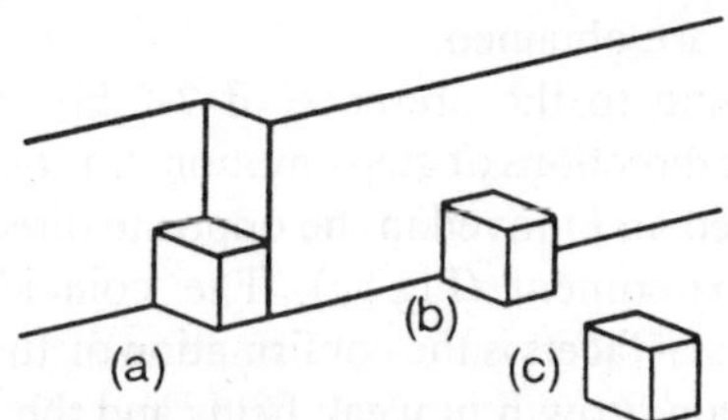

FIGURE 11. Nucleation process during the first-order phase transition: (a) - 1-D nucleus, (b) - 2-D nucleus, (c) - 3-D nucleus.

At large oversaturation degrees the crystal is growing from the melt through nucleation at the phase boundary.[67,68] The obtained crystals become shapeless due to the chaotic character of this process.

The decreasing of oversaturation degree leads to the suppression of the probability of **2-D** nucleation. Nevertheless the growth of new phase is continued. It is common to explain this fact by **1-D** nucleation at the steps on the phase boundary and layer by layer growth of the crystal by the motion of this steps along the boundary (Fig.12). As a result phase boundary become regular shaped and the crystals with regular facets are obtained.[70,71]

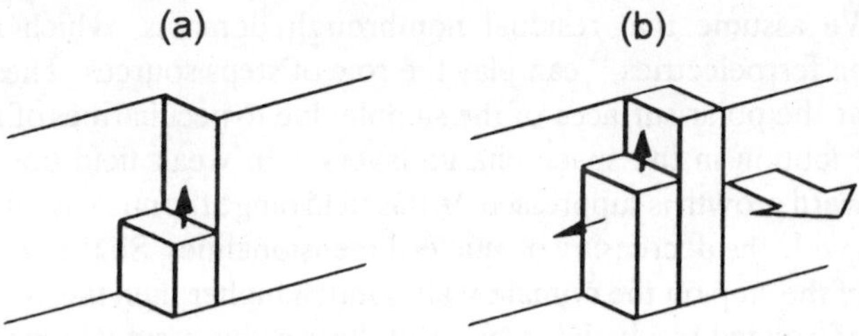

FIGURE 12. One-dimensional nucleation at the steps of the phase boundary and layer by layer growth of new phase.

5.2. Sideways Domain Wall Motion

Let us show that just the same mechanisms drive the domain evolution during switching in the model uniaxial crystals. One can assume by the analogy with the first-order phase transition at small oversaturation that in weak field layer by layer growth of domains of preferable orientation takes place. In **PGO** the trigonal anisotropy of the surface energy lead to the existence of three distinguished directions of preferable steps motion. It is easy to show that in this case six plane domain walls must be formed (Fig. 4b). Thus the regular hexagonal domains are obtained.

In **GMO** owing to the presence of 2-fold symmetry axis there are only two preferable directions of steps motion. So, two parallel plane domain walls must be formed and moved in the opposite directions (Fig. 5b). This is observed in the experiment (Fig.5a). The coincidence of domain wall orientation with crystal facets is the confirmation of the unity of the processes governing the domain growth in weak fields and the crystal growth at small oversaturation.[70,71]

5.3. Influence of Residual Domains

The main problem which must be solved for using the mechanism of layer by layer growth in our case is the origin of the steps on the domain walls. In the theory of crystal growth it is commonly understood that the steps are arising as a result of intersection of screw dislocation with the phase boundary.[72] These steps provide spiral growth of crystal surface. On analogy with crystal growth Nakamura[73] assumed that domain wall motion is due to spiral step growth, produced by screw dislocation normal to the wall. But Miller and Weinreich[74] showed, that this mechanism required unrealistic number of dislocations for the explanation of experimental data.

We assume that residual nonthrough domains, which are always existing in ferroelectrics,[13] can play the role of steps sources. These domains arise near the polar surfaces of the sample due to peculiarities of the internal field distribution in the space charge layers.[75] In weak field the velocity of their forward growth is suppressed. In this field range the nucleation probability increase with the decreasing of nuclei dimensionalities. So the velocity of the growth of the step on the domain wall is much higher then the velocity of the growth of isolated residual domain. But these stable residual domains can be intersected by the moving domain wall during switching. Immediately they are transformed into lugs on the wall, which rapidly grow through the sample and form the steps on the wall suitable for layer by layer growth (Fig. 13).

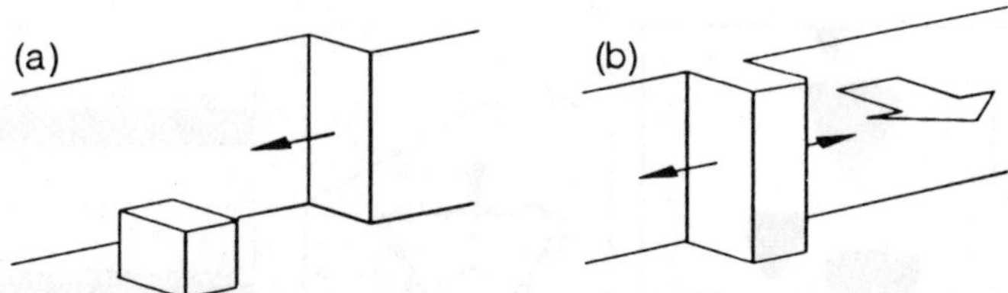

FIGURE 13. The role of residual nonthrough domains as the sources of steps on domain wall during switching.

The estimations show that density about 10^3 - 10^4 mm^{-2} of residual domains is enough for realization of the layer by layer growth. This quantity correlates with the number of domains arising in **PGO** during the switching in strong fields.[73] It must be noted that in any field range the existence of residual domains accelerate the switching process.

5.4. Delay Time and Switching in Steps

It is necessary to take into account existence of the delay time between the moment of electric field switching on and the starting of the nucleation process.[77] The value of delay time is higher for the nuclei with greater dimensionalities $\mathbf{t_d(1\text{-}D) < t_d(2\text{-}D) < t_d(3\text{-}D)}$. If the time of field action, which is equal to the pulse duration or the switching time, is less than delay time for any kind of nuclei, the kinetics of domain structure qualitatively changes.[62,78] This effect can be easily demonstrated in **PGO**. Under the action of train of short strong field pulses (duration less than the time needed for the arising of through domains) the domain structure is changed by sideways motion of domain walls, as in weak fields.[78] During the growth domains acquire the shape of scalene triangles, whose sides are parallel to the planes [100] (Fig.14).

In order to explain these peculiarities we must take into account that in this case both mechanisms of domain wall motion take place one after another (Fig.14). During the strong field action the **2-D** nuclei are formed on the wall. Their orientation is determined by the anisotropy of surface energy. These nuclei formed the steps on domain walls. In the pause between pulses in weak field the walls are smoothed as a result of steps motion along the domain walls. In this case it is easy to show that the walls must move in three directions, orthogonal to the directions of motion of hexagonal domain walls. Thus triangular domains, whose sides are rotated with regard to the sides of hexagonal domains, are formed (Fig.14b).

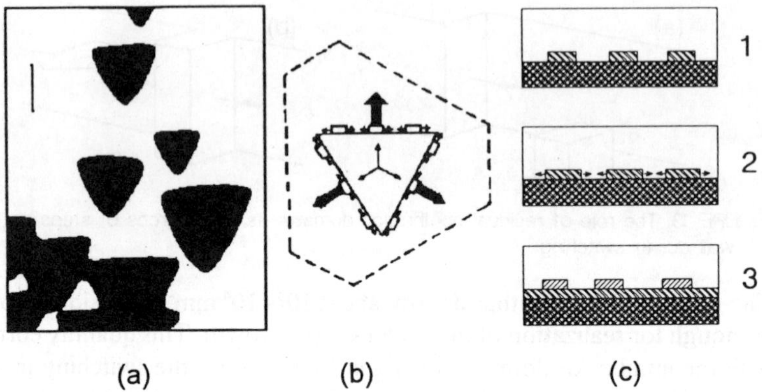

FIGURE 14. The forming of triangle domains in PGO under the action of the train of strong field short pulses: (a) - the domain pattern; (b) - schematic growing of domain growth, the dashed line - the facets of single crystal; (c) - main stages of domain growth during switching: 1 - arising of nuclei in strong field pulse, 2 - steps motion in weak field between pulses, 3 - arising of nuclei in the next strong field pulse.

6. LOCAL INTERNAL ELECTRIC FIELD

It is clear that nucleation probability at the given place and consequently the rate of domain growth depends on the local strength of internal field. In ferroelectric capacitor the field is determined not only by the electric potential difference between electrodes, but also by the sum of depolarization field of bound charges and screening fields produced by charges on electrodes (external screening) and bulk charges (bulk screening).

6.1. Depolarization Field

In ferroelectric samples with finite sizes in polar direction the increase of the spontaneous polarization magnitude due to the temperature change (including the phase transition) is always accompanied by arising of additional electric field producing by bound charges. In real ferroelectric single crystals these charges appear in several places: firstly, near the polar surfaces of the sample; secondly, near the encountering (head to head or tail to tail) domain walls; and at last, in the bulk, if the spontaneous polarization is heterogeneous. It must be noted that the last effect is extremely strong in ceramics. In any case the bound charge produce electric field which try to change the direction of spontaneous polarization, the so-called depolarization field $\mathbf{E_d}=\mathbf{P_s}/\varepsilon\varepsilon_0$. The strength of this field can be extremely high. For example, in ferroelectric capacitor preparing from **PGO** in single domain state at RT depolarization field is about 10^8 V/m. So the spontaneous backswitching must take place after the end of external field pulse. As a result, without field ferroelectric capacitor would be divided into small domains. It is easy to show that such multidomainization reduce the depolarization field strength.

The same problem exist during polarization reversal, because in any case domain kinetics goes through the stages of the arising of great number of nonthrough domains with charged domain walls. The special longitude spike-like shape of such domains minimize the depolarization field.[74,79]

From this point of view it is interesting to study the charged domain walls of large area, which can be arise, for example, in **PGO** after phase transition (Fig. 15).[80-83] One can see that such walls have a very complicated nonplanar dentate shape with a great number of narrow irregular dents/spikes.

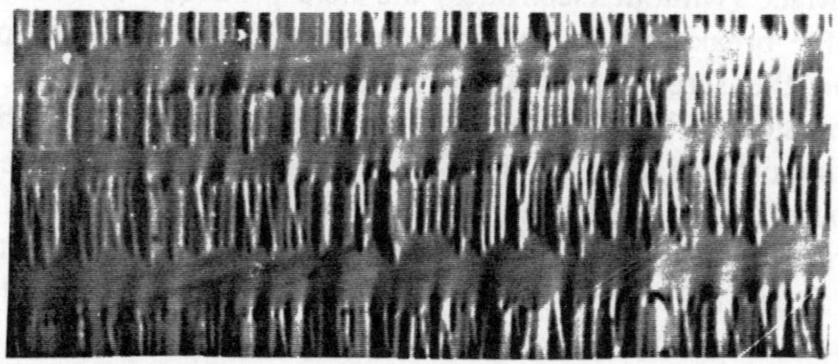

FIGURE 15. Charged incountering domain walls in as-grown domain structure in PGO. Visualizing by using the cleavage method.[82,83]

The velocity of domains growth (nucleation probability near the domain wall) depends upon shape and sizes of the domains, especially for nonthrough ones with charged domain walls. This effect is very important for the arising of **3-D** nuclei and their forward growth through the thickness of the sample.

Ferromagnetics display the similar behavior, but their fundamental difference from ferroelectrics is due to the fact that in the latter these fields can be compensated through the redistribution of charges (screening effect).

6.2. External Screening Effect

The most fast and effective way to compensate depolarization field is external screening by forming of compensating charges at the polar faces of samples.

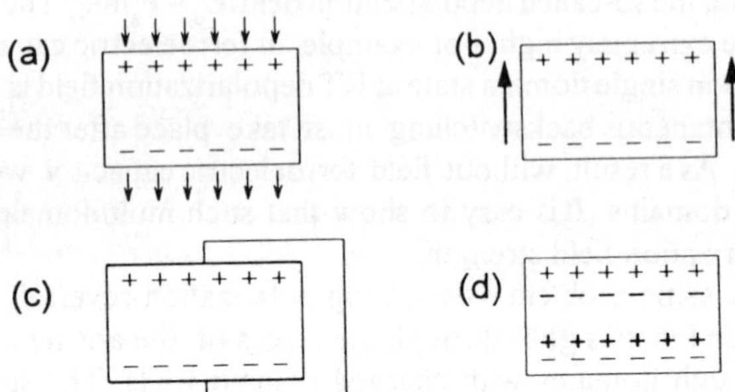

FIGURE 16. Mechanisms of screening processes: a - external screening by breakdown in external medium or electron emission; b - external screening by surface conductivity; c - external screening in ferroelectric capacitor; d - bulk screening.

This process can be realized by several ways (Fig. 16). For the sample with free surface (without electrodes) the field produced by bound charges can cause the breakdown in external medium. If this sample is placed in the vacuum the screening is achieved through field emission of electrons (exoemission).[84,85] The conductivity along the lateral facets of the sample can be of importance, also. In the ferroelectric capacitor, whose polar faces are coated by electrodes, the screening process occurs through the redistribution of charges on electrodes due to the current in external circuit.

It is necessary to take into account that ferroelectric capacitor possesses the surface layers with special properties.[10,12] At present the assumption about their dielectric nature is rather popular. Then there are effective dielectric surface layers in ferroelectric capacitor, the so-called dielectric gaps[9] of thickness **L** and dielectric permittivity ε_L (Fig.17). As a result, the layers of bound charges are formed at some distance from the electroded surface. The existence of these dielectric gaps leads to such peculiarities in screening effects, as incomplete compensation of depolarization field by external screening, especially in the vicinity of the domain walls.

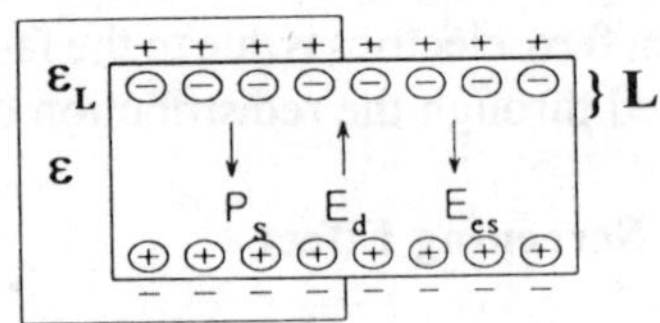

FIGURE 17. The ferroelectric capacitor with dielectric gaps.

Let us consider ferroelectric capacitor with single planar neutral domain wall as a model for the evaluation of the internal field distribution. There are spatial regions with incomplete external screening on both sides near the wall. One can approximate the charge distribution by two narrow strips of opposite charges just at the wall (Fig. 18). Their widths are equal to the thickness of dielectric gap and the surface charge density is about the value of spontaneous polarization. It is possible to evaluate the field distribution by summing up the fields from different strips and by using formula for the average field at the edge of the strip capacitor:

$$E(A) = \sigma/\pi\varepsilon\varepsilon_0 d\ [2\mathrm{arctg}(A/d) + A/d\ \ln(1+d^2/A^2)] \qquad (1)$$

where **A**, **d** - width and thickness of the capacitor, σ - surface charge density. The space distribution of total internal field for stationary or slowly moving domain wall is plotted in Fig. 18 (curve 1). It is seen that field strength reaches its maximum value near the domain wall.

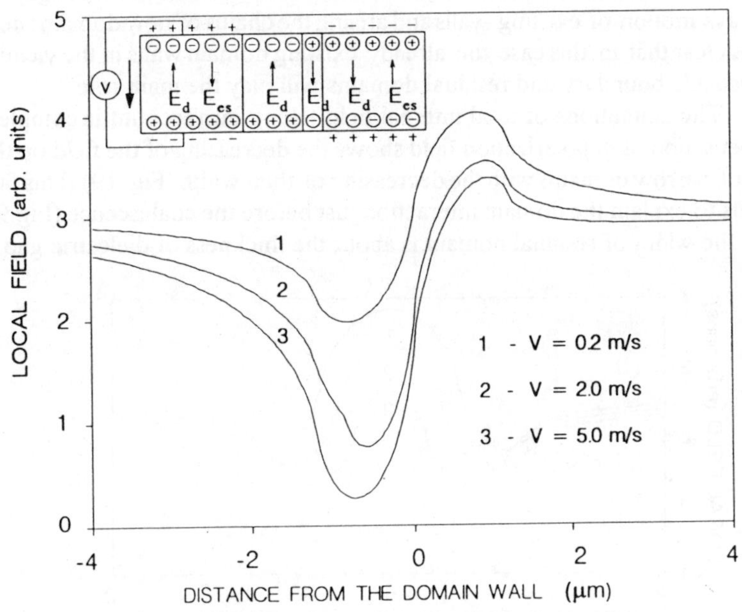

FIGURE 18. The calculated space distribution of the total internal field near the moving plane domain wall. In insert - peculiarities of the external screening near the domain wall during switching.

During switching the additional noncompensated bound charge, the so-called trail, appears behind the moving wall, decreasing switching field and slowing walls motion.[86] This effect is due to retardation nature of external screening. One can take it into account assuming some asymmetry in charged distribution near the moving domain wall (Fig. 18). The width of this effective trail is determined by the product of domain wall velocity and time constant of external circuit. The space distribution of total internal field becomes asymmetrical, but even in this case internal field reaches its maximum in front of the wall (Fig. 18). Such behavior makes it possible to explain the arising of chain-like domain structure in superstrong field (Fig. 7 and 8).[60]

One can observe another peculiarity of domain kinetics in superstrong field, namely the suppression of the arising of new domains. This effect can be explained if we take into account the delay time - the interval between the moment of voltage applying and the moment when the nuclei start to form. It is worth to mention that this delay time increases with increasing of nuclei dimension. The ratio of delay time to the time of external field action (switching time) is very important. If switching time becomes less than delay time for **3-D** nuclei the polarization reversal will be achieved through the sideways motion of existing walls and arising the chains of new domains near it. It is clear that in this case the already existing domain walls in the vicinity of electrode boundary and residual domains will play the main role.

The estimations of total internal field taking into account in complete compensation of depolarization field shows the decreasing of the field on the walls of narrow domains with the decreasing of their width (Fig. 19). This fact permits to explain the domain interaction just before the coalescence (Fig.9), when the width of residual domain is about the thickness of dielectric gap.

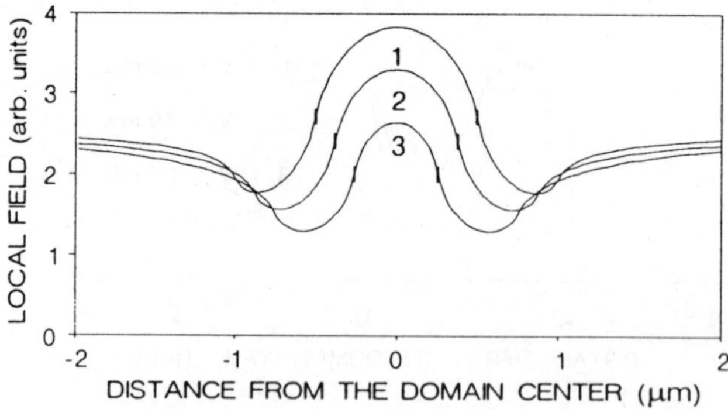

FIGURE 19. The calculated space distribution of the total internal field for the narrow strip domains. Domain width: 1 - 1.0 μm, 2 - 0.7 μm, 3 - 0.5 μm

6.3. Bulk Screening Effect

It was shown that in short-circuited capacitor with dielectric gaps the depolarization field is compensated by external screening in the first instance. Nevertheless, even in single domain state after finishing of this process, the residual depolarization field remains in the bulk $\mathbf{E_{dr} = 2P_s L/\varepsilon_L \varepsilon_o d}$. The bulk screening is the only way to compensate this field. There are several variants of the bulk screening: redistribution of space charges or reorientation of dipolar defects produced in the crystal by doping or by irradiation.[87]

The standing at constant conditions leads to screening of depolarization field due to redistribution of carriers in the bulk and formation of space charges bound by traps. The time constants of the bulk screening are extremely long (about 10^2 s in **PGO** and 10^4 s in **GMO** at RT).[88] If the bulk screening is completed one can obtain the unipolarity of switching process and even complete backswitching after the end of field pulse.[89] The field produced by space charges can be measured as a bias field of hysteresis loop[88,90] $\mathbf{E_b = 0.5(E_c^+ + E_c^-)}$, where $\mathbf{E_c}$ - coercive field. The measured value of bias field is 10^5 V/m. The bulk screening field changes its value and sign at domain walls. It was shown that bulk screening occur not only in the static domain structure but even during polarization reversal. The local strength of bulk screening field during cyclic switching is equal to time averaged field in the chosen place of the crystal.[77,91] It is clear from above consideration that bulk screening considerably alters the domain kinetics especially in multidomain state.

The additional peculiarities of sideways domain wall motion take place during the switching from the multidomain state: in weak fields the wall shifts to some distance from initial position and stops; after the field is switched off spontaneous backward motion of the wall is observed (Fig. 20).[92]

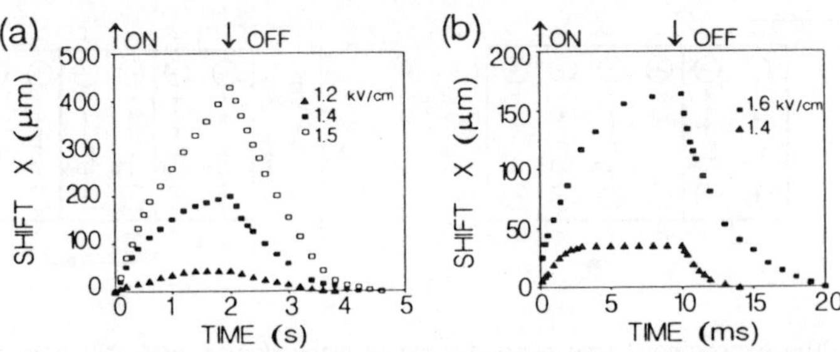

FIGURE 20. The time dependencies of plane domain wall shift from initial position in various electric fields: (a) PGO and (b) GMO.

In order to explain this behavior we must add bulk screening field $\mathbf{E_{bs}}$ to already considered model. One must taken into account the existence of additional pair of charged planes situated at the distance about effective bulk screening length $\mathbf{L_s}$ from the surface (Fig. 21). This field is practically constant during switching because the time constant of bulk screening $\mathbf{t_{bs}}$ is much longer than the switching time. We obtain for the internal field at the domain wall $\mathbf{E_{iw}}$ the following dependence upon the value of wall shift from equilibrium position[92]

$$E_{iw} = E_{ex} + k\, E_{dr}\, F\,(x/2L_s) \qquad (2)$$

where $\mathbf{F(A) = 1/\pi\ [2arctgA + A\ ln(1 + A^{-2})]}$; **k** - screening factor. In both model crystals all experimental data of domain wall motion in external field and theoretical results, obtained under considered assumptions, are in good agreement (Fig. 21).[92]

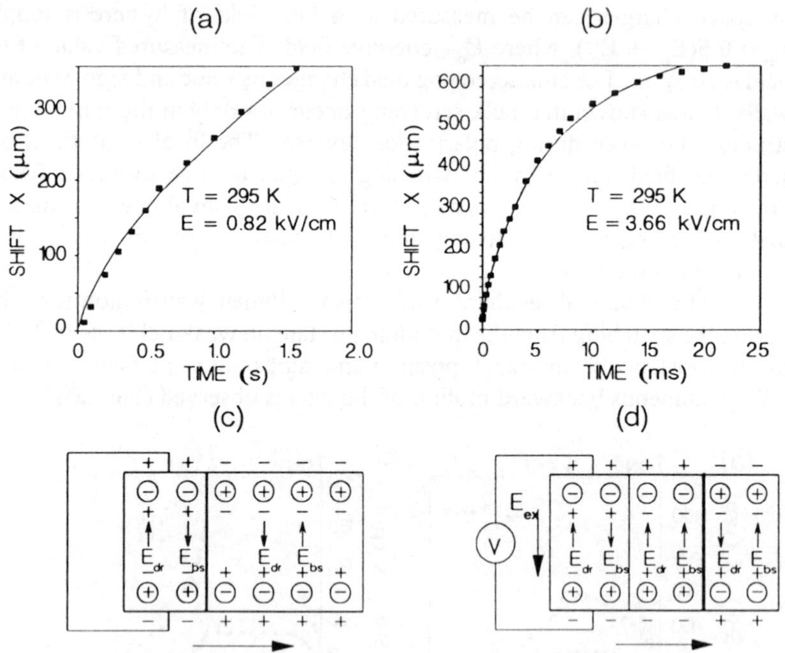

FIGURE 21. The experimental time dependencies of plane domain wall shift from stable position (dots) approximated by theoretical curve: (a) - PGO, (b) - GMO. The scheme of external and bulk screening of depolarization field during the wall motion from the stable position: (c) - initial state, (d) - during switching.

«Broad domain boundary». Recently it was shown, that switching in steps of **PGO**, under the train of short pulses of strong field, was accompanied by formation of triangle domains.[62,78] We had found that applying of superstrong field short pulses (duration is less than delay time) to the sample with multidomain structure led additionally to the arising of the chains of nonthrough domains in the vicinity of parent domain wall.[66] These domains grew in polar direction and coalesced with parent domain. Due to the very slow growth of these domains and arising of new chains of domains near the previous ones the transition region, so-called «broad domain boundary» **(BDB)**, consisting of a great number of nonthrough domains was forming with sideways enlargement (Fig. 22a).[66] Using the train of pulses with additional weak negative bias it is possible to choose switching mode which provides the increasing only of the size of **BDB**.

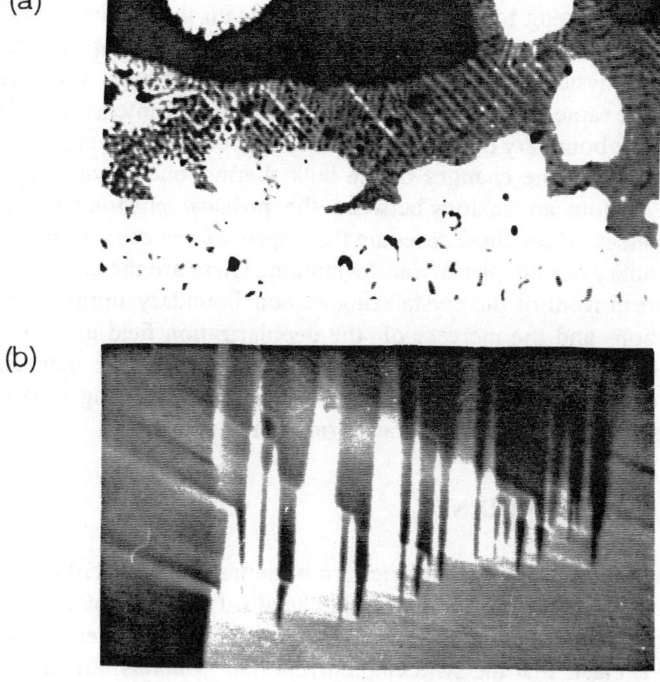

FIGURE 22. Broad domain boundary in PGO: (a) - domain pattern visualizing by optical method. View in the polar direction, (b) - nonthrough domains in broad domain boundary visualizing by cleavage method. Cleavage surface is parallel to the polar direction.[82,83]

The formation time of **BDB** is about 15-20 min. Its width reached hundreds microns. The density of nonthrough domains in this region is about 10^6 cm^{-2}. It must be noted that nonthrough domains are arranged in regular structure with the symmetry corresponding to the crystal one. By means of optical method and by cutting along the planes closed to the cleavage planes it was confirmed that all nonthrough domains in **BDB** are formed on the anode and grew to the cathode (Fig. 22).

Fractal-type domain structure. During switching from single domain state by the train of short pulses of superstrong field the wide region of nonthrough domains arise and organize very complicated domain structure of fractal type.[66,71] The spatial distribution density of domains had been carried out in terms of fractal theory.[93] It was shown that dependence **N(R)**, where **N** - number of domains, **R** - distance from cluster center, could be described by power law relation with index $\mathbf{D_F}$, the so-called fractal dimensionality. Its value lies in interval from 1,59 to 1,74.

It is interesting to compare this pattern with the results of explosive crystallization of amorphous films.[94,95] Their similarity is not of accidental nature. The physical processes governing these fractal-type structures are practically the same. During explosive crystallization the intensive nucleation near the phase boundary occurs due to the special effects of heat «screening» and as a result to the changes of the bulk thermoconductivity. But more important to note an analogy between the physical phenomena defining these processes. Both these cases are the display of remote action effect of phase boundary during phase transformation. There are the increase of the temperature in front of the crystallizing region boundary during explosive crystallization and the increase of the depolarization field in front of the domain boundary during switching in superstrong field. It demonstrates ones more the usefulness of proposed approach considering polarization reversal as an example of phase transformation phenomena.

7. TRANSIENT CURRENT

Up to now the information extracted from the transient current data in thin films is limited because the traditional mathematical treatment of experimental data commonly yield only field and temperature dependencies of switching time.[10] It is clear, that the switching current data contains the information about domain structure kinetics: arising, growth and coalescence of domains.[59,60] It is reasonable to propose that with adequate theoretical description the kinetic parameters of the main stages of domain evolution can be extract even from these integral measurements.

7.1. Problems of the Classical Approach

The theoretical description of the switching process[96,97] is usually based on Kolmogorov-Avrami (K-A) statistical theory[98,99] developed initially for the description of metal crystallization in infinite media (e.g. in the approximation that the average distance between the centers of crystallization is much less than the crystallizing media sizes). The most consistent application of K-A theory for the description of switching kinetics and transient current shape in ferroelectrics was given by Ishibashi and Takagi.[6]

Recently there were made attempts to modify K-A approach for the description of the behavior of real finite objects.[100,101] It was pointed out that the effects due to the finite sizes are very important in polycrystal films and ceramics. In such objects the reconstruction of domain structure goes simultaneously and independently in individual grains of extremely small sizes. The averaging over the great number of independent experimental realizations favours the application of statistical description in this case.

Two limiting situations are considered usually by this statistical model. Following classification of Kolmogorov[98] they are: firstly, α-model - in which the number of domains is progressively increasing (the nucleation probability per volume α usually remains constant during switching) and, secondly, β- model - when all domains involved in switching are arising instantaneously at the very beginning of the process with the density β per volume. In both cases domain wall velocity **v** is taken to be constant. Within these approximations the main formula describing the variation of the fraction of the area occupied by unswitched domains runs as follows:

$$q(t) = \exp[-c\alpha\, v^n\, t^{n+1}/(n+1)] \qquad \text{for } \alpha\text{-model} \qquad (3)$$

$$q(t) = \exp(-c\, b\, v^n\, t^n) \qquad \text{for } \mathbf{b}\text{-model} \qquad (4)$$

where **n** - the dimensionality of the kinetic process, depending upon the type of domain structure (**n** = 1 or **1-D** process for lamellar domains, **n** = 2 or **2-D** process for isolated cylindrical domains and **n** = 3 or **3-D** process for bulk ones), **c** - the shape constant.[98]

While describing the experimental transient current pulses in ferroelectric films by these expressions the fractal values of the kinetic process dimensionalities **n** are usually obtained.[5,102] Moreover it appears impossible to describe the decaying part of the current pulse by these formulas. It is apparent because the proposed theory did not account for the fact that at the instant the growing domain touches the sample boundaries there occurs the geometrical catastrophe. The domain ceases to grow in one of the directions - in other words the abrupt change of the shape constant occurs. This process drastically changes the shape of transient current. Such effect is the most pronounced while observing the growth of wedge-like or lens-like domains (at the high anisotropy of domain wall velocity) which is rather typical for thin ferroelectric films.[103,104]

7.2. Switching in finite media

Let us consider the polarization reversal in thin square plate (with area $\mathbf{S = l^2}$) oriented perpendicular to polar axis. We can consider this kinetic process as two-dimensional (**2-D**) because the domain growth velocity in polar direction is always considerably greater then in others. Let us take into account that during any experiment there exists limited time interval between consecutive measurements $\mathbf{\Delta t}$. Thus for any value of the anisotropy of the velocity of sideways domain growth $\mathbf{A = v_x / v_y}$ we can consider the process as the isotropic movement of the domain walls occurring on the discrete matrix $\mathbf{L_x \times L_y}$, where $\mathbf{L_x = l/(v_x \Delta t)}$ and $\mathbf{L_y = l/(v_y \Delta t)}$.

Let us consider at first the β-model with the appropriate number of the simultaneously arising domains $\mathbf{N_0 = \beta_0 l^2}$, which are growing in the shape of rhombus with high anisotropy $\mathbf{A>>1}$ ($\mathbf{L_x << L_y}$). One can choose the **1-D** line with the length $\mathbf{L_y}$ (i-th row of the matrix) situated at the distance $\mathbf{\Delta i = |L - i|}$ from the middle of the matrix (Fig. 23).

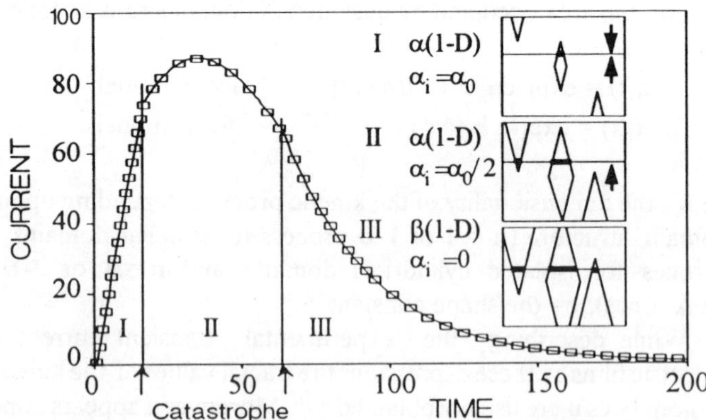

FIGURE 23. Computer simulation of switching current for definite line of 2-D matrix for β-model. Squares - results of computer experiments, Solid curve - approximation by Eq. (3) and (5). In insert - scheme of domain kinetics at the different stages of switching process.

The switching process in such a line satisfies all the requirements needed for the application of K-A theory. As it is seen in Fig.23 the number of the domains in the considered line is increasing when the domains from another regions reach it. So the switching process in the line occurs according to α-model.

The limitations on the matrix sizes in **x** direction lead to the abrupt changes of the intensity of domains arising in the line per time during switching. At the beginning $\alpha_i(t) = \alpha_0 = N_0/(2AL^3) = \beta/L$, where $2L = L_x$, but then it undergoes two catastrophic abrupt changes:

$$\alpha_i(t) = \begin{cases} \alpha_0, & \text{for} \quad 0 < t < L - \Delta i \\ \alpha_0/2, & \text{for} \quad L - \Delta i < t < L + \Delta i \\ 0, & \text{for} \quad L + \Delta i < t \end{cases} \qquad (5)$$

The initial value of α undergoes twofold decrease as a result of the first catastrophe and after the second - even the type of the model is changed. In other words the **α(1-D)→β(1-D)** geometrical phase transition occurs.

In order to verify the validity of proposed description the computer simulation of the switching in **2-D** matrix for β-model was carried out with the random distribution of growing rhombic domains. For every step the variation in the fraction of the area occupied by arising domains **ΔS/Δt** (the analog of transient current) was averaged over great number of independent realizations (computer experiments). It is seen in Fig. 23 that the switching in definite line (**1-D** part of the sample) is approximated extremely good by Eq. (3) with $n = 2$ and two catastrophic changes of parameters in accordance with Eq. (5).

For the time dependence of the fraction of the area occupied by unswitched domains in the whole matrix we have:

$$q(t) = 2\prod_{i=1}^{L} q_i(t) = \prod_{i=1}^{L} \exp[-c\alpha_i(t)\, v^n t^n] = 2\exp[-\sum_{i=1}^{L} c\,\alpha_i(t)\, v^n t^n] \qquad (6)$$

It is seen that **β(2-D)** kinetic process is the sum of local **α(1-D)** processes. As a result we obtain the formula similar to Eq. (4) in which the density of the domains per square **β(t)** effectively diminishes during switching (in other words the gradually decreasing of shape constant obtain):

$$\beta(t) = 2\sum_{i=1}^{L} \alpha_i / L = \beta_0 (1 - t/t_c) \qquad (7)$$

where $t_c = 2L$.

It is seen that during switching in restricted area with high anisotropy of sideways motion of domain walls there are two important peculiarities: at first, the number of the domains involved in the β(**2-D**) process effectively diminishes and, at second, the dimensionality of the kinetic process catastrophically changes at the appropriate moment $\mathbf{t_c}$. It means that β(**2-D**)$\rightarrow$ β(**1-D**) transition occurs. The contribution of these peculiarities diminishes with the increasing of effective sizes and in the limit (**L** $\rightarrow\infty$) the well-known K-A formula for β-model is obtained.

It is possible to show that the expression analogous to Eq. (7) can be used for the description of the transient current shape for any model situations: arbitrary value of anisotropy (account for the limitations in both directions) and α-model as well.

For α-model:

$$i(t) = (2\,P_s\,l^2\,n\,t^n/t_{0\alpha}{}^{n+1})\{1 - [1+1/(n+1)]\,t/t_m\}\exp[-(t/t_{0\alpha})^{n+1}(1 - t/t_m)] \quad (8)$$

for β-model:

$$i(t) = (2\,P_s\,l^2\,n\,t^{n-1}/t_{0\beta}{}^n)[1 - (1 + 1/n)\,t/t_m]\exp[-(t/t_{0\beta})^n\,(1 - t/t_m)], \quad (9)$$

where $t_{0\alpha} = [c\alpha v^{n+1}/(n+1)]^{-1/(n+1)}$, $t_{0\beta} = (c\beta v^n)^{-1/n}$, P_s - spontaneous polarization. In this case the decay time $\mathbf{t_m}$ (the characteristic time of the process of the effective decreasing of domain number) is the function of anisotropy, shape constant and nucleation intensity. In the above considered case for β-model with high anisotropy $\mathbf{t_m}$ is equal to $\mathbf{t_c}$.

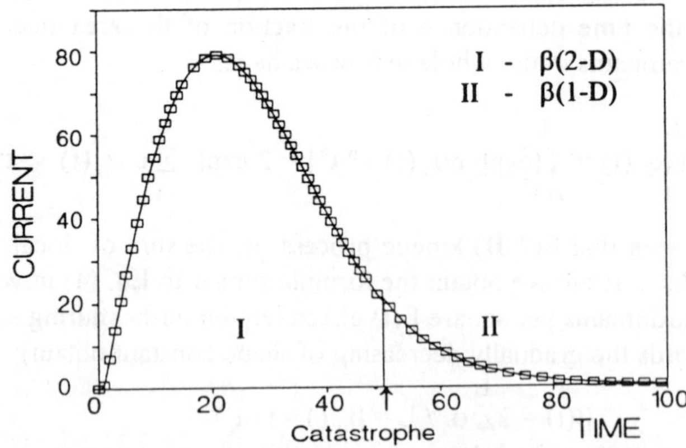

FIGURE 24 Computer simulation for β-model with anisotropic domain sideways growth. The points - results of computer experiment, the solid curve - approximation by Eq. (9).

7.3. Computer Experiments

The computer simulations of the switching process was carried out on the **2-D** matrix of 500×500 unit cells. The averaging of current pulses over great number of realizations (more than 1000) was held for obtaining statistically reliable results corresponding to the experimental conditions in polycrystal thin films. The computer experiments were done for α- and β-models in the wide range of the model parameters. The various shapes of growing domains with different orientation relative to the «sample» boundaries were examined as well.

The results of computer experiments for α- and β-models are approximated extremely good by the Eq. (8) and (9) with integer values of kinetic process dimensionalities. It is important that these formulas succeeded in describing the decaying part of the pulses which is impossible within the framework of the classical approach. The typical results for β-model with isotropic domain growth are shown in the Fig. 24. The existence of the geometrical catastrophes is clearly seen in both cases.

7.4. Experimental results

The investigations of the switching were done in thin plates of model single crystal lead germanate **PGO** (with optically distinguished domains) for the experimental testing of the method. Two different types of initial domain structure were prepared: the single domain state (Fig. 25a) and the controlled maze structure with high density of domain walls (Fig. 25b).[62] The main stages of domains' evolution obtained as a result of mathematical treatment of transient current data are in qualitative agreement with results of direct observations of domain kinetics.

The mathematical treatment allows us to show the qualitative and quantitative difference in the domains kinetics in these two cases. Switching from single domain state demonstrate the **β(2-D)→β(1-D)** geometrical transition in the decaying part of the current pulse. It is necessary to mention that the first catastrophe in Fig.25a is due to the relatively large dimension of the sample in polar direction (**d** = 130 μm) and never can be realized in thin films. During switching from the maze domain structure we reveal another type of geometrical transition **β(1-D)→α(2-D).** This fact enables us to state that the change from the maze to isolated cylindrical domains occurs. All predicted transformations of the types of domain structures are confirmed by direct real-time visualization of the domain kinetics.

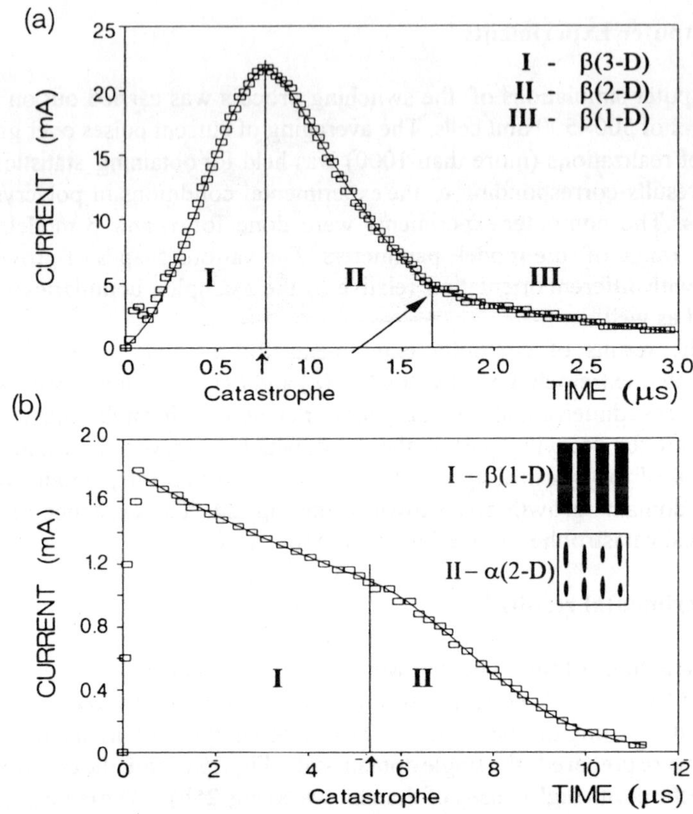

FIGURE 25 The transient current for PGO single crystal during switching from: (a) - single domain state, (b) - the meza domain structure, approximated by Eq. (8) and (9). In insert - scheme of the domain structure at the different stages of switching.

The width of maze elements in our experiments was about few microns and it was difficult to observe the momentary domain patterns by optical methods. The proposed method was very effective even in this case for extracting the information about the «geometry» (type) of the domain structure and its changing during switching. It must be stressed that in contrast to any method of visualization of momentary domain patterns the proposed method has practically no resolution limits.

The method has been applied with much success for the studying of thin polycrystal films of PLT and PZT (Fig. 26). The **β(2-D)→β(1-D)** and **α(2-D)→α(1-D)** transitions during switching are rather typical in thin films.

This kind of transitions was discussed above in details and confirmed by computer simulation. The considered transformations can be interpreted as the result of anisotropic growth of residual domains in finite media. The predicted domain evolution is in accordance with the results of the TEM observation of static domain patterns in thin films.[103,104]

Within proposed mathematical treatment we can obtain such important parameters of domain structure evolution as: kinetic process dimensionality **n**, anisotropy of domain growth **A** and characteristic times t_0, t_m and t_c. The knowledge of such parameters in turn makes it possible to describe the reconstruction of the domain structure in terms of the number of arising domains per volume (α and β) and their sideways velocities (v_x and v_y). The obtain parameters are depended upon the voltage and temperature. They are changing considerably during aging and degradation.

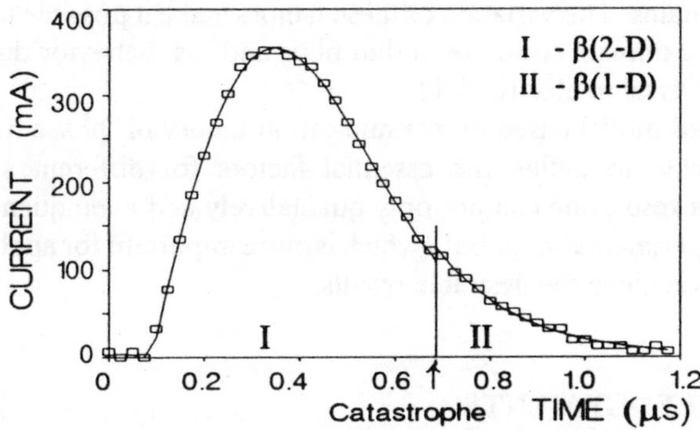

FIGURE 26a The transient current for PLT thin film produced by magnetron sputtering approximated by Eq. (9)

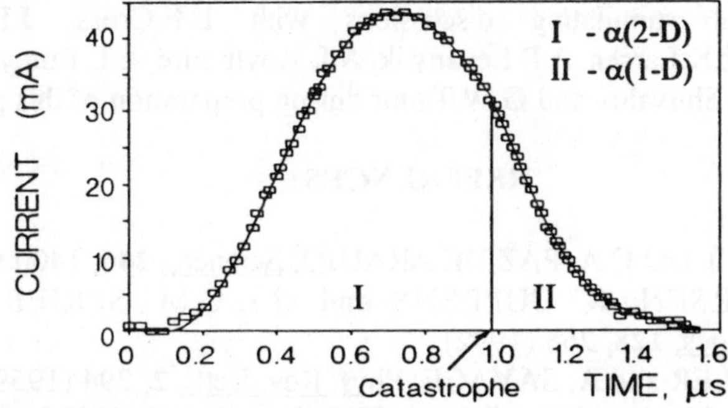

FIGURE 26b The transient current for PZT thin film produced by laser ablation approximated by Eq. (8).

8. CONCLUSION

In this paper we considered the nonequilibrium process of domain structure evolution. For the description of numerous experimental results in model ferroelectrics with controlled parameters we proposed the simple and clear physical model developed for the description of phase transition. The specific of ferroelectrics lies in the fact that nucleation rate is drived by the local electric fields. These fields are of retardation nature because the charged subsystems can not be rearranged immediately. As a result domain structure kinetics in ferroelectrics depends upon the relation between the time of field action and delay time; time constant of external screening and bulk one; sample thickness and screening length. These parameters depend also on many factors: value of spontaneous polarization, bulk and surface layer conductivity, dielectric permittivity and time constant of external circuit, so as concentration of remnant domains. The variation of these factors make it possible to control the ferroelectric domain structure in thin films and its behavior during fast polarization reversal in electric field.

Proposed model based on the nucleation theory of phase transitions makes it possible to define the essential factors for different switching situations. As a result one can not only qualitatively and even quantitatively interpretate experimental date but (which is more important for applications) predict how to achieve the desirable results.

9. ACKNOWLEDGEMENTS

This work was supported in part by Russian Foundation of Fundamental Research under Grant No.93-02-2451. It is a pleasure to acknowledge the many helpful stimulating discussions with L.E.Cross, J.Fousek, V.M.Fridkin, P.K.Larsen, A.P.Levanyuk, A.L.Roytbourd, E.L.Rumyantsev, J.F.Scott, L.A.Shuvalov and G.W.Tailor during preparation of this paper.

REFERENCES

1. J.F. SCOTT and C.A. PAZ DE ARAUJO, Science, **246**, 1400 (1989); P.K. LARSEN, R. CUPPENS and G.A.C.M. SPRIERINGS, Ferroelectrics, **128**, 265 (1992).
2. R.C. MILLER and A. SAVAGE, Phys. Rev. Lett., **2**, 294 (1959).
3. V.V. SHAKMANOV, S.I. YAKUNIN, G.V. SPIVAK, N.V. VASILYEVA and V.Sh.MELAMED, Krystallographiya, **17**, 351 (1972).

4. W.J. MERZ, Phys. Rev, **95**, 690 (1954).
5. K. DIMMLER, M. PARRIS, D. BUTLER, S. EATON, B. POULIGNY, J.F. SCOTT and Y. ISHIBASHI, J.Appl. Phys., **61**, 5467 (1987).
6. Y. ISHIBASHI and Y. TAKAGI, J. Phys. Soc. Jap., **31**, 506 (1971).
7. V.P.SKRIPOV and V.P.KOVERDA, Spontaneous Crystallization of Overcool Liquids (Nauka, Moscow, 1984).
8. Ya.B. ZELDOVITCH, ZHETF, **12**, 525 (1942).
9. V.M.FRIDKIN, Photoferroelectrics (Springer-Verlag, Berlin, Heidelberg, New York, 1979).
10. E.FATUZZO and W.J.MERZ, Ferroelectricity (North-Holland Publishing Company, Amsterdam, 1967).
11. M.E.LINES and A.M.GLASS, Principles and Application of Ferroelectrics and Related Materials (Clarendon Press, Oxford, 1977).
12. J.C.BURFOOT and G.W.TAYLOR, Polar Dielectrics and their Applications (MacMillan Press LTD, 1979).
13. F.IONA and G.SHIRANE, Ferroelectric Crystals (Pergamon Press, Oxford-London, 1962).
14. A.G.CHYNOWETH, J.Appl.Phys., **27**, 78 (1956).
15. F.AIKAWA, J.HATANO, M.DATE, T.FURUKAWA and H.FUTAMA, J.Phys.Soc.Jap., **58**, 1063 (1989).
16. J.C. BURFOOT and R.V. LATHAM, Brit.J.Appl.Phys., **14**, 933(1963).
17. A. HADNI and R. THOMAS, Phys.Stat.Sol., **31A**, 71 (1975).
18. Ch.D. TRAN, X. GERBAUX and A. HADNI, Ferroelectrics, **33**, 31 (1981).
19. R.V. LATHAM, Brit. J. Appl. Phys., **18**, 1383 (1967).
20. R.LE BIHAN, Ferroelectrics, **97**, 19 (1989).
21. R.LE BIHAN, J.L.CHARTIER and L.JEAN, Ferroelectrics, **13**, 475 (1976).
22. A.RODER and C.SCHEIDING, Crys. Res. Technol., **20**, 1489 (1985).
23. A.A.SOGR and V.Z.BORODIN, Izv.AN SSSR, Ser.Fiz., **41**, 1498 (1977).
24. Y. FURUHATA and K. TORIYAMA, Phys.Lett., **23**, 361 (1973).
25. V.P. KONSTANTINOVA, N.A. TIKHOMIROVA and M. GLOGAROVA, Ferroelectrics, **20**, 259 (1978).
26. N.A. TIKHOMIROVA, L.A. SHUVALOV, A.I. BARANOV, A.R. KARASEV, L.I. DONTSOVA, E.S. POPOV, A.V. SHILNIKOV and L.G. BULATOVA, Ferroelectrics, **29**, 51 (1980).
27. L.I. DONTSOVA, N.A. TIKHOMIROVA and L.A. SHUVALOV, Ferroelectrics, **97**, 87 (1989).
28. A.P. MALOZEMOFF and J.S. SLONCZEWSKI, Magnetic Domain Walls in Bubble Materials (Academic Press, New York, 1979).
29. T.H. O'DELL, Ferromagnetodynamics (MacMillan Press LTD, 1981).

30. H. BLATTNER, W. KANZIG, W. MERZ and H. SUTTER, Helv. Phys. Acta., **21**, 207 (1948).
31. B. MATTHIAS and A. HIPPEL, Phys.Rev., **73**, 1378 (1948).
32. E.A. LITTLE, Phys. Rev, **98**, 978 (1955).
33. W.J. MERZ, J.Appl.Phys., **27**, 8, 938 (1956).
34. J.A. HOOTON and W.J. MERZ, Phys.Rev., **98**, 409 (1955).
35. A. SAVAGE and R.C. MILLER, J.Appl. Phys., **31**, 1546 (1960).
36. J. KOBAYASHI, N. YAMADA and T. NAKAMURA, Phys. Rev. Lett., **11**, 410 (1963).
37. R. ABE, J.Phys.Soc.Jap., **13**, 244 (1958).
38. T. MITSUI and J. FURUICHI, Phys. Rev., **95**, 558 (1954).
39. S.E. CUMMINS, J. Appl. Phys., **36**, 1958 (1965).
40. S.E. CUMMINS, Ferroelectrics , **1**, 11 (1970).
41. T. NAKAMURA, T. KONDO and A. KUMADA, Phys. Lett., **36A**,141 (1971).
42. J.P. DOUGHERTY, E. SAWAGUCHI and L.E. CROSS, Appl. Phys. Lett., **20**, 364 (1972).
43. H. IWASAKI, K. SUGII, N. NIIZEKI and H. TOYODA, Ferroelectrics, **3**, 157 (1972).
44. R.C. MILLER and A. SAVAGE, J.Appl. Phys., **32**, 714 (1961).
45. D.P. CAMERON, IBM J. Res. Developm., **1**, 2 (1957).
46. E.I. SPERANSKAYA, Zh. Neorg. Khim., **4**, 421 (1960).
47. H. IWASAKI, K. SUGII, T. YAMADA and N. NIIZEKI, Appl. Phys. Lett., **18**, 444 (1971).
48. S. NANAMATSU, H. SUJIAMA, K. DOI and Y. KONDO, J. Phys. Soc. Jap, **31**, 616 (1971).
49. R.E. NEWNHAM and L.E. CROSS, Endeavour, **33**, 18 (1974).
50. A.P.LEVANYUK and D.G.SANNIKOV, Uspehi Fiz.Nauk, **112**,561 (1974).
51. K. AIZU, A. KUMADA, H. YUMOTO and S. ASHIDA, J. Phys. Soc. Jap., **27**, 511 (1969).
52. J. SAPRIEL, Phys. Rev. B., **12**, 5128 (1975).
53. R.B. FLIPPEN and C.W. HAAS, Solid State Communs., **13**, 1207 (1973).
54. I.M.SHEPHERD and J.R.BARKLEY, Solid State Communs., **10**, 123 (1972).
55. A.N. ALEKSEEV, A.L. PROKLOV and N.A. TIKHOMIROVA, Krystallografiya, **32**, 1189 (1987).
56. A.N.ALEKSEEV and A.L.PROKLOV, Izv.AN SSSR, Ser.Fiz., **50**, 407 (1986).
57. V.Ya. SHUR, A.L. GRUVERMAN, V.P. KUMINOV and N.A. TONKACHYOVA, Ferroelectrics, **130**, 341 (1992).
58. V.Ya. SHUR, V.V. LETUCHEV, E.L. RUMYANTSEV and T.B. CHARIKOVA, Fiz. Tverd. Tela., **28**, 2829 (1986).

59. V.Ya. SHUR, A.L. GRUVERMAN, N.Yu. PONOMAREV, E.L. RUMYANTSEV and N.A.TONKACHYOVA, JETP Lett, **53**,615(1991).
60. V.Ya. SHUR, A.L. GRUVERMAN, N.Yu. PONOMAREV, E.L. RUMYANTSEV and N.A.TONKACHYOVA, Integrated Ferroelectrics, **2**, 51 (1992).
61. V.Ya. SHUR, A.L. GRUVERMAN and E.L. RUMYANTSEV, Ferroelectrics, **111**, 123 (1990).
62. V.Ya. SHUR, A.L. GRUVERMAN, V.V. LETUCHEV, E.L. RUMYANTSEV and A.L. SUBBOTIN, Ferroelectrics, **98**, 29 (1989).
63. V.Ya.SHUR, V.V. LETUCHEV and E.L. RUMYANTSEV, Fiz. Tverd. Tela, **26**, 2510 (1984).
64. A. KUMADA, Phys. Lett., **30A**, 186 (1969).
65. V.Ya. SHUR, V.V. LETUCHEV, E.L. RUMYANTSEV and T.B. CHARIKOVA, Zh. Techn. Fiz., **55**, 1666 (1985).
66. V.Ya. SHUR, A.L. GRUVERMAN, N.Yu. PONOMAREV and N.A. TONKACHYOVA, Ferroelectrics, **126**, 371 (1992).
67. W.K. BURTON, N. CABRERA and F.C. FRANK, Phil. Trans. Roy. Soc., **243**, 299 (1951).
68. A.A.CHERNOV, in Modern Crystallography (Nauka,Moscow,1980), **3**, 7-232.
69. J.W. CAHN, Acta Metallurgica, **8**, 554 (1960).
70. V.Ya.SHUR and E.L.RUMYANTSEV, Ferroelectrics, 142, 1-7 (1993).
71. V.Ya.SHUR and E.L.RUMYANTSEV, Ferroelectrics (in press).
72. F.C. FRANK, Disc. Faraday Soc., **5**, 48 (1949).
73. T. NAKAMURA, J. Phys. Soc. Jap., **9**, 425 (1954), **15**, 1379 (1960).
74. R.C. MILLER and G. WEINREICH, Phys. Rev., **117**, 1460 (1960).
75. V. JANOVEC, Czechosl. J. Phys., **9**, 468 (1959).
76. V.Ya. SHUR,V.V. LETUCHEV,Yu.A.POPOV and V.I.SARAPULOV, Krystallografiya, **30**, 945 (1985).
77. V.Ya. SHUR, V.V. LETUCHEV, E.L. RUMYANTSEV and I.V. OVECHKINA, Fiz. Tverd. Tela, **27**, 1232 (1985).
78. V.Ya. SHUR, V.V. LETUCHEV, E.L. RUMYANTSEV and I.V. OVECHKINA, Fiz. Tverd. Tela, **27**, 1585 (1985).
79. M. HAYASHI, J.Phys.Soc.Jap., **33**, 616 (1972).
80. V.Ya. SHUR, Yu.A. POPOV, E.L. RUMYANTSEV, A.L. SUBBOTIN and V.V. VSHIVKOVA, Izv. AN SSSR, Ser. Fiz., **49**,1860 (1985).
81. V.Ya. SHUR, A.V. GURIEV, L.V. BUNINA, A.L. SUBBOTIN and Yu. A. POPOV, Fiz. Tverd. Tela, **30**, 3143 (1988).
82. V.Ya. SHUR, A.L. SUBBOTIN and V.P. KUMINOV, Ferroelectrics, **140**, 101 (1993).

83. V.Ya. SHUR, E.L. RUMYANTSEV and A.L. SUBBOTIN, Ferroelectrics, **140**, 305 (1993).
84. G.I. ROSENMAN, V.A. OKHAPKIN, Yu.L. CHEPELEV and V.Ya. SHUR, JETP Lett., **39**, 477 (1984).
85. G.I. ROSENMAN, V.V. LETUCHEV, Yu.L. CHEPELEV, O.V. MALYSHKINA, V.Ya. SHUR and V.P. KUMINOV, Appl. Phys. Lett., **56**, 689 (1990).
86. M.E.DROUGARD and R.LANDAUER, J.Appl. Phys., **30**, 1663 (1959).
87. P.V. LAMBECK and G.H. JONKER, Ferroelectrics, **22**, 729 (1978).
88. V.Ya. SHUR, Yu.A. POPOV and N.V. KOROVINA, Fiz. Tverd. Tela, **26**, 781 (1984).
89. V.Ya. SHUR, V.V. LETUCHEV and I.V. OVECHKINA, Fiz. Tverd. Tela, **26**, 3474 (1984).
90. A.P. LEVANYUK and A.S. SIGOV, Defects and Structural Phase Transitions (Gordon & Breach Science Publishers, Amsterdam, 1988).
91. V.Ya. SHUR, A.L. GRUVERMAN, E.V. KOROVINA, M.Z. ORLOVA and L.V. SHERSTOBITOVA, Fiz. Tverd. Tela., **30**, 299 (1988).
92. V.Ya. SHUR, A.L. GRUVERMAN, V.P. KUMINOV and N.A. TONKACHYOVA, Ferroelectrics, **111**, 197 (1990).
93. R. JULLIEN, Comm. Cond. Mat. Phys., **13**, 177 (1987).
94. H.J. LEAMY, W.L. BROWN and G.K. CELLER, Appl. Phys. Lett., **38**, 137 (1981).
95. N.M. BOGDANOV, V.P. KOVERDA, V.N. SKOKOV, V.P. SKRIPOV and A.A. DICK, Krystallografiya, **33**, 1251 (1988).
96. E. FATUZZO, Phys.Rev., **127**, 1999 (1962).
97. H.H. WIEDER, J.Appl.Phys., **31**, 180 (1960).
98. A.N.KOLMOGOROV, Izv.Acad.Nauk USSR, Ser. Math., **3**, 355 (1937).
99. M. AVRAMI, J.Chem.Phys., **7**, 1103 (1939), **8**, 212 (1940), **9**, 177 (1941).
100. H.M. DUIKER and P.D. BEALE, Phys. Rev. B, **41**, 490 (1990).
101. Y. ISHIBASHI and H. ORIHARA, J.Phys.Soc.Jap., 61, 1919 (1992), 61, 4650 (1992).
102. J.F. SCOTT, L. KAMMERDINER, M. PARRIS, S. TRAYNOR, V. OTTENBACHER, A. SHAWABKEH and W.F. OLIVER, J.Appl. Phys., **64**, 787 (1988).
103. M. HUFFMAN, J. ZHU and M.M. AL-JASSIM, Ferroelectrics, **140**, 191 (1993).
104. M. KLEE, U. MACKENS and A.DE VEIRMAN, Ferroelectrics, **140**, 211 (1993).

CHAPTER 7

ELECTRODE MATERIALS FOR FERROELECTRIC THIN FILM CAPACITORS AND THEIR EFFECT ON THE ELECTRICAL PROPERTIES

Husam N. Al-Shareef and Angus I. Kingon
Department of Materials Science & Engineering, N.C. State University, Raleigh, North Carolina 27695-7919

1. INTRODUCTION

It has become clear over the past few years during the development of thin film ferroelectric devices that the electrodes play a major role in determining the device properties and performance. In this chapter we discuss the selection and processing of electrode materials, as well as the effect of electrodes on the properties of the ferroelectric film capacitors. In some cases the effect of the electrodes is via modification of the microstructures of the films grown on the bottom electrodes, and in some cases the electrodes directly control the properties. Most of the discussion will be of the lead zirconate titanate (PZT) system because it has been the most extensively studied system among the perovskite family for memory applications.

There are a number of criteria which affect the choice of the electrodes. We present a partial list here to emphasize the complex role of the electrodes:

1) The electrode must have sufficiently low resistance
2) It must be chemically compatible with the semiconductor, and not react with the ferroelectric.

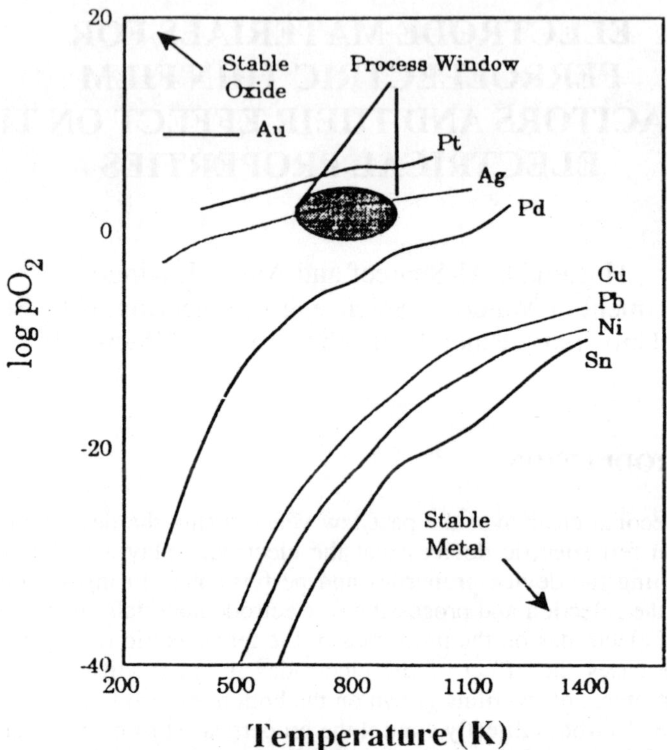

FIGURE 1. Schematic showing boundary of stability of a range of metals with respect to their oxides (Reproduced from Ref. 1).

3) It must display adequate adhesion to both ferroelectric and underlying structure.
4) It must be morphologically stable under the process conditions.
5) It may need to act as a diffusion barrier for oxygen or other species.
6) It must provide suitable interfacial electronic properties,
7) It may be necessary to utilize the electrode to control the film microstructure or orientation.

Most of the ferroelectric materials which have been studied and are under consideration for thin film devices are oxides. This immediately limits the choice of electrodes, particularly the bottom electrode, as the oxides must be deposited or processed under oxidizing conditions. This is obvious from a consideration of Fig.1, which shows the conditions of temperature and oxygen partial pressure for the stability of a range of candidate metals relative to their oxides.[1] Thus the choice is limited to a few noble metals. In addition, silver and gold react with perovskites during processing, having solid solubility in the perovskite lattice. This effectively eliminates them for use as bottom electrodes, and even as top electrodes if there is any "back-end" post processing which involves temperatures above about 550°C. Therefore, platinum has remained as the only viable **metal** for use as bottom electrode for ferroelectric thin film devices. It is these issues which has recently led to a substantial amount of research on conducting oxide electrodes as an alternative. In this chapter we review results for both platinum and oxide electrodes. A discussion covering a slightly broader range of electrode combinations may be found in an earlier publication.[2] There is a further issue regarding electrode selection which must be mentioned. This relates to the device density and size in the case of ferroelectric thin films which are integrated with semiconductor devices (i.e. for ferroelectric non-volatile memories and DRAMs). A typical low density FRAM[3] is shown schematically in Fig. 2. It can be seen that the ferroelectric capacitor is located to the side of the semiconducting FET, and the bottom electrode is deposited on top of an insulating oxide layer, typically SiO_2. In contrast, a typical high density (> 4Mb) DRAM is shown schematically in Fig. 3. In this case the bottom electrode needs to make and maintain contact directly down to the underlying silicon device. In the case shown in Fig. 3 this is via a polysilicon or tungsten plug. This turns out to be a difficult problem as it is necessary to prevent silicide formation, and to prevent the formation of interfacial oxides via oxygen diffusion through the Pt.[4] Solutions to this problem therefore require a conducting barrier below the Pt bottom electrode. At the time of writing, this issue has not been fully resolved, and the choice of barrier provides an upper temperature or thermal budget limit to the processing steps.[4] The subject is not discussed further in this chapter, but rather we deal with the simpler and more extensively studied case of the bottom electrode being deposited on underlying structures which may be insulating.

The rest of the chapter is divided into the following sections: section 2 in which we discuss electrode processing, and we include the effects on the ferroelectric film nucleation and growth; and section 3 in which we discuss the effects of electrodes on device properties.

2. PROCESSING AND RELIABILITY OF ELECTRODE MATERIALS

In this section the electrode materials which have been studied for use with $Pb(Zr_xTi_{1-x})O_3$ (PZT) ferroelectric thin films are grouped into three

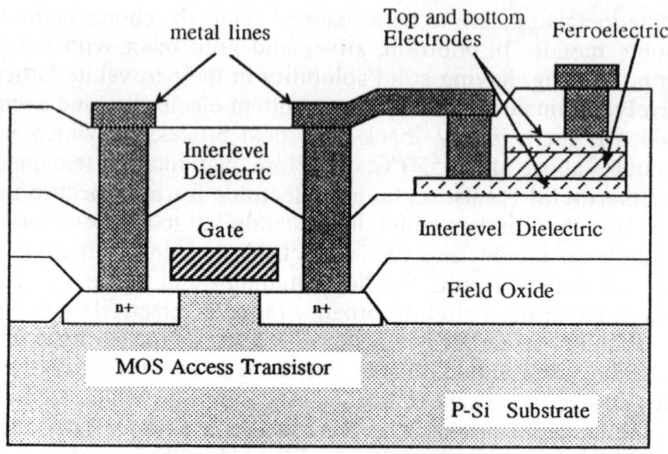

FIGURE 2. Schematic of a low density ferroelectric non-volatile memory (FRAM) showing the ferroelectric capacitor situated above an insulator. (Reproduced from Ref. 3).

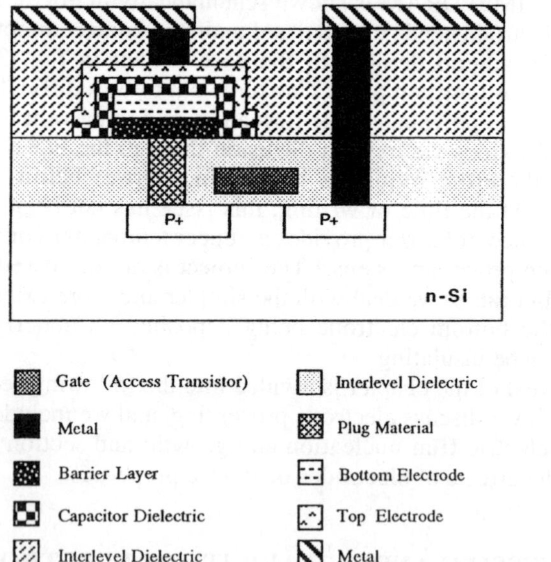

FIGURE 3. Schematic of a high density DRAM showing electrical contact between bottom electrode and a plug, via a barrier layer.

categories: metallic, simple oxide, and complex oxide electrodes. In each category, several issues will be discussed. These include deposition methods, stability of the electrode material, its effect on perovskite PZT nucleation, orientation, and microstructure, and its interaction with the PZT thin film.

2.1. Metal Electrodes

The most commonly used metal electrode has been platinum, although other metals such as palladium and gold have also been used. Methods used for platinum film deposition have predominantly been sputtering and e-beam evaporation. Table 1 gives a listing of the deposition methods and conditions reported by researchers using Pt as an electrode material for ferroelectric thin films. Since the adhesion of Pt to SiO_2 is poor, an adhesion layer has typically been used. The thicknesses and the types of adhesion layers are also included in Table 1.

In the references listed in Table 1 the films were often post-deposition annealed at various temperatures and in different annealing atmospheres. The purpose of these experiments was to simulate the ferroelectric thin film deposition. The Pt/Ti electrode system deposited by these various methods generally has two major instability problems. These are hillock formation and Pt/Ti interaction[2,5-8], and they are discussed in turn in the following subsections.

TABLE 1. Techniques and Conditions used for Pt deposition.

Deposition Method	Deposition Temp. (°C)	Thickness of Pt (Å)	Adhesion Layer Thick., Å	Ref.
Ion Beam Sputt.	RT-600	500-4000	Ti (200-600)	2,5
DC Mag. Sputt.	RT	1500	Ta,Ti (500)	6
e-beam Evapor.	25-250	250-2000	Ti (50-2000)	7
Sputtering	RT	170-1000	Ti (0-200)	8
RF Sputtering	RT	1400-2400	Ti (70-1000)	9
DC Sputtering	RT	2000	Ti (1000)	10

2.1.1. Hillock formation

The first instability problem observed in the Pt/Ti system was that of hillock formation. Hillocks are thin elevated features which protrude from the metal surface(Pt in this case) during thermal cycling. Such hillocks can lead to capacitor shorts as reported earlier.[2,5,8] An example of the scale of such hillocks is shown in the TEM micrograph of Fig. 4. Notice that the height of these hillocks is in the same range (even higher)

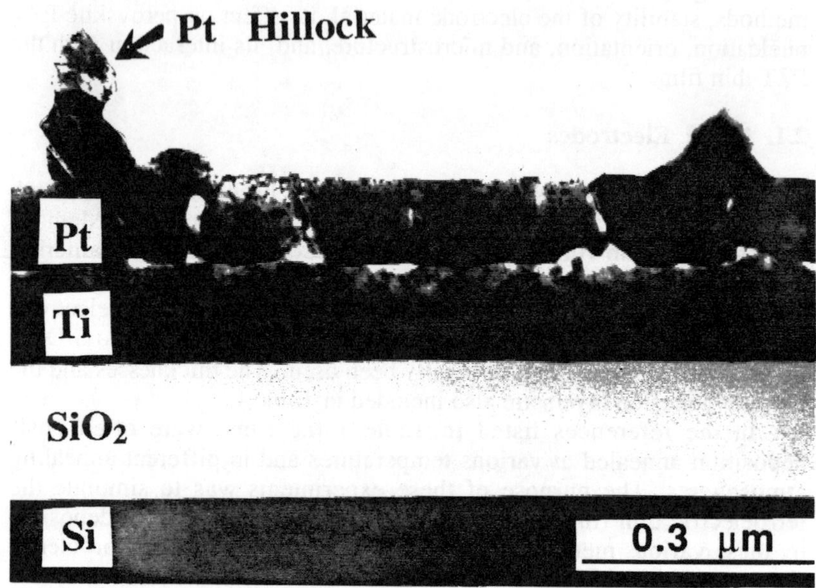

FIGURE 4. TEM micrograph showing a typical Pt hillock in a film grown on Ti/SiO_2/Si. (Reproduced from Ref. 7).

as a typical ferroelectric thin film, which explains why formation of such hillocks is a major concern.

Our understanding of hillock formation in Pt electrodes has markedly improved in the last year. The available data clearly suggest that hillock formation is related to stress relief in the Pt films. The morphological changes which take place in a Pt film during heat treatment are best understood by monitoring stress evolution in the films during heat treatment. Of course this has to be coupled with examination of film surface morphology using techniques such as electron microscopy or atomic force microscopy. A typical stress versus temperature curve is shown in Fig. 5 for a Pt thin film sputter-deposited on SiO_2/Si. A summary of our understanding of hillock formation in Pt films is presented below with the aid of Fig. 5.

1) Notice that compressive stress in the Pt film increases with temperature up to around 400°C (segment A-B). This compressive stress results mainly from the thermal expansion coefficient mismatch

between the Pt film and the underlying Si substrate. The initial film stress at room temperature (stress after deposition) can be either compressive or tensile. This will depend on the deposition technique and conditions. For example, sputtered films tend to have compressive stress while evaporated films tend to have tensile stress.

2) As the temperature is further raised above 400°C, mobility of the Pt atoms increases and rapid stress relief by hillock formation occurs along the B-C segment of the curve. It is in fact well known[11] that compressive

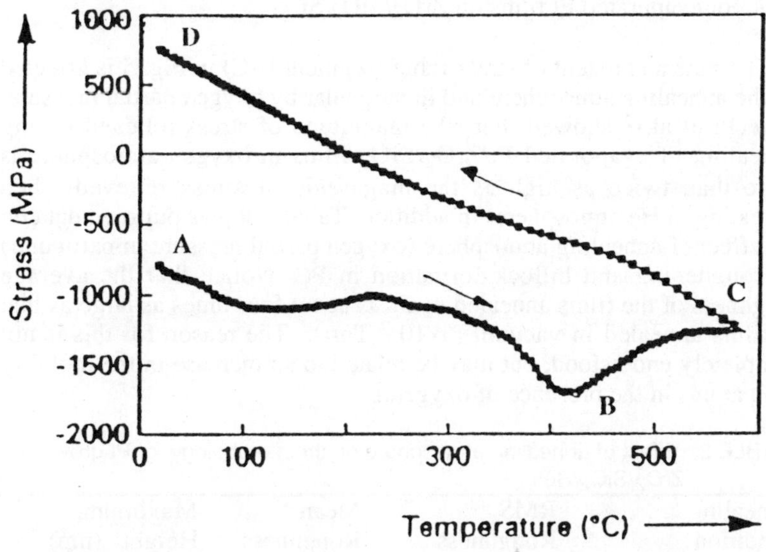

FIGURE 5. Stress as a function of temperature in a Pt film grown on $ZrO_2/SiO_2/Si$ (Reproduced from Ref. 8).

stress relief occurs by hillock formation. The stress at point C on such curves can be tensile or compressive and this may depend on factors such as initial film stress and the degree of thermal expansion mismatch between the Pt film and underlying substrate. For instance, Kotecki et al.[12] have shown that Pt films grown on SiO_2 and $ZrO_2/SiO_2/Si$ under identical condition have tensile and compressive stress, respectively, at point C in their stress versus temperature curves.

3) Once stress relief occurs (along segment B-C), then cooling and heating the Pt film afterwards causes film stress to change in an essentially linear fashion (segment C-D in Fig. 5). This portion of the curve corresponds to stress caused by thermal expansion mismatch. This is important as it explains why some authors reported an absence of hillocks after annealing at higher temperature.

4) Hillocks are metastable. In other words, as the film is heated past point C in Fig. 5 (the point were stress relief and hillock formation is complete), hillocks disappear. This was recently shown by Summerfelt et al.[13] for evaporated Pt films on $ZrO_2/SiO_2/Si$.

5) The rate and extent of stress relief (segment B-C) in Fig. 5 is affected by the annealing atmosphere and in particular by oxygen partial pressure. Kotecki et al.[12] showed that the magnitude of stress relieved during annealing of evaporated $Pt/ZrO_2/SiO_2$ films in oxygen atmosphere is more than twice as high as the magnitude of stress relieved when annealing in He atmosphere. In addition, Table 2 shows our own data on the effect of annealing atmosphere (oxygen partial pressure in particular) on roughening and hillock formation in Pt. Notice that the average roughness of the films annealed in air is about four times as large as that of films annealed in vacuum ($5x10^{-7}$ Torr). The reason for this is not completely understood, but may be related to an increase in the mobility of Pt atoms in the presence of oxygen.

TABLE 2. Effect of annealing atmosphere on the morphology of Pt grown on $ZrO_2/SiO_2/Si$).

Annealing Condition	RMS Roughness (nm)	Mean Roughness (nm)	Maximum Height* (nm)
700°C, 40 min. in air	14.8	11.6	132
700°C, 40 min. in vacuum ($5x10^{-7}$ Torr)	3.87	2.51	95.4

*The maximum height correlates with the largest hillocks.

6) The roughening of Pt films due to grain growth has not been studied. More studies on grain growth in Pt and how it fits into the overall issue of stability of the Pt electrodes are needed.

Changing the Pt deposition technique, deposition conditions, annealing atmosphere, and film thickness affects the magnitude of certain parameters in Fig. 5 (e.g. initial film stress and onset of stress relief). However, the features of the stress versus temperature curve shown in Fig. 5 are essentially what is observed for Pt films grown under a variety of conditions (this holds for Pt grown on substrates with a lower thermal expansion coefficient than Pt, e.g. Si or SiO_2/Si).

2.1.2. Pt/Ti interactions

The Pt/Ti interaction involves "intermixing" of the two elements and the oxidation of the Ti. It is important to note that by the term "intermixing" we do not necessarily mean elemental interdiffusion with alloy or compound formation. It is usual for the Ti to oxidize to TiO_x during annealing, and the Pt and TiO_x remain phase segregated although they can move their positions relative to one another as discussed below. Their have been reports of Pt_3Ti alloy formation[10], although these are likely to occur only under conditions of low oxygen partial pressure. In terms of analysis techniques such as Auger depth profiling or Rutherford Backscattering Spectrometry (RBS) analysis, interface roughening and elemental interdiffusion can not be differentiated from one another. Interpretation therefore requires an additional technique such as cross-sectional transmission electron microscopy (TEM).

The intermixing of Pt and Ti during growth of ferroelectric thin films on Pt/Ti electrode has been a subject of numerous studies. Olowolafe et al.[10] used Auger electron spectroscopy and Rutherford backscattering to study the interdiffusion of Pt/Ti bilayers and the effect on crystallization of PLZT thin films. Their samples were 2000Å Pt on 1000Å Ti on oxidized Si substrate. The Pt/Ti layers were dc sputter-deposited at room temperature and were further annealed in O_2 and N_2 atmospheres at various temperatures. For samples annealed to 800°C for 30 minutes in O_2, Pt_3Ti and TiO_2 compounds were formed with a thin Pt film remaining on the surface. For samples annealed in N_2, Pt_3Ti and Ti_5Si_3 formed. Fig. 6 shows Auger depth profiles of as-deposited and annealed Pt/Ti bilayers. The annealing atmosphere was oxygen and the annealing time was 30 minutes.

Takemura et al.[6] deposited Pt/Ti bilayers directly on Si by DC magnetron sputtering and further annealed them at 500-800°C in vacuum and in oxygen atmosphere. For Pt(50nm)/Ti(150nm)/Si annealed in vacuum at about 600°C, interdiffusion between Pt and Ti and between Ti and Si was observed. In addition, Si reached to the surface of the structure and reacted with the Pt forming a Pt silicide. In the case of structures annealed in oxygen, interdiffusion and oxidation of the

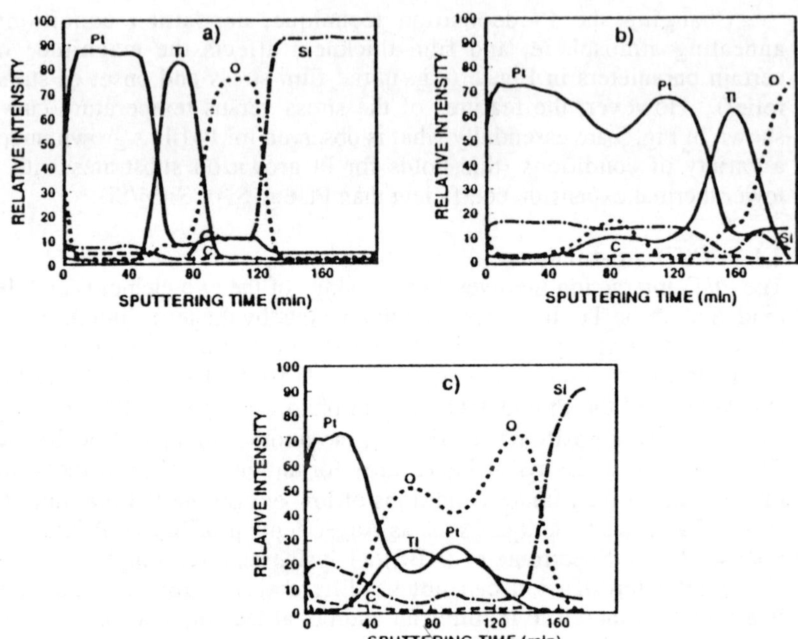

FIGURE 6 AES depth profiles of Pt/Ti/SiO_2/Si films with the Pt layer (a) as-deposited, (b) annealed at 500°C and at (c) 800°C for 30 minutes in oxygen (Reproduced from Ref. 10).

titanium and silicon occurred. The authors did the experiment with a Ta adhesion layer, and a summary of their results is shown in Fig. 7.

The intermixing of Pt/Ti was also observed in our laboratory on ion beam sputter-deposited Pt/Ti bilayers deposited on SiO_2/Si substrates. However, we found that replacing the Ti layer by a partially oxidized Ti layer (TiO_x) significantly reduces the interaction between the Pt and Ti layer. Figs. 8 and 9 show RBS spectra of a Pt/Ti/SiO_2/Si and of a Pt/TiO_x/SiO_2/Si layered structures before and after annealing at 550°C for 90 minutes in air. The Pt itself was deposited at 200°C using ion beam sputter-deposition. Notice that intermixing of the Pt and Ti occurs after the annealing step only for the sample having Pt deposited on Ti; the sample with Pt deposited on TiO_x shows negligible intermixing. This observation was recently confirmed by Sreenivas et al.[14] TEM analysis

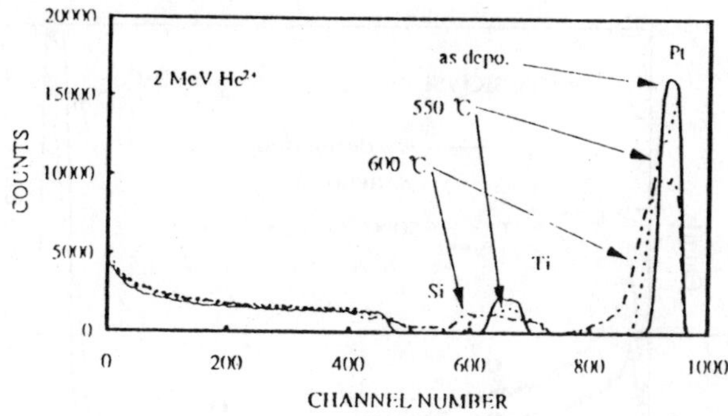

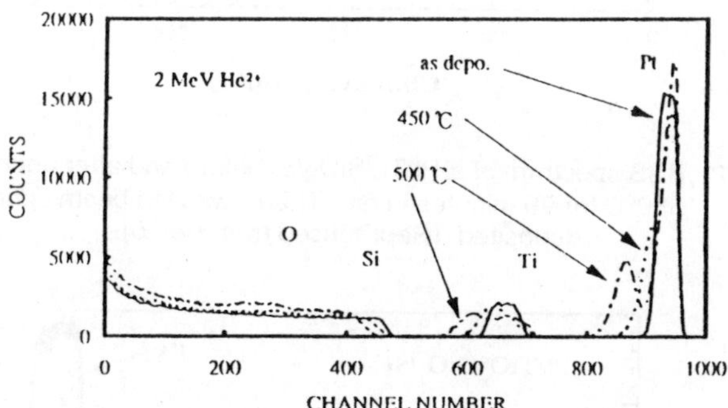

FIGURE 7 RBS spectra showing the effect of annealing Pt/Ti bilayers (a) in vacuum and (b) in oxygen, at various temperatures. (Reproduced from Ref. 6).

of our samples showed that the "intermixing" corresponded only to interface roughening in the case where the Pt was deposited at elevated temperatures and was therefore large-grained and dense. If the Pt was deposited at room temperature then TiO_x grains could be found at the top interface after annealing.

Bruchhaus et al.[9] reported important studies on the interdiffusion of Pt/Ti bilayers deposited on SiO_2/Si by RF sputtering. They observed that during the heat treatment of the Pt/Ti bilayer, oxidation induced migration of the Ti layer through Pt to the surface occurs, as shown in Fig. 10. The loss of Ti degraded adhesion to the underlying SiO_2/Si substrate. By using thicker Ti (about 1000Å) adhesion improved, but oxidation of Ti at the Pt surface was still observed. Eichorst et al.[7] also reported intermixing of the Pt and Ti in their evaporated Pt/Ti thin films.

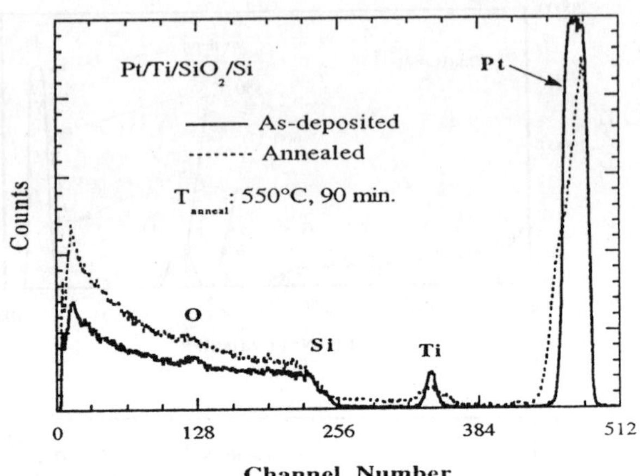

FIGURE 8. RBS spectrum of a Pt/Ti /SiO_2/Si before and after annealing at 550°C for 90 minutes in air. The Pt was ion beam sputter-deposited. (Reproduced from Ref. 5).

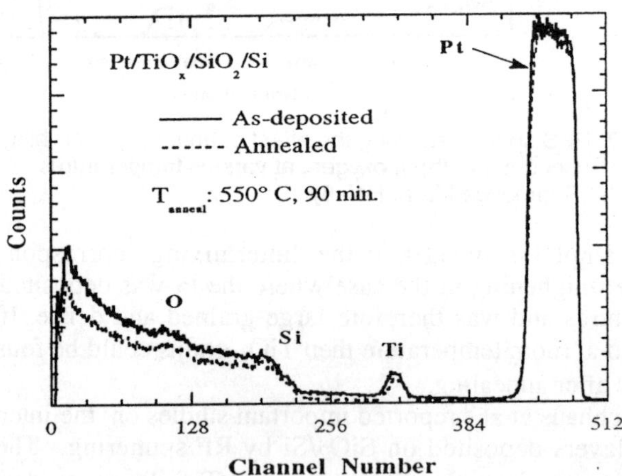

FIGURE 9. RBS spectrum of a Pt/TiO_x /SiO_2/Si before and after annealing at 550°C for 90 minutes in air. The Pt was ion beam sputter-deposited. (Reproduced from Ref. 5).

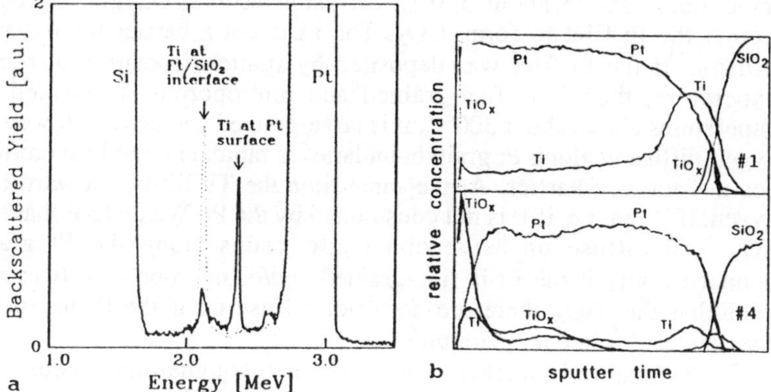

FIGURE 10. (a) RBS spectrum and (b) Auger depth profile showing the diffusion of Ti to the surface of Pt and its subsequent oxidation. The broken line in the RBS spectrum and the top Auger profile (#1) are for the Pt/Ti /SiO_2/Si structure before annealing at 450°C for 2 hours. (Reproduced from Ref. 9).

They noticed that the Pt peaks in the x-ray diffraction patterns shifted with annealing, indicating the possible formation of a Pt-Ti alloy.

From the above data, it is clear that intermixing of the Pt/Ti bilayer and subsequent oxidation of Ti occur during PZT growth (or annealing). This may increase the effective resistivity of the bottom electrode and therefore degrade capacitor performance. Similar to hillock formation, the annealing atmosphere seems to affect the extent of Pt/Ti interaction. In addition, the microstructure of the Pt layer probably plays a major role in the extent of Ti diffusion into the Pt. Tison and Drobeck[15] found that Ti is the faster diffusing species in Ti/Pt, Ti/Pd, and Ti/Au couples during vacuum annealing. They also showed that Ti diffuses along the grain boundaries of the other elements. Therefore, minimizing the extent of Ti interdiffusion into the Pt may be partially controlled by controlling the Pt microstructure. Thus, one should attempt depositing Pt layers using conditions that produce a dense, large-grained microstructure and conditions which do not favor the formation of columnar grains.

Our understanding of the interactions in the Pt/Ti system may therefore be summarized as follows:

1) There is a large chemical potential gradient between the Ti layer under the Pt, and oxygen in the ambient (or in the ferroelectric film). At

temperatures above about 350°C, diffusion occurs. Oxygen diffuses through the Pt film to form TiO_x. The Pt is not a barrier for oxygen diffusion. If the Pt film was deposited by sputter deposition at room temperature, then it is fine-grained and microporous. If formed at temperatures above about 300°C, it is large-grained and dense. However, oxygen diffusion along Pt grain boundaries is rapid, and the Pt therefore does not act as a barrier. At the same time the Ti diffuses toward the oxygen, if it can, i.e. if it is not constrained by the Pt. We deduce that the TiO_x will diffuse up as second phase grains along the Pt grain boundaries only if the Pt is fine-grained and/or microporous. It is our belief that the TiO_x therefore does not diffuse up if the Pt has been deposited at elevated temperatures.

2) Substantial surface and interface roughening occurs in conjunction with the diffusion which results from the oxygen activity gradient.

3) Deposition of TiO_x as the adhesion layer substantially reduces the chemical potential gradient, and therefore the interface roughening or "intermixing".

4) This roughening due to interdiffusion adds to the roughening due to the sources described previously, namely hillock formation and grain growth.

2.1.3. Effect of Pt/Ti on PZT film microstructure and orientation

There are numerous studies which have shown that the electrode has a significant effect on the ferroelectric film microstructure. We have selected some of these for discussion. Spierings et al.[8] suggested that Pt hillocks in Pt/Ti bilayers acted as nucleation sites for the perovskite phase in their sol-gel PZT thin films. They compared the microstructure of two PZT thin films grown on as-deposited (room temperature deposited) Pt/Ti and on annealed Pt/Ti (the exact annealing temperature was not reported). They found that PZT films grown on annealed Pt/Ti were single phase and homogeneous, while the PZT films grown on room temperature deposited Pt/Ti consisted of perovskite rosettes surrounded by a pyrochlore-type matrix.

More recently Hase et al.[16] proposed a mechanism where the TiO_x present on the surface and in the grain boundaries of the Pt layer acts as a nucleation site for perovskite PZT. Therefore, films deposited on Pt/Ti were dense and homogeneous due to high nucleation site density. In contrast, films grown on Pt without the Ti layer had a low nucleation site density. This resulted in PZT films grown on Pt with inhomogeneous microstructure, consisting of perovskite islands (rosettes) surrounded by a pyrochlore-type matrix. A similar observation was made by Sreenivas

et al.[14] It should be noted however, that not all groups have experienced difficulties in achieving single phase perovskite on Pt electrodes. In the case of sol-gel processing in our own laboratory, the films deposited on Pt bilayers or on Pt/TiO_x are generally single phase and their microstructures (grain size, density, orientation) depend on the substrate type, the technique used, and the deposition conditions.

The crystallization of the perovskite phase on Pt/Ti electrodes has been studied by several authors.[17,18,19,20,21] It is generally observed that crystallization of perovskite PZT is usually preceded by the formation of a pyrochlore-type phase. Figs. 11 and 12, show the crystallization sequence and the microstructural evolution of PZT thin films deposited by the sol-gel process[21] on evaporated Pt/Ti bilayers. As can be seen in Fig. 11, the formation of the perovskite phase is preceded by the formation of a pyrochlore-type phase. This is essentially what has been observed by other researchers.[17-21]

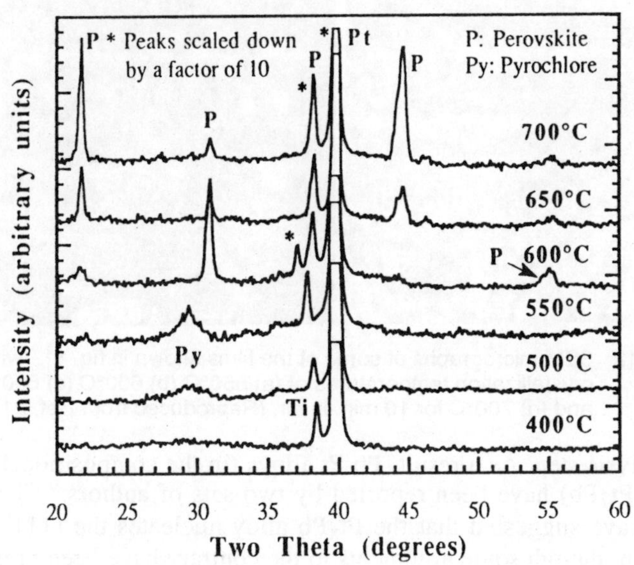

FIGURE 11. X-ray diffraction patterns of $Pb(Zr_{0.53}Ti_{0.47})O_3$ thin films grown on $Pt/Ti/SiO_2/Si$ by the sol-gel process. Films were crystallized at 400°C to 700°C for 10 min. in air. (Reproduced from Ref. 21).

In the case of sol-gel processing of PZT on Pt lower electrodes, it is important to note that the oxygen activity in the film can be low during

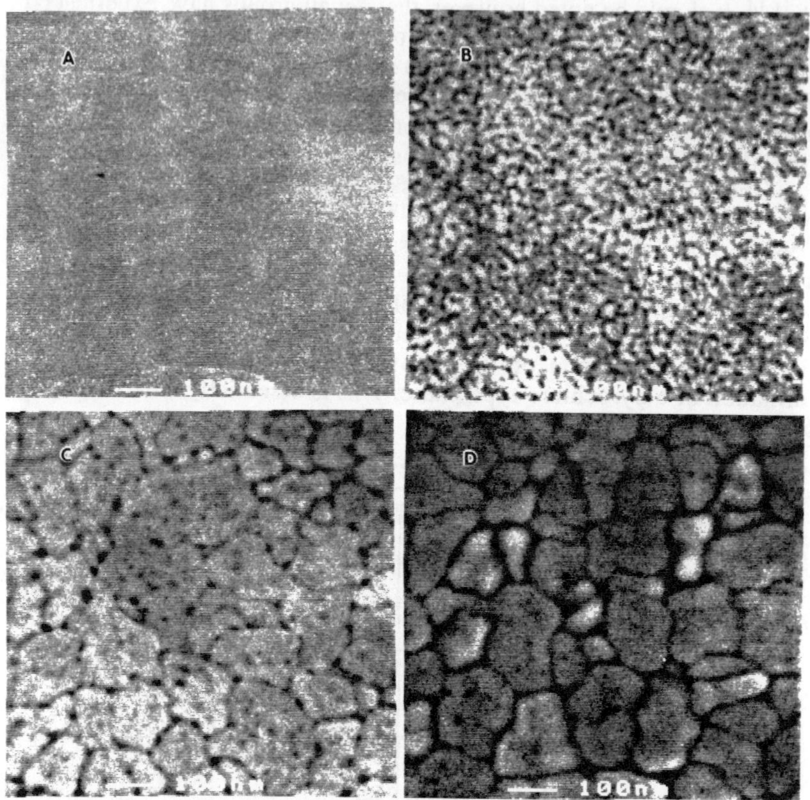

FIGURE 12. SEM micrographs of some of the films shown in fig. 11, with crystallization temperatures of (a) 550°C (b) 600°C (c) 650°C and (d) 700°C for 10 min. in air. (Reproduced from Ref. 21).

the pyrolysis step. As a result, Pb-Pt alloys (in the compositional range Pt_5Pb to Pt_7Pb) have been reported by two sets of authors. Chen and Chen[22] have suggested that the Pt_xPb alloy nucleates the (111) PZT orientation, though some arguments to the contrary have been presented by Brooks et al.[23] In addition, excess PbO has also been suggested to control PZT film orientation.

Evidence for excess PbO at the film/electrode interface enhancing the (100)c PZT orientation was presented by Chen and Chen[22]. Auciello et al.[24] showed evidence for excess PbO enhancing the (101)t orientation in a pulsed laser deposition process. In summary, the Pt bottom electrode affects the PZT film nucleation and orientation in a variety of ways. Care must be taken to control these effects.

Finally, it should be mentioned that by growing the PZT film on epitaxial Pt electrodes, nearly epitaxial PZT films can be achieved. For instance, our group[5] has grown epitaxial Pt films on (100) MgO substrates by ion beam sputter-deposition at 500°C. An electron diffraction pattern of such film is shown in Fig. 13. As will be discussed later, PZT films with better ferroelectric properties can be produced by using such epitaxial bottom electrodes.

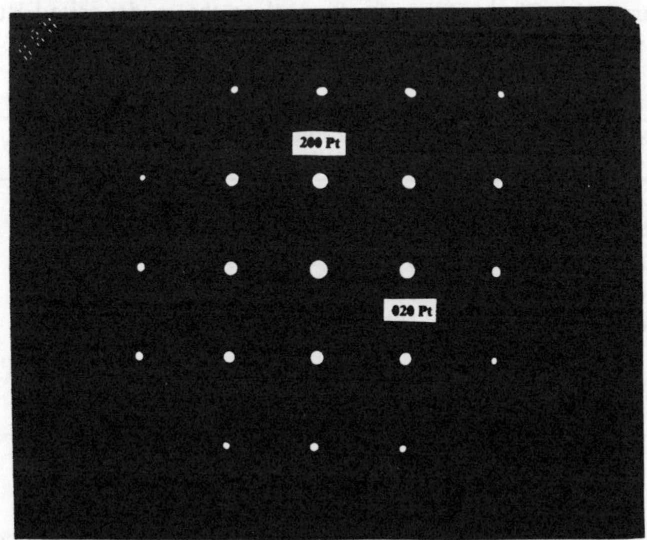

FIGURE 13 Electron diffraction pattern of a Pt thin film which was grown on (100) MgO by ion beam sputter-deposition at 500°C. (Reproduced from Ref. 13).

2.2. Simple Oxides

RuO_2 is by far the most studied simple oxide electrode for use with ferroelectric thin films for non-volatile memory applications. RuO_2 has the rutile crystal structure with a primitive tetragonal lattice having a c/a ratio of 0.69.[25] More recently, $SrRuO_3$[26] has been used as an electrode material for PZT thin films. As will be discussed in the electrical property section, the use of RuO_2, $SrRuO_3$ and other oxide electrodes has been the key to solving the fatigue problem of PZT ferroelectric capacitors. In this section, we review issues relating to the growth of RuO_2 and its effect on the orientation and microstructure of PZT thin films.

The methods used to deposit RuO_2 have mostly been sputtering and chemical vapor deposition. Table 3 lists the methods of RuO_2 thin film deposition reported by researchers. Ruthenium oxide is much more stable against morphological changes than Pt. For example, ruthenium oxide thin films do not have a hillock formation problem. In fact, RuO_2 thin films deposited in our lab under similar conditions of a Pt which had hillocks were smooth: the average surface roughness of such films was around ± 20 Å. Even after annealing the as-deposited RuO_2 thin films at temperatures of 800°C, they remained smoother than Pt.

TABLE 3. Deposition methods of RuO_2.

Deposition Technique	Deposition Temp. (°C)	Resistivity (μΩ-cm)	Substrate	Ref.
CVD	575	90	SiO_2/Si	27
RF Reactive Sputtering	RT	150-180	SiO_2/Si	28
DC Magnetron Sputt..	RT-500°C	500-40	SiO_2/Si	29
Ion Beam Sputtering	NA	NA	SiO_2/Si	30
Ion Beam Sputtering	200-500		SiO_2/Si, MgO	2,5
CVD	350	50	SiO_2/Si, $SrTiO_3$	31

2.2.1. Electrical resistivity

The resistivity of RuO_2 is higher than that of Pt and this may be a small disadvantage. However, we are able to get RuO_2 with resistivities as low as 56 μΩ-cm. De Keijser et al.[31] reported a resistivity of 50 μΩ-cm for their CVD grown RuO_2 thin films. In addition, in making an actual memory it is only necessary to have the ruthenium oxide in contact with the ferroelectric capacitors; the wiring from one cell to the other can still be done using metal lines. However, compared to other exotic oxides such as LSCO and YBCO (which will be discussed later), RuO_2 has lower resistivity.

2.2.2. Microstructure and orientation of PZT films grown on RuO_2

The growth and electrical properties of PZT thin films deposited on RuO_2 compared to films grown on Pt electrodes has also been a subject of numerous investigations.[30,32-37] From such studies it is clear that the electrode material directly affects the nucleation, the microstructure (grain size, phases present, orientation), and the electrical properties of PZT thin films. In our sol-gel and ion beam sputter-deposited PZT thin films, we observe that nucleation of the perovskite phase is more difficult on RuO_2 than it is on Pt. In fact, second phases are always observed in films deposited on RuO_2. In addition, the films grown on RuO_2 tend to have larger grain size (1-2 μm) than film deposited on Pt (0.1-0.2 μm). This difference results because the nucleation site density is much higher in films grown on Pt than in films grown on RuO_2. The microstructural differences between PZT films grown on Pt and those grown on RuO_2 electrodes result in difference in their electrical properties as will be discussed later in this chapter.

The crystallization sequence of PZT thin films grown on RuO_2 was studied and the results are shown in Figs. 14 and 15. Similar to the films grown on Pt, crystallization of the perovskite phase is preceded by formation of a pyrochlore-type phase. However, as shown in the SEM micrographs of Fig. 15 and the TEM micrograph in Fig. 16, a pyrochlore-type phase still exists in the films even after annealing at 750°C for 10 minutes. Notice also that the number of perovskite phase

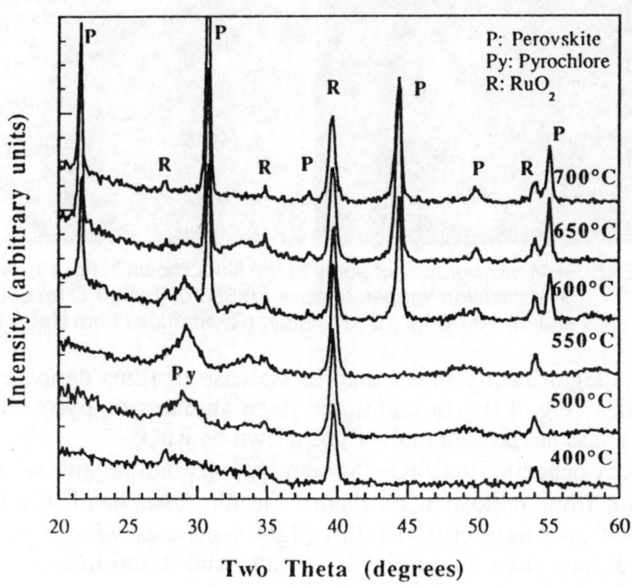

FIGURE 14 X-ray diffraction patterns of $Pb(Zr_{0.53}Ti_{0.47})O_3$ films grown on $RuO_2/SiO_2/Si$ by the sol-gel process. Films were crystallized at 400°C to 700°C for 10 min. in air. (Reproduced from Ref. 21)

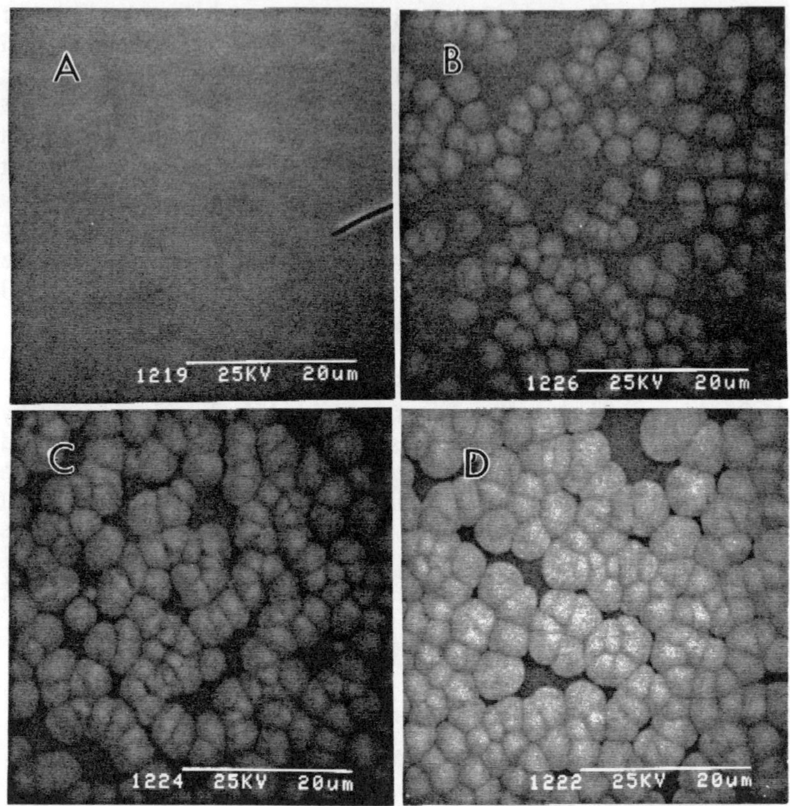

FIGURE 15 SEM micrographs of some of the films shown in fig. 14, with crystallization temperatures of (a) 550°C (b) 600°C (c) 650°C and (d) 700°C for 10 min. in air. (Reproduced from Ref. 21).

nuclei is significantly lower than in the case of films deposited on Pt electrodes (Fig. 12). In addition, large amounts of pyrochlore-type second phase are present in the films grown on RuO_2.

Films deposited on SiO_2/Si were polycrystalline and so were the PZT thin films deposited on them. On the other hand, RuO_2 films deposited on single crystal (100) MgO substrates were highly (101) oriented and they maintained an epitaxial relationship with the substrates. This orientation results because it leads to a relatively low lattice mismatch 5%. A selected area electron diffraction pattern of a RuO_2/MgO thin film is shown in Fig. 17. The PZT films deposited on such substrates were usually (101) oriented.[5]

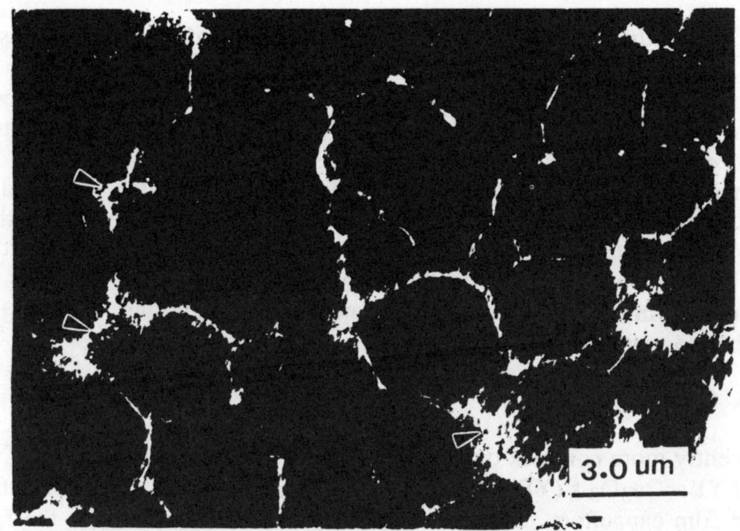

FIGURE 16. TEM plan view of sol-gel $Pb(Zr_{0.53}Ti_{0.47})O_3$ film grown on RuO_2/MgO substrates at 700°C, 10 min. in air.

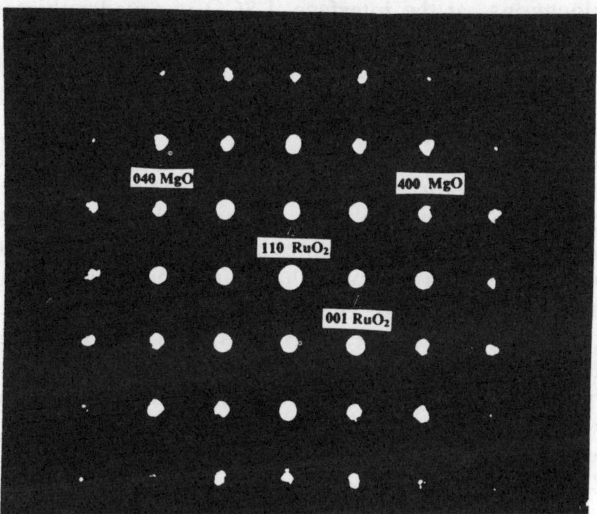

FIGURE 17. SAD diffraction pattern of an RuO_2 thin film grown on (100) MgO. (Reproduced from Ref. 21).

Recently, another relatively simple oxide ($SrRuO_3$)[26] has been used as an electrode material for growth of PZT thin films. $SrRuO_3$ is conducting, but its reported room temperature resistivity of (340 μΩ-cm) is several time larger than that of RuO_2. However, $SrRuO_3$ has better lattice matching to PZT (2.7%)[26], and this may lead to reduced defect densities near the film electrode interface. Studies on the growth and microstructural characterization of PZT thin films grown on this electrode are more scarce than the data available on films grown on Pt or RuO_2 electrodes. However, $SrRuO_3$ appears to be a candidate as an electrode material for nonvolatile memory applications. $SrRuO_3$ is stable up to 900°C in oxidizing or inert gas atmospheres.[38]

2.3. Complex Oxides

Recently more complex oxide electrodes such as $La_{1-x}Sr_xCoO_3$ (LSCO) and $YBa_2Cu_3O_{3-\delta}$ (YBCO) have been used as electrodes for ferroelectric thin film capacitors. Interest in these electrodes resulted from the fact that they have perovskite-based crystal structures with good lattice matching with the PZT ferroelectric thin film. The lattice matching may be important since it results in a lower density of point defects such as misfit dislocations at the interface. LSCO is psuedocubic and has a lattice parameter of approximately 3.83Å.[39] The a-b plane of YBCO has lattice parameters of 3.82 and 3.88Å[40], respectively. In comparison, tetragonal perovskite PZT has lattice parameters of 4.04Å and 4.14Å for the a-axis and c-axis, respectively.[41]

LSCO is the most conductive perovskite oxide with a resistivity about a 100 μΩ-cm[42] for bulk samples measured at room temperature. The resistivity also depends on the Sr content in the films. YBCO, on the other hand, has a resistivity two times higher[42] than LSCO. The oxygen content of these two oxides will change depending on the annealing temperature and atmosphere. This will in turn affect their resistivity. However, the oxygen sublattice in LSCO is more stable than the oxygen sublattice in YBCO[42] (i.e. less susceptible to oxygen loss than). Nonstoichiometry in YBCO and LSCO have been a subject of numerous studies.[43,44] Another important issue which little is known about is the possibility of chemical reactions taking place between the PZT film and the oxide electrodes. These complex oxides are also more difficult to produce with controlled stoichiometries due to the presence of several components. The deposition methods and conditions used to deposit LSCO and YBCO for use as an electrode material of PZT thin films are listed in Table 4.

PZT thin films grown on LSCO and YBCO were epitaxial due to the good lattice matching between those electrodes and the PZT thin film. X-ray diffraction revealed that YBCO and LSCO films deposited on single crystal (100) MgO are (001) and (100) oriented[48], respectively. In contrast, YBCO and LSCO films deposited on oxidized silicon were polycrystalline. In these previous sections we have discussed several issues relating to the viability of metal and oxide materials as electrodes for PZT thin film capacitors. These issues are summarized below.

TABLE 4. Deposition conditions of LSCO and YBCO.

Technique	Deposition Temp. (°C)	Resistivity (μΩ-cm)	Substrate	Ref.
Laser Ablation, LSCO	600	200	SiO_2/Si, MgO	45
Laser Ablation, LSCO	675	100	(100) MgO	42
PECVD, LSCO	700	200	(100) MgO	46
Ion beam Sputt., YBCO	680	450	(100) MgO	47
Laser Ablation, YBCO	650	600	(100) MgO	48

1) It was observed that oxide electrodes are in general more resistant to morphological changes than are metal electrodes. (i.e. smaller hillock formation, adhesion, interdiffusion). Oxide electrodes do not form hillocks as Pt does. However, using a thin Pt layers will in general reduce the problem of hillock formation in sputtered Pt thin films. In addition, oxide electrodes do not present an adhesion problem as Pt and other noble metals do. Therefore, no additional adhesion layer is required. Use of an adhesion layer as was the case in Pt/Ti resulted in interdiffusion of Ti into Pt and in an increase in the bottom Pt electrode resistivity. In order to minimize the Ti interdiffusion into Pt, use of a TiO_x adhesion layer instead of Ti is an effective approach.

2) Metal electrodes have the advantage that their resistivity is much lower than oxide electrodes. Resistivity of the electrode material should be as low as possible for optimum device speed.

3) Metal electrodes also have the advantage of being easier to deposit than oxide electrodes. In addition, etching and patterning of the metal electrodes is likely to be easier than patterning complex oxide electrodes such as LSCO and YBCO.

4) The ability to grow single phase perovskite PZT on a given electrode material will depend on the deposition technique. For instance, while growth of single phase perovskite PZT on Pt electrodes can be achieved

by ion beam sputter-deposition and sol-gel processing, it is generally not possible to achieve single perovskite phase on Pt using laser ablation deposition. The exception to this is the case where the PZT films are deposited at low temperatures using pulse laser deposition followed by a post-deposition annealing treatment. Another example is that in all deposition techniques used in our laboratory (sol-gel, ion beam sputter-deposition, and laser ablation deposition) growth of PZT films on RuO_2 electrodes is always accompanied by pyrochlore formation. These observations suggest that atom type and arrangement in the bottom electrode material affect the nucleation of the perovskite phase. In addition, for the vapor phase techniques, the sticking coefficient of lead plays a role in determining the phases that will nucleate on a given bottom electrode.

5) For high density integrated devices, the bottom electrode will need to be connected directly to the underlying Si via a plug. This will require a conducting barrier layer under the bottom electrode.

The choice of the most suitable electrode material will be dictated not only by the factors discussed above, but also by the effect such electrode has on the short term and long term properties of the ferroelectric capacitor. Ultimately, it may be the effect of the electrode material on capacitor properties that determines its viability. In the next section, we discuss the properties of PZT film capacitors with metal and oxide electrodes. Data will be presented for films deposited by a variety of techniques.

3. EFFECT OF THE ELECTRODE MATERIAL ON THE ELECTRICAL PROPERTIES

Electrical testing of thin film capacitors with metal and oxide electrodes has revealed large differences especially in long term properties. This section will be divided into three parts. In the first, the short term properties will be discussed. In the second part the effect of the electrode type on fatigue and retention will be discussed. In the third part the effect of electrode type on DC leakage and dielectric breakdown behavior will be presented.

3.1. Short Term Properties.

By short term properties we mean the basic variables obtained from a hysteresis measurements. These include remanent polarization, saturation polarization, and the coercive field. These variables provide important information about the quality of the film and provide initial evaluation of its viability for use in ferroelectric memories. Tables 5-7 show the short term properties of PZT thin films deposited on Pt, RuO_2, and LSCO electrodes by selected deposition techniques. It should be emphasized that this is a partial list.

Several comments can be made about the data listed in Tables 5-7. All samples have sufficient polarization to make them acceptable for use in memory devices. The films were also fully switched at or below 5 volts. Our sol-gel derived films on RuO_2 electrodes were actually well saturated at 3-4 volts for film thicknesses of about 2000Å. Films deposited on single crystal substrates tended to have higher polarization values due to their preferred orientation. Oriented films would be preferred over polycrystalline films for memory applications. However, single crystals such as MgO and Al_2O_3 cannot be used as a substrate if these ferroelectric capacitors are to be integrated in Si-based integrated circuits. Recently, researchers have used buffer layers on Si such as $Bi_4Ti_3O_{12}$[62], which always grows c-axis oriented regardless of the substrate used, to grow oriented PZT thin films. Highly oriented or epitaxial PZT films, however, can provide useful information and help understand several phenomena of ferroelectric thin films. The coercive field data should be viewed with caution since the coercive fields depend on film thickness. In general however, the films deposited on the oxide electrode had lower resistivity and higher leakage currents. The resistivity data is not included in the tables, but it will be discussed in a later section.

TABLE 5. Properties of PZT capacitors with Pt electrodes.

Deposition Technique	Substrate	Pr $\mu C/cm^2$	Ps $\mu C/cm^2$	Ec kV/cm	Ref.
Sol-gel	SiO_2/Si	14	30	100	32
	MgO	40	50	100	32
Sol-gel	SiO_2/Si	26	45	43	49
Sol-gel	SiO_2/Si	13	32	33	50
Ion Beam Sputtering	MgO	30	58	200	51
DC Sputtering	SiO_2/Si	8	30	75	52
RF Sputtering	SiO_2/Si	10	48	39	53
Laser Ablation	SiO_2/Si	32	50	130	54
Laser Ablation	SiO_2/Si	10	32	60	48
Laser Ablation	SiO_2/Si	17	25	40	55
MOCVD	MgO	40	50	40	56
PECVD	SiO_2/Si	12	16	54	57

TABLE 6. Properties of PZT with RuO_2† and $SrRuO_3$# electrodes.

Deposition Technique	Substrate	Pr μC/cm²	Ps μC/cm²	Ec kV/cm	Ref.
Sol-gel†	SiO_2/Si	14	35	35	58
	MgO	30	50	50	59
Sol-gel†	SiO_2/Si	14	34	30	33
Ion Beam Sputter Dep.†	MgO	20	35	150	60
Off-axis Sputtering#	$SrTiO_3$	27	68	70	26

TABLE 7. Properties of PZT films with LSCO† and YBCO# electrodes.

Deposition Technique	Substrate	Pr μC/cm²	Ps μC/cm²	Ec kV/cm	Ref.
Sol-gel†	MgO	50	60	80	61
Laser Ablation†	MgO	30	35	75	48
Laser Ablation†	SiO_2/Si	20	30	40	48
Laser Ablation†	BT/SiO_2/Si	20	45	35	62
Laser Ablation#	MgO	30	48	120	48

3.2 Effect of Electrode Material on Fatigue

Fatigue or loss of polarization upon repeated cycling of the ferroelectric capacitor has been a subject of many studies. It was initially observed that PZT thin films grown on Pt electrodes underwent severe polarization fatigue during capacitor cycling. Several models have been proposed to explain the phenomenon of fatigue[63-69], but in each case experimental data has been obtained which puts the postulated hyposthesis into question. It is generally believed that oxygen vacancies, which happen to be the most mobile species in perovskite PZT, contribute to fatigue of the ferroelectric capacitors. At the same time, these models were being debated, data began to appear which showed that the use of an oxide electrodes in place of the metallic (mainly platinum) electrodes reduced or eliminated the polarization fatigue of PZT capacitors . These oxides included RuO_2[30,33,34], $SrRuO_3$[26], $YBa_2Cu_3O_{7-\delta}$[48,70], and $La_{1-x}Sr_xCoO_3$ (LSCO)[45,48,62]. This emphasized the important role that the electrode material plays in controlling film properties. There were several attempts to improve the fatigue of

Pt/PZT/Pt capacitors by modifying the PZT composition (e.g. changing Zr/Ti ratio[71,72,73,74], donor doping[75-76], and film lead content[74]). Thus, despite all these attempts, only marginal improvement in the fatigue behavior of Pt-based PZT capacitors was achieved. Using oxide electrodes was the key to solving the fatigue problem in PZT capacitors. However, early in 1992, Symetrix corporation announced their discovery that certain layered perovskite ferroelectrics ($SrBi_2Ta_2O_9$ and similar compounds[77]) show negligible polarization fatigue with Pt electrodes.

An example which illustrates the difference in fatigue behavior of PZT thin films with metal and oxide electrodes is shown in Fig. 18. There is clearly an advantage to using the oxide electrode in terms of improving capacitor fatigue behavior. The reason why PZT films with oxide electrodes have improved fatigue is still a subject of controversy. We have attempted to separate the contributions of film microstructure and film/electrode interface to fatigue by performing the following experiment. Capacitors with four different electrode configurations were fabricated. These are shown in Fig. 19. The first two capacitors ($Pt/PZT/Pt/SiO_2/Si$ and $RuO_2/PZT/Pt/SiO_2/Si$) were actually made by depositing Pt and RuO_2 top electrodes next to each other on the same $PZT/Pt/SiO_2/Si$ film.

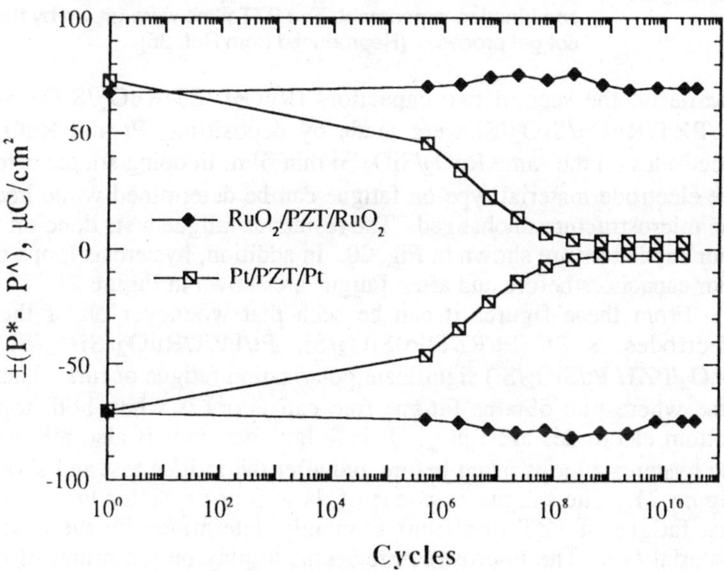

FIGURE 18. Fatigue curve of PZT films with Pt and RuO_2 electrodes.

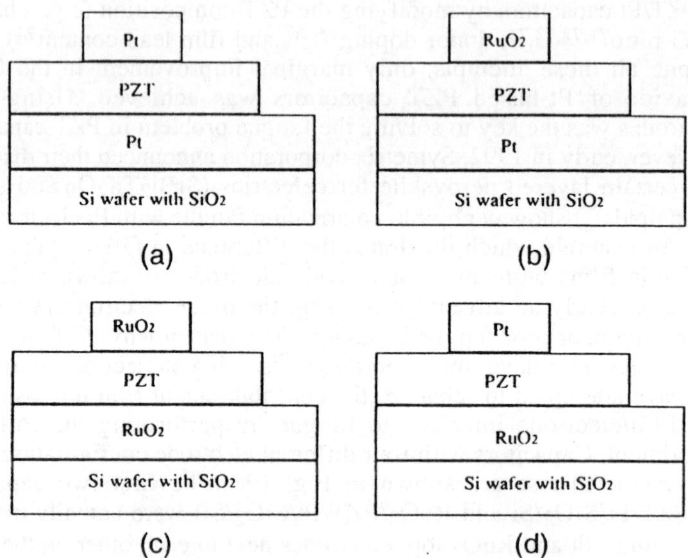

Figure 19. A schematic of the four capacitor types used in the electrode combination experiment. The PZT films were grown by the sol-gel process. (Reproduced from Ref. 36).

Similarly, the second two capacitors (RuO_2/PZT/RuO_2/SiO_2/Si and Pt/PZT/RuO_2/SiO_2/Si were made by depositing Pt and RuO_2 top electrodes on the same RuO_2/SiO_2/Si thin film. In doing so, the effect of the electrode material type on fatigue can be determined while keeping the microstructure unchanged. The results of fatigue tests done on these four capacitors are shown in Fig. 20. In addition, hysteresis loops of the four capacitors before and after fatigue are shown in Figure 21.

From these figures it can be seen that whenever on of the two electrodes is Pt (Pt/PZT/Pt/SiO_2/Si, Pt/PZT/RuO_2/SiO_2/Si, and RuO_2/PZT/ Pt/SiO_2/Si) significant polarization fatigue occurs. The only case where one obtains fatigue-free capacitors is when both top and bottom electrodes are RuO_2. This fatigue behavior is also reflected in the hysteresis loops taken before and after the fatigue test and shown in Figure 21. The fatigue response of these four capacitor types suggest that fatigue of PZT thin films is mainly determined by the electrode material type. The microstructure seems to play only a minor effect in controlling fatigue. Evidence for the effect of microstructure effect on fatigue has been manifested in at least two cases. One is the observation

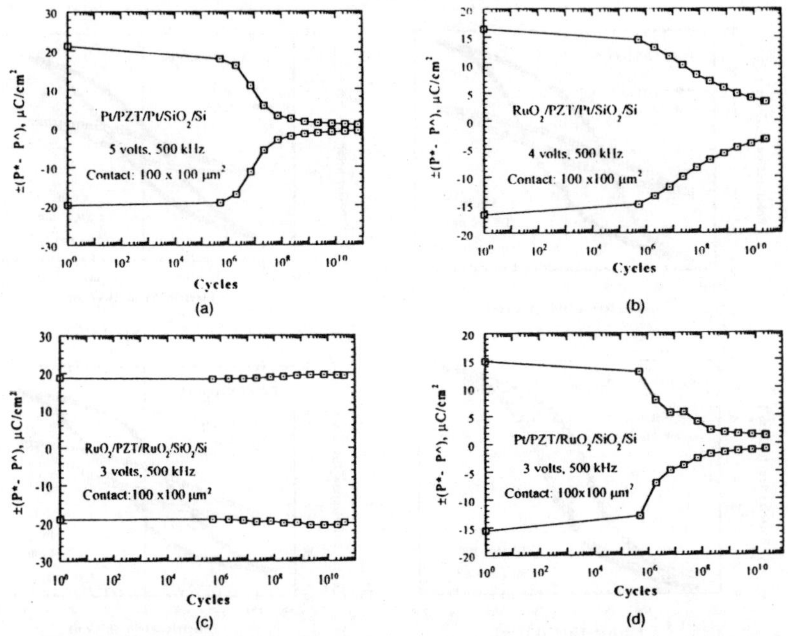

FIGURE 20. Fatigue characteristics of the four capacitor types shown in Fig. 19. (Reproduced from Ref. 36).

that epitaxial films can in some cases fatigue less than polycrystalline ones.[32] The second, is the observation that sol-gel films annealed at various temperatures show different fatigue rates.[74]

In most fatigue models[63-65] it is suggested that fatigue results due to the buildup of oxygen vacancies near the electrode/film interface. However, the exact reason for the migration and accumulation of oxygen vacancies near the film/electrode interface is a subject of controversy. If in fact oxygen vacancy accumulation near the film electrode interface is responsible for fatigue, it is certainly tempting to postulate that the oxide electrode acts as a sink for the oxygen vacancies, thereby reducing or eliminating fatigue. More experiments are still needed to confirm this postulate.

3.3 Effect of Electrode Material on DC leakage

While using oxide electrodes essentially alleviates the fatigue problem, capacitors with oxide electrodes tended to have low resistivity compared

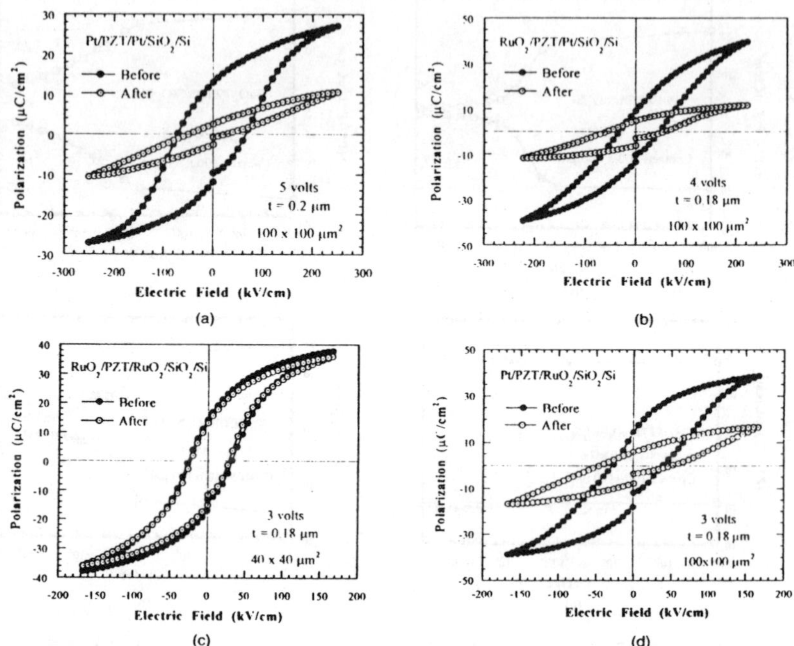

FIGURE 21. Hysteresis loops before and after the fatigue test of the four capacitor types shown in Fig. 19. (Reproduced from Ref. 36).

to capacitors having Pt electrodes. The effective resistivity of PZT thin films with oxide electrodes (measured using the RT66A ferroelectric tester) are listed in Table 8. Notice that the resistivities of films with Pt electrodes are one to three orders of magnitude larger than resistivities of films with oxide electrodes. This difference between the resistivities of PZT films with Pt and oxide electrodes is also reflected in the DC leakage current characteristics of the capacitors as will be discussed shortly. It is necessary to understand the reason for the high leakage currents of PZT capacitors with oxide electrodes and to reduce them to acceptable levels. A low leakage current and acceptable dielectric breakdown properties are essential for the successful implementation of PZT thin films into commercial devices. We have undertaken much work on comparing the characteristics of leakage current of PZT capacitors with Pt and RuO_2 electrodes.[36,37] Similar studies were carried out by Yoo et al.[78]

TABLE 8. Effective resistivities of PZT films with various electrodes.

Deposition Method	Electrode Material	Resistivity ($\mu\Omega$-cm)	Ref.
Sol-gel	Pt//Pt	$5x10^{10}$	32
Laser ablation	Pt//Pt	$1x10^{10}$	48
Ion beam Sputt.	Pt//Pt	$2x10^{10}$	51
Laser ablation	LSCO//LSCO*	$1\text{-}5x10^9$	62
Laser ablation	LSCO//LSCO	$2x10^9$	48
Laser ablation	YBCO//YBCO	$5x10^8\text{-}5x10^9$	70
Laser ablation	Pt//YBCO	$3x10^{10}$	48
Sol-gel	RuO_2//RuO_2	$1.3\ x10^7$	30
Sol-gel	RuO_2//RuO_2	$10^6\text{-}10^9$	21

* PZT film is lanthanum doped.

As an example, results of leakage current measurements performed on the four capacitor types shown in Fig. 19 are presented below. When a DC voltage is applied to a PZT thin film capacitor, the current response usually shows three regions in the time spectrum. These regions are transitory, steady state, and resistivity degradation and/or dielectric breakdown. We have already presented discussions of all three regions[37,79,80] in earlier papers. The effect of electrode material on the current under DC bias in each of these three regions is discussed next.

Upon initial application of a DC bias, the response current of a typical ferroelectric film capacitor decreases with time (see Fig. 22). The I-V or J (current density) versus E (electric field) follows a power law dependence of the form:

$$J \propto t^{-n}$$

This behavior is known as the "Curie-von Schweidler" or "Universal law" behavior. It is important to note that the current in this region can be observed in both charging and discharging cycles. It is therefore not true leakage, but corresponds to long range polarization. The mechanisms are still under active debate. However, preliminary data suggest that the process is bulk controlled[81] by the film rather than the electrodes. Furthermore, the power law dependence implies that greater care should be exercised in determining the true leakage mechanisms: for example no I-V or (J-E) data should be presented without first measuring and showing I-t data at selected voltages. Charge and discharge curves should always be measured to differentiate polarization from true leakage current. The transition region is observed at low fields unless "swamped" by the leakage currents.

Fig. 22 and 23 show the I-t curves and the corresponding I-V curves of the four capacitor types shown in Fig. 19. From Fig. 22 it can be seen that for capacitors with Pt bottom electrode ($Pt/PZT/Pt/SiO_2/Si$ and $RuO_2/PZT/Pt/SiO_2/Si$) the leakage currents are low and it is difficult to reach steady state at room temperature, regardless of what top electrode is used. On the other hand, capacitors with RuO_2 bottom electrode ($RuO_2/PZT/RuO_2/SiO_2/Si$ and $Pt/PZT/RuO_2/SiO_2/Si$) have relatively high leakage currents and the I-t curves are almost flat irrespective of the top electrode material type. Fig. 23 also shows that the capacitors with same bottom electrode have similar I-V characteristics regardless of the top electrode material type. At low voltages, the I-V curves of all capacitors usually show ohmic behavior and negligible polarity dependence. This is a characteristic of bulk conduction.

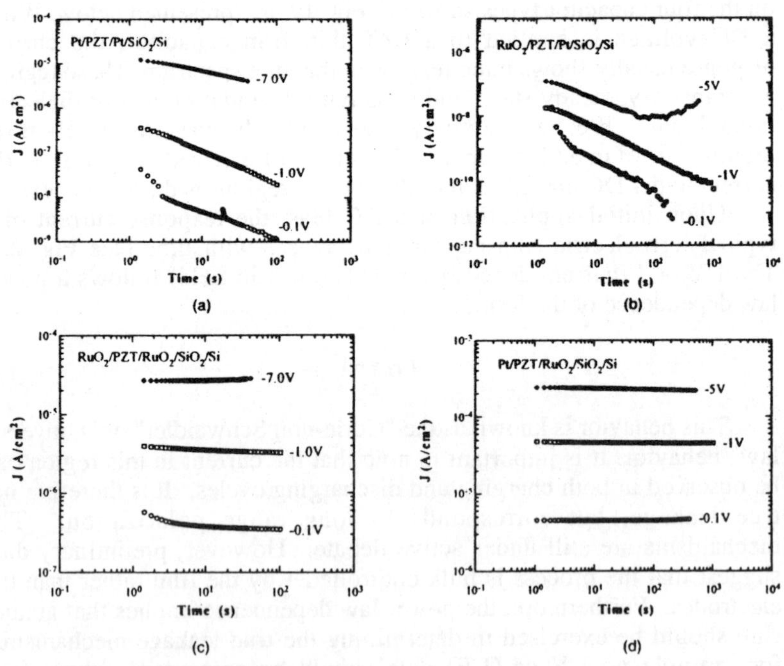

FIGURE 22. I-t curves of the four capacitor types shown in Fig. 19. (Reproduced from Ref. 36).

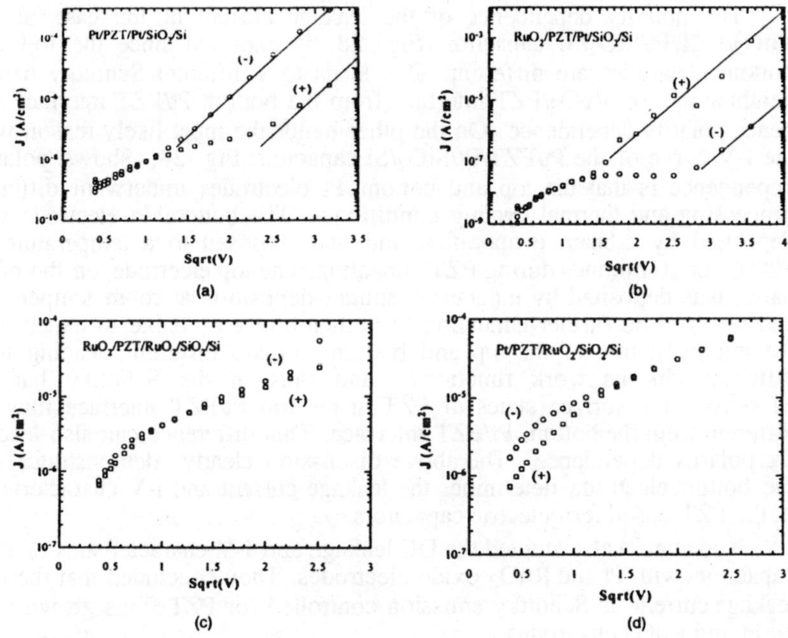

FIGURE 23. I-V curves of the four capacitor types shown in Fig. 19. (Reproduced from Ref. 36).

At higher voltages, a new leakage mechanism sets in the case of capacitors with Pt bottom electrode ($Pt/PZT/Pt/SiO_2/Si$ and $RuO_2/PZT/Pt/SiO_2/Si$). The electrical current has strong polarity dependence and a plot of log (J) vs $\sqrt{V}$ shows a straight line, as shown in Fig. 23. While other leakage current mechanisms such as Poole-Frenkel emission in the bulk may also give a linear plot of log(J) vs $\sqrt{V}$, they are less likely to result in the observed polarity dependence. The most reasonable explanation, therefore, is that the current in the high voltage range is Schottky emission controlled for capacitors with Pt bottom electrode ($Pt/PZT/Pt/SiO_2/Si$, and $RuO_2/PZT/Pt/SiO_2/Si$). However, measurements such as the temperature dependence of leakage currents are needed to confirm the leakage current mechanism.

The polarity dependence of the leakage current in the case of the RuO_2/PZT/Pt/SiO_2/Si capacitor, Fig. 23, is expected since the top and bottom electrodes are different. This leads to a different Schottky barrier height at the top RuO_2/PZT interface from the bottom Pt/PZT interface and hence polarity dependence. On the other hand, the most likely reason why the I-V curve of the Pt/PZT/Pt/SiO_2/Si capacitor, Fig. 23, shows polarity dependence is that the top and bottom Pt electrodes underwent different processing and thermal cycling conditions. The bottom Pt electrode was deposited by e-beam evaporation and was exposed to a temperature of 700°C for 10 minutes during PZT annealing. The top electrode, on the other hand, was deposited by magnetron sputter-deposition at room temperature and received no further annealing. It is therefore reasonable to expect that the microstructure of the top and bottom Pt are different, leading to a difference in the work functions and thus in the Schottky barrier. Moreover, the surface states of PZT at the top Pt/PZT interface may be different from the bottom Pt/PZT interface. This difference can also lead to the polarity dependence. The above discussion clearly demonstrates that the bottom electrode determines the leakage current and I-V characteristics of the PZT-based ferroelectric capacitors.

Yoo et al.[78] also studied the DC leakage and I-V characteristics of PZT capacitors with Pt and RuO_2 oxide electrodes. They concluded that the true leakage current is Schottky emission controlled for PZT films grown both on Pt and RuO_2 electrodes.

It is then obvious that while the oxide electrodes do solve the problem of polarization fatigue, they generally result in PZT capacitors with large leakage currents. We have recently used electrodes which include both Pt and RuO_2[82] "hybrid electrodes" to produce PZT capacitors with both good resistance to polarization fatigue and low leakage current.

Dielectric breakdown can also display a dependence on the electrode type. An example of this phenomenon in PZT has been published by Chen et al.[80]

4. ACKNOWLEDGMENTS

We gratefully acknowledge the contributions of our colleagues including Drs. Orlando Auciello, Daniel Lichtenwalner, Xiaohua Chen, and Y.L. Chen and Mr. Manoj Vellakail, and Cem Basceri, as well as useful discussions with Dr. Scott Summerfelt (TI), and Dr. Dave Kotecki (IBM). We also thank all the authors who kindly gave us permission to include figures from their work in this chapter. This research has been supported by ARPA under contract N00014-93-1-0591.

5. REFERENCES

1. B. GNADE, presented at NATO Advanced Study Institute on Electroceramic Thin Films, Sapri, Italy, (June,1994).
2. P.D. HREN, S. H. ROU, H. N. AL-SHAREEF, K.D. GIFFORD, O. AUCIELLO, and A. I. KINGON, Integrated Ferroelectrics, **2**, 311(1992).
3. H. YOSHIMORI, H. NAKANO, T. MIHARA, H. WATANABE, C. A. PAZ DE ARAUJO, and L. D. MCMILLAN, ISIF 4 Proceedings, 75 (1992).
4. A. I. KINGON, presentation at Symposium I2, MRS Fall Meeting, Boston, Nov. 30, 1994. To be published in MRS Symposium Proceedings, Ferroelectric Thin Films IV, 1995.
5. H. N. AL-SHAREEF, K.D. GIFFORD, S. H. ROU, P.D. HREN, O. AUCIELLO, and A. I. KINGON, Integrated Ferroelectrics, **3**, 321(1993).
6. K.TAKEMURA,T.SAKUMA,S. MATSUBARA, S.YAMAMICHI, H. YAMAGUCHI, and Y. MIYASAKA, ISIF 4 Proceedings, 481 (1992).
7. D. J. EICHORST , T. N. BLANTON, C. L. BARNES, and L. A. BOSWORTH, Integrated Ferroelectrics, **4**, 239 (1994).
8. G. A. C. M. SPIERINGS, J. B. A. VAN ZON, M. KLEE, and P. K. LARSEN, ISIF 4 Proceedings, 280 (1992). Also, presented at ISIF 6, 1994 (Monterey, CA).
9. R. BRUCHHAUS, D. PITZER, O. EIBL, U. SCHEITHAUER, and W. HOESLER, MRS Symp. Proc., **243**, 123 (1992).
10. J. O. OLOWOLAFE, R. E. JONES, A. C. CAMPBELL, P. D. MANIAR, R. I. HEGDE, AND C. J. MOGAB, MRS Symp. Proc., **243**, 355 (1992).
11. F. M. D'HEURLE, International Materials Reviews, **34**, 53 (1989).
12. D. KOTECKI, private communication.
13. S. SUMMERFELT, presentation at Symposium I2, MRS Fall Meeting, Boston, Nov. 30, 1994. To be published in MRS Symposium Proceedings, Ferroelectric Thin Films IV, 1995.
14. K. SREENIVAS, IAN REANEY, T. MAEDER, and N. SETTER, J. Appl. Phys., **75**, 232(1994).
15. T. C. TISONE and DROBECK, J. Vac. Sci. & Tech., **9**, 271 (1971).
16. T. HASE, T. SAKUMA, K. AMANUMA, T. MORI, A. OCHI, and Y. MIYASAKA, presented at ISIF 6, Monterey, CA (March,1994).
17. C. K. KWOK, S. B. DESU, and L. KAMMERDINER, MRS Symp. Proc., **200**, 83(1990).
18. B. A. TUTTLE, R. W. SCHWARTZ, D. H. DOUGHTY, and J. A. VOIGT, MRS Symp. Proc., **200**, 159 (1990).
19. C. LIVAGE, A. SAFARI AND L.C. KLEIN, ISAF 8 Proceedings, 444(1992).

20. J. CHEN, K. G. BROOKS, K. R. UDAYAKUMAR, and L. E. CROSS, MRS Symp. Proc., **243**, 33(1992).
21. H. N. AL-SHAREEF, K. R. BELLUR, O. AUCIELLO, and A. I. KINGON, Ferroelectrics, **152**, 85 (1994).
22. S.Y. CHEN and I.W. CHEN, J. Am. Ceram. Soc., **77**, 2332 (1995).
23. K.G. BROOKS, I. M. REANEY, R. KLISSUSKA, Y. HUANG, L. BURSILL, and N. SETTER, J. Mat. Res., **9**, 2540 (1994).
24. O. AUCIELLO, L. MANTESE, J. DUARTE, X. CHEN, S. H. ROU, A. I. KINGON, A. F. SCHREINER, and A. R. KRAUSS, J. Appl. Phys. , **73**, 5197(1993).
25. W. D. RYDEN, A. W. LAWSON, and C.C. SARTAIN, Phys. Rev. B, **1**, 1494 (1970).
26. C. B. EOM, R. B. VAN DOVER, J. M. PHYLLIPS, D. J. WERDER, C. H. CHEN, R. J. CAVA, R. M. FLEMING, and D. K. FORK, Appl. Phys. Lett., **63**, 2570 (1993).
27. M. L. GREEN, M. E. GROSS, L. E. PAPA, K. J. SCHNOES, and D. B. BRASEN, J. Electrochem. Soc., **132**, 2677 (1985).
28. E. KOLAWA, F.C. TSO, W. FLICK, X. A. ZHAO, E. T-S PAN, and M. A. NICOLET, Thin Solid Films, **173**, 217 (1989).
29. Q. X. JIA, Z. Q. SHI, K. L. JIAO, and W. A. ANDERSON, Thin Solid Films, **196**, 29 (1991).
30. S. BERNSTEIN, T. Y. WONG, Y. KISLER, and R. TUSTISON, J. Mat. Res., **8**, 12(1993).
31. M. DE KEIJSER, G. J. M. DORMANS, and P. J. VAN VELDHOVEN, Integrated Ferroelectrics, **5**, 221 (1994).
32. K. R. BELLUR, H. N. AL-SHAREEF, O. AUCIELLO, and A. I. KINGON, ISAF 8 Proceedings, 448 (1992).
33. D. VIJAY, C. K. KWOK, W. PAN, I. K. YOO, and S. B. DESU, ISAF 8 Proceedings, 408 (1992).
34. O. AUCIELLO, K. D. GIFFORD, AND A. I. KINGON, Appl. Phys. Lett., **64**, 2873 (1994).
35. H. N. AL-SHAREEF, K. R. BELLUR, O. AUCIELLO, and A. I. KINGON, Integrated Ferroelectrics, **5**, 185 (1994).
36. H. N. AL-SHAREEF, A. I. KINGON, X. CHEN, K. R. BELLUR, and O. AUCIELLO, J. Mat. Res., **9**, 2968 (1994).
37. X. CHEN, H. N. AL-SHAREEF, K. R. BELLUR, and A. I. KINGON, Ferroelectrics, **151**, 133 (1994).
38. W. BENSCH, H. W. SCHAMALE, and A. RELLER, Solid State Ionics, **43**, 171 (1990).
39. Powder Diffraction File - Inorganic Phases, ed. by W. F. Mc Clure, File number 36-1394.
40. Powder Diffraction File - Inorganic Phases, ed. by W. F. Mc Clure, File number 38-1433.
41. Powder Diffraction File - Inorganic Phases, ed. by W. F. Mc Clure, File number 33-784.

42. J. CHEUNG, P. E. MORGAN, and R. NEURGAONKAR, ISIF 4 Proceedings, 158 (1992).
43. J. MIZUSAKI, S. YAMAUCHI, K. FUEKI, and A. ISHIKAWA, Solid State Ionics, **12**, 119 (1984).
44. J. Mizusaki, Y. Mima, S. Yamauchi, K. Fucki, and H. Tagawa, J. of Solid State Chemistry, **80**, 102 (1988).
45. R. Dat, D. J. Lichtenwalner, O. Auciello, and A. I. Kingon, Appl. Phys. Lett. **64**, 2873(1994).
46. J. ZHANG, G. CUI, D. GORDON, P. VAN BUSKIRK, and J. STEINBECK, MRS Symp. Proc., **310**, 249 (1993).
47. C.N. SOBLE, II, S. H. ROU, D. J. LICHTENWALNER, T. M. GRAETTINGER, R. WOOLCOTT, JR., O. AUCIELLO, and A. I. KINGON, Appl. Phys. Lett., **61**, 1844 (1992).
48. D. J. LICHTENWALNER, R. DAT, O. AUCIELLO, and A. I. KINGON, Ferroelectrics, **152**, 97 (1994).
49. R.W. SCHWARTZ, R. A. ASSINK, and T. J. HEADLY, MRS Symp.Proceedings, **243**, 264(1993).
50. G. TEOWEE, J. M. BOULTON, S. C. LEE, and D. R. UNLMANN, MRS Symp. Proc., **243**, 255 (1992).
51. K.D. GIFFORD, H. N. AL-SHAREEF, O. AUCIELLO, and A. I. KINGON, MRS Symp. Proc. **243**, 191 (1992).
52. RAINER BRUCHHAUS, H. HUBER, D. PITZER, and W. WERSING, ISIF 3 Proceedings, 453 (1991).
53. S. B. DESU, C. H. PENG, L. KAMMERDINER, and P. J. SCHUELE, MRS Symp. Proc., **200**, 319 (1990).
54. S. B. KRUPANIDHI, D. ROY, N. MAFFEI, and C. J. PENG, ISIF 3 Proceedings, 100(1991).
55. R.E. LEUCHTNER, K. S. GRABOWSKI, D. B. CHRISEY, and J. S. HORWITZ, ISIF 3 Proceedings, 72 (1991).
56. Y. SAKASHITA, T. ONO, H. SEGAWA, K. TOMINAGA, and M. OKADA, J. Appl. Phys., **69**, 8532 (1991).
57. W. T. PETUSKI and D. A. RICHARDSON, ISIF 3 Proceedings, 571 (1991).
58. H. N. AL-SHAREEF, K. R. BELLUR, O. AUCIELLO, and A.I. KINGON, Thin Solid Films, **256**, 73 (1995).
59. H. N. AL-SHAREEF, K. R. BELLUR, O. AUCIELLO, and A.I. KINGON, unpublished results.
60. K. D. GIFFORD, O. AUCIELLO, and A. I. KINGON, Presented at ISIF 6, (March, 1994).
61. H. N. AL-SHAREEF, K. R. BELLUR, O. AUCIELLO, and A. I. KINGON, unpublished results.
62. R. RAMESH, T. SANDS, and V. G. KERAMIDAS, J. Electronic Materials, **23**, 19 (1994).

63. H. M. DUIKER, P. D. BEALE, J. F. SCOTT, C. A. PAZ DE ARAUJO, B. M. MELNICK, J. D. CUCHIARO, and L. D. MCMILLIAN, J. Appl. Phys., **68**, 5783(1992).
64. I.K. YOO and S.B. DESU, Phys. Status Solidi A , **133**, 565 (1992).
65. J. F. SCOTT, C. A. PAZ DE ARAUJO, B. M. MELNICK, L. D. MCMILLAN, and R. ZULEEG, J. Appl. Phys. **70**, 382 (1991).
66. K. AMANUMA, T. MORI, T. HASE, T. SAKUMA, A. OCHI, and Y. MIYASAKA, Proc. Sixth US-Japan Seminar on Dielectric and Piezoelectric Ceramics, 29 (1993).
67. T. MIHARA, H. WATANABE, H. YOSHIMORI, and C. A. PAZ DE ARAUJO, Presented at ISIF 5, 1993.
68. W. PAN AND C. F. YUE, Ceramic Transactions, **25**, 385 (1991).
69. W. L. WARREN, D. DIMOS, B. A. TUTTLE, R. D. NASBY, and G. E. PIKE, Appl. Phys. Lett., **65**, 1018(1994).
70. R. RAMESH, W. K. CHAN, B. WILKENS, T. SANDS, J. M. TARASCON, V. G. KERAMIDAS, D. K. FORK, J. LEE, and A. SAFARI, Appl. Phys. Lett., **61**, 1537(1992).
71. H. WATANABE, T. MIHARA, H. YASHIMORI, and C. A. PAZ DE ARAUJO, ISIF 3 Proceedings, 139 (1992).
72. S. D. BERNSTEIN, Y. KISTER, J. M. WHAL, S. E. BERNACKI, and S. R. COLLINS, MRS Symp. Proc., **243**, 373 (1992).
73. G. TOEWEE, J. M. BOULTON, M. N. ORR, C. D. BAERTLEIN, R. K. WADE, D. P. BIRNIE, and D. R. UHLMAN, MRS Symp. Proc., **310**, 423 (1992).
74. H. N. AL-SHAREEF. K. R. BELLUR, O. AUCIELLO, and A. I. KINGON, Thin Solid Films, **252**, 38 (1994).
75. J. CHEN, M. P. HARMER, and D. SMYTH, ISAF 8 Proceedings, 111(1992).
76. H. WATANABE, T. MIHARA, H. YASHIMON, and C. A. PAZ DE ARAUJO, ISIF 4 Proceedings, 346 (1992).
77. C. A. PAZ DE ARAUJO, J. D. CUCHIARO, M. C. SCOTT, and L.D. MCMILLAN, International Patent Application, International Publication No. WO 93/12542, 1993.
78. I. K. YOO AND S. B. DESU, ISAF 8 Proceedings, 255 (1992).
79. X. CHEN, A. I. KINGON, L. MANTESE, O. AUCIELLO, and K. Y. HSIEH, Integrated Ferroelectrics, **3**, 355 (1993).
80. X. CHEN, A. I. KINGON, K. BELLUR, and O. AUCIELLO, Integrated Ferroelectrics, **5**, 59(1994).
81. R. WASER, presented at NATO Advanced Study Institute on Electroceramic Thin Films, Sapri, Italy, (June, 1994).
82. H. N. AL-SHAREEF, K. R. BELLUR, O. AUCIELLOand A. I. KINGON, Appl. Phys. Lett., **66**, 239 (1995).

PART II

WET CHEMICAL SYNTHESIS AND DEPOSITION

CHAPTER 8

METAL ALKOXIDES - PRECURSORS FOR FERROELECTRIC MATERIALS

N.Ya. Turova
M.V. Lomonosov Moscow State University, Moscow, Russia

M.I. Yanovskaya
L.Ya. Karpov Institute of Physical Chemistry, Moscow, Russia

1. INTRODUCTION

During the last 10 years, decomposition of metal alkoxides $M(OR)_n$ became widely used as a synthetic route to oxide materials. In general, different ways of decomposition of alkoxides leading to oxides are possible:

$$Zr(OR)_4 \begin{cases} \nearrow ZrO_2 + C_nH_{2n} + R_2O + ROH + \ldots \\ \rightarrow ZrO_2 + CO_2 + H_2O \\ \searrow ZrO_2 \cdot nH_2O \xrightarrow{t^o} ZrO_2 + H_2O \end{cases}$$

The first two reactions of pyrolysis (decomposition in plasma and oxidation) give high purity oxides and only gaseous by-products. Volatility of alkoxides allows their application for preparation of films by vapor decomposition (MOCVD). However, it is indeed the third reaction which proved to be most interesting for oxide preparation. This process, viz. the hydrolytic decomposition of alkoxide solutions in organic solvents accompanied by precipitation of hydroxides which form oxides on dehydration, has eventually become the basis for sol-gel technology of oxides.

Sol-gel method can be successfully used in preparation of pure oxides for electronics, optics, ceramics for engines and utilization of radioactive wastes, filters in chemical and food industries. Among important materials which are prepared by this method one should mention ferro- and dielectrics, ion-conducting refractory materials, membranes, protective, decorating films as well as films with special optic and dielectric properties. During the last few years this method has been also applied for synthesis of high-temperature superconductors (HTSC) (see scheme 1).

Application of metal alkoxides in the synthesis of complex oxide materials usually ensures high chemical homogeneity of the products, which is its special advantage in comparison with precipitation from aqueous solutions. In the latter case, precipitation of different M_2O_n occurs at different pH values; moreover, the coexistence in aqueous solutions of M^{n+} with SiO_4^{2-}, PO_4^{3-}, BO_3^{3-} anions is, generally speaking, impossible. Low temperatures of phase formation and sintering as well as considerable shortening of sintering time (in comparison with oxides prepared by alternative techniques) prevent violation of stoichiometry caused by the loss of the most volatile components during thermal treatment.

Presently, the application of alkoxides is becoming an industrial route to oxides. In accordance with the data of the review[1], 800 patents on application of sol-gel technology in the synthesis of oxides were issued and disclosed only in Japan in the period from 1978 through 1986 (the number of patent applications was considerably increased around 1980 and after). Since 1981 several series of International conferences on sol-gel chemistry and applications are held on a regular basis: 1) International Workshop on Glasses and Glass Ceramics from Gels (1981, 1983, 1985, 1987, 1989, 1991); 2) International Conference on Ultrastructure Processing of Ceramics, Glasses and Composites (1983, 1985, 1987, 1989, 1991); 3) Materials Research Society Symposium on Better Ceramics

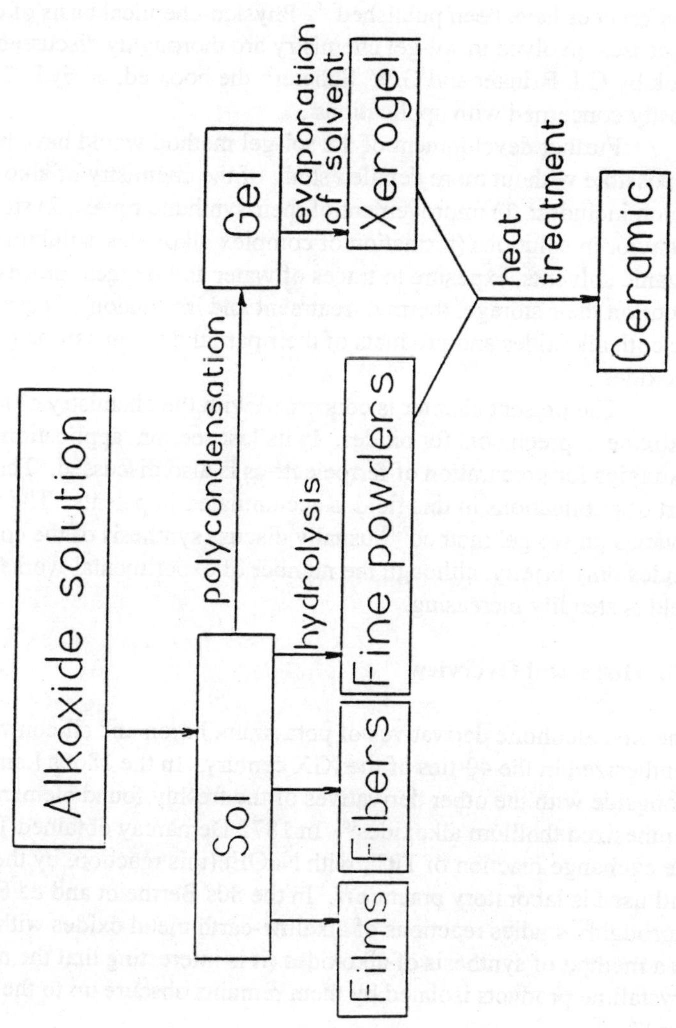
Alkoxide solution
Sol
polycondensation
Gel
evaporation of solvent
Xerogel
hydrolysis
Films
Fibers
Fine powders
heat treatment
Ceramic

SCHEME 1

Through Chemistry (1984, 1986, 1988, 1990). Materials of the Conferences have been published[2-4]. Physico-chemical basis of complex processes involved in sol-gel chemistry are thoroughly discussed in the book by C.J. Brinker and G.W. Scherer[5]; the book edited by L.C. Klein is mostly concerned with applications[6].

Further development of the sol-gel method would have been impossible without more detailed study of the chemistry of alkoxides, which includes: 1) improvement of their synthetic routes; 2) study of their behavior in solutions (formation of complex alkoxides, solubility in organic solvents, exposure to traces of water and oxygen, processes which occur on their storage, thermal treatment and irradiation); 3) structural data for both alkoxides and products of their partial decomposition on the way to oxides.

The present chapter is concerned with the chemistry of metal alkoxides - precursors for oxides. In its last section, application of alkoxides for preparation of ferroelectrics is also discussed. The major part of publications in this field is accumulated in patents. The numerous reviews on sol-gel method[7-15] usually discuss synthesis of the complex oxides only briefly, although the number of experimental works in this field is steadily increasing.

1.1. Historical Overview

The first alcoholic derivatives of potassium, boron and silicon were synthesized in the 40-ties of the XIX century. In the 1860s Lamy alongside with the other derivatives of the freshly found element synthesized thallium alkoxides[16]. In 1875 Demarcay obtained $Ti(OEt)_4$ by the exchange reaction of $TiCl_4$ with NaOEt (this reaction, by the way, is still used is laboratory practice)[17]. In the 80s Berthelot and de Forcrand thoroughly studies reactions of alkaline-earth metal oxides with alcohols as a method of synthesis of alkoxides (it is interesting that the nature of the crystalline products isolated by them remains obscure up to the present day)[16].

Tishchenko's dissertation "On Action of Aluminum Amalgam on Alcohols (Al Alkoxides, their Properties and Reactions)," published in St. Petersburgh in 1899 may be considered as a first systematic study of metal alkoxides which even now has not lost its importance[18]. In this work a few general principles of synthesis of alkoxides have been elaborated; certain regularities in the changes of physico-chemical properties in homologous series and with ramification of the radical (from primary to secondary and tertiary) have been studies; general mechanisms of thermal decomposition have been demonstrated; $Al(OR)_3$ were also suggested as reagents in the aldehyde condensation reaction with the ester formation, the so-called ester-condensation well-known today as the Tishchenko reaction. For a few decades which followed, alkoxides (mostly Na, Al, Ti) remained of interest only for specialists in organic synthesis so far as they were employed as reagents for reduction of carbonyl compounds, catalysts in disproportionation of aldehydes, condensations and polymerisations. Among these works, one of the most interesting dealt with the synthesis of bimetallic alkoxides $M[M'(OR)_n]_m$ by Meerwein and Bersin[16]. They named the new class synthesized by them "Meerwein alkoxosalts" by analogy with the less stable $M[M'(OH)_n]_m$. Another interesting new class - much more reactive than alkoxides themselves - alkoxohalogenides (for example, ClMgOEt) - has also been studied by Meerwein and Schmidt[16]. Considerable contribution to the development of synthetic and physico-chemical methods has been made by the studies of uranium (IV-V) alkoxides, performed during the World War II in Los Alamos by Gilman and coworkers as a part of the Manhattan Project (the goal was to find new volatile compounds necessary for separation of isotopes)[19]. An important step in the chemistry of alkoxides was made in 1950 by Bradley and his coworkers. The characteristic feature of their works is a very broad scope of elements from III to VIII group of the Periodic Table, including lanthanides and actinides. As a result, basic laboratory synthetic routes have been developed. A large number of homologs (including derivatives with highly branched R) synthesized allowed Bradley to formulate the main principle regularity which determines the structures of alkoxides: the molecule is prone to have a minimum degree of association which provides the metal atom with maximum possible coordination number. Taking into account this principle rule, it became possible to make suggestions concerning the structures of associates $[M(OR)_n]_m$ on the basis of the results of molecular weight determinations[20,21]. These works stimulated to a considerable extent the application of the whole complex of modern techniques to physico-chemical studies of metal alkoxides. The direct X-ray structural investigation performed most recently proved to be especially fruitful. Their results not only changed considerably many of the widely accepted views on the structures of alkoxides but often led to considerable revision even of their empirical formulae, thus explaining peculiarities in the chemical and physical behavior of these rather unusual compounds (see below for some details).

At present, we see a new upsurge of interest in the chemistry of alkoxides. It is primarily associated with the search for precursors for synthesis of materials. However, systematic studies of the compounds of this class have been performed only in several laboratories. Bradley at Queen Mary College (London, Great Britain) continues his research of alkoxides, in particular oxoalkoxides of the III group metals[22]. His former student Mehrotra, with his many coworkers in India (University of Rajasthan, Jaipur), is engaged in the synthesis of partially substituted mono-, di-, and even trimetallic alkoxides[23].

In France, Hubert-Pfalzgraf (Universite de Nice) carries out very interesting preparative and structural investigations of Y, Ln, Bi, Nb, Pb, Cu, and other alkoxides and methoxyethoxides (including oxoalkoxides) as precursors for HTSC and piezoelectrics[7,24]. Caulton (Indiana University, Bloomington, Indiana, USA) in his attempt to find new volatile alkoxides of alkaline, alkaline earth metals, Y, La, Ce, bimetallic alkoxozirconates has synthesized and characterized the above mentioned alkoxides and solved a series of exceptionally interesting molecular structures[25]. In the same university, the former student of Bradley, Chisholm, has carried out his classical investigations of alkoxides of I.o and W (III, IV) with metal-metal multiple bonds[26].

For more than 30 years in the Moscow State University (Russia), we have been studying chemistry of simple and bimetallic alkoxides (alkoxoaluminates, -titanates, -zirconates, -niobates, -molybdates, -tungstates) and alkoxocomplexes $M(OR)_nX_m$, where X = Hal,O. Alongside with preparative and structural investigations, we widely use the classical technique of physico-chemical analysis (solubility diagrams in di- and tri-component systems), which allows us to specify the compositions of alkoxide complexes, and to establish the field of existence of the particular compounds in solutions. The practical goal of our work lies in the choice of such homolog among the alkoxides of every particular metal whose properties would prove to be optimal for the oxide material technology. Another important objective of our work is to find the most convenient was of synthesis of such alkoxides[27, 28]. One of the techniques developed by us - electrochemical synthesis[29] - has been used for commercial production of transition metal alkoxides in the Research and Design Institute of Monomers (VNIPIM, Tula, Russia). Synthesized alkoxides are applied for preparation of oxides; these investigations we carry out in the Karpov Institute of Physical Chemistry (Moscow)[30]. These works give the scientific base for the development of sol-gel technology in Russia.

For more than 50 years, hydrolysis of metal alkoxides has been used for preparation of oxide catalysis and absorbents with highly developed surfaces, stable sols and gels of hydroxides as well as for application of oxide coatings[31]. Even before World War II in the State Optical Institute in Leningrad (Russia), techniques for preparation of the SiO_2 - TiO_2 coatings found industrial applications in Schott Glasswerke Co., Ltd. at first for the large scale production of automotive rear-view mirrors, and afterwards for anti-reflection and sun-shielding windows of different compositions (see review by Dislich[33]). Nowadays, works in this direction are being intensively continued in the same company[34, 35], as well as in many other laboratories. Different application techniques (dipping, spinning, spraying) of alkoxide solutions which were used first of all for application of optical films have been successfully extended to application of films for opto- and acoustoelectronics, antistatic, decorating, protective, refractory, ferroelectric films, HTSC.

At present, it is clear that metal alkoxides are indispensable for thin films applications. In this case, advantages of sol-gel method, such as high purity of oxides, low temperatures of their formation, preservation of stoichiometric composition of the starting solution in the oxides formed alongside with the certain versatility and ease of processing techniques - are most clearly seen. On the other hand, relatively high cost of metal alkoxides, which is traditionally considered to be the main obstacle for their extensive application, in this case, does not play any significant role because of small quantities of precursors necessary.

Another rapidly developing application of alkoxides is preparation of fibers[36]. Fibrous gels are drawn from the sols partially hydrolyzed alkoxide solutions at room temperature and then are converted to glass or ceramic fibers by heating. The main compositions of fibers are as follows:

SiO_2, TiO_2, TiO_2-ZrO_2, ZrO_2, Al_2O_3, and lately HTSC[37]. Fibers are produced on a commercial scale by a few Japanese companies: Asahi Glass Co. (SiO_2)[38], Sumimoto Chemical Co., Ltd. (Al_2O_3)[39].

The same method of slow controlled hydrolysis of alkoxide solutions is used for production of thin (30-100 A) sheets of Al_2O_3 by Mitsubishi Mining & Cement Co., Ltd.[40]. Traditionally, sol-gel technology has been used for preparation of special Al_2O_3, SiO_2, TiO_2-based glasses without melting. Gel, which is formed on slow hydrolysis with polymerization of sols is dried and annealed. The most difficult point in production of monoliths is to prevent development of cracks in the process of drying the gel. Several methods are used to prevent this: very slow stepwise thermal treatment[41, 42], supercritical drying[43] (especially for the preparation of thermal insulating materials), and addition to the solutions of drying control chemical additives (DCCA), the most commonly used additive is formamide[44]. The latter technique has been, for instance, successfully applied for optical filters based on SiO_2 doped by rare-earths. Despite the fact that the mechanism of action of these agents is not yet clear, introduction of DCAA opens a new way of modifying the properties of sols and the future oxide materials.

The above discussed applications of sol-gel method concern mostly the SiO_2-based materials, obtained from $Si(OEt)_4$. At present, more than 40% of patents on sol-gel chemistry are concerned with SiO_2-based materials; however, the circle of the compositions is gradually extending. Different oxides in the form of fine powders for applications in capacitor, dipiezoelectric and refractory ceramics have been synthesized. Investigations in this direction begun by the pioneering works of Mazdiyasni performed in the late 60s and early 70s in the Air Force Wright Aeronautical Laboratories, Ohio, USA[45, 46]. He studied the synthesis of fine powders ($BaTiO_3$, $SrTiO_3$, PZT, PLZT, ZrO_2-Y_2O_3) which are of the most practical interest even today. Mazdiyasni also demonstrated the main advantages of application of alkoxides for synthesis of oxides: high purity, chemical homogeneity of the oxides obtained, their high reactivity and ability to sinter in dense ceramic bodies at temperatures far lower than usual.

During the last decade these methods of oxides preparation are being extensively studied in Japan. A large number of complex silicates, aluminates, stannates, germanates, vanadates have been synthesized by

Yamaguchi[47]. Ferroelectric materials from metal alkoxides are obtained by Ozaki with coworkers[9]. According to[48] more than 100 different kinds of fine powders for electronic ceramics have been synthesized by this method to date. High purity fine $BaTiO_3$ for multilayer capacitors, mullite - Al_6Si_2O13, stabilized zirconium dioxide (ZrO_2 + 4-9% Y_2O_3)[49] are examples of industrial products manufactured by this method.

It is also noteworthy that in addition to this relatively new application alkoxides (first of all Si, Al, Ti derivatives) are traditionally consumed in a large scale by technology of materials, where they are used as hydrophobic, cross-linking agents for polyhydroxocompounds, epoxide and polyether resins. The products of their partial hydrolysis and thermolysis - polymetaloxanes - are the components of the heat resistant layers.

2.1. General Characteristics of Alkoxides - Synthesis

Metal alkoxides are the products of substitution of H atoms in the molecules of alcohols ROH by metals; they are formal analogs of hydroxides, but usually much more thermo-stable. Properties of $M(OR)_n$ are determined by the electro-negativity of the metal atom, acidity of the alcohol (pK_{ROH} = 15-19), as well as by the nature of the alkyl radical. The second two factors lead to considerable variety in the properties of alkoxides in comparison with hydroxides. Thus, even within the series of alkaline alkoxides the methylates are definitely ionic, whereas the t-butylates exhibit properties of molecular organic compounds resembling organic molecules. Methoxides are soluble in MeOH, liquid ammonia (but not hydrocarbons), their solutions are electroconductive, they crystallize in the polymeric networks analogous to LiOH (anti-PbO-type), decompose without melting at > 200°C[16]. On the other hand, M^IOBu-t are soluble in hydrocarbons, ethers, and other non-polar solvents; their crystals are built of oligomeric molecular associates, e.g. $[KOBu\text{-}t]_4$ in the form of cubane-like molecules, can be vaporized without decomposition[50]. Variations of the properties in the series of homologs are not so pronounced in the case of other metals (n ≥ 3); however, all methylates have, in fact, polymeric structures (more stable than alkaline derivatives due to the higher covalency of M-O bonds), they are insoluble in organic solvents, do not melt and, as a rule, do not exist in the gaseous form.

The recent progress in the sol-gel technology (especially in the field of ferroelectric and HTSC films) is associated with substitution of metal alkoxides by methoxyethoxides - $M(OC_2H_4OR)_n$ (R = Me, Et) - which are the derivatives of monoethers of ethylene glycol. The second oxygen atom of the ligand ensures the formation of the chelate cycles

$$\begin{array}{ccc} & \overset{R}{\diagup} & \\ \diagdown\!\!\overset{|}{\underset{|}{M}}\!\!\overset{\swarrow O}{\underset{\diagdown O}{}} & & \begin{array}{l} - CH_2 \\ \quad | \\ - CH_2 \end{array} \end{array}$$

which hinders the association of the molecules and results in considerably higher solubility and volatility of methoxyethoxides in comparison with the alkoxides. Besides, methyoxyethoxides are somewhat more stable to hydrolysis and oxidation; especially stable metal hydroxide sols are formed in the mixed H_2O - $CH_3OC_2H_4OH$ systems.

Other alkoxide derivatives which are much more stable than alkoxides them selves, are the products of partial substitution of OR groups by carboxylate- or acetylacetonate-groups $M(OR)_mX_{n-m}$ (X = RCOO, acac); their molecules also involve chelate cycles. Quite a lot of such products have been obtained by Mehrotra with coworkers[23].; they are usually liquids or non-crystalline solids; however, their individuality in many cases seems doubtful. Recently, due to their application for preparation of oxides, they have attracted considerable interest, some crystalline products of the derivatives of this class have been isolated and their crystal structures established.

All metal alkoxides are sensitive to moisture: they decompose even in the presence of traces of water in solvents or in the atmosphere, the products of their partial decomposition usually contain O(oxo) groups. Therefore, all operations with alkoxides are to be performed in inert atmosphere of nitrogen or argon, dried over P_2O_5, zeolite, or other moisture absorbents; only absolute solvents should be used (some considerations concerning preparation of absolute alcohols are discussed below).

At present alcoholic derivatives are known, in fact, for all elements of the Periodic Table, including transuraniums and xenon. The following metal alkoxides are important for preparation of ferroelectrics: alkaline, alkaline-earth, magnesium, lead, rare earth, titanium, zirconium, niobium, tantalum, molybdenum, tungsten. However, ferroelectric materials usually contain different elements as dopants, thus practically all metal alkoxides are used for preparation of ferroelectric materials.

Synthesis of metal alkoxides has been studies for more than a century by different researchers, numerous synthetic methods have been suggested. In the 50s in the course of the systematic studies Bradley suggested a few general techniques[27]. In the present review we would like to differentiate the methods which found certain interest for technological applications and those which are mostly used in the laboratory practice.

2.1.1. Reactions of metals and alcohols

$$M + nROH \rightarrow M(OR)_n + n/2\ H_2\uparrow$$

This method may be used only in the case of the most active metals, with the values of the standard electrode potential in the range of -3.0 ÷ -2.3 V (alkaline and alkaline-earth metals). These metals usually react readily with the first representatives of the homological series of alcohols; however, in order to bring reaction to completion it is important to dissolve the alkoxides obtained, that is why usually a considerable excess of alcohols is used together with heating of the reaction mixture. Aikoxides crystallize usually in the form of solvates, e.g. LiOEt.2EtOH[51], NaOPr-I.3i-PrOH[16], KOMe • eMeOH[52], $Mg(OMe)_2$ • 4MeOH, $Ca(OEt)_2$ • 4EtOH, $Ba(OEt)_2$ • 4EtOH[53]. High temperatures of pyrolysis of $M(OR)_n$) determine the necessity of direct synthesis of non-solvated alkoxides (which are far more reactive and find applications, for instance, as catalysts). One of such methods involves reaction of metal with the stoichiometric amount of alcohol in an inert solvent such as xylene or dioxane. The solvated alkoxide first formed in the process reacts with the new portion of metal while the non-solvated alkoxides are precipitated[54].

In the other variant of this method liquid ammonia is used as a solvent, e.g.:

$$Na + NH_3(liquid) \xrightarrow{-e} [Na(NH_3)_6]^+ \xrightarrow[-NH_3]{+ROH} NaOR\downarrow + 1/2H_2\uparrow$$

This reaction is especially important for the synthesis of the derivatives of higher unsaturated and polyatomic alcohols[16, 55].

Intensification of reactions of metals with alcohols is usually performed by amalgamation with mercury or its salts:

$$2M + Hg^{2+} \longrightarrow M/Hg + M^{n+}$$

Production of NaOR by Mathienson Alkali Works is based, for instance, on the reaction of sodium amalgam (formed as a result of the electrolysis of aqueous NaCl solution with the mercury cathode) with alcohol, NaOR • nROH is isolated from the solutions[16].

In addition to mercury activation of the reactions of metals with alcohols can be performed by I_2, Br_2, $MHal_n$, $M(OR)_n$; good results were obtained when two activation agents were added simultaneously, e.g. $HgCl_2 + I_2$ or $HgCl_2 + Hg(OAc)_2$. Under this condition the following metal alkoxides have been synthesized: $Be(OEt)_2$[56], $Al(OR)_3$ ($Al(OPr\text{-}I)_3$ and $Al(OBu\text{-}s)_3$ are commercially produced by this method)[57], $M(OR)_3$ (M = Y, La, Ln)[45, 58]. The mechanism of this activation lies in formation in the solutions of oxonium complexes $HgI_2 \cdot nROH$ which readily react with the metal. The product of this reaction $MHAL_n(OR)_m$ further interacts with metal to liberate $HgI_2 \cdot nROH$, and so on. The value of the M^{n+}/M-solid is not changed, however over voltage is removed.

It is noteworthy that solubility of the metal alkoxide formed in many cases is even more important for the reaction than the values of the M standard electrode potentials or the mobility of H atoms of the alcohol. The following examples illustrate this statement. Despite higher acidity of MeOH in comparison with EtOH (pK = 15.5 and 16 respectively, the insoluble $Ca(OMe)_2$ is formed very slowly in comparison with the soluble derivatives $Ca(OEt)_2$ or $Mg(OMe)_2$ (both of them crystallize from solutions as solvates)[56]. Aluminum readily reacts with I-PrOH with the formation of the highly soluble $Al(OPr\text{-}I)_3$ even in the absence of the catalyst (pK_{ROH} = 18, Al^{3+}/Al_{solid} = -1.66 V). On the other hand, polymeric $Al(OMe)_3$ and $Al(OEt)_3$ are formed only on prolonged refluxing of the metal with alcohol in xylene (140°C) in the presence of $HgCl_2$ and I_2[57]. Bonding of the polymeric alkoxide in a soluble complex allows to perform reaction of metals with alcohols even for the elements whose standard potentials are close to zero, e.g.:

$$Cu + ROH + Am \longrightarrow Cu(OR)_2 \bullet nAm \bullet mROH$$

$$(R = Me, Et; Am = En, Dipy, Phen)^{59}$$

2.1.2. Anodic oxidation of metals As far as only limited scope of alkoxides can be obtained by direct interaction of metals with alcohols it seemed promising to use anodic oxidation for synthesis of the less active metals. Besides, some of the more active metals (e.g., In) dissolve in alcohols so slowly that application of electrochemical synthesis in these cases is also quite essential. The first attempt in this direction has been made by Szilard in 1906: he performed electrolysis of NaOMe, solutions in MeOH with Cu or Pb anodes[60]. For many years, however, this idea was not receiving the appropriate attention; it was not until 1970 that Monsanto Chem Co. disclosed electrochemical synthesis of Si, Ge, Ti, Zr, Ta alkoxides by anodic dissolution in alcohols in the presence of electroconductive additives (NH_4 Cl or R_4NCl)[61]. In 1974 Lehmkuhl and coworkers elaborated a continuous process of synthesis of insoluble $M(OR)_2$ (M = Fe, Co, Ni; R = Me, Et, n-, t-Bu) in alcohols with $[Bu_4N]$ Br or NaBr as additives[62]. In the early 80s we suggested and introduced into laboratory practice a series of electrochemical techniques for synthesis of the following metal alkoxides: $M(OR)_n$, M = Sc, Y, La, Ln, Ti, Zr, Hf, Nb, Ta, Ni, Cr(III); R = Et, I-Pr, C_2H_4OMe; $MO(OR)_4$, M = Mo, W; R = Me, Et, I-Pr, OC_2H_4OMe as well as methoxyethoxides $M(OC_2H_4OMe)n$, M = Be, Cu, Fe(III), Co, Ni, Cr(III); $[Bu_4N]Br$ or LiCl were used as additives. In contrast to application of NH_4Cl with these additives the process does not cease. The yields of the alkoxides are 60-95%[29, 63-67]. Limitation of the method is associated with refinement of some metals on the cathode which is due to reduction of alkoxohalogenides of metals with low standard potentials formed on the first stage of the reaction. In these cases it is possible to isolate only those alkoxides which are insoluble in electrolyte. Thus among bismuth alkoxides the only derivative which could be synthesized by electrochemistry is polymeric methoxide[68]. Recently, synthesis of the following insoluble derivatives has been also disclosed: $Ni(OR)_2$[66, 69], $U(OEt)_4$[70], $Cu(OR)_2$, R = Me, C_2H_4OMe[67], Am, Ph[71], $Sn(OMe)_2$[72]. It is obvious that when methoxyethanol is used as a reaction medium the potential of the system M^{n+}/M_{solid} is considerably changed. For instance, while anodic oxidation of Fe in aliphatic alcohols leads to polymeric $Fe(OR)_2$ (quite sensitive to traces of oxygen)[62], quite stable soluble $Fe(OC_2H_4OMe)_3$ has been obtained from solutions in methoxyethanol[73], the latter is obviously a very convenient precursor for preparation of Fe_2O_3-based materials.

An important advantage of the described technique lies in its simplicity: metals as starting materials are much more convenient in comparison with anhydrous $MHal_n$ (which are in major cases difficult to access); the consumption of organic solvents is much smaller which also decreases contaminations associated with hydrolysis of the products. Electrochemical method allows to create a highly efficient, continuous technology of alkoxides which yields no wastes or contaminations (its only b-product is hydrogen).

2.1.3. Reactions of oxides or hydroxides with alcohols

$$M^{I}OH + ROH \rightleftharpoons M^{I}OR + H_2O$$

$$M^{II}O + ROH \rightleftharpoons M^{II}(OR)_2 + H_2O$$

Both reactions are used for commercial production of alkaline and alkaline-earth alkoxides from very cheap raw materials. As far as the alkoxides thus formed are soluble in alcohols both processes are reversible. Thus, application of these reactions is expedient in the case of higher alcohols with the boiling temperature higher than 100°C (which allows to remove water from the reaction sphere). When low-boiling alcohols are used the reaction time increases greatly, water is eliminated in the form of an azeotrope with benzene. Sometimes hydrophilic agents (e.g., CaC_2) are added; alkoxides may be also precipitated by addition of acetone (NaOEt)[16, 74].

This method is especially important for synthesis of thallium (I) alkoxides. The most frequently used is TlOEt, obtained as follows:

$$TlOH + EtOH \rightleftharpoons TlOEt + H_2O$$

$$2\ Tl + EtOH + 1/2\ O_2 \longrightarrow TlOEt + TlOH$$

Thallium ethoxide is an oily liquid; in TlOEt-EtOH system separation takes place; alkoxide is accumulated in the bottom (95% TlOEt) while TlOH and water are accumulated in the upper layer. Thus in this case separation of alkoxide from water does not require any special measures[75]. Alkaline metals react in analogous way with alcohols which do not mix with water. For instance, reaction of KOH with Et_2CHOH (pentanol-3) at 120°C also results in formation of two layers; the upper alcoholic layer contains 40% of KOR and 2% of KOH while the lower (aqueous) contains 54% KOH[76].

Application of the reaction of alcoholysis of oxides for synthesis of alkoxides of the metals of III-VIII groups is restricted to "covalent" oxides with low values of lattice activation energies. Usually these are derivatives of M in the higher oxidation states and their interaction with alcohols is complicated by oxidation-reduction processes, which can be illustrated by the following examples:

$$V_2O_5 + 6ROH \xrightarrow{t} 2VO(OR)_3 + 3H_2O\uparrow \text{ (azeotrope with benzene}$$

$$R = Et - i\text{-}Am,\ \text{orange liquid, b.p. } 108\text{-}155/2\,mmHg^{77,78}$$

$$CrO_3 + 2t\text{-}BuOH \xrightarrow{(CCl_4)} \underset{\text{yellow liquid}}{CrO_2(OBu\text{-}t)_2} + \underset{\text{red liquid}}{Cr_2O_5(OBu\text{-}t)_2}$$

$$+ \underset{\text{(in the upper layer)}}{Cr_3O_8(OBu\text{-}t)_2 + H_2O^{79}}$$

$$MoO_3 + 3[CH_3-CH(OH)-]_2 \xrightarrow[\text{b.p.} > 180°C]{} CH_3-CH(O)-CH(O)-CH_3 \cdot MoO_2 \bullet 2R(OH)_2 + H_2O$$ [80]

$$Re_2O_7 + 2t\text{-}BuOH \xrightarrow{[CCl_4]} 2ReO_3(OBu\text{-}t) + HReO_4$$ [81]

colorless crystals, decompose -4°C

$$Os^{VIII}O_4 + (-CH_2OH)_2 \xrightarrow[\text{b.p.}197°C]{} Os^{VI}O[(-CH_2O)_2]_2 + H_2O\uparrow + \ldots$$ [82]

$$H_2XeO_4 + t\text{-}BuOH \xrightarrow{(H_2O)} XeO_2(OH)(OBu\text{-}t) + H_2O$$

$$\downarrow KOBu\text{-}t$$

$$K[XeO_3(OBu\text{-}t)] \bullet t\text{-}BuOH$$ [83]

In conclusion, it should be mentioned that the reactions discussed in the present section are very important for the drying of alcohols[16]. Application of sodium alkoxides, which has been employed in old techniques, can hardly be recommended because the equilibrium of the reaction $NaOEt + H_2O \rightarrow NaOH + EtOH$ is shifted to the left. (The equilibrium constant of the reaction $EtOH + OH^- \rightarrow EtO^- + H_2O$ at 25°C is equal to 0.68[84, 85], which means that in the 0.1 M solution of NaOH in 99% EtOH, 96% of Na is present in the form of NaOEt). There are unfortunately no available thermodynamic data on the equilibria of alkaline-earth metals with water; however, determination of water in alcohols after refluxing over $Ca(OR)_2$ or $Mg(OR)_2$ allows to recommend these alkoxides for the purpose of absolutizing of alcohols. Presumably, low solubility of the hydrolysis products - $MO_x(OR)_{2-2x}$ ensures shift of the equilibrium towards formation of alcohols.

2.1.4. Alcoholysis of MX_n

$$MX_n + nROH \longrightarrow M(OR)_n + nHX$$
$$X = H, Alk, C\equiv C, \equiv N, NH_2, NR_2, N(SiR_3)_2 + SH, \ldots$$

The main idea of these processes lies in the interaction of the "active" hydrogen atom of the alcohols with the "anions" of hydrides alkyls, acetylides, nitrides, amides, dialkylamides, trialkylamides, sulfides, etc., with formation of compounds where H atom is bonded by a strong covalent bond (usually gaseous HX)[16]. Alkaline hydrides of the most active metals (K, Rb, Cs) are used in order to slow down the reaction, sometimes it is necessary simply to avoid explosion.

On the other hand, in order to accelerate alcoholysis of NaH and KH in solutions of benzene or THF 2.2.1. - 2.2.2. - cryptands are used (which effectively chelate alkaline metal into a rather stable complex)[86]. Quite stable, volatile perfluorotert-butoxides were first obtained in reactions of LiH or NaH with $(CF_3)_3COH$ (distillation temperatures at atmospheric pressure 218 and 232°C respectively)[87]. Li and Na hydrides are used as cheaper (in comparison with metals) raw materials for production of the corresponding alkoxides[88].

Alcoholysis of bimetallic hydrides, and primarily $MAlH_4$, is used in the quantitative analysis for the volumetric determination of "active" hydrogen in alcohols as well as for the synthesis of $M^IAlH_n(OR)_{4-n}$ - mild, selective reducing agents for different classes of organic compounds. Among them $MAlH(OBu\text{-}t)_3$ - the only individual compounds isolated in solid state have special functions[89].

Alcoholysis of amides is widely used for synthesis of alkaline alkoxides. Reaction of M^INH_2 with aromatic alcohols is performed in liquid ammonia: $MNH_2 + PhCH_2OH \rightarrow MOCH_2Ph\downarrow + NH_3$ [16, 90]. This reaction was, for the first time, applied for the synthesis of $U(OR)_4$ by Gilman[19], afterwards Bradley and Thomas[21] demonstrated the universal character of this reaction for synthesis of $Cr(OR\text{-}t)_4$, $V(OR)_4$, $Sn(OR)_4$, $Nb(OBu\text{-}t)_5$ which could not be synthesized by other methods. Afterwards Chisholm used this reaction for the synthesis of clusters of the new type $M_2(OR\text{-}I)_6$ (M = Mo, W)[26]. By-products of this reaction -R_2NH are characterized by low boiling points, and thus can be easily eliminated from the reaction mixtures; the yields are quantitative. For preparation of the most difficult for synthesis alkoxides ditrimethylsilylamides $M[N(SiMe_3)_2]$ are used. Due to the bulky and branched ligand these components are always monomeric, highly soluble in non-polar organic solvents and are more reactive than $M(NR_2)_n$. This method was used for synthesis of a series of alkoxides of Mn(II), Cr(II), Pb(II), as well as methoxyethoxides of Cd, Bi, etc.,[91-95]. Alcoholysis of alkyl- and silylamides is a quite universal synthetic route to metal alkoxides in laboratory practice; however, the starting materials themselves are hard to synthesize: they are obtained in very poor yields not quite easily (a few % in the exchange reactions of MCl_n with lithium amides. That is why, if possible, it is always much better to synthesize $M(OR)_n$ directly from chlorides (e.g., $W(OR)_6$, $V(OR)_4$ [96,97]).

In conclusion, we would like to mention alcoholysis of MR_n (accompanied by evolution of hydrocarbons). However, as far as alkanes are not readily available they are used for synthesis of alkoxides very rarely. Only few examples are known: CuOMe[98], $Mn(OPr\text{-}I)_2$ [99], $V(OBu\text{-}t)_4$ [100], $Cr(OR)_2$ [101]. On the other hand, alcoholysis of alkanes is frequently performed to obtain alkoxoalkanes $Mr_n(OR)_{n-m}$, which are used as catalysts in polymerization processes[102]. The latter are quite stable to alcoholysis, and cannot be transferred to alkoxides.

2.1.5. *Exchange reactions of metal halogenides with alkaline alkoxides or ammonia*

$$MHal_n + nM^IOR \longrightarrow M(OR)_n + nM^IHal$$

This method hitherto is the most commonly used for synthesis of metal alkoxides, the reaction is usually performed in alcoholic solutions. That is why at first some considerations on interaction of metal halogenides with alcohols should be made. It has been emphasized by Tishchenko[18] that dissolution of $ZnCl_2$ in alcohols is accompanied by vigorous reaction leading to formation of $ZnCl_n(OR)_{2-n}$, evolution of HCl and further interaction in accordance with the following scheme:

$$ROH + HCl \longrightarrow RCl + H_2O$$

This reaction is catalyzed by $ZnCl_n(OR)_{2-n}$. Solvolysis is characteristic of halogenides as strong Lewis acids and its rate is increasing in the series primary < secondary < tertiary. Indeed, interaction of $ZnCl_2$ with EtOH or $SiCl_4$ with t-BuOH results in formation of $MO_2 \bullet nH_2O$[23]. The following examples illustrate the dependence of the rate of the reaction on the valency of metal (the ionic character of M-Cl bond) and the pK of alcohol:

$$\left.\begin{array}{l} VCl_3 + ROH \longrightarrow VCl_3 \bullet nROH \\ VCl_4 + ROH \longrightarrow VOCl_2 \bullet nROH \end{array}\right\} \quad \begin{array}{l} R = Me,\ Et,\ i\text{-}Pr \\ n = 3;4 \end{array}$$

$$VOCl_3 + EtOH \xrightarrow[{[CCl_4]}]{} VOCl_2(OEt) \xrightarrow[EtOH]{} VOCl(OEt)_2$$

$$VOCl_3 + PhOH \longrightarrow VOCl(OPh)_2 \quad [pK = 10]$$

$$VOCl_3 + p\text{-}ClC_6H_4OH \longrightarrow VO(OC_6H_4Cl\text{-}p)_3 \quad (pK = 8)$$ [103,104]

Despite the fact that the reactions are accompanied by formation of oxochlorides with the partial substitution of chlorine atoms by alkoxide groups they should be avoided in synthesis of alkoxides. Alkoxylation of $MHal_n$ is performed by NaOR or LiOR or by ammonia in alcoholic solution (formally "ammonium alkoxide"). Reaction with NaOR is accompanied by formation of NaCl which is almost completely insoluble in organic solvents and is, therefore, used in the synthesis of soluble alkoxides. One of the first examples of application of such reactions, has not lost its importance as a method of synthesis to this day:

$$TiCl_4 + 4NaOR \xrightarrow{(ROH)} Ti(OR)_4 + 4NaCl\downarrow$$

For many years a lot of studies have been performed in order to improve this method by different ways of component introduction, separation of the products, lowering of the temperature, etc. In all cases the yield of the alkoxide did not exceed 70%[17, 105]. We succeeded in considerable improvement of the yields of $M(OR)_n$ (M = Ge, Ti, La, Ln, Bi, V, Nb, Mo, W, Fe) in these reactions by complete blocking of halogenide alcoholysis: the latter were very slowly introduced either without solvent or in a form of solution or suspension in an inert organic solvent (ether, hydrocarbon, alkylhalogenide) in reaction with the alcoholic solution of NaOR[63, 65, 67, 68, 106, 107]. The process is complicated by the formation of bimetallic complexes $NaM_n(OR)_p$ or $NaM_n(OR)_pCl_q$ which may be decomposed by prolonged refluxing or stirring of the reaction mixture after the addition of the reagents is complete. In the case of $Ti(OR)_4$ due to the low stability of alkoxotitanates it proves possible to bring reaction to completion, however the attempt to synthesize $Zr(OR)_4$ by this reaction is doomed to failure: $NaZr_2(OR)_9$ are quite stable and may even be distilled in vacuo without decomposition[23]. Alkoxochlorides and especially $M(OMe)_nCl_m$ which are formed in this reaction are also usually very stable. Interestingly, that YCl_3 and $NdCl_3$ react with 3 moles of NaOR to give $Y_3(OBu\text{-}t)_8Cl \cdot 2Thf$ and $Nd_6(OPr\text{-}I)_{17}Cl$, while $LaCl_3$ in the same conditions yield $La_3(OBu\text{-}t)_9 \cdot 2Thf$[108, 109]. Rare example of alkoxylation of fluorides with NaOR are the syntheses of $U(CEt)_5$ and $U(OEt)_6$ [110, 111].

In order to synthesize insoluble polymeric alkoxides (primarily methoxides) LiOR is usually used, e.g.

$$UCl_4 + 4LiOMe \xrightarrow{(MeOH)} U(OMe)_4\downarrow + 4LiCl^{19}$$

$$MCl_2 + 2LiOMe \xrightarrow{(MeOH)} M(OMe)_2\downarrow + 2LiCl$$

$$M^{II} = Cr,\ Mn,\ Co,\ Ni,\ Cu^{112}$$

Solubility of LiCl in MeOH, EtOH and n-BuOH is 30.4, 19.6, and 13.9%, respectively. That is why after the refluxing of the reaction mixture and washing off the precipitate with alcohol alkoxides free from LiCl are obtained. However, this reaction in many cases is also complicated by formation of bimetallic complexes. Thus, in accordance with the data[66] the synthesis of $Ni(OR)_2$ suggested by Mehrotra[113] leads, in fact, to $Li_nNi(OR)_{1.8+n}Cl_{0.2}$ (R = Me, Et). In the synthesis of alkoxides highly soluble in hydrocarbons (with branched or chelated radicals) both LiOR or NaOR may be used. The product is usually extracted from the mixture with LiHal or NaHal by hexane, Et_2O or other low-boiling solvents, e.g.:

$$BiCl_3 + 3NaOC_2H_4OMe \xrightarrow[2h]{(Thf)} Bi(OC_2H_4OMe)_3 + 3NaCl\downarrow$$

$$BiBr_3 + 3NaOBu\text{-}t \xrightarrow[24\,h]{(Thf)} Bi(OBu\text{-}t)_3 + 3NaBr\downarrow^{24,93}$$
$$[sublim.\,80°/10^{-2}]$$

$$CuCl + LiOCH(t\text{-}Bu)_2 \xrightarrow[48\,h]{Et_2O} Cu\ OCH(t\text{-}Bu)_2 + LiCl\downarrow^{114}$$

Ammonia in alcoholic solvent is used for alkoxylation much more frequently, than the alkaline alkoxides. This method which was first patented by Nelles in 1939[21,105] remains hitherto the main technological method for production of titanium alkoxides:

$$TiCl_4 + 4ROH + 4NH_3 \longrightarrow Ti(OR)_4 + 4NH_4Cl\downarrow$$

$$R = Me, Et, n\text{-}, i\text{-}Pr, n\text{-}, i\text{-}Bu, n\text{-}, i\text{-}Am, n\text{-}C_6H_{13}, n\text{-}C_8H_{18}^{115}$$

Chloride may be also fixed by $HCONH_2$ or $HCONMe_2$: NH_4Cl is soluble in amides (and nitriles), while $Ti(OR)_4$ are isolated as the bottom layer. However, reactions with gaseous NH_3 in alcohol-benzene (toluene, cyclohexane, CCl_4) mixture are by far the most economical. Other solvents are added to decrease the solubility of NH_4Cl in alcohols (which comprises 3.3% in MeOH, 1.5% in EtOH). Higher concentration of hydrocarbons results in very fine precipitates which cannot be filtered off. It has already been mentioned above that high extent of solvolysis does not allow to perform synthesis of $Ti(OR\text{-}t)_4$ in the conditions described. In order to overcome this obstacle $TiCl_4$ is introduced in reaction with alcohols in the alkaline medium in the form of $TiCl_4 \bullet 8NH_3$ (which crystallizes on bubbling of NH_3 through the solution of $TiCl_4$ in benzene or cyclohexane). On refluxing of this complex with ROH (R = t-Bu, t-Am, $CH_2 = CH\text{-}CH_2$, Et) $Ti(OR)_4$ were obtained in high yields[105, 115].

Application of NH_3 instead of M^IOR allows to overcome formation of bimetallic complexes, thus by this method $Zr(OR)_4$ and $Hf(OR)_4$ were first synthesized[21]. Alkoxides of different elements Si, Ge, V, Nb, Ta, Sb, Mo, W, U, Fe, Re, Etc., were also synthesized in this way. Sometimes instead of NH_3 tertiary amines were used:

$$2ReOCl_4 + 8MeOH + 8R_3N \longrightarrow [O{=}Re(OMe)_2]_2[\mu\text{-}O][\mu\text{-}OMe)_2 + 8[R_3NH]Cl^{116}$$

Application of this method is only limited by high stability and presumably low solubility in alcohols of the complexes of chlorides with ammonia, e.g. $LnCl_3 \bullet 3NH_3$. Besides interaction of $MHal_n$ with "ammonium alkoxides" should not lead to compounds with higher basicity, e.g. $Th(OR)_4$ (its alcoholic solution is alkaline). Such alkoxides are synthesized by reaction of $ThCl_4$ with NaOR.

In one of the variants of this reaction Bradley suggested to use instead of $MHal_4$ their complexes with pyridinium chloride:

$$(PyH)_2ZrCl_6 + 6NH_3 + 4ROH \longrightarrow Zr(OR)_4 + 6NH_4Cl\downarrow + 2Py$$

Such a substitution allows to exclude alcoholysis of the strong Lewis acid and to use a precursor which is much more stable than the chloride; it has been synthesized by reaction:

$$ZrOCl_2 \bullet 8H_2O + 4HCl + 2Py \xrightarrow{[EtOH]} [PyH]_2ZrCl_6 + 9H_2O^{117}$$

This method is the only possible way of the synthesis of Ce(OR) and $Pu(OR)_4$ as far as the corresponding $MHal_4$ have not been described[21].

In addition to $M^{I}OR$ halogenides may also react with other alkoxilation agents. For instance, Mo and W alkoxides and alkoxohalogenides may be synthesized via silicon alkylalkoxides or B, Nb, P alkoxides[118, 119]:

$$MoF_6 + Me_nSi(OMe)_{4-n} \longrightarrow MoF_x(OMe)_{6-x} + nMeF$$

$$WCl_6 + Me_3Si(OMe) \rightarrow WCl_3(OMe)_3 \rightarrow WCl_2(OMe)_4 \xrightarrow{+2NaOMe} W(OMe)$$

$$WF_6 + B(OMe)_3 \longrightarrow WF_5(OMe) + BF_3 + MeF$$

Besides $MHal_n$ carboxylates (and especially acetates) are sometimes used in the synthesis of alkoxides. Their advantage as precursors for these syntheses lies in considerably low stability of alkoxoacetates in comparison with alkoxohalogenides. Indeed, reactions

$$M(OAc)_2 + 2NaOMe \xrightarrow{MeOH} M(OMe)_2\downarrow + 2NaOAc \quad (M = Zn, Cd, Hg)$$

$$M(OAc)_2 + 2LiOR \xrightarrow{EtOH} M(OEt)_2\downarrow + 2LiOAc \quad (M = Cd, Hg)$$

can be performed due to precipitation of polymeric alkoxides[120]; whereas in the synthesis of $Pb(OR)_2$ it is NaOAc, which is insoluble in alcohols:

$$Pb(OAc)_2 + 2NaOR \xrightarrow{(ROH)} Pb(OR)_2 + 2NaOAc\downarrow^{27}$$

$$R = i\text{-}Pr,\ Bu$$

$$Pb(OAc)_2\downarrow + 2NaOBu\text{-}t \xrightarrow[(PhMe)]{} Pb_4O(OBu\text{-}t)_6 + t\text{-}Bu_2O + 2NaOAc\downarrow^{94}$$

The use of halogenides in the same conditions leads to $MHal_n(OR)_{2-n}$ which do not decompose in the excess of $M^{I}OR$, or under refluxing.

Patent[121] deals with the general method of preparation of the stable alkoxide solutions (which is used in technology); it comprises a heterogeneous reaction of Zn, Cd, Ln, In, Pb, Sn, Zr, Sb, Bi, Mn carboxylates in alcoholic solutions with NH_3 or ethanolamines with subsequent purification by ionic exchange[121].

The use of complex cerium nitrate (IV) (which is easily available in comparison with $(PyH)_2CeCl_6$) in the synthesis of $Ce(OR)_4$ proved quite fruitful:

$$(NH_4)_2Ca(NO_3)_6 + \left[\begin{matrix} NH_3 + ROH \\ NaOR \end{matrix}\right. \xrightarrow{(ROH)} Ce(OR)_4 + \left[\begin{matrix} NH_4NO_3 \\ NaNo_3 \end{matrix}\right. + NH_3$$

$$(R = Me,\ Et,\ i\text{-}Pr,\ \text{-}CH_2\text{-}CH_2\text{-})^{122}$$

2.1.6. Interaction of alkoxides with other alcohols or esters (reactions of alcoholic exchange or transesterification)

$$M(OR)_n + nR'OH \rightleftharpoons M(OR')_n + nROH$$

$$M(OR)_n + nR''COOR' \rightleftharpoons M(OR')_n + nR''COOR$$

This method is especially important for the preparation of homologs which are hardly accessible by other techniques. According to the data of 1H NMR spectroscopy the first step of the reaction lies in addition to the free alcohol and presumably destruction of the existing $M(OR)_n$ oligomers[21]. There may be two reasons which allow to shift the equilibrium in the desired direction: lower boiling point of the liberating alcohol allowing to distill it off (possibly as an azeotrope with benzene) or precipitation of the newly formed alkoxide. Tishchenko was the first to use this method for the synthesis of insoluble $Al(OMe)_3$ as well as the derivatives of allyl alcohol and glycol (which do not react with Al-amalgam) from Al(OPR-$I)_3$. The advantages of this method are mild reaction conditions, the absence of the side reactions, associated with the presence of $MHal_n$ and HCl. The use of benzene as a solvent allows to control the substitution (for the synthesis of $M(OR)_m(OR')_n$) or use the stoichiometric quantities of ROH for synthesis of $M(OR)_n$; besides, it allows to lower the reaction temperature, which is especially important in the case of the high-boiling alcohols.

Afterwards Bradley and Mehrotra demonstrated on different examples that high-boiling alcohols may be substituted by low-boiling alcohols; however, in this case considerable excess of R'OH is necessary alongside with repeated treatment (with complete evacuation of the solvent)[21]. This reaction is, in fact, used for synthesis of all methoxides (because of their polymeric character and minimal boiling temperature of MeOH in comparison with other alcohols), and for the derivatives of unstable alcohols which in other cases may undergo decomposition. For instance, trimethylsilanolates $M(OSiMe_3)_n$, the formal analogues of t-butoxides, may be obtained only in this way (usually from ethoxides or isopropoxides). Substitution of OR groups in the polymeric alkoxides is especially difficult to perform. Thus refluxing of $Zn(OMe)_2$ for 120-130 hours with fractionation allows only to obtain $Zn(OR)_n(OMe)_{2-n}$, where R = Et, t-Bu[23]. However, it proved possible to obtain $Be(OMe)_2$ and $Cd(OEt)_2$ from the insoluble $Be(OEt)_2$ and $Cd(OMe)_2$ after the refluxing of the reaction mixture only for a few hours[120, 123]. Substitution by OR groups with branched radicals (especially tertiary) is never complete, presumably because of high stability of the partially substituted alkoxides, e.g.:

$$2Al(OPr\text{-}i)_3 + 4\,t\text{-}ROH \rightarrow [Al(OR\text{-}t)_2(OPr\text{-}i)]_2 + 4\,i\text{-}PrOH \quad (R = Bu,\ Am)$$

$$Zr(OEt)_4 + 3\,t\text{-}BuOH \rightarrow Zr(OBu\text{-}t)_3(OEt)$$

$$Ta(OMe)_5 + 4\,ROH \rightarrow Ta(OR)_4(OMe) + 4\,MeOH \quad (R = i\text{-}Pr,\ t\text{-}Bu)$$

$$2Mg[Al(OEt)_4]_2 + 6\,t\text{-}BuOH \rightarrow Mg[Al(OBu\text{-}t)_3(OEt)]_2^{23}$$

Apart from alcohols, esters are also used as alkoxylating agents for $M(OR)_n$. If the boiling points of the introduced and eliminated alcohols are the same (e.g., I-PrOH and t-BuOH, b.p. 82.4°C), application of AcOBu-t allows to obtain t-butoxides by distilling off the low-boiling ester:

$$Zr(OPr\text{-}i)_4 + 4AcOBu\text{-}t \longrightarrow Zr(OBu\text{-}t)_4 + 4AcOPr\text{-}i$$
$$\text{b.p. }98^\circ C \qquad\qquad \text{b.p. }89^\circ C$$

In many cases application of esters in contrast to alcohols results in complete substitution of OR-groups, which (according to Bradley) is due to the stronger coordination of the carbonyl oxygen in comparison with the alkoxide oxygen on the first stage of the reaction[21]. Application of esters in the synthesis of trialkylsilanolates is expedient due to their high stability in comparison with R_3SiOH which readily undergo polymerization and oxidation:

$$Ti(OR)_4 + 4AcOSiR_3 \longrightarrow Ti(OSiR_3)_4 + 4AcOPr\text{-}i$$ [23]

2.1.7. Redox reactions - Oxidation of organometallic compounds

Only a few of rather numerous reactions of oxidation of alkoxides by oxygen, peroxides, halogens and other compounds may be used for synthesis of $M(OR)_n$ (with considerable yields). These reactions are known for uranium alkoxides[19, 20], e.g.:

$$5U(OEt)_4 + O_2 \longrightarrow [U(OEt)_5]_2 + UO_2 \quad (\text{yield } 90\%)$$

$$U(OEt)_4 \xrightarrow{Br_2} U(OEt)_4Br \xrightarrow{NaOEt} [U(OEt)_5]_2$$

$$[U(OEt)_5]_2 \xrightarrow[(-NaBr)]{NaOEt\ +\ [PhCH_2O\text{-})} U(OEt)_6 \quad (\text{yield } 50\%)$$

Alkoxides of Mo and W with the metal-metal bonds are readily oxidized to M(VI) alkoxides:

$$Mo^{III}_2(OR)_6 + 2O_2 \longrightarrow 2MoO_2(OR)_2 + 2RO\bullet \quad (R = i\text{-}Pr,\ t\text{-}Bu)$$

$$Mo^{IV}_2(OR)_8 + O_2 \longrightarrow 2MoO(OR)_4 \ (R = i\text{-}Pr,\ neo\text{-}Am;\ t\text{-}Bu\text{-}conc.) > 0.25M/\ell)$$

$$W_2(OR)_6 + O_2 \longrightarrow W_2O_3(OR)_6^{124}$$

Oxidation of $Cr(OR)_3$ or $Cr(C_6H_6)_2$ (e.g., by t-butylperoxide) is the only way to synthesize $Cr(OR\text{-}t)_4$ - one of the unique compounds of Cr(IV); they are quite stable to hydrolysis and oxidation, presumably because of high stability of the tetrahedral molecules with all the central atoms hindered by t-radical. On the other hand, primary and secondary ROH reduce $(Cr(OR)_4$ to $Cr(OR)_3$[20, 125, 126].

Reduction reactions are primarily characteristic of the compounds of Cr, Mo, and W(VI). The UV-irradiation of the alcoholic solutions of $(NH_4)_2CrO_4$ or $C_6H_6Cr(CO)_3$ results in $Cr(OR)_3$[127, 128]. Reaction of pinacol $[(CF_3)_2C\text{-}OH]_2$ with aqueous solutions of K, Cs or Et_4N chromates leads to alkoxochromates (V) $MCrO(RO_2)_2$; they are stable to hydrolysis due to chelating of the ligand[129]. For W alkoxides reduction of very stable $W(OAr)_6$ by K (in the presence of Reney nickel) to $W(OPh)_5$ and $W(OC_6H_4Me\text{-}p)_5$ is of practical interest[130]. $Ti(OR)_3$ (R = Et, Pr, Bu) were obtained by reduction of $Ti(OR)_4$ by Na in alcoholic solutions (i.e. by hydrogen in statu nascendi) or by K in ether[105]. Recently, $Ce_4O(OPr\text{-}I)_{13}$ has been obtained in the $(\text{-}CH_2OMe)_2$-I-PrOH mixture by irradiation of $Ce(OPr\text{-}I)_4 \bullet$ I-PrOH by the visible light[131]. Partial reduction of alkoxides occur in the course of electrolysis of $M(OR)_n$, and is observed in the processes of electrochemical synthesis of titanium and niobium alkoxides: cathode reduction leads to colored $M(OR)_{n-1}$. When electrolysis with Nb anode and Pt cathode has been performed oxoalkoxoniobates $NbO(OR)_3$ (R = Me, Et, Bu), the products of oxidation of the $Nb(OR)_4$, were isolated[132]. Although the reduction products are highly unstable (they cannot be used for synthesis of oxides) their formation in the course of electrochemical synthesis should be taken into consideration, because their subsequent oxidation by O_2 (or even its traces) leads to oxoalkoxides.

Butlerov and Francland in the XIX century made an attempt to synthesize alkoxides by oxidation of MR_n. However, the decomposition of the first formed peroxides

$$ZnR_2 + O_2 \longrightarrow Zn(OOR)_2$$

is a rather complicated process, alkoxide is only one of the products (its formation is usually registered by the presence of alcohols in the hydrolysis products of the reaction mixture)[133].

In conclusion of this part of the chapter, we would like to give some practical recommendations concerning the methods of absolutizing of alcohols and synthesis of metal alkoxides which are most frequently used in preparation of ferroelectrics. Alcohols are usually absolutized by the most readily available metal alkoxides Mg, Ca, Al. These alkoxides are synthesized by reactions of metals with absolute alcohols, introduced into the alcohol which is to be absolutized (H_2O content ≤ 1%); refluxing of the reaction mixture for 4-6 hours is performed, then absolute alcohol is distilled off with a delegmator. 5-8 g of Mg or Al or ~ 20 g of Ca is necessary for absolutizing of 1ℓ of alcohol. $Mg(OMe) \cdot 4MeOH$ is convenient for absolutizing of MeOH, $Ca(OEt)_2 \cdot 4\,EtOH$ (or $Mg[Al(OEt)_4]_2$) - for EtOH; $Al(OR)_3$ - for BuOH and I-PrOH, $Mg(OC_2H_4OMe)_2$ - for $MeOC_2H_4OH$. In the two latter cases it is also useful to reflux the alcohols thus obtained with zeolites (with the pore size < 5A) (as far as the products of partial hydrolysis of alkoxides are not completely eliminated from the reaction sphere). The content of H_2O in the alkoxides prepared is ≤ 0.02% (which has been determined by Fischer technique).

For synthesis of solutions containing two or more alkoxides for ferroelectric applications it is very convenient to introduce alkaline and alkaline earth metals in the form of metals in the alcoholic solutions of $M(OR)_n$ (M = Ti, Zr, Nb, etc.; R = Et, I-Pr, Bu, C_2H_4OMe).

Table 1 lists the homologs for each particular alkoxide which seem most useful for applications: the easiest for synthesis, soluble in parent alcohol or methoxyethanol (the most universal solvent now used for preparation of alkoxide solutions).*

The electrochemical synthesis of $M(OR)_n$ (method 2) is carried out in a cell without diaphragm with an anode made from the corresponding metal, while Pt or stainless steel are used as cathodes in absolute alcohols with U = 110-220 V, current density 0.1-0.4 A/cm.^{2} • $[R_4N]$ Br, LiCl or LiBr are used as conductive additives (in their presence the process does not slow down). The yields of the products are 60 - 95%. This method is used both for synthesis of reagents $M(OR)_n$ where M = Y, Ti, Zr, Hf, Nb, Ta, Mo, W, Cu as well as solutions containing a few alkoxides prepared for this films applications in the Institute of Monomers (Russia)**

3. PHYSICO-CHEMICAL PROPERTIES OF ALKOXIDES

3.1. Variations in Homologous Series

High donor activity of the OR group oxygen atom determines possibility of formation of the M-OR-M bridges which are responsible for association of molecules into oligo- and polymeric structures. The main factor which determines all the physico-chemical characteristics of the alkoxides is the degree of association. The latter usually decreases in the homologous series from methoxides to ethoxides, and from the derivatives of normal ROH to branched primary and then secondary and tertiary R. The branched tertiary radicals present considerable sterical hindering for oxygen atom in formation of new donor-acceptor bonds, therefore $M(OR\text{-}t)_n$ always demonstrate minimal degree of association among their homologues; they are the most soluble and volatile.

---

* $Cu(OR)_2$ can be dissolved only due to formation of complexes with amines $H_nN(C_2H_4OH)_{3-n}$

** VNIPIM, prosp. Lenina, 106-a, Tula, 300600, Russia. Distributor of alkoxides in North America and Asia, James F. Scott.
FAX: (719) 598-3437 (USA) Symetrix International, Inc., 5055 Mark Dabling Blvd., Suite 100, Colorado Springs, CO 80918, USA.

TABLE 1. Synthesis and properties of some metal alkoxides

Alkoxide	Synthesis technique	Appearance	M.p., °C	B.p., °C/mmHg	References
$Mg(OMe)_2 \bullet 4MeOH$ 1		colorless cryst.	decomp. 40°	--	[53]
$M(OC_2H_4OMe)_2$, M=Mg, Sr, Ba	1	viscous liquids	--	--	[67]
<<$M(OPr\text{-}I)_3$>> = $M_5O(OPr\text{-}I)_{13}$ M=Sc,Y,La,Ln	2;5	colorless cryst.	120(La)	Subl. $150/10^{-2}$ (Sc)	[29,63,58,134]
$M(OC_2H_4OMe)_3$ M=Sc,Y, La,Ln	1;2	colorless cryst.	--	decomp.	[67,24]
$Pb(OPr\text{-}I)_2$	5	colorless cryst.	decomp. > 100	--	[27]
$Pb(OC_2H_4OMe)_2$	4	colorless cryst.	--	$160/10^{-4}$ (subl. with decomp.)	[92]
$Ti(OPr\text{-}I)_4$	2;5	colorless cryst.	~20 n_p^{10} 1.467 d_4^{20} 0.971	49/0.1	[29,105,135]
$Ti(OBu\text{-}n)_4$	2;5	light-yellow liquid	n_D^{20} 1.493 d_4^{20} 0.993	142/0.1	[29,105]
$Ti(OC_2H_4OMe)_4$	2	colorless liquid	d_4^{18} 1.191 n_D^{18} 1.508	182/3	[1105]
$Zr(OPr\text{-}I)_4 \bullet I\text{-}PrOH$	2;5	colorless cryst.	135 (with decomp.)	160/0.1	[23,28,117,29, 135]
$Zr(OC_2H_4OMe)_4$	2;6	colorless liquid	--	$150/10^{-4}$	[67]

Alkoxide	Synthesis technique	Appearance	M.p., °C	B.p., °C/mmHg	References
$Hf(OC_2H_4OMe)_4$	2;6	-"-	--	$150/10^{-4}$	[67]
$Bi(OEt)_3$	5	colorless cryst. (dark on irradiation)	decomp. 105	--	[68]
$Bi(OC_2H_4OMe)_3$	5;6	colorless cryst. or liquid	45-48	$150/10^{-4}$	[67,93]
$Nb(OPr\text{-}I)_5$	2;5	colorless cryst.	79	60-70/0.1 (subl.)	[29,136,137]
$Nb(OBu\text{-}n)_5$	2;5	yellow liquid	--	197/0.15	[29,137]
$Nb(OC_2H\ OMe)_5$	2;6	-"-			
$Ta(OPr\text{-}I)_5$	2;5	colorless cryst.	84 (subl.)	122/0.1	[29,137]
$Ta(OBu\text{-}n)_5$	2;5	yellow liquid	--	217/0.15	[29,137]
$MoO_2(OC_2H_4OMe)_2$	6	colorless cryst.	102 (with decomp.)	$>100/10^{-4}$	[67,138]
$MoO(OEt)_4$	2;5	yellow liquid	--	decomp. 140	[65]
$WO(OC_2H_4OMe)_4$	2	colorless liquid	--	$>100/10^{-4}$	[67]
$WO(OEt)_4$	2	colorless cryst.	126	subl. $125/10^{-2}$ decomp. 170	[64,139]
$Cu(OMe)_2$	2;5	blue cryst. powder	--	decomp.	[71,140,141]
$Cu(OC_2H_4OMe)_2$	2	-"-	--	decomp.	[67,140]

These regularities may be demonstrated using LiOR as an example. Methoxide and ethoxide crystallize from solutions in the form of LiOR•2ROH, they are soluble in alcohols (solubilities of LiOMe and LiOEt at 20°C are 12.0 and 16.4%, respectively)[51]. Their solutions are electroconductive (the values of λ for solutions with the concentration of 3.10^{-3} mol/l are as follows: LiOMe - 92, LiOEt - 40, LiOPr - 20)[142]. Heating results in desolvation of LiOR•2ROH under the solution at 28 and 55°C for ROH - at 40 and 110°C, respectively. Figure 1 demonstrates that desolvation changes the sign of ΔH of dissolution to the opposite: from endothermal for LiOR•2ROH, to exothermal for LiOR. The difference can be easily explained by the difference in the structures of precipitates. LiOMe has a rigid polymeric structure analogous to LiOH (see 3.2), whereas crystal structure of the solvate is built of oligomeric molecules linked by relatively weak bonds. LiOMe is not volatile, LiOEt can be sublimated at $150°C/10^{-2}$ mmHg; LiOPr-I sublimes at $100°C/5.10^{-2}$ mmHg, and in contrast to methoxide and ethoxide it is soluble in nonpolar organic solvents (benzene, cyclohexane, etc.). LiOBu-t sublimes readily (sublimation temperature $110°C/10^{-1}$ mmHg), there are fragments of hexameric molecules in its mass-spectra; different authors give different values of molecular association (2-11) for LiOPr-I and the only value (6) - for LiOBu-t. Chemical shifts in the ^1H NMR spectra of LiOBu-t considerably exceed the corresponding values for free t-BuOH, which testifies to the more covalent character of Li-O bond even in comparison with O-H[16, 143]. LiOBu-t is readily soluble in hydrocarbons and ether but, in fact, does not dissolve in the parent alcohol (1.3%). Although there are no data concerning crystal structure of LiOBu-t its hexameric molecules are presumably analogous to $(NaOBu\text{-}t)_6$ or enolate $[Me_3C\text{-}C(=CH_2)OLi]_6$[144, 145]: the hexagonal prisms with the alternating Li and O vertexes.

In all the homologous series methoxides have peculiar properties - they are polymeric, non-soluble in organic solvents, non-volatile and do not melt. Sometimes such properties are also characteristic of other alkoxides of normal structures (especially ethoxides and propoxides).

In the homologous series of the oligomeric $M(OR)_n$ all parameters which are associated with the degree of ordering, or molecular packing density, such as melting temperatures, density and the like decrease regularly. Boiling and sublimation temperatures as well as ΔH of evaporation increase in the same series together with the increase of molecular weight; besides the effect of chain entanglement also takes place, which influences the changes of viscosity, (e.g., in the series of $Ge(OR)_4$, see Fig. 2). In the homologous series of $Ge(OR)_4$ and $Ti(OR)_4$ refractive indexes change in different directions; however, molecular refraction (MR_D) as usual increases with the increase of molecular weight. The influence of branching of R on the changes in boiling points may be illustrated in the series of isomeric amyloxides of Ti, Zr, Nb, Ta[149] and Al[150, 151] (Fig. 3). In the series of primary $M(OR)_n$ boiling points decrease with branching of R if the degree of association "n" remains constant (4 for $Al(OR)_3$, 1 for $Ti(OR)_4$). Interestingly, the boiling point of $M(OCH_2CMe_3)_n$ - neoamyloxides in all the series remains as low as in the case of $M(OAm\text{-}t)_n$ (for both isomers of $Al(OR)_3$ n=2). In the case of $M(OR\text{-}s)_n$ boiling points of all three isomers are very close to one another and decreases greatly for $M(OR\text{-}t)_n$, although the degree of association in this case, in fact, does not change (it is 1 for the derivatives of Ta, Nb, Ti and - 2 for Al and Zr). Thus, in this case the temperature of transition into gaseous phase turns out to be more sensitive to steric hindrance than to the degree of association.

It is clear that branching of the radical should lead to the increase of the solubility of alkoxides in organic solvents (especially nonpolar); however practically no quantitative data are available. The reasons for such situation have been discussed below (see 3.3).

The state of alkoxides in the gaseous form seems very important in connection with the problem of purification of alkoxides. The mass-spectra of the volatile products testify to the existence of associates which usually have the same size as those in condensed state (according to the data of X-ray structural analysis[28]).

3.2. Types of Molecular Structures

The data on structures of oligomeric and polymeric alkoxides are fundamental properties of compounds which present practical interest in connection with the problems of solubility in organic solvents (polar and nonpolar), solvation, etc.

All structural investigations have been performed during the last 30 years; however, they widened considerably recently. To date, about a few

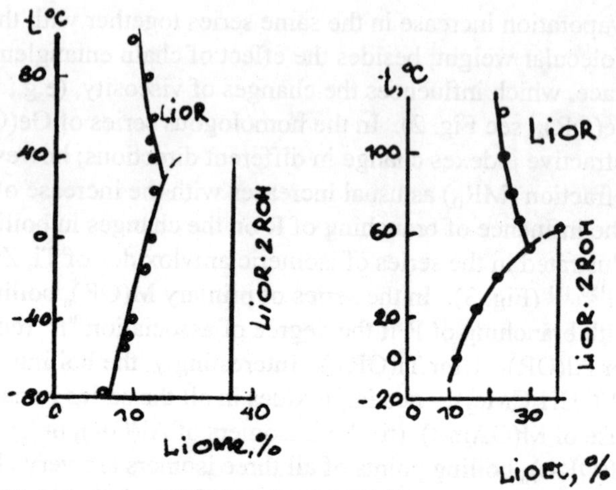

FIGURE 1. **Solubility in the system LiOH-ROH (R = Me, Et).[51]**

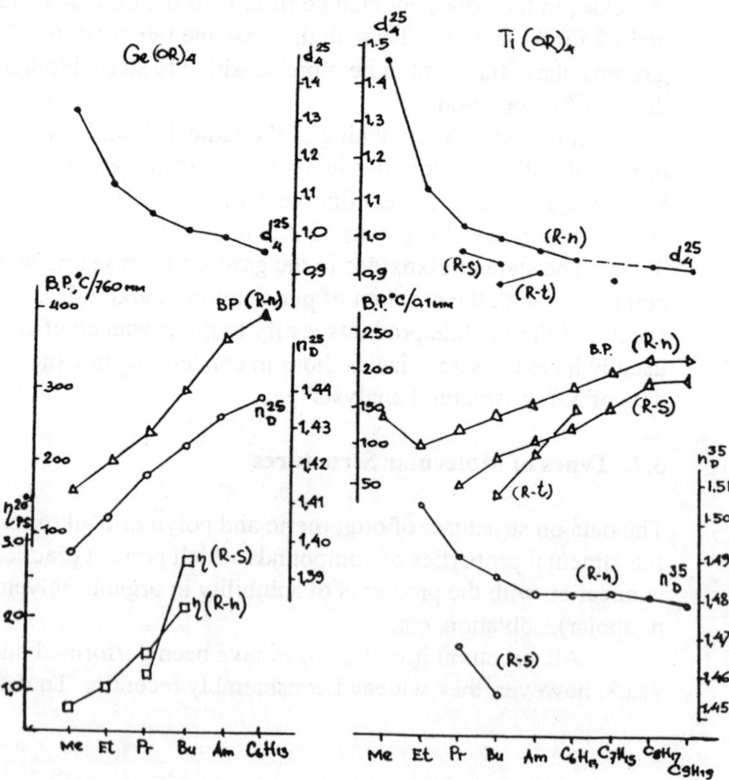

FIGURE 2. **Variations of boiling points (b.p.), refractory indices (n_D) and viscosity (η) in the homologous series $Ga(OR)_4$ and $Ti(OR)_4$.[23,105,146-148]**

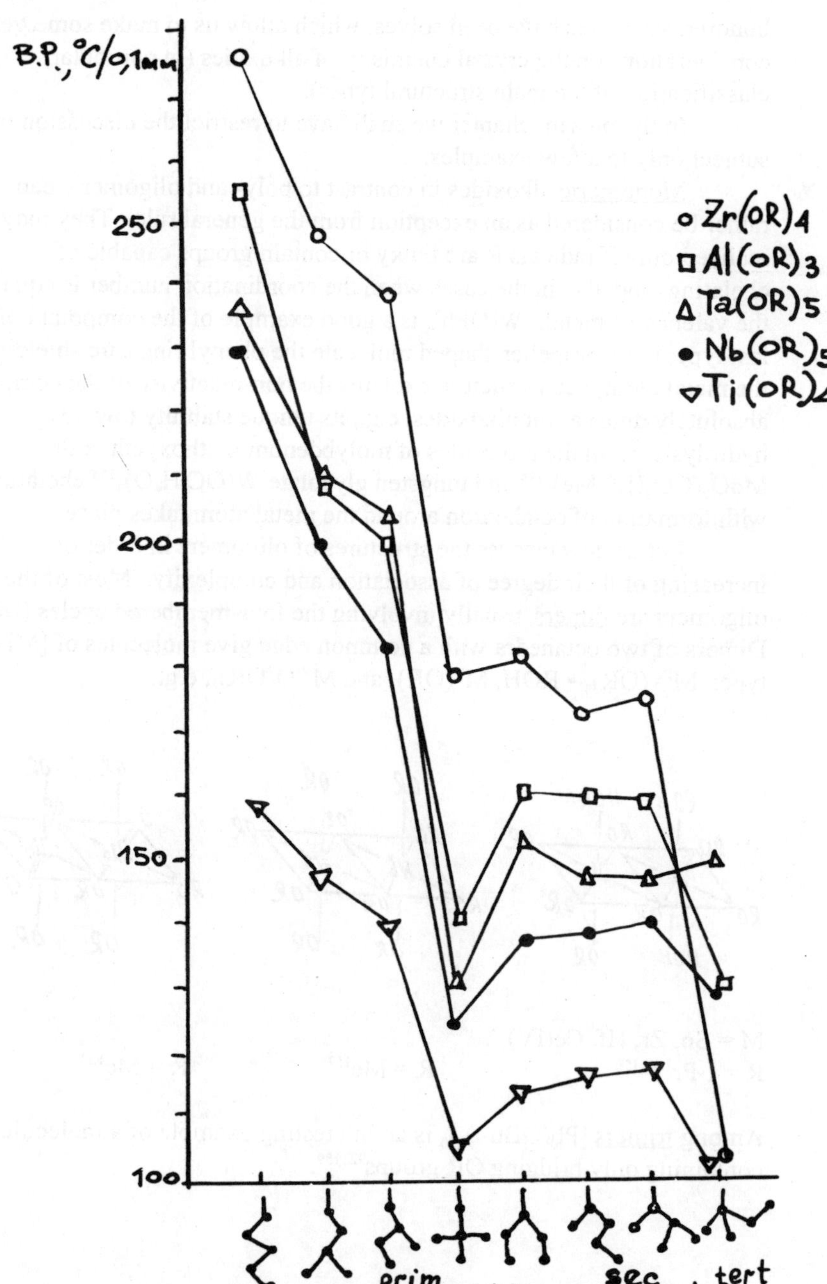

FIGURE 3. Variation of boiling points in the series of isomeric amyloxides with ramification of R.[23]

hundred structures have been solves, which allow us to make some general considerations on the crystal chemistry of alkoxides (in particular, classification of the main structural types).

In the present chapter we shall have to restrict the discussion of the subject only to a few examples.

Monomeric alkoxides in contrast to poly- and oligomeric can rather be considered as an exception from the general rule. They may be realized only if radicals R are bulky or contain groups capable of chelating, and also in the cases when the coordination number is equal to the valency of metal. $W(OPh)_6$ is a good example of the compound of the last type; in its propeller shaped molecule the phenyl rings are shielding the metal atom; such structure explains the non-reactivity of the compound absolutely unusual for alkoxides, e.g., its unique stability towards hydrolysis[152]. In the molecules of molybdenum methoxyethoxide $MoO_2(CO_2H_4OMe)_2$[138] and tungsten glycolate $W(OC_2H_4O)_3$[153] chelating with formation of octahedron around the metal atom takes place.

Let us now discuss the structures of oligomers in order of increasing of their degree of association and complexity. Most of the oligomers are dimers, usually involving the four-membered cycles $[M_2O_2]$. Dimers of two octahedra with a common edge give molecules of $[ML_5]_2$ type: $M^{IV}(OR)_4 \cdot ROH$, $M^V(OR)_5$ and $M^{VI}O(OR)_4$, e.g.:

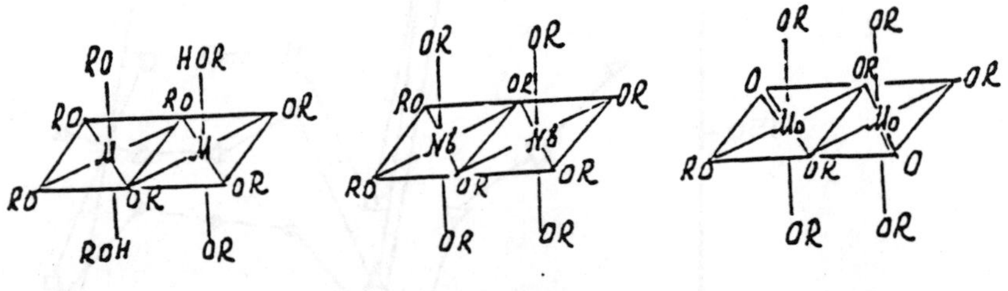

M = Sn, Zr, Hf, Ce(IV)
R = I-Pr[154, 156] R = Me[157] R = Me[158]

Among trimers $[Pb(OBu\text{-}t)_2]_3$ is an interesting example of a molecule containing only bridging OR groups[92, 159]:

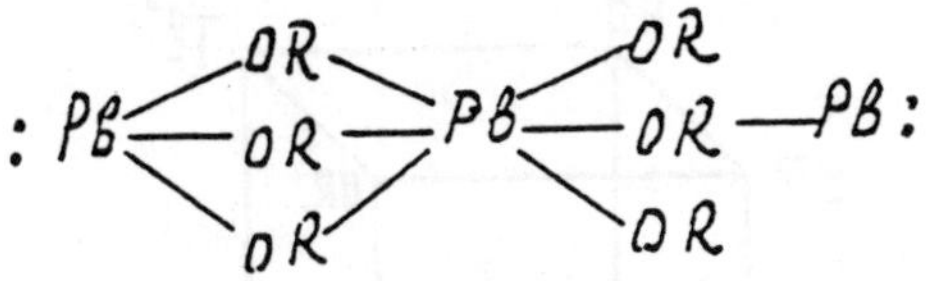

All tetrameric molecules may belong to one of two possible types which may be represented by different combination of the two pairs of octahedra, combined by a common edge. In one case the lines between the two centers of the octahedra of each of the pairs are parallel; whereas in the other case they are perpendicular to each other. As a result, in the first case a "planar" complex of four octahedra lying on the faces is formed, in the second case the cubane-like associate is formed. Indeed, in the tetrameric titanium methoxide and ethoxide $[Ti(OR)_4]_4$, as well as in $[Ti(OMe)(OEt)_3]_4$ with the structures of the first type all four Ti atoms are located in one plane[148, 160, 161].

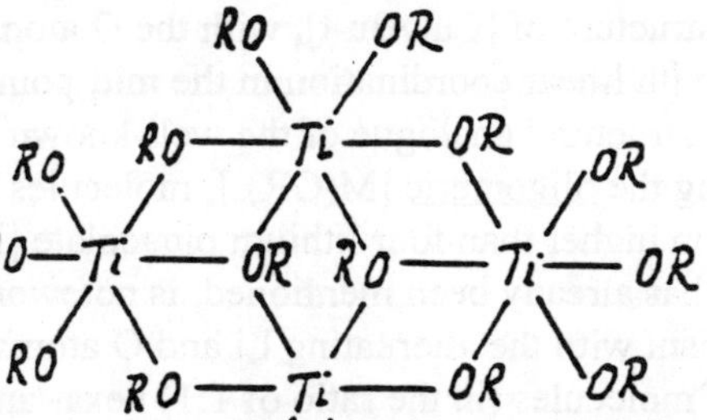

$[Nd(OPr\text{-}I)_3 \bullet I\text{-}PrOH]_4$ [152], the methoxyethoxide cation $[Mg_4(OC_2H_4OMe)_6(MeOC_2H_4OMe)_2]^{2+}$ [163] and the $[W(OEt)_4]_4$ cluster[164] have the analogous structures.

The cubane-like structures of alkoxides have been predicted by Sidgwick on the basis of determination of molecular weight and parahor of the oily TlOEt[23].

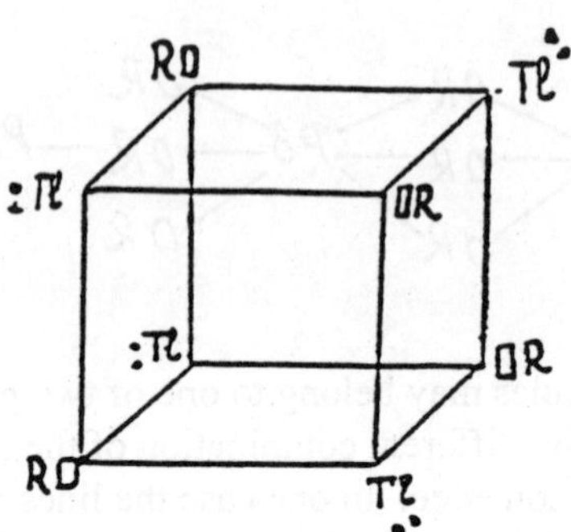

Afterwards this structure has been confirmed in the crystalline $[TlOMe]_4$[165], $[TlOSiPh_3]_4$ [166], as well as in t-butoxides and silanoxides of alkaline metals - $[MOBu\text{-}t]_4$, M = K, Rb, Cs [50], $[MOSiMe_3]_4$, M = Li, Na, K, Rb, Cs [167]. The cubanes are presumably more stable than "planar" octahedra arrangements: tetramers of alkaline metals in contrast to $[Ti(OR)_4]_4$ exist both in gaseous phase and in solutions.

Another type of the tetranuclear alkoxides - square - has been found in the structure of $[CuOBu\text{-}t]_4$ with the O atoms in its vertexes and Cu(I) atoms with linear coordination in the mid-points of the edges. This molecule is a structural analogue of the well-known anion $[Cu_4O_4]^{4-}$ [168].

Among the oligomeric $[M(OR)_n]_x$ molecules with the degree of polymerization higher than four lithium pinacolate $[LiO\text{-}C\,(=CH_2)(t\text{-}Bu)]_6$, which has already been mentioned, is noteworthy. Its molecule is a hexagonal prism with the alternating Li and O atoms in its vertexes[145]. Two types of molecules (in the ratio of 1:1) hexa- and nonamers coexist in the very interesting structure of NaOBu-t). The first is an analogue of the Li complex, while the second may in the same terminology be described as a hexagonal bipyramid with truncated apical vertexes. In the gaseous phase, however, only $[NaOBu\text{-}t]_6$ exist[144]. The new type of structures has been found recently among methoxyethoxides. Cyclic decamer $Y(OC_2H_4OMe)_3$ contains 7-coordinated Y-atoms in the following elementary units[24]:

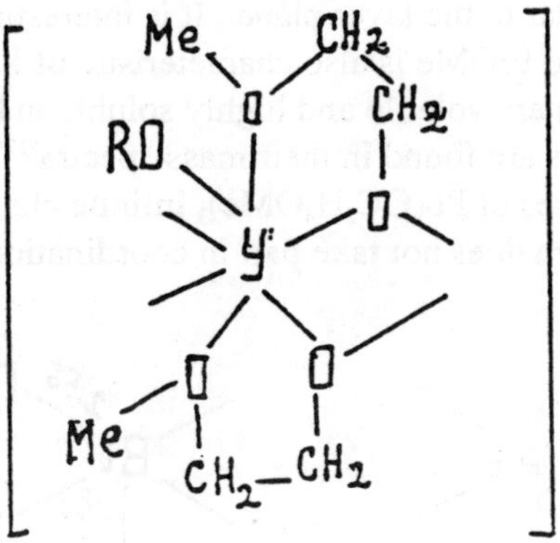

Ca and Cd methoxyethoxides are built of the planar 9-nuclear molecules $[M(OC_2H_4OMe)_2]_9\,(MeOC_2H_4OH)_2$ [95, 169], which may be described as "further development" of the $Ti(OR)_4$ tetrameric motive or a fragment of the polymeric layer $[Cd(OR)_2]_\infty$ (see below).

High donor activity of the O atom of methoxides together with the minimal dimensions of R ensure the maximal coordination capacity of the M atom in $M(OMe)_n$; practically all methoxides - either crystalline or amorphous - are <u>polymeric</u>. Sometimes the peculiarities of methoxides concern even their color. For instance, $Ni(OMe)_2$ with octahedral Ni atoms in its polymeric structure is yellow-green in contrast to the violet $Ni(OR)_2$ (R = Et, Pr, I--Pr) with the tetrahedrally coordinated Ni atoms[66]. The violet color of the polymeric $Cr(OMe)_2$ also differs from other $Cr(OR)_2$ which are orange[91].

X-ray powder investigations of methoxides (and sometimes ethoxides) of alkaline[52, 70, 171], bivalent (Mg, Ca, Sr, Ba, Cd, Pb, Ni)[53, 27, 66, 120, 172] and trivalent (Al, Cr, Fe)[173] elements revealed that they crystallize in the layered structures of hydroxides - LiOH (layers of tetrahedra $[MO_{4/4}]$), $Mg(OH)_2$ (layers of octahedra $[MO_{6/3}]$) and $Al(OH)_3$ (octahedra $[AlO_{6/2}]$)[***]. The layers of the methoxide groups are located in the crystal lattice between the $[MO_n]$ layers in accordance with the [MOMeMeO ...] motive. The analogous structures of $M(OEt)_2$ are usually more distorted in the

*** For $Mg(OR)_2$ and $M(OR)_3$ the crystalline phases existing in the amorphous matrix are discussed.

direction perpendicular to the layer plane. It is interesting that the layered structure analogous to LiOMe is also characteristic of K, Rb and Cs isopropoxides, which are volatile and highly soluble in hydrocarbons (only monomeric fragments are found in their mass-spectra)[174].

In the structures of $Pb(OC_2H_4OMe)_2$ infinite chains exist, in both cases the ether O atom does not take part in coordination[92, 93]:

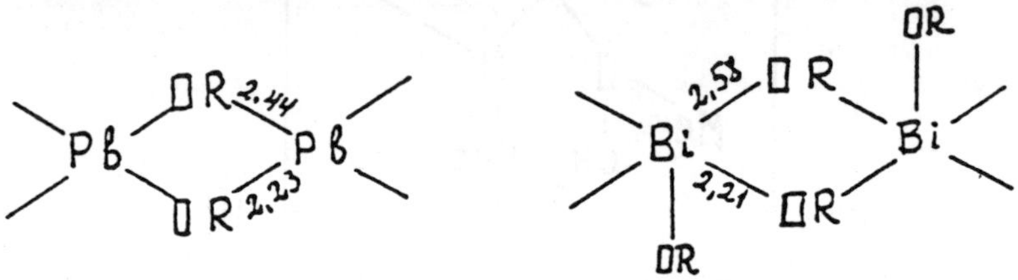

Solubility in methoxyethanol and hydrocarbons testifies to readily occurring depolymerization, the saturation of the coordination sphere of the metal atom is presumably achieved by the ether O atom and is accompanied by chelating.

Thus, the polymeric structure of the alkoxide does not at all mean that $M(OR)_n$ is absolutely inert. The character of the M-OR bond also plays an important role. The polymeric layers of ionic M^IOMe can be destroyed easily by alcohol, liquid NH_3, and other polar solvents with subsequent solvation which makes the process exothermic. The donor-acceptor bonds in the structures of M^IOPr-I or Pb and Bi methoxyethoxides are to considerable extent covalent (considerable difference in the length of two M-O bridges has been observed in $M(OC_2H_4OMe)_2$) and split in the presence of the nonpolar molecules of the solvent. On the other hand, Al, Cr, Fe methoxides and ethoxides which form the most stable M-O bonds in $[MO_{6/6}]$ octahedra (which do not demonstrate any trend to undergo solvolysis in the alcoholic solutions) turn out to be absolutely inert. They can be dissolved only by formation of complexes or by partial destruction with formation of oxo-bridges.

3.3. Variations in the Molecular Composition - "Coordination Polymerism"

The above described structural types characterize crystalline alkoxides. At the same time, amorphous state is much more typical of metal alkoxides. The reason for this fact lies in non-homogeneity of the molecular composition. For more than 20 years a lot of examples of non-reproducibility of the properties of some metal alkoxides (melting points, IR- and mass-spectra) have been frequently discussed in literature. Aluminum ethoxide and isopropoxide[57], $Ga(OPr\text{-}I)_3$ [175], $Ti(OMe)_4$ [176] already became classical examples. It has been estimated by us that variations in the constants (melting points, solubility values, temperatures of transitions into gaseous state, color intensity and even aggregative state) for the samples with different "history" - synthesis temperature, the nature of the solvent, the storage time ("age") are typical of almost all metal alkoxides of the III - VIII groups. By means of mass-, IR-Raman-spectroscopies, cryoscopy and X-ray diffraction we have demonstrated that alkoxides have a trend to form a few types of associates (or different size and structure), which are sometimes present in one and the same sample, we gave this phenomenon the name of "coordination polymerism." Studies of different alkoxides of Al, Ti, Zr, Hf, Nbo^{3+}, MoO^{4+}, MoO_2^{2+}, WO_2^{2+}, Fe, Ni [28, 65-67, 106, 107, 132, 177, 181] allowed us to estimate the following general regularities: (1) The degree of association increases with storage time ("aging") and under the action of alcohols, which is accompanied by the increase of the melting points, lowering of solubility in alcohol, decoloration. The enhanced temperature of gas phase transition increases the low molecular weight fraction in the fragmentation products in the mass-spectra. (2) Oligomerization occurs when $[M(OR)_n]_x$ are dissolved in aromatic hydrocarbons or after melting and distillation. The oligomerization products demonstrate lower values for all thermal constants, higher solubility in alcohol, their color is more pronounced. The mass-spectra are characterized by ions with high molecular weights, corresponding to oligomers which also exist in the condensed matter. The knowledge of these regularities allows to vary the properties of the samples (see scheme 2).

We assume that the basis of "coordination polymerism" lies in sterical instability of oligomeric alkoxides, which is caused by incomplete coordination of the metal atom. The reason for thermodynamic instability of the crystalline and amorphous phases lies in the tendency of the metal atom to build up its coordination polyhedron by polymerization, isomerization, etc.

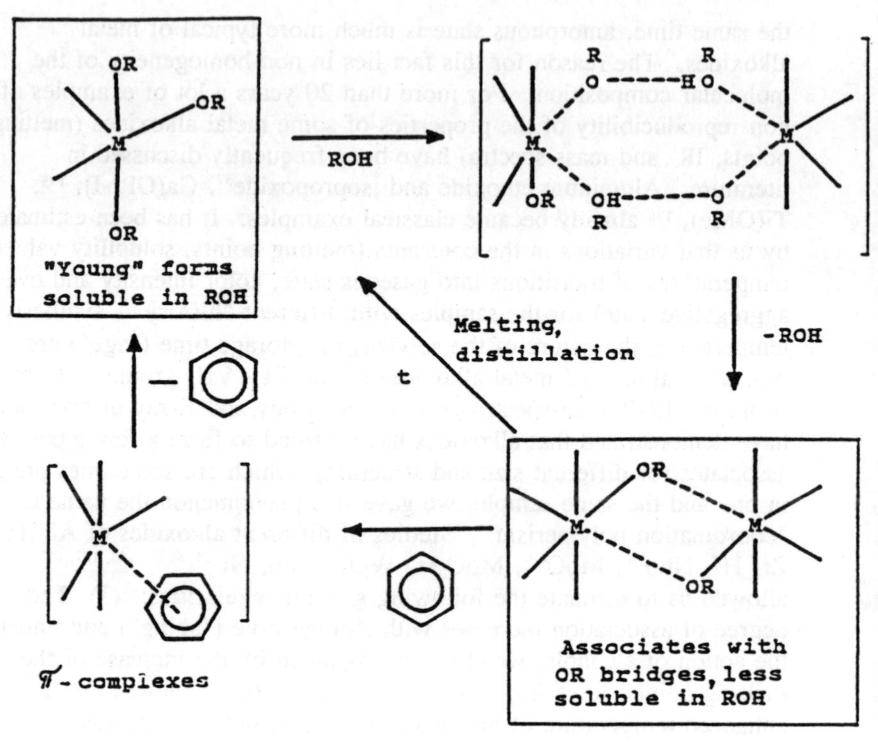

OR
OR
M
OR
OR
"Young" forms
soluble in ROH
ROH
R
O
R
HO
M
M
OR
OH
O
R
-ROH
Melting,
distillation
t
OR
M
M
OR
Associates with
OR bridges, less
soluble in ROH
π - complexes

SCHEME 2

One of the most illustrative examples is to behavior of $Al(OPr-I)_3$. Its freshly distilled liquid samples contain trimeric molecules, with the molecular structure analogous to $Al_3Cl_5(OPr-I)_4$. The structure of the latter molecules has been solved by means of the X-ray structural study[177]. Storage or interaction of the melt with alcohols yields crystals with the maximal m.p. 140°C, they are built of tetramers $[Al(OR)_3]_4$. "Aging" process which involves the increase of melting point and degree of association with time may be represented in the form of the diagram trimer-tetramer (Fig. 4). The trimer has a 5-coordinated Al atom, while in the tetramer Al has octahedral coordination. Different molecular forms behave towards alcohol in different manners (Fig. 5). Liquid $[Al(OR)_3]_3$ is dissolved in alcohol limitless; however, in the presence of ROH quick transition trimer → tetramer occurs, the solubility decreases down to the equilibrium values for the tetramer (curve 1). Solubility of the crystals of tetramer increases up to the same value (curve 3); for the samples containing both forms at first extraction of the trimer takes place, which is accompanied by a maximum on curve 2. Similarly, in the course of interaction of liquid $Zr(OPr-I)_4$, with alcohol, the values of solubility as well as m.p. of the solid depend on the quantity of the introduced sample (i.e. the metastable trimer, which is present in the solution). Afterwards (for several days) solubility decreases down to the equilibrium values which correspond to precipitate - tetrameric molecules (m.p. 135°C) [25]. The above discussed phenomena explain the well-known fact that solubility in solutions of metal alkoxides which are obtained in the course of their syntheses (mother liquor) is much higher in comparison with the values obtained on dissolution of solid samples. IR and Raman spectroscopy data for liquid and solid samples of $Zr(OPr-I)_4$ trimers and teramers allow to suggest that their structures are analogous to associates of $Ti(OEt)_4$ [28, 178, 180] (Fig. 6).

"Polymerism" is especially pronounced in the case of methoxyethoxides. The other O atom gives one more opportunity for filling of the coordination sphere of the M-atom - by chelating: in accordance with the data of structural analysis both O atoms may be mono-, bi-, and tridentate, as well as non-coordinated. These factors give supplementary opportunities for "polymerism." One of its consequences is that most of methoxyethoxides are liquids, crystals may be obtained in extremely poor yields from hydrocarbons.

The second factor which influences greatly the oligomerization processes is the presence in the samples of oxoalkoxides - the products of partial decomposition of alkoxides. Recently, very many structures of oligomeric oxoalkoxides have been described (see below). Thermal treatment of $[Al(OEt)_3]_\infty$ at 240°C or its prolonged refluxing with benzene results in partial dissolution with formation of $[AlO_n(OR)_{3-2n}]_m$, $n << 1$. Iron ethoxide behaves very much the same. The structures of the oligomeric molecules which in fact exist in solutions of "$M(OEt)_3$" may be reconstructed from the data of X-ray structural analysis of oxocomplexes $M_{10}O_4(OEt)_{22}$, precipitating from these solutions (if oxo-groups are regarded as the places of condensation). Molecules $[Al(OEt)_3]_n$, n = 3, 4, and 5, which have been earlier registered in the mass-spectra, may be distinguished in Fig. 7[181].

It is also noteworthy that usually in the description of the chemistry of the particular metal alkoxide it may be difficult to distinguish between the two factors - the variety of molecular composition on the one hand and the presence of oxoalkoxides in the samples on the other hand. The reversibility may be considered in some cases as such a criterion: the possibility of multiple reproduction of the properties of the oligomeric samples (e.g., $Al(OPr\text{-}I)_3$, $Zr(OPr\text{-}I)_4$, etc.) testifies to invariability of the

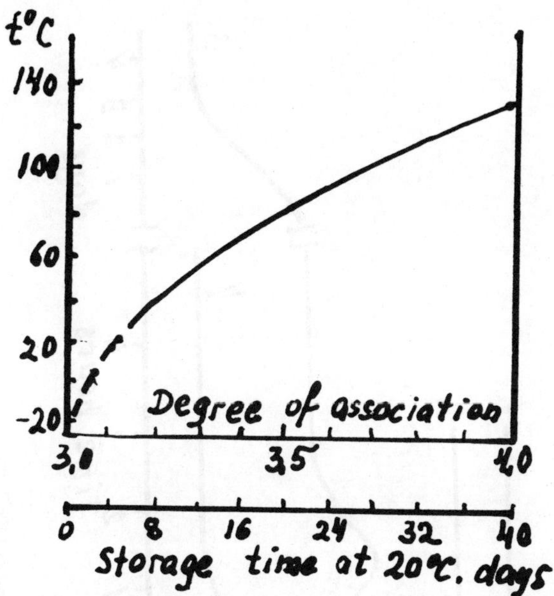

FIGURE 4. Structure of $Al(OPr\text{-}i)_3$-oligomers and thermodynamic diagram for the system $[Al(OR)_3]_3$ - $[Al(OR)_3]_4$.[177]

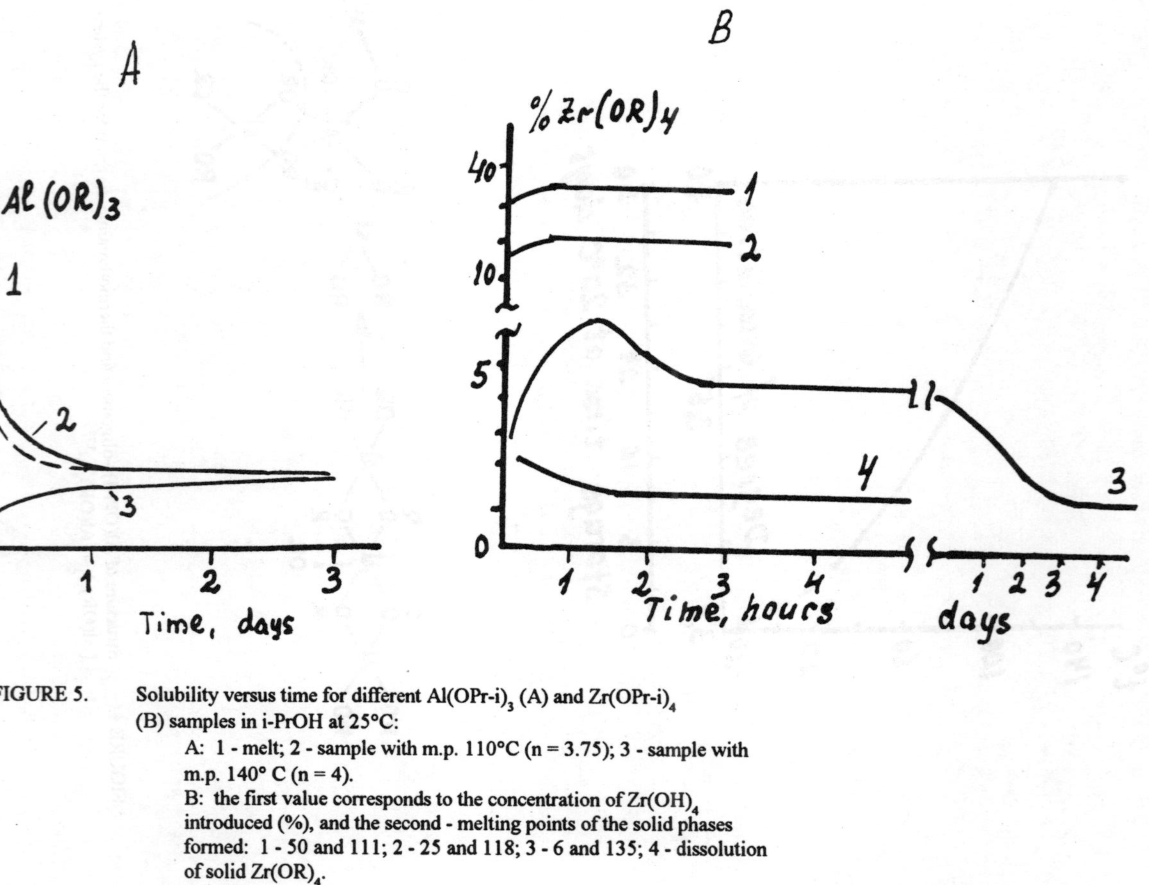

FIGURE 5. Solubility versus time for different $Al(OPr\text{-}i)_3$ (A) and $Zr(OPr\text{-}i)_4$ (B) samples in i-PrOH at 25°C:

A: 1 - melt; 2 - sample with m.p. 110°C (n = 3.75); 3 - sample with m.p. 140° C (n = 4).

B: the first value corresponds to the concentration of $Zr(OH)_4$ introduced (%), and the second - melting points of the solid phases formed: 1 - 50 and 111; 2 - 25 and 118; 3 - 6 and 135; 4 - dissolution of solid $Zr(OR)_4$.

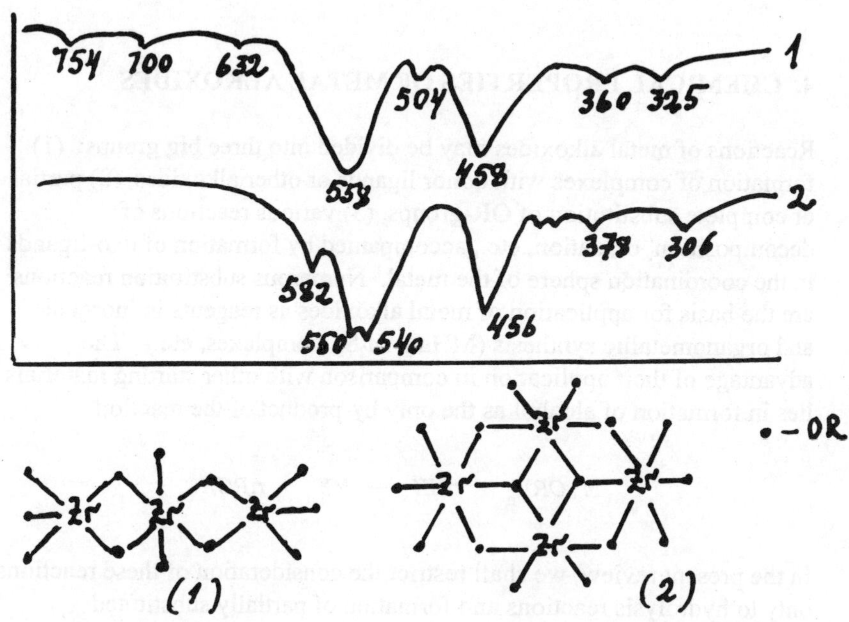

FIGURE 6. IR spectra of liquid $Zr(OPr\text{-}i)_4$ - (1) and its suspension in nujol - (2) the anticipated structures of trimer and tetramer.

chemical composition and dominating of the first factor. On the other hand, irreversible polymerization may occur in both cases.

4. CHEMICAL PROPERTIES OF METAL ALKOXIDES

Reactions of metal alkoxides may be divided into three big groups: (1) formation of complexes with donor ligands or other alkoxides, (2) partial or complete substitution of OR-groups, (3) various reactions of decomposition, oxidation, etc., accompanied by formation of oxo-ligands in the coordination sphere of the metal. Numerous substitution reactions are the basis for application of metal alkoxides as reagents in inorganic and organometallic synthesis ($MHal_n$, M_2S_n, complexes, etc.). The advantage of their application in comparison with other starting materials lies in formation of alcohol as the only by-product of the reaction

$$M(OR)_n + nHX \longrightarrow MX_n + nROH$$

In the present review, we shall restrict the consideration of these reactions only to hydrolysis reactions and formation of partially substituted $M'(OR)_mX_{n-m}$, X = RCOO, acac etc. - the derivatives which are used in sol-gel preparation of oxides. Reactions of partial decomposition of alkoxides is of special practical interest in association with the mechanism of formation of oxides from alkoxides.

4.1. Complexation - Bimetallic Alkoxides

Formation of solvates with alcohols is in particular characteristic of ionic M^IOR and $M^{II}(OR)_2$ (with $M(OPr\text{-}I)_4 \cdot I\text{-}PrOH$, M = Sn, Zr, Hf, Ce being the only exceptions among the multivalent metals). Solvates with ethers, amines, phosphines, etc., form monomeric or oligomeric "covalent" $M(OR)_n$, the solvating molecules complete the coordination sphere of the metal atom. However, the OR bridges are exceptionally active donors which makes solvates, in fact, rare in the chemistry of alkoxides.

Bimetallic alkoxides (the products of complexation of two $M(OR)_n$ with different metal atoms) attracted special attention since they were first

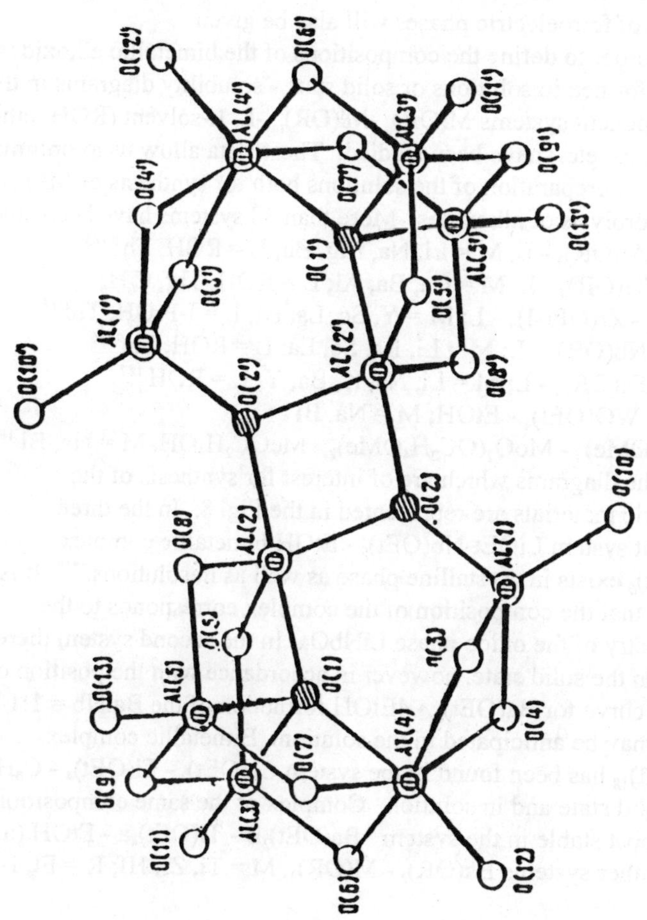

FIGURE 7. Molecular structures of $M_{10}O_4(OEt)_{22}$, M = Al, Fe.[181]

described in 1929 by Meerwein. Recently, interest to these compounds increased due to their application for preparation of complex (bimetallic) oxides. The modern state of chemistry of these compounds has been recently reviewed[182]; therefore, in this chapter, we shall discuss only the mutual solubility and complexation in the systems containing two alkoxides a few examples of the crystalline structures of $MM_m'(OR)_n$ - precursors of ferroelectric phases will also be given.

In order to define the compositions of the bimetallic alkoxides, which are formed in solutions or solid state - solubility diagrams in the three-component systems $M(OR)_n$ -$M'(OR)_m$ -L, L-solvent (ROH, ether, hydrocarbons, etc.) have been studied. These data allow us to optimize the conditions of preparation of the solutions both for synthesis of $MM'(OR)_n$ and for hydrolysis of alkoxides. More than 30 systems have been studied:
$M(OR)_n$ - $Al(OR)_3$ - L, M = Li, Na, Mg, Ba; L = ROH, Thf[183]
$M(OR)_n$ - $Ti(OR)_4$ - L, M = Ca, Ba, Al; L = ROH, Thf, C_6H_6 [184]
$M(OPr\text{-}I)_n$ - $Zr(OPr\text{-}I)_4$ - L; M = Y, Sc, La, Al; L = I-PrOH, Thf[185]
$M(OR)_n$ - $Nb(OR)_5$ - L; M = Li, Ba, Sc, La; L = ROH, Thf[186]
$M(OR)_n$ - $Fe(OR)_3$ - L; M = Li, Na, K, Ba, Y; L = ROH[187]
$M(OEt)_n$ - $WO(OEt)_4$ - EtOH; M = Na, Bi
$M(OC_2H_4OMe)_n$ - $MoO_2(OC_2H_4OMe)_2$ - $MeOC_2H_4OH$, M = Na, Bi[188]
Some of the diagrams which are of interest for synthesis of the ferroelectric materials are represented in the Fig. 8. In the three-component system LiOEt-$Nb(OEt)_5$ - EtOH bimetallic complex $LiNb(OEt)_6$ exists in crystalline phase as well as in solutions.**** It is very important that the composition of the complex corresponds to the stoichiometry of the oxide phase $LiNbO_3$. In the second system there is no complex in the solid state, however in accordance with the position of the solubility curve for $Ba(OEt)_2$ • 4EtOH formation of the Ba:Nb = 1:1 complex may be anticipated in the solution. Bimetallic complex $CaTi_4(OEt)_{18}$ has been found in the system $Ca(OEt)_2$ - $Ti(OEt)_4$ - C_6H_6**** both in solid state and in solution. Complex of the same composition is also the most stable in the system $Ba(OEt)_2$B - $Ti(OEt)_4$a - EtOH (as well as in the other systems $Ba(OR)_2$ - $M(OR)_4$, M = Ti, Zr, Hf; R = Et, I-Pr[182,

---

**** Composition of the complex in solution may be determined by extrapolation of the solubility curve of LiOEt• 2EtOH to the side of the triangle LiOEt - $Nb(OEt)_5$ (see ref. 189).

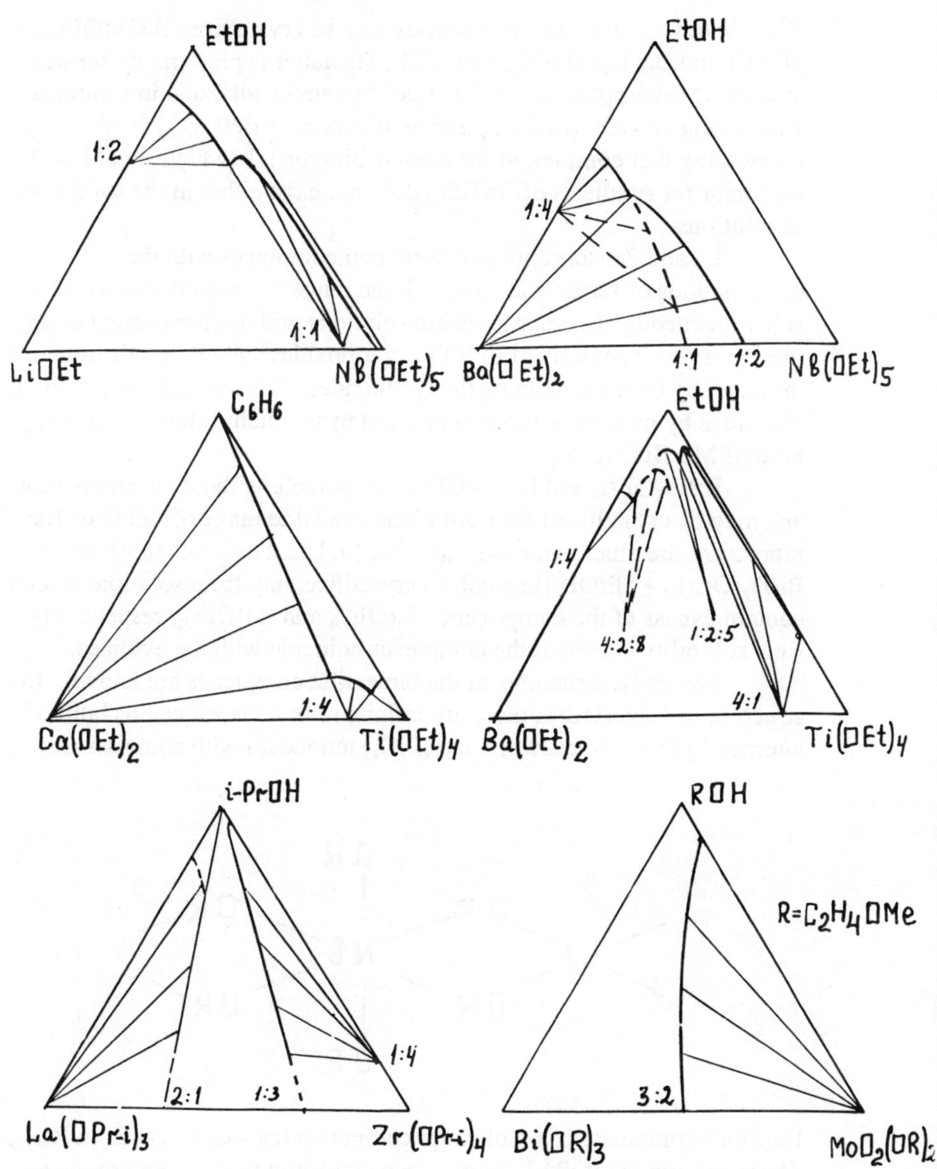

FIGURE 8. Solubility at 20°C in the systems $M(OR)_n$-$M'(OR)_m$-L (mass %).

[190]). However, two more compounds may be crystallized $BaTi_2(OEt)_{10}$ • $5EtOH$ and $Ba_4Ti_2O(OEt)_{14}$ • $8EtOH$. The latter is presumably formed as a result of decomposition of the "true" bimetallic alkoxide in solutions with strong basicity (in the presence of excess of $Ba(OEt)_2$)[191]. It is noteworthy that complex of the composition of 1:1 (which should be a precursor for synthesis of $BaTiO_3$) does not exist either in the solid state or in solutions.

La and Zr isopropoxides form two complexes with the compositions of La:Zr = 2:1 and 1:3, the range between these two sections is homogeneous. In order to obtain solutions with higher content of Zr (necessary for synthesis of PLZT) isopropoxides should be substituted, for instance, by better soluble methoxyethoxides. Similarly, substitution of alkoxides by methoxyethoxides allowed us to obtain solutions with any ratio of Mo:Bi (Fig. 8f).

$LiNb(OEt)_6$ and $BaTi_4(OEt)_{18}$ are soluble without decomposition and may be crystallized from solutions in a wide range of Li:Nb or Ba:Ti ratios. On the other hand, $CaTi_4(OEt)_{18}$ (in benzene or alcohol) and $BaTi_2(OEt)_{10}$ • $5EtOH$ (in alcohol) crystallize only from solutions which contain excess of the component - $Ti(OEt)_4$ and $Ba(OEt)_2$, respectively. Real solubility values of the complexes coincide with the evtonics.

Presently, structures of the bimetallic complexes are known. In the structure of $LiNb(OEt)_6$, there are infinite screw-shaped chains built of the alternating $[NbO_6]$ octahedra and $[LiO_4]$ tetrahedra with common ribs[191]:

OR
OR OR
Li Nb
OR OR
OR

Barium tetratitanate is a molecular compound (as well as $Ba[Zr_2(OR)_9]_2$). There are two $[Ti_2(OR)_9]$ groups - two octahedra with a common face in its composition. Each is bonded to the central eight-coordinated Ba atom via two μ_2 - OR and two μ_3 -OR groups[190, 191, 193].

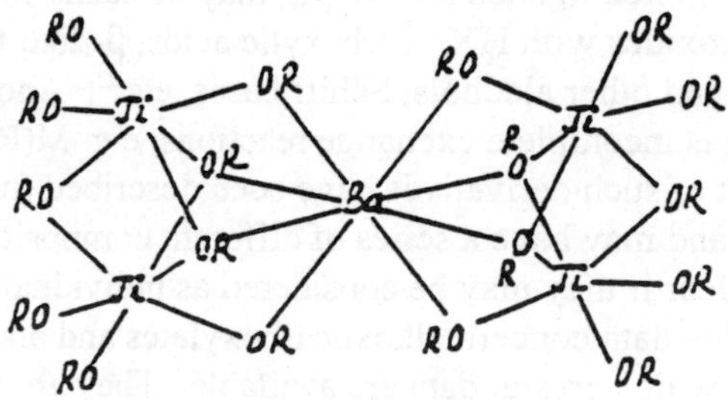

Dititanate may be considered as an ionic complex. In its cation a "half" of another molecule is bonded to the central Ba atom, other 4 places in the coordination sphere of Ba atom are occupied by the molecules of the solvating alcohol:

$$\left[(ROH)_4Ba(\mu_3-OR)_2(\mu-OR)_2Ti_2(\mu-OR)(OR)_4\right]^+\left[H(OR)_2\right]^{-193}$$

$Sr_2Ti(OPr\text{-}I)_8$ • 5i-PrOH can be crystallized from the solutions of Sr and Ti isopropoxides. In its structures $[Sr_2TiO_3]$ cycles with three octahedra with common faces exist; two μ_3 - OR groups are located below and over the plane of the cycle.

It is clear from the above described structures that neither of these compounds contains elements of the future oxide phase and thus cannot be its direct precursor.

Besides the bimetallic complexes which are formed in the systems $InSn(OBu\text{-}t)_3$ and $M^{II}Pb(OBu\text{-}t)_{32}$ M^{II} = Sr, Ba are worth mentioning. They can be formed either by complexation of two alkoxides or in exchange reactions with chlorides. Their molecules consist of infinite chains containing no terminal groups[194]:

$$: In(\mu-OR)_3Sn: \qquad : Pb(\mu-OR)_3BA(\mu-OR)_3Pb:$$

4.2. Substitution Reactions - Hydrolysis

Alkoxides with mixed ligands $M(OR)_nX_m$ may be easily formed in reactions of alkoxides with HX - carboxylic acids, β-dike-tones, alkanolamines and other alcohols, Schiff bases, etc.[23]. Another way of their formation is incomplete exchange reactions, e.g. $M(RCOO)_n$ + NaOR (see 2.5). A lot of such derivatives have been described (usually they are not crystalline and may have a series of different compositions); however, it is not quite clear if they may be considered as individual compounds. The most reliable data concern alkoxocarboxylates and alkoxo-β-dicetonates for which crystal data are available. They are interesting as oligomeric precursors soluble in organic solvents to a greater extent than the corresponding $M(OR)_n$. Thus $[Cr(OMe)(acac)_2]_2$, $[Sn(OPr\text{-}I)_3(acac)]_2$, $[Al(OPr\text{-}I)_2(acac)]_3$, $[Cu(OMe)(AcO)]_4$ form di- tri-, and tetrameric molecules containing $M(\mu\text{-}OR)_2M$ fragments[195, 198]. It is noteworthy that $Cr(OR)_3$ and $Cu(OR)_2$ are polymers. Great interest has been expressed to the recently synthesized "ferric wheel" $[Fe(OMe)_2(CH_2ClCOO)]_{10}$ - planar cyclic dimer with the diameter of ~ 12 A[199]. Bimetallic complex $[BaCu(\mu_3\text{-}OR)_2(acac)_2ROH]_2$ - precursor for HTSC - has been synthesized by reaction of $[Cu(OC_2H_4OMe)(acac)]_2$ with $Ba(OC_2H_4OMe)_2$ [200]. $Co_2Zr_2(OPr\text{-}n)_{10}(acac)_2$ [201] and $Pb_2Zn_2(OR)_4(AcO)_4$ [202] are complexes of the same type. The latter is the derivative of MeOCH(Me)OH (isomer of methoxyethanol) and is formed on alcoholysis of two acetates.

Naturally hydrolysis is the most important reaction among substitution reactions, insofar as it is the basis for application of metal alkoxides in the synthesis of oxides. Hydrolysis which may be represented by the general scheme:

$$M(OR)_n + NH_2O \longrightarrow M_2O_n \bullet xH_2O + nROH$$

comprises in fact two processes - substitution of the alkoxide groups by hydroxides and the subsequent condensation (see review by Livage[203]). On the first stage water molecules which are stronger donors than ROH substitute ROH in the coordination sphere of $M(OR)_n \bullet xROH$. Then intramolecular migration of the proton which results in appearance of hydroxide group in the place of alkoxide group occurs. On the next stage alongside with the increase of the number of the hydroxide groups association which is due to hydrogen bonds ("oxolation") and subsequent condensation ("olation") takes place:

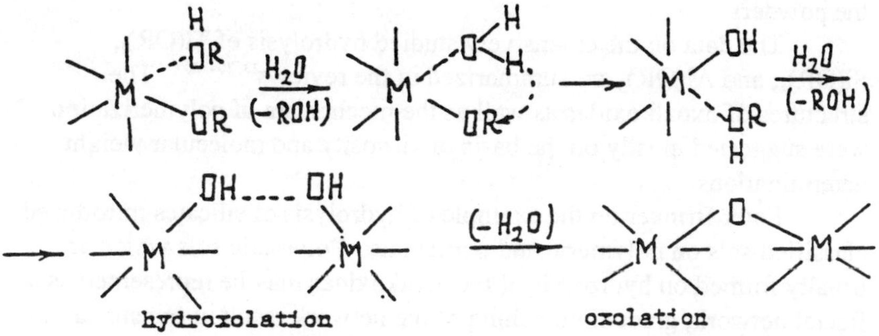

hydroxolation oxolation

Prevailing of certain particular forms (sometimes containing residual OR- and ROH) depends on the nature of the metal, H_2O:ROH ratio in the system, the initial $M(OR)_n$ concentration and the conditions of processing (t°C, mixing rate, the consequence of introduction of the reagent in the system, etc.). Three types of bridges may exist in the structure of the hydroxide thus formed:

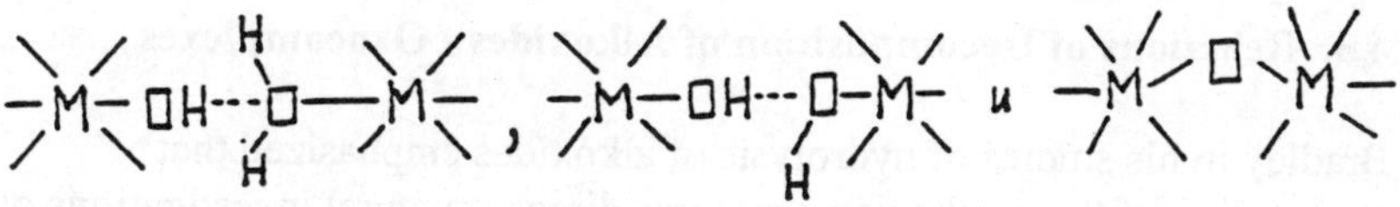

The properties of the samples are determined by their ratio. Hydroxides prepared on hydrolysis of metal alkoxides differ from all other $M_2O_n \bullet xH_2O$ in the maximal content of OH-groups; consequently they easily form colloid systems (sols and gels) and demonstrate very high reactivity. Dehydration occurs at very low temperature and is accompanied by formation of oxide powders with very small particles (~ 100 A) highly developed surface, consider able chemical activity. They may be amorphous or crystalline, frequently metastable phases are formed. Variations of the dehydration temperature allow to vary the properties of the powders.

The data on the extensively studied hydrolysis of $Ti(OR)_4$, $Si(OR)_4$, and $Al(OR)_3$ are summarized in the reviews[23, 204, 205]. The structures of oxoalkoxides as well as the mechanism of polymerization were suggested mostly on the basis of viscosity and molecular weight determinations.

Later Brinker on the example of hydrolysis of silicates introduced classified sols on polymeric and particulate. Polymeric sols which are usually formed on hydrolysis of metal alkoxides may be represented as a fractal network, gradual branching of the network result in transformation of sol → gel. Particulate systems have identifiable dense primary particles with the size > 1 nm.

An interesting example of formation of sols of different types is hydrolysis of Mo and W alkoxides. In accordance with the trend of tungsten to form W-O-W bridges hydrolysis of $WO(OEt)_4$ leads to gradual branching of the network in the colloid solution. On the other hand, hydrolysis of $MoO_2(OEt)_2$ results in formation of the structures analogous to polymolubdate anions, which contain alongside with the bridging terminal oxo-groups (Mo = 0)[30].

4.3. Reactions of Decomposition of Alkoxides - Oxocomplexes

Bradley in his studies of hydrolysis of alkoxides emphasized that explanation of the mechanisms requires direct structural investigations of the intermediate hydrolysis products - oxoalkoxies[204]. The results of the first structural investigations of the so-called "first hydrolysis products" $Ti_7O_4(OEt)_{20}$ [206-208] $Nb_8O_{10}(OEt)_{20}$ [20] and $Zr_{13}O_8(OMe)_{36}$ [209] are usually discussed in all the reviews concerning hydrolysis of alkoxides. It is noteworthy that all oxo-ligands in these molecules are bridging (even Nb atom has no terminal Nb = 0 groups) and are usually regarded as the points of condensation of two hydroxide groups. Recently, many new oxoalkoxides have been characterized by means of X-ray structural analysis. Moreover, many derivatives which were earlier described as alkoxides turned out to contain oxo-ligands in their structures (usually only few OR groups are substituted by O atoms). For example, "isopropoxides" of rare earth elements and In should, in fact, be represented by the formula $M_5O(OPr\text{-}I)_{13}$, M = Sc, Y, La, Yb, Nd[22, 24, 63]; crystalline "ethoxides" of Al and Fe were found to be $M_{10}O_4(OEt)_{22}$ [181]. Frequently oxoalkoxides are crystallized in the form of solvates which makes impossible to distinguish between them and $M(OR)_n$ by chemical analysis. Although many authors continue to regard oxoalkoxides as the products of uncontrollable hydrolysis of alkoxides conditions of their synthesis in many cases testify to quite different mechanisms of decomposition of alkoxides.

One of the most interesting recent results is the identity of single crystals $Ti_7O_4(OEt)_{20}$ and $Nb_8O_{10}(OEt)_{20}$ isolated by two different methods: by hydrolysis and under the conditions of complete isolation from traces of moisture[132, 208]. One of the paths of their formation may lie in elimination of ether which has been first described by Bradley for niobium t-butoxides.

$$Nb(OEt)_5 + 5\,t\text{-}BuOH \xrightarrow[{[-EtOH]}]{} NbO(OBu\text{-}t)_3 + Nb_2O(OBu\text{-}t)_8 + t\text{-}Bu_2O$$

Although ether in this case has not been registered the heterolytic splitting of O-R bond with the subsequent attack of the carbocation on the M-OR bond of the other OR-group has been suggested[210]:

$$\text{M(OR)(OR)} \longrightarrow \text{M}\cdots\overset{\delta-}{\text{O}}\text{—}\overset{\delta+}{\text{R}}\ \ \text{OR} \longrightarrow \text{M=O} + R_2O$$

The analogous processes are characteristic to a considerable extent of the derivatives of tungsten and especially molybdenum: among the derivatives of $Mo(OR)_6$ series only methylate has been isolated, $MoO(OR)_4$ tend to decompose with formation of $MoO_2(OR)_2$. An attempt to isolate from solutions, bimetallic alkoxides of Mo and Na (1:1), leads to decomposition with precipitation of highly crystalline complex oxide Na_2MoO_4; R_2O in these cases were registered by means of GLC[65, 188].

Another possible way of formation of oxo-groups in the coordination on sphere of M lies in exceptionally active interaction of $M(OR)_n$ - derivatives of alkaline, alkaline-earth and some other metals with oxygen which leads to peroxides and hydroperoxides on the first step (see 2.7[133]}, their decomposition gives water alongside with numerous other products.

Relations between different reactions of formation of oxo-products even before hydrolysis stage needs careful investigation. Nevertheless, it is clear that mono- and bimetallic oxoalkoxides play very important role and are to be considered as real precursors of oxides. Therefore, conditions of their crystallization as well as structural characterization seem important.

The trinuclear isopropoxides - $M_3(\mu_3 - 0)\ (\mu_3 - OR)\ (\mu - OR)_3\ (OR)_6$ (M = Mo, W) (with oxo- and one OPr-I- group under and over the center of the M_3 triangle) have been synthesized on oxidation of $M_2(OR)_6$ by O_2 and $WO(OR)_4$ respectively[211]. The tetranuclear $[Pb_4(\mu_4 - 0)\ (\mu - OSiPh_3)_6]$. C_6H_6 is formed on alcoholysis of $PbCp_2$ in ether. In its tetrahedral molecules six OR bridges are located on the ribs[212]. The isopropoxides of In and rare earth elements have been synthesized by different techniques (1, 2, 4 and 5). They are in fact pentanuclear molecules $M_5(\mu_5 - 0)\ (\mu_3 - OR)_4\ (\mu - OR)_4\ (OR)_5$. The $[M_5]$ square bipyramid contains O-atom in its center, four μ_3 - OR - over the centers of the four faces and four μ - OR - along the ribs of the basis, each M atom has one terminal OR-group[22, 24, 63]. Square bipyramid is also the center of $Ba_5(\mu_5 - 0)\ (OPh)_8\ (PhOH)\ (Thf)_8$ molecule. This complex is formed in reaction of barium with phenol[213]. Barium "methoxyethoxide" proved in fact to be oxoalkoxide $Ba_6(\mu_6 - 0)\ (OR)_{10}\ (ROH)_4$ ($R = C_2H_4OMe$) with $[OBa_6]$ octahedron[217]. Short refluxing of the alcoholic solutions of $Ca(OEt)_2$ even in the presence of traces of O_2 results in the crystalline $Ca_6(\mu_4 - 0)_2\ (\mu_3 - OEt)_4\ (OEt)_4 \bullet$ 14EtOH. The molecule consists of two cubanes $[Ca_4O_4]$ with a common face $[Ca_2(O\text{-}oxo)_2]$ [193]. Crystallization of $Pb_6\ (\mu_3 - OPr\text{-}I)_4$ occurs from solutions "$Pb(OPr\text{-}I)_2$" (synthesized by methods 2 and 5).

Its octahedral molecule is isostructural to $[3PbO \cdot H_2O]_2$ [214]. Formation of the oxo-complex in this case is presumably the result of elimination of ether which is especially characteristic of $Pb(OR)_2$ [27, 94].

$Ti_7O_4(OEt)_{20}$ which has been always considered as the first hydrolysis product of $Ti(OEt)_4$ has been recently isolated also as a result of the prolonged heating (under enhanced pressure) of the solution of etoxide in absolute alcohol[208]. Its planar centrosymmetric molecule may be regarded as a product of condensation of the trimer and tetramer $Ti(OR)_4$ [207].

The other well-known hydrolysis product $Nb_8O_{10}(OEt)_{22}$ [20] is also crystallized on storage from the liquid $NbO(OEt)_3$ [132].

The products of thermal oligomerization of $M(OEt)_3$ (M = Al, Fe) can be represented as decameric molecules $M_{10}(\mu_4 - 0)_2\ (\mu_3 - 0)_2\ (\mu - OEt)_{14}\ (OEt)_8$ which consist of two pentamers with two planar μ_3 - 0 (oxo) and $[Al_2O_2]$ cycle (Fig. 7).

$Zr_{13}O_8(OMe)_{36}$ was obtained on hydrolysis of $Zr(OMe)_4$ in alkaline solutions. The central Zr atom is coordinated by 8 oxogroups, which bond it to 12 peripheral Zr atoms with the coordination number 7 - $Zr(\mu_4 - 0)_2\ (\mu - OR)_4\ (OR)$ [209].

Among the oligomeric oxoalkoxides $Ti_{16}\ (\mu_4 - 0)_4\ (\mu_3 - 0)_8\ (\mu - 0)_4\ (\mu - OEt)_{16}\ (OEt)_{16}$ is the largest known molecule. This compound is formed on the prolonged thermolysis of $Ti(OEt)_4$ in alcoholic solution[208].

Dissolution of the oxoalkoxides in alcohols is presumably accompanied by oligomerization, which never allows to obtain equilibrium values of solubility (compositions of the molecules in solutions and in solid state differ). Nevertheless some of oxoalkoxides exist in gaseous state, which has been proved by mass-spectrometry ($Al_{10}O_4(OEt)_{22}$, $Y_5O(OPr\text{-}I)_{13}$).

In conclusion it is worth mentioning that synthesis of all oxoalkoxides (excluding three hydrolysis products) has been performed under complete isolation of moisture, thus they are formed by non-hydrolytic mechanisms. Oxoalkoxides are also very well known among derivatives with mixed ligands $MO_m(OR)_nX_p$ [215-??].

A few examples which concern complex oxoalkoxides - precursors for ferroelectric materials are given below.

In the molecule $LiNbO(OEt)_4 \cdot EtOH$ (which has been earlier interpreted as the partial hydrolysis product $LiNb(OH)\ (OEt)_5$ [219]) the terminal oxo atom of the Nb = 0 group is bonded via H-bond with Li atom:

It is quite clear from the scheme that even slight thermal treatment results in break-off of the molecule of solvating alcohol with formation of Li-O-Nb bridge.

Crystalline $[BaTiO(OPr\text{-}I)_4 \bullet 7/8\ I\text{-}PrOH]_4$ with the Ba:Ti ratio 1:1 crystallizes from solutions of Ba and Ti isopropoxides in isopropenol. There are tetrameric molecules with the central "cubane" $[Ba_4O_4]$ in its structure (Fig. 9). $[Ti(OR)_4]$ fragments with the 5-coordinated Ti atoms are bonded to three Ba atoms with μ_2 - and μ_3-OR groups. The molecules have Ba-O-Ti bridges with exceptionally short Ti-O bonds (1.71, 1.79 A) [220]. The complex may, in fact, be described as a combination of pairs of anions and cations $[Ba(ROH)_4]^{2+}$ $[TiO(OR)_4]_4^{2-}$ in contrast to $BaTi_4(OR)_{18}$. In solutions it may presumably exist in the form of separate ions or associates with the Ba:Ti ratio of 1:1.

The oxoethoxoniobate $[BaNbO(OEt)_5 \bullet 3/2EtOH]_4$ resembles the above described molecule greatly. It also contains Ba-O-Nb bridges and may be interpreted as $[Ba(ROH)_{1.5}]_4^{2+}$ $[NbO(OR)_5]_4^{2-}$ [193].

$Pb_6Nb_4(\mu_4 - 0)_4 (OEt)_{24}$ is especially important molecule as a precursor for synthesis of ferroelectric complex niobates[221].

5. SYNTHESIS OF FERROELECTRIC MATERIALS FROM METAL ALKOXIDES

Preparation of complex oxides from metal alkoxides includes preparation

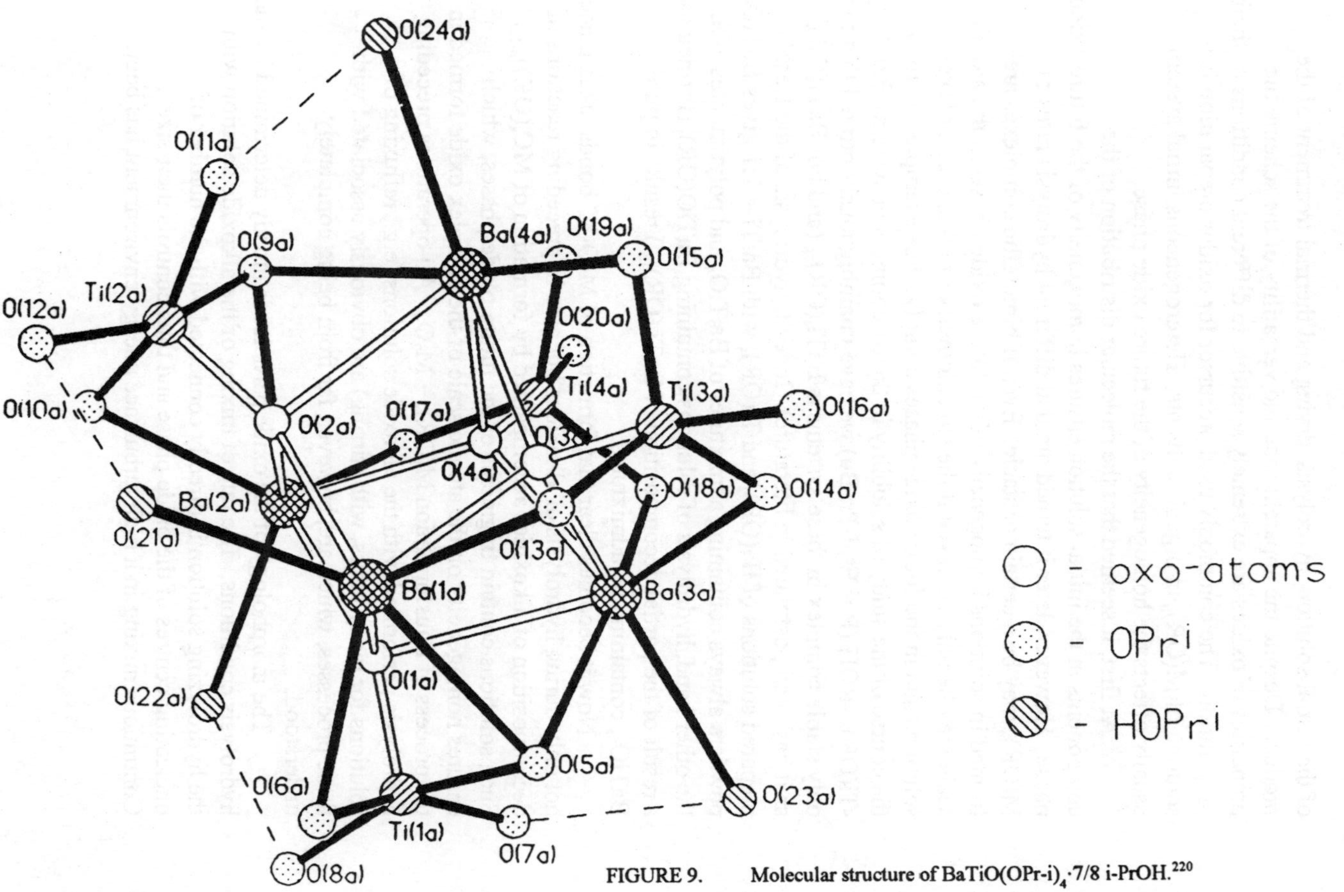

FIGURE 9. Molecular structure of $BaTiO(OPr\text{-}i)_4 \cdot 7/8$ i-PrOH.[220]

of the initial solution, hydrolysis, drying and thermal treatment of the product. Despite the apparent ease and versatility of the scheme the properties of oxides are extremely sensitive to different conditions of their preparation. The commonly used precursor for oxides preparation is a solution of $M(OR)_n$ in organic solvents, Heterogeneous initial system usually affects the homogeneity of the future oxide phase.

At first, it seemed that the molecular distribution of the components in the initial solution ensures homogeneity of the future oxide phase. However, later it turned out that different hydrolysis rates of $M(OR)_n$ may be a certain obstacle. Even if bimetallic complexes are formed in solutions homogeneity of the future oxide phase on the atomic scale may be only achieved if the stoichiometry of alkoxo-complexes in solutions and in the future oxide phases comply. For example, recently in the course of the study of solubility in the tri-component systems $Ba(OR)_2$ -$Ti(OR)_4$ -ROH (R = Et, I-Pr, Bu) we have unambiguously proved that the only stable complex in these systems is $BaTi_4(OR)_{18}$ (and not $BaTi(OR)_6$, as it has been postulated in literature). Indeed, hydrolysis of the freshly prepared solutions of $Ba(OR)_2$ and $Ti(OR)_4$ with Ba:Ti = 1:1 gives $BaTiO_3$ powders always containing admixtures of Ba_2TiO_4 and polytitanates. On the other hand, hydrolysis of solutions containing $BaTiO(OR)_4$ (formed as a result of the partial decomposition of $BaTi_4(OR)_{18}$) results in pure $BaTiO_3$, containing no admixtures.

Now it becomes clear that formation of M-O-M' bonds takes place not only during hydrolysis of alkoxides but also in preceding reactions of decomposition of alkoxides accompanied by formation of $MO_x(OR))_{n-2x}$. Such solutions contain fragments of the future oxide phases, which ensures homogeneity on the atomic scale of the complex oxide formed in the process of transformation $M(OR)_n \rightarrow M_2O_n$. All operations preceding hydrolysis performed with the alkoxide solutions (e.g., refluxing of solutions for many hours, with stirring) are obviously associated with these processes, which are, however, far from being completely understood.

The morphology of the oxide particles is mostly determined by the hydrolysis conditions. The rate of mixing of the alkoxide solution with the hydrolyzing solution is directly connected with the number of nucleating centres of the oxide phase and thus controls their size. Continuous mixing in a high turbulence energy environment has been successfully used in the synthesis of $BaTiO_3$ and, undoubtedly, will be fruitful for synthesis of many other complex oxides.

There is, however, another way of conducting hydrolysis - gradual addition of water (usually in a mixture with an organic solvent), which leads to the formation of sol which then convert into gel. Its slow drying gives glasses, while quick drying results in powders, the so-called xerogels. In this case, particle size distribution as well as their shape is determined by the structure of sol and the extent of branching of the polymeric network of the gel. In this type of hydrolysis chemical additives (usually organic acids, glycols, β-diketones) play very important role. Their addition results in partial substitution of OR-groups by acetic or acetylacetonate groups, condensation with formation of oxo-bridges and elimination of ester. The number of the alkoxide groups taking part in the subsequent hydrolysis is thus considerably limited. This method has been successfully used, e.g., for synthesis of uniform $SrTiO_2$ [254]. Thus, it is possible to change qualitatively the properties of the oxide materials obtained by variation of the hydrolysis conditions.

Ozaki[9] in his review classified perovskites according to conditions of their crystallization. He estimated the following three possibilities of crystallization of perovskites: (1) directly in the course of hydrolysis; (2) in one step on thermal treatment of the amorphous hydrolysis products, and (3) as a result of the solid-state reactions between the first crystallized oxides. At present, it became evident that the careful choice of processing conditions (which include pre-hydrolysis and hydrolysis stages) allows to obtain most of the enumerated by Ozaki perovskites without thermal treatment after hydrolysis. If thermal treatment is, nevertheless, necessary it is very important to choose the atmosphere (air, oxygen or oxygen-water vapor flow).

Preparation of the initial solution becomes an especially important stage for solutions used for application of films. The most popular solvent in these cases is methoxyethanol, which is more and more widely used instead of aliphatic alcohols. It is really hard to overestimate the role of this solvent which has only recently been introduced into sol-gel technology. Special properties of methoxyethanol as a solvent for film-forming solutions need careful consideration; however, one of its advantages lies in its ability to dissolve alongside with alkoxides, other organic and inorganic derivatives of metals, which extend greatly the potentials of sol-gel method.

Some examples of application of metal alkoxides for synthesis of ferroelectrics - $BaTiO_3$-based materials, complex niobates and tantalates, PZT and PLZT-materials are discussed below.

5.1. Barium Titanate and $BaTiO_3$-based Solid Solutions

Barium titanate and $BaTiO_3$-based materials certainly are among most commonly used oxides for electronics. However, it is the $BaTiO_3$ powder of extremely high quality (in respect of its purity, stoichiometry, particles morphology) which is required for most of the modern applications. These are particularly those characteristics which may be considerably improved by application of alkoxides. Thus, it is of no surprise that a lot of publications are concerned with the synthesis of $BaTiO_3$ from alkoxides. The first works on preparation of $BaTiO_3$ in reaction of $Ba(OH)_2$ with $Ti(OR)_4$ date back to 1955[227]; in the late 60s preparation of $BaTiO_3$ by hydrolysis of the solution of alkoxides has been studied by Mazdiyasni[238]. In the last few years (1986-1990) we witness a new upsurge of interest to the synthesis of $BaTiO_3$ (see Table 2) which is presumably caused by attempts of industrial application of these methods.

All the suggested techniques may be divided in accordance with the method of introduction of barium in the system into three groups - application of salts (organic or inorganic), $Ba(OH)_2$, and finally - $Ba(OR)_2$.

Barium salts BaX_2 ($X = Cl^-, Br^-, CH_3COO^-$) do not react with organic solutions of $Ti(OR)_4$; this fact explains why in the products of thermal treatment of the hydrolysis products $BaCO_3$ is registered, its decomposition is complete only by 1200°C. It is thus quite natural that the reaction product contains alongside with $BaTiO_3$ admixtures of different complex oxides of $BaO-TiO_2$ system together with oxohalogenides (in the case when barium halogenides are used as starting materials). Introduction of barium in the form of solution of $Ba(OAc)_2$ in acetic acid allows to obtain film-forming sols 267. Their recent study by means of small-angle X-ray scattering, IR- and Raman spectroscopy suggests that the Ba and Ti species coexist in the sol and gel separately. Thermolysis at 200-400°C results in formation of $BaCO_3$ and TiO_2; crystallization of $BaTiO_3$ occurs only after heating to 650°C as a result of solid-state interaction 226.

The following reaction which results in crystallization of $BaTiO_3$ from solutions is probably one of the simplest ways to synthesize not only $BaTiO_3$ but also a lot of other perovskites:

$$Ba(OH)_2 \cdot 8H_2O \quad + \quad Ti(OR)_4 \quad \xrightarrow{t \geq 85^\circ C} \quad BaTiO_3$$

(aqueous solution) or crystalline (alcoholic solution) or without solvent (crystalline)

This process is studied in France by Boilot and coworkers (together with Thomson LCC)[229], and in a number of companies, such as E.I. Dupont de Nemours (USA)[230], Philips (FRG)[235]. The properties of $BaTiO_3$ thus obtained, and especially its stoichiometry and morphology are extremely sensitive to different conditions of the process (the way of introduction and mixing of the components), their ratio, isolation and drying conditions, application of surfactants. Careful choice of all of the above mentioned conditions allows to obtain submicron powders with uniform particles, mean size of ~0.5 μm and specific area ~ 10 m^2/g. Usually, ceramic materials sintered from these powders are characterized by the values of $\epsilon \approx 3000$; $tg\delta \approx 0.02$ values are $\epsilon \approx 4050 - 5400$ [229] and 8500 - 9500 [230] at $tg\delta$ 0.02 - 0.03.

Despite the considerable interest of technologists, and many publications on this method, the mechanism of $BaTiO_3$ formation in accordance with the above mentioned reaction remains vague. Boilot[229] following Flaschen[227] suggested formation of $[Ti(OH)_6]^{2-}$ ion in this reaction. It is noteworthy, on the other hand, that hydroxotitanates have never been registered either in aqueous solutions or in solid state. The absorption mechanism suggested in[233] seems much more probable. Initially, the hydrolysis of $Ti(OR)_4$ takes place forming a TiO_2 gel in which hydroxyl groups are absorbed giving a negative charge to the particles. Later, the alkaline earth cations diffuse inside the particles to counterbalance the negative charge. Finally, a condensation mechanism produces the three-dimensional perovskite structure expelling water molecules. This mechanism, by the way, allows us to explain the difference in the temperatures of formation of barium and strontium titanates (at 80 and 20°C, respectively) by higher mobility of Sr^{2+}.

TABLE 2. Synthesis of $BaTiO_3$ and $BaTiO_3$-based solid solutions

Material, some characteristics	Starting reagents	Crystallization temperature	References
$BaTiO_3$-powder with admixtures of $BaCO_3$ Ba_2TiO_4, BaX_2 (X = Hal)	[BaX_2 + HX + $Ti(OEt)_4$ + EtOH] (X = Cl, Br, I, CH_3COO)	600 - 1200	[222,223]
$BaTiO_3$, $BaZrO_3$ - powders, sintering into dense ceramic at 1000°C	[$Ba(CH_3COO)_2$ + CH_3COOH + ROH] + [$Ti(OEt)_4$ and/or $Zr(OPr)_4$]	700 - 1000	[224-226]
$BaTiO_3$ - powders (cubic phase), particles < 1 μm, S=12-14 m^2/g, sintering at 1115°C, ϵ = 3000-8000; tgδ = 0.02 (t = 20°C; f = 11 kHz)	$Ti(OR)_4$ - pure or solution in ROH (R=I-Pr, Bu) + ↓$Ba(OH)_2 \cdot 8H_2O$ or [$Ba(OH)_2$ + H_2O] or [↓$Ba(OH)_2$ + MeOH]	85	[227-235]
Ba_xSr_{1-x} $(Ti_uZr_vHr_w)$ O_3 (O ≤ x ≤ 1; u + v + w = 1), powders	Analogous to the above	-"-	[236,237]
$BaTiO_3$ - powders, particles < 1 μm, S = 10 m^2/g, sintering at 1200°C, ϵ ≃ 2500; tgδ = 0.02 (t = 20°C, f = 1 kHz), multilayer ceramic capacitors	$Ba(OR)_2$ + $Ti(OR)_4$ (R = Et, I-Pr, Bu, T-Am) + ROH or C_6H_6, or $CH_3OC_2H_4OH$	80 - 800	[238-247]
$Ba_xSr_{1-x}Ti$ Zr_{1-y} (O ≤ x, y ≤ 1), powders	Analogous to the above	800	[248]
$BaTiO_3$, powder (tetragonal phase)	$Ba(OH)_2$ + $Ti(OR)_4$ (R = I-Pr, Bu) + xylene or $Et_2C_6H_4$ → $BaTiO_2(OR)_2$	800	[249]
-"-	$Ba(OPr\text{-}I)_2$ + $Ti(OPr\text{-}I)_4$ + I-PrOH → $BaTiO(OPr\text{-}I)_4 \cdot 7/8$ I-PrOH	800	[220,245]

Material, some characteristics	Starting reagents	Crystallization temperature	References
$BaTiO_3$ doped with Nb, Ta, Ln - oxides; ceramics with uniform grains ~ 1μm	1) $Ba(OR)_2 + Ti(OR)_4 + M(OR)_n$ (M = Nb, Ta, Ln; R = I-Pr, Bu, Am) 2) $\downarrow BaTiO_3 + [M(OR)_n + ROH]$	800	[250-253]
$SrTiO_3$ - powders with spherical particles 0.4 ÷ 2.9 μm	$Ti(OPr\text{-}I)_3(OOCR') + Sr(OR)_2$	--	[254]
$BaO\text{-}xTiO_2$ - powders, capacitor and microwave ceramics	$Ba(OR)_2 + xTi(OR)_4 + ROH$	800 - 1200	[255-259]
$BaTiO_3$, $SrTiO_3$ - dielectric films	$Ba(OR)_2 + Ti(OR)_4 + ROH$ (R = Me, Et, I-Pr, Ba, $CH_3OC_2H_4$)	600 - 800	[260-264]
$BaTiO_3$, $SrTiO_3$ - films 1.2 - 1.8 μm thick, photoelectrodes	$M(OPr\text{-}I)_2$ (Mr, Sr, Ba) + $Ti(OPr\text{-}I)_4$ + acac + CH_3COOH	600 - 1000	[265,266]
$BaTiO_3$, $SrTiO_3$ - ferroelectric films	$Ba(CH_3COO)_2 + Ti(OR)_4 + CH_3COOH$ (R = I-Pr, Bu)	1000	[267,268]

Suggestion of the absorption mechanism allowed to improve considerably the morphology of powders by separation of two stages - hydrolysis of $Ti(OR)_4$ and subsequent interaction of hydrolysis product with aqueous $Ba(OH)_2$.

More than 20 years ago Mazdiyasni suggested preparation of $BaTiO_3$ by hydrolysis of $Ti(OAM\text{-}t)_4$ (obtained by alcoholic exchange from $Ti(OPr\text{-}I)_4$ and $Ba(OPr\text{-}I)_2$. Solutions containing Ba and Ti alkoxides were refluxed for several hours, and then slowly hydrolyzed, which lead to crystallization of $BaTiO_3$. After calcination at 800°C the powder was characterized by high purity, uniformity and sintered into dense ceramic with $\epsilon = 2500$ (t = 20°C, f = 1 kHz) at 1300°C. This remarkable technique, which should be considered as a first brick in the chemistry of preparation of complex oxides from metal alkoxides, leaves a lot of unsolved questions. It remains in particular strange, what determined such an unusual choice of alkoxides. In accordance with the modern data formation of $BaTi(OR)_6$ as a result of refluxing of Ba- and Ti-containing solutions seems highly improbable. Complex of such composition does not at all exist (see 4.1). No information is given about the hydrolysis conditions, such as H_2O : $M(OR)_n$ or temperature, while in the following works it has been shown that these conditions control the phase composition of the powder[246].

Interaction between $Ba(OPr\text{-}I)_2$ and $Ti(OPr\text{-}I)_4$ in I-PrOH has been thoroughly studied by Kirby[245]. He was the first to have isolated single crystals with the ratio Ba:Ti = 1:1 (he, however, did not ascribe any precise composition to these crystals). On their thermolysis tetragonal $BaTiO_3$ phase is formed. X-ray structural investigations of the crystals has demonstrated that their composition if $BaTiO(OPr\text{-}I)_4 \cdot 7/8i\text{-}PrOH$ [220]. Fragments of this structure containing Ba-O-Ti bridges which are true precursors of $BaTiO_3$ are accumulated in solution during refluxing of the alkoxide solutions.

The following scheme 3 summarized all the processes taking place under different conditions suggested in techniques for $BaTiO_3$ preparation.

Sometimes, crystalline $BaTiO_3$ prepared by hydrolysis of the alkoxide solutions in the form of dispersion without drying was successfully used to make a ceramic slip formulation, which was then cast as a ceramic green body for microcapacitors 1-50 μm thick.

First $BaTiO_3$ films from alkoxide solutions were also obtained

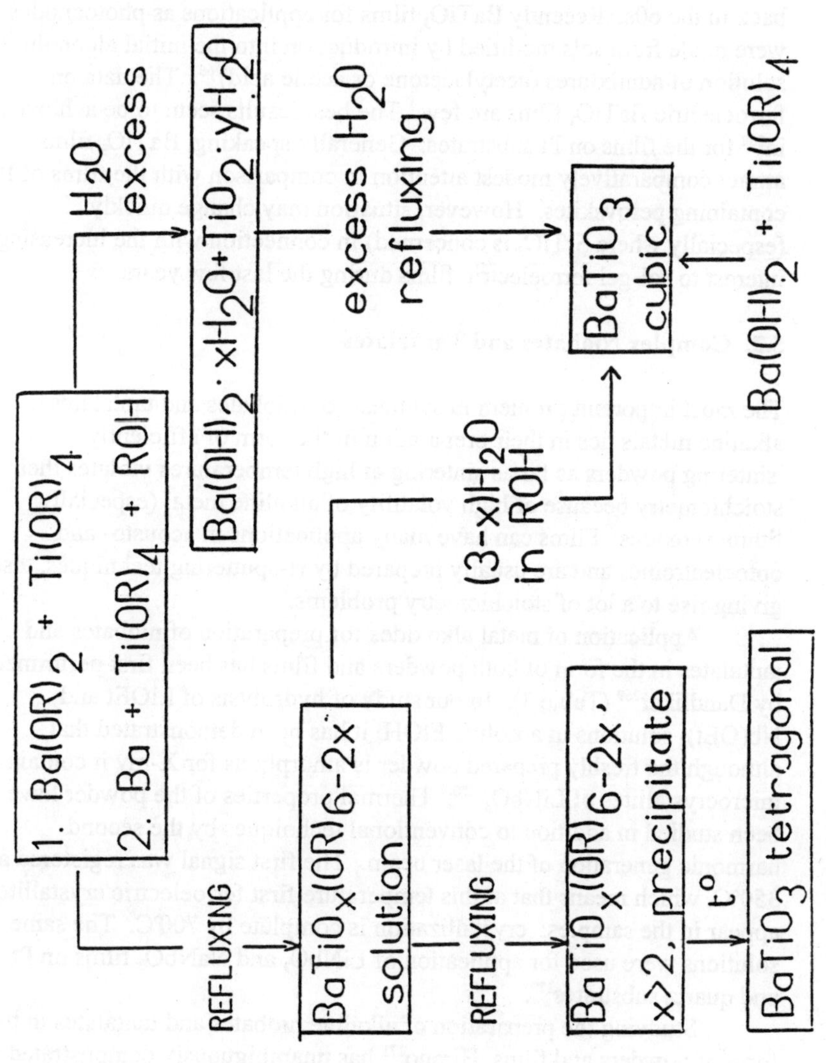
1. Ba(OR)2 + Ti(OR)4
2 Ba + Ti(OR)4 + ROH
REFLUXING
BaTiOx(OR)6-2x
solution
REFLUXING
BaTiOx(OR)6-2x',
x>2, precipitate
t°
BaTiO3 tetragonal
H2O
excess
Ba(OH)2·xH2O+TiO2·yH2O
excess H2O
refluxing
(3-x)H2O
in ROH
BaTiO3
cubic
Ba(OH)2 + Ti(OR)4

SCHEME 3

back in the 60s. Recently $BaTiO_3$ films for applications as photoanodes were made from sols modified by introduction into the initial alcoholic solution of admixtures (acetylacetone or acetic acid)[266]. The data on ferroelectric $BaTiO_3$ films are few. The best results seem to be achieved in[268] for the films on Pt substrates. Generally speaking, $BaTiO_3$ films attract comparatively modest attention in comparison with the films of Pb-containing perovskites. However, situation may change quickly (especially where $SrTiO_3$ is concerned) in connection with the increasing interest to sol-gel ferroelectric films during the last few years.

5.2. Complex Niobates and Tantalates

The most important problem in synthesis of miobates and tantalates of alkaline metals lies in their preparation in the form of efficiently sintering powders as far as sintering at high temperatures violates their stoichiometry because of high volatility of alkaline metal (especially lithium) oxides. Films can have many applications in acousto- and optoelectronics and are usually prepared by rf-sputtering techniques, also giving rise to a lot of stoichiometry problems.

Application of metal alkoxides for preparation of niobates and tantalates in the form of both powders and films has been first performed by Dandliker[269] (Table 3). In our study of hydrolysis of LiOEt and $Nb(OEt)_5$ solutions in absolute EtOH, it has been demonstrated that although the freshly prepared powder is amorphous for X-ray it contains microcrystallites of $LiNbO_3$ [270]. Thermal properties of the powder have been studied in addition to conventional techniques by the second harmonic generation of the laser beam. The first signal was registered at 350°C, which means that at this temperature first ferroelectric crystallites appear in the samples: crystallization is complete by 700°C. The same solutions were used for application of $LiNbO_3$ and $NaNbO_3$ films on Pt and quartz substrates[270].

Studying the preparation of alkaline niobates and tantalates in the form of powders and films, Hirano[271] has unambiguously demonstrated the necessity of refluxing of the initial alkoxide solutions before hydrolysis or film application. Only such solutions allowed to prepare pure niobates containing no admixtures. The explanation for this fact Hirano sees in formation of $LiNb(OEt)_6$ in solutions during their refluxing. However, in accordance with[186] complex of this composition crystallizes on combining of the components. On the other hand, on refluxing or storage of these solutions oxoethoxoniobates are formed. (The structure of the crystalline $LiNbO(OEt)_4 \bullet EtOH$ has been solved in[219]). Calcination of the hydrolysis products and films was performed in a flow of oxygen mixed with water vapor (to facilitate complete hydrolysis and crystallization). Crystallization of $NiNbO_3$ occurs in these conditions at very low temperature of 250 - 300°C.

Films were obtained from solutions partially hydrolyzed after refluxing. The choice of concentration and thickness of the layer which undergoes calcination is very important for preparation of films without cracks and bubbles. Oriented $LiNb_xTa_{1-x}O_3$ ($0 \leq x \leq 1$) were obtained on the respectively oriented sapphire substrates (012), (110), or (001). However, on Si ((100) and (111) planes) oriented films could not be obtained. Afterwards it has been suggested to use instead of solutions in ethanol much more stable solutions in methoxyethanol[274].

Other oxide phases of the system Li_2O - Ta_2O_5 can be synthesized from alkoxide solutions with different Li(OEt) : $Ta(OEt)_5$ ratios[279].

Barium- and strontium-containing perovskites $AB_{1/3}C_{2/3}O_3$ (A = Ba, Sr; B = Zn, Mg, Co, Ni; C = Nb, Ta) centrosymmetric pseudocubic phases are now used as the base for a relatively new class of resonators with high Q-factor. Dielectric characteristic of ceramic materials are affected dramatically by the presence of admixtures of foreign phases, which is rather a serious problem as far as niobates and especially tantalates sinter only at extremely high temperatures (1400 - 1500°C). Pb-containing complex oxides - $PbZn_{1/3}Nb_{2/3}O_3$; $PbFe_{1/2}Nb_{1/2}O_3$ - are ferroelectric materials with high dielectric constants and a high electrostrictive strain coefficients. Relatively low sintering temperatures make these materials promising for electronics applications.

For a few years synthesis of complex niobates and tantalates from metal alkoxides has been studied by Matsushita Electric Industrial Co., Ltd. (Japan)[282-284]. In the first attempts to synthesize $BaZn_{1/3}Ta_{2/3}O_3$ crystalline inorganic compounds of Ba and Zn were suspended in the alcoholic solution of $Ta(OEt)_5$, dispersion was hydrolyzed and calcined.

TABLE 3. Synthesis of complex niobates and tantalates

Materials	Starting reagents	Crystallization temperature	References
$Li_xNa_{1-x}Nb_yTa_{1-y}O_3$ (O ≤ x, y ≤ 1) powders, ceramics, polycrystalline and oriented films	M(OEt) + $M'(OEt)_5$ (M = Li, Na; M' = Nb, Ta) in EtOH	400 - 600	[269-273]
$LiNbO_3$ - films	LiOEt + $Nb(OEt)_5$ + $CH_3OC_2H_4OH$	500	[274]
$KTa_{1-x}Nb_xO_3$ (O ≤ x ≤ 1), ceramics	KOR (R = Me, Et) + $M(OEt)_5$ (M = Nb, Ta) in EtOH	700 - 750	[275]
$LiMO_3$ (M = Nb, Ta), powders	Separate hydrolysis of [LiOEt + EtOH] and [$M(OEt)_5$, (M = Nb, Ta) + EtOH]	1000 - 1150	[276]
$Li_xNa_{1-x}NbO_3$ (0.03 ≤ x ≤ 0.17), powders	$Nb(OEt)_5$ + NaOEt + LiOAc in EtOH	600	[277]
$LiNbO_3$ - fibers	LiCl + $Nb(OEt)_5$ + EtOH	450 - 600	[278]
$LiTaO_3$, $LiTa_3O_8$, Li_3TaO_4, β - Ta_2O_5, powders	Different [LiOEt] : [$Ta(OEt)_5$] ratios in EtOH	500 - 800	[279]
$Sr_xBa_{1-x}Nb_2O_6$ (O ≤ x ≤ 1), powders, ferroelectric films on Si	$Ba(OR)_2$ + $Sr(OR)_2$ + $Nb(OR)_5$ + ROH	700 - 1000	[280,281]
$BaZn_{1/3}Ta_{2/3}O_3$ - powders	↓MO, $M(OH)_2$, MX_2 (M = Ba, Zn; X = Hal) in [$Ta(OEt)_5$ + EtOH]	> 1100	[282]
-"-	$Ba(OEt)_2$ + $Ta(OEt)_5$ + $Zn(acac)_2$ + EtOH	~ 1030	[283]
$AB_{1/3}C_{2/3}O_3$ (A = Ca-Ba; B = Mg, Zn, Co, Ni; C = Nb, Ta) - powders	[$A(OH)_2$ + H_2O] + [$B(OR)_2$ + $C(OR)_5$ + ROH]	≥ 95	[191,284,285]

Materials	Starting reagents	Crystallization temperature	References
$AB_{1/3}C_{2/3}O_3$ (A = Sr, Ba; B = Mg, Zn; C = Nb, Ta) - powders	$A(OPr\text{-}I)_2 + B(OEt)_2 + C(OPr\text{-}I)_5 + C_6H_6$	800 - 1000	[286]
-"- powders, multilayer ceramic capacitors	$A(OR)_2 + B(OR)_2 + C(OR)_5$ (R = Et, I-Pr, Bu) + mixture (C_6H_6 + I-PrOH) or I-PrOH	85 - 100	[287,288]
$PbM_{1/3}Nb_{2/3}O_3$ - powders	$[Pb_4O(OPr\text{-}I)_6 + Mg[Nb(OEt)_6]_2 + EtOH]$ or $[Pb(OP\text{-}t)_2 + Mg[Nb(OEt)_6]_2 + EtOH$ + xylene	≥ 700	[289]
-"- powders, films	$Pb(CH_3COO)_2 + Mg(OEt)_2 + Nb(OEt)_5 + CH_3OC_2H_4OH$	> 700	[290,291]
$PbMg_{1/3}Nb_{2/3}O_3$ - $PbTiO_3$ solid solutions	$[Ti(OBu)_4 + Zr(OBu)_4 + Nb(OBu)_5 + BuOH] + [Pb(CH_3COO)_2 \cdot 3H_2O + Mg(CH_3COO)_2 \cdot 4H_2O]$	--	[292]
$PbZn_{1/3}C_{2/3}O_3$ (C = Nb, Ta) - ceramics	$Pb(CH_3COO)_2 + Zn(CH_3COO)_2 + C(OPr\text{-}I)_5$ (C = Nb, Ta) + $CH_3OC_2H_4OH$		[293]
$PbFe_{1/3}Nb_{1/3}O_3$ - powders, ceramics	$[Fe(OBu)_3 + Nb(OEt)_5 + C_6H_6] + PbO$	850	[294]
-"-	$Nb(OEt)_5 + Fe(acac)_3 + Pb(OAc)_2 \cdot 3H_2O + AcOH$	--	[295]

Uniform distribution of Ba and Zn in this gel was achieved by vigorous stirring of the reaction mixture. High degree of dispersion and reactivity of $Ta_2O_5 \bullet H_2O$ ensures formation of complex oxide at temperatures ~ 150°C lower than in the case of solid-state reactions. Then all three elements were dissolved in the organic solvent Ba and Ta - in the form of alkoxides, while Zn - in the form of its most available derivative - acetylacetonate. However, $Zn(acac)_2$ does not hydrolyze and reacts with the alkoxide hydrolysis products only on the thermolysis step, thus formation of the complex oxide in this case also occurs only as a result of the solid-state reaction between the components at ~ 1030° C.

According to the patent data[286] complex oxides of this class may be obtained by hydrolysis of the alcoholic solutions containing all three alkoxides. Crystallization in this case occurs at considerably lower temperature (800 - 1000°C). Recently, complex niobates and tantalates were obtained in the form of crystalline phases directly in the course of hydrolysis of the alkoxide solution (analogous to preparation of $BaTiO_3$). Hydrolysis of the alkoxide solutions in the mixed C_6H_6-I-PrOH solvent was carried out by the excess of water ~ 80°C with the subsequent stirring of the reaction mixture at this temperature for several hours. However, full understanding of the correlation between the hydrolysis conditions and the extent of crystallinity has not yet been achieved. Crystalline products obtained directly by hydrolysis may be employed for preparation of dispersions used for sintering of ceramic sheets.

The synthesis of complex perovskites by alkoxide-hydroxide route (studied by now in detail in the case of $BaTiO_3$) seems to be rather promising

$$Ba(OH)_2 + 1/3Mg[Nb_2(OR)_6]_2 \xrightarrow{t \geq 80°C} BaMg_{1/3}Nb_{2/3}O_3$$

The main problem in this process lies in the introduction of $M(OR)_2$ (M = Mg, Zn, Co, Ni) into the alcoholic solution of niobium or tantalum alkoxides, because their synthesis presents certain difficulties. In 191, 285 Mg was dissolved in the alcoholic solutions of niobium or tantalum alkoxides (which activate reaction of Mg with alcohols; see above). However, dissolution of Zn, Co, Ni under this condition cannot be performed.

Lead-containing perovskites $PbMg_{1/3}Nb_{2/3}O_3$ may be synthesized by hydrolysis of the alcoholic solutions containing all three elements in the form of alkoxides; however $Pb(OR)_2$, which is not so easy to synthesize, may be substituted by $Pb(OAc)_2$. In this case, however, methoxyethanol should be used as a solvent.

5.3. PZT and PLZT Solid Solutions

First studies on the synthesis of PZT and PLZT materials from metal alkoxides date back to the early 70s. They dealt primarily with preparation of powders for sintering of the electro-optic PLZT ceramics[304]. In fact, all elaborated methods are based on the hydrolysis of titanium and zirconium alkoxides (see Table 4). Although Mazdiyasni suggested to hydrolyze solution containing all four elements as alkoxides in most of the following studies lanthanum alkoxides have not been used; instead of that, hydrolysis of the solution in organic solvent has been performed by aqueous solution of lanthanum acetate. Different lead derivatives were also used: lead alkoxides, dissolved in alcohols together with titanium and niobium alkoxides, PbO-powder which was vigorously stirred in the alcoholic solution of titanium and zirconium alkoxides during the hydrolysis stage, or aqueous solution of lead acetate, which was used as an agent for hydrolysis.

Application of all four alkoxides allowed to obtain powder with the purity of 99.95%, which sintered into dense optically transparent ceramics at 1050 - 1175°C without hot pressing. However, for the large scale synthesis another technique, employing PbO and aqueous $La(OAc)_3$ has been used[314]. The following investigations demonstrated that hydrated Ti and Zr hydroxides which are formed on hydrolysis of corresponding metal alkoxides are extraordinarily active and react readily with suspended PbO to form amorphous matrix with uniform distribution of elements[316]. Crystallization of PZT from this matrix occurs in one stage at 600°C. The sintered ceramic is characterized by the absence of composition fluctuations and residual porosity and thus demonstrates very good optical properties. Introduction of lead in the form of its acetate does not give very good results as far as decomposition of $Pb(OAc)_2$ occurs only on calcination and the perovskite phase is thus formed only as a result of the

TABLE 4. Synthesis of $PbTiO_3$, PZT, and PLZT materials

Material	Starting reagents	References
$PbTiO_3$ - powders	$Pb(OR)_2 + La(OR)_3 + Ti(OR)_4 + Zr(OR)_4$	[290]
PZT - powders	(R = n-Pr, I-Pr, Bu) in ROH and $CH_3OC_2H_4OH$	[291-302]
PZT - films		[303]
PLZT - powders, ceramics		[304]
$PbTiO_3$ - powders	$[Ti(OPr\text{-}I)_4 + Zr(OBu)_4 + ROH] + [Pb(NO_3)_2 + La(NO_3)_3 + H_2O]$ or	[305]
PZT - powders	$Ti(OPr\text{-}I)_4 + [ZrO(NO_3)_2 + Pb(NO_3)_2 + La(NO_3)_3 + H_2O]$	[306]
PLZT - powder, ceramics		[307,308]
PZT - powders, ceramics	$\downarrow PbO + [Zr(OBu)_4 + La(OPr\text{-}I)_3 + I\text{-}PrOH]$	[309,310]
PLZT - powders		[311]
PZT - powders	$[Ti(OBu)_4 + Zr(OBu)_4$ - without solvent, or in BuOH + crystalline $Pb(CH_3COO)_2 \bullet 3H_2O$ or $[Pb(CH_3COO)_2 + H_2O]$	[312]
PLZT - ceramics	$\downarrow PbO + [Ti(OR)_4 + Zr(OR)_4 + ROH] + [La(CH_3COO)_2 + H_2O]$ (R = I-Pr, Bu)	[313-316]
	$[Zr(OBu)_4 + Ti(OBu)_4 + BuOH] + [Pb(CH_3COO)_2 + La(CH_3COO)_3 + H_2O]$	[318]
$PbTiO_3$ - films for optics	$[Ti(OPr\text{-}I)_4 + acac] + [Pb (CH_3COO)_2 + AcOH]$	[319]
$PbTiO_3$ - powders	$[Ti(OR)_4 + Zr(OR)_4 + Pb(R'COO)_2$ in I-PrOH or hydrocarbons (R = Bu, Pr)	[320] [321]
PZT - powders		
PZT - films		[322]

solid-state reactions between the components.

Comparison of different commercially available ZrO_2 samples in sintering of PZT has demonstrated that crystallization of PZT occurs in one step only is samples prepared by hydrolysis of alkoxides are used[334].

Despite good laboratory results alkoxides have not yet found industrial application for sintering of piezoelectric ceramics, presumably because of their high cost. On the other hand, the last few years are characterized by the growth of interest to the synthesis of PZT from metal alkoxides in the form of thin films. As far as this book contains special chapter concerning application of PZT films by sol-gel method (see the Chapter by S.K. Dey) we shall not discuss this subject in the present review. We would only like to emphasize that after the first attempts to use different precursors all the latest techniques use methoxyethanol as a solvent. This solvent dissolves both $Ti(OR)_4$ and $Zr(OR)_4$ as well as $Pb(OAc)_2$ which react in solutions with formation of complex alkoxoacetates. Moreover, commercial $Pb(CH_3COO)_2 \bullet 3H_2O$ can be readily dehydrated by recrystallization from methoxyethanol. However, all chemical processes which comprise preparation of the precursing sol are far from being completely understood. One of the first investigations in this field has been performed by Payne[335]. The composition of complexes in methoxyethoxide solutions has been studied by means of mass-spectra and NMR spectroscopy. This application of metal alkoxides seems to be at present one of the most important in the synthesis of ferroelectric materials.

REFERENCES

1. Y. NOMIYA, J. Cerm. Soc. Jap. Int. Ed., **97**, No. 4, 21 (1989).
2. J. Non-Cryst. Solids, **48** (1982); **63** (1984); **82** (1986); **100** (1988); **121** (1990).
3. Editors L.L. HENCH, J.D. MACKENZIE AND D.R. ULRICH, Ultrastructure Processing od Ceramics, Glasses and Composites (Wiley, N.Y., 1984, 1986, 1988).
4. Editors C.J. BRINKER, D.E. CLARK, D.R. ULRICH AND J.B. ZELINSKI, Better Ceramics Through Chemistry (North-Holland, N.Y., 1984; Pittsburgh, PA., 1986, 1988, 1990).
5. C.J. BRINKER AND G.W. SCHERER, Sol. Gel. Science (Academic Press, Inc., Boston, 1980).
6. Editor L. KLEIN, Sol-Gel Technology for Thin Films, Fibers, Preforms, Electronics and Specialty Shapes (Noyes, Park Ridge, N.Y., 1988).
7. L.G. HUBERT-PFALZGRAF, New J. Chem., **11**, 663 (1987).
8. D.R. ULRICH, J. Non-Cryst. Solids, **121**, 465 (1990).
9. Y. CZAKI, Ferroelectrics, **49**, 285, (1983).
10. P. RAVINDRANATHAN, S. KOMARNENI, S.C. CHOI, A.S. BHALLA AND R. ROY, Ferroelectrics, **87**, 133 (1988).
11. C. SANCHEZ AND J. LIVAGE, New J. Chem., **14** (1990).
12. B.J.J. ZELINSKI AND D.R. UHLMANN, J. Phys. Chem. Solids, **45**, 1069 (1984).
13. S. SAKKA, Treatise on Materials Science and Technology, **22**, 129 (1982).
14. H. DISLICH, J. Non-Cryst. Solids, **57**, 371 (1983).
15. N.YA. TUROVA AND M.I. YANOVSKAYA, Inorg. Mater. (Izv. AN SSSR, Neorg. Mater.), **19**, 615 (1983).
16. N.YA. TUROVA AND A.V. NOVOSELOVA, Uspekhi Khimii, **34**, 385 (1965).
17. E. DEMARCAY, Compt. Rend., **80**, 51 (1975).
18. W.E. TISCHTSCHENKO, Zhurn. Russ. Fiz. Khim. Obsch., **31**, 694, 784 (1899).
19. H. GILMAN, et al., J. Amer. Chem. Soc., **78**, 4285, 4287, 4289, 6027, 6030 (1956); **79**, 4921 (1957).

20. D.C. BRADLEY, in progress in Inorg. Chem., **2**, edited by F.A. Cotton (Interscience Publ., N.Y., 1960), p. 304; Adv. in Inorg. Chem. and Radiochem., **15**, 259 (1972).
21. D.C. BRADLEY, in progress in Inorg. Reactions, **2**, edited by W.L. Jolley (Interscience Publ., N.Y., 1965) p. 226.
22. D.C. BRADLEY, H. CHUDZYNSKA, et. al., Polyhedron, **9**, 719 (1990).
23. D.C. BRADLEY, R.C. MEHROTRA AND D.P. GAUR, Metal Alkoxides, (Academic Press, London, 1978).
24. L.G. HUBERT-PFALZGRAF, et. al., J. Chem. Soc. Chem. Commun., **1989**, 1846; **1990**, 695, 301; Inorg. Chem., **28**, 263 (1989).
25. K.G. CAULTON, et. al., Inorg. Chem., **29**, 3126 (1990); **30**, 1720, 1723, 2317 (1991); J. Chem. Soc. Chem. Commun., **1990**, 1750.
26. M.H. CHRISHOLM, et. al., Faraday Symp. Soc., **1980**, N14, 194; in progress in Inorg. Chem., **29**, (NY, 1982), p. 1; Inorg. Chem.: Toward 21st Century Symp., Bloomington, IN, 1982 (Washington, D.C. 1983) p. 243; J. Solid State Chem., **57**, 120 (1985); Chem. Soc. Rev., **14**, 69 (1985); Polyhedron, **5**, 25 (1986); New J. Chem., **11**, 459 (1987).
27. A.V. NOVOSELOVA, N. YA. TUROVA, et. al., Izv. Akad. Nauk SSSR, Neorg. Mat., **15**, N6; 950 (1979); **18**, N2, 271 (1982).
28. N.YA. TUROVA, et. al., Russ. J. Inorg. Chem., **25**, 1188, 3263 (1980); Sov. J. Coord. Chem., **6**, 221 (1980), **8**, 346 (1982); **9**, N6, 430, 685 (1983); **11**, 366, 867, 945 (1985); **12**, 440, 619 (1986); **13**, 586 (1987); **15**, 867 (1989).
29. V.A. SHREIDER, E.P. TUREVSKAYA, N.I. KOZLOVA, AND N.YA. TUROVA, Inorg. Chim. Acta., **13**, L.73 (1981).
30. M.I. YANOVSKAYA, N.YA. TUROVA, et. al., J. Mater. Sci. Lett., **6**, 274 (1987); J. Mater. Sci., **23**, 395 (1988); J. Non-Cryst. Solids., **124**, 155 (1990); Ferroelectrics, **105**, 39 (1990).
31. W. GEFFCKEN AND E. BERGER, Ger. Pat., **736**, 411 (1939).
32. N.V. SUYKOVSKAYA, Chemical Methods of Preparation of Thin Transparent Films, (Khimiya, Leningrad, 1971).
33. H. DISLICH, Angew Chem. Int. Ed. Engl., **10**, 363 (1971).
34. H. DISLICH AND E. HUSSMANN, Thin Solid Films, 77, 129 (1981).

35. H. DISLICH, Sol-Gel Technology for Thin Films, Fibers, Preforms, Electronics and Specialty Shapes, edited by L.C. Klein (Noyes Publication, Park Ridge, NJ, 1988) p. 50.
36. S.SAKKA, Ibid., p. 140.
37. S. SAKKA, H. KOZUKA, AND T. UMEDA, J. Ceram. Soc. Japan, **96**, (1988), p. 463.
38. H. OTA, New Glass, **8**, 25 (1988).
39. S. HORIKIRI AND Y. ABE, Ceramics Data Book, (Kogyo-Seihin-Gijutsu-Kyokai, 1980), p. 332.
40. K. YAMADA, FC Report, **8**, 281 (1988).
41. M. YAMANE AND S. OKANO, Yogyo-Kyokai-Shi, **87**, 434 (1979).
42. B.E. YOLDAS, J. Mater. Sci., **14**, 1893 (1979).
43. J. FRICKE, J. Amer. Chem. Soc., **78**, 4285, 4287, 4289, 6027, 6030 (1956); **79**, 4921 (1957).
44. S. WALLACE AND L.L. HENCH, Mat. Res. Soc. Symp. Proc., **32**, 47 (1984).
45. K.S. MAZDIYASNI AND C.T. LYNCH, C.r. J. Etud Solids, Finement Div. (1967), p. 9.
46. K.S. MAZDIYASNI, et. al., J. Amer. Ceram. Soc., **52**, 523 (1965); **54**, 539, 548 (1971); **55**, 633 (1972).
47. O. YAMAGUCHI, et. al., J. Amer. Ceram. Soc., **66**, 169 (1983); **68**, 110 (1985); **69**, 36 (1986); **72**, 1914 (1989).
48. Application of Ultra Fine Powders, (Nikkan Kogyo-Sinbunsha, 1986), p. 76.
49. Y. OZAKI, Seramikkusu, **21**, 102 (1986); Jap. Sci. and Technol., **25**, 227, 43 (1984).
50. E. WEISS, H. ALSDORF, H. KUHR, AND H. GRUTZMACHER, Ber, **101**, 3777 (1968); Angew. Chem., **79**, 816 (1967).
51. N.YA. TUROVA AND A.V. NOVOSELOVA, Bull. Acad. Sci. USSR, Div. Chem. Sci., **1970**, 709.
52. P.J. WHEATLEY, J. Chem. Soc., **1960**, 4270.
53. N.YA. TUROVA, B.A. POPWKIN AND A.N. NOWOSELOWA, Z. Anorg. Allg. Chem., **365**, 100 (1969).
54. N. FISHER AND S.M. MCELVAIN, J. Am. Chem. Soc., **56**, 1768 (1934).

55. E. CHABLAY, Compt. Rend., **140**, 1343 (1905); **153**, 819, 953 (1911); **154**, 1507 (1912); Amer. Chim., [9], **8**, 145, 167, 179, 185 (1917).
56. N.YA. TUROVA, A.V. NOVOSELOVA, AND K.N. SEMENENKO, Russ. J. Inorg. Chem., **4**, 997 (1959).
57. R. PIE KOS, Uspekhi Khimii, **37**, 647 (1968).
58. L.M. BROWN AND K.S. MAZDIYASNI, Inorg. Chem., **9**, 2783 (1980).
59. V.N. KOKOZEY, et. al., Ukr. Khim. Zhurn., **44**, 92! (1978); **46**, 1157 (1980).
60. B. SZILARD, Z. Electrochem., **12**, 393 (1906).
61. T.G. TRIPP, Ger. Offen. Pat., 2.005.835 (1970); 2.121.732 (1972); Brit. Pat., 1.246.032 (1971); Fr. Pat., 2.091.229 (1972).
62. H. LEHMKUHL AND W. EISENBACH, Ann., **1975**, 672; Ger. Offen. Pat. 2.349.561 (1974); Canad. Pat. 1.024.466 (1978).
63. N.I. KOZLOVA, N. YA. TUROVA AND E.P. TUREVSKAYA, Sov. J. Coord. Chem., **8**, 339 (1982).
64. S.I. KUCHEKO, N.YA. TUROVA AND V.A. SHREIDER, Zh. Obshch. Khim., (Russ.) **55**, 2412 (1985).
65. N.YA. TUROVA, V.G. KESSLER, et. al., ibid. **60**, 113 (1990); Sov. J. Coord. Chem., **15**, 867 (1989).
66. T.V. ROGOVA, N.YA. TUROVA AND B.V. ZHADANOV, Sov. J. Coord. Chem., **11**, N6, 448 (1985).
67. N.YA. TUROVA, E.P. TUREVSKAYA, et. al., Russ. J. Inorg. Chem., **97**, N1, 89 (1992).
68. S.I. KUCHEIKO, V.G. KESSLER AND N.YA. TUROVA, Sov. J. Coord. Chem., **13**, 586 (1987).
69. W. SCHLAFER, K. ANDREAS AND E. KRETSCHMER, Pat. DDR 158.410 (1983).
70. M.T. VILHENA, A.M.T.S. DOMINGOS AND A. PIRES DE MATOS, Inorg. Chim. Acta, **95**, 11 (1984).
71. J.C. BANAIT AND P.K. PANIL, Synth. React. Inorg. Met-Org. Chem., **16**, 1217 (1986); **18**, 309 (1988); Bull. Electrochem., **5**, 264 (1989).
72. Jpn. Kokai, 60-159 183 (1985).
73. L.I. SOLOVJEVA, Private Commun.
74. Pat. DDR 246.988 (1987).

75. E.P. TUREVSKAYA, N.YA. TUROVA AND A.V. NOVOSELOVA, Russ. J. Inorg. Chem., **20**, 838 (1975).
76. Pat. USA 2.796.443 (1957).
77. W. PRANDTL AND L. HESS, Z. Anorg. Chem., **82**, 103 (1913).
78. N.F. ORLOV AND M.G. VORONKOV, Bull. Acad. Sci. USSR, Div. Chem. Sci., **1959**, 933.
79. W.J. BEHR AND J. FUCHS, Z. Naturforsch., **28B**, 597 (1973).
80. M.W. BISHOP, G. BUTLER, J. CHATT, et. al., J. Chem. Soc. Dalton Trans., **1979**, 1843.
81. C. RINGEL AND G. BODEN, Z. Anorg. Chem., **393**, 65 (1972).
82. R.J. COLLIN, J. JONES AND W.P. GRIFFITH, J. Chem. Soc. Dalton Trans., **1974**, 1904.
83. B. JASELSKIS AND J.P. WARRINER, J. Amer. Chem. Soc., **91**, 210 (1969).
84. E.F. CALDIN AND G. LONG, J. Chem. Soc., **1954**, 3737.
85. J. MURTO, Suomen Kem., **35**, N9, B157 (1962).
86. R. LE GOALLER, M.A. POSOUINI AND J.L. PIERRE, Tetrahedron, **36**, 237 (1980).
87. R.E.A. DEAR, W.B. FOX, et. al., Inorg. Chem., **9**, 2590 (1970).
88. E.V. CHETVERIKOV, M.V. KOROTKOVA, et. al., V All-Union Conference on Non-aqueous Solutions, (Rostov, 1985), p. 149.
89. N.YA. TUROVA, M.I. KARPOVSKAYA (YANOVSKAYA), et. al., Inorg. Chim. Acta, **21**, 157 (1977); Sov. J. Coord. Chem., **4**, N6, 907 (1978).
90. G.F. WHEITE, A.B. MORRISON AND E.G. ANDERSON, J. Am. Chem. Soc., **46**, 961 (1924).
91. B. HORVATH, E.G. NORWATH, et. al., Z. Anorg. Chem., **449**, 41 (1979); **457**, 51 (1979).
92. S.C. GOEL, M.Y. CHIANG AND W.E. BUHRO, Inorg. Chem., **29**, 358, 4640 (1990).
93. M.C. MASSIANI, R. PAPIERNIK AND L.G. HUBERT-PFALZGRAF, Polyhedron, **10**, 437 (1991).
94. R. PAPIERNIK, L.G. HUBERT-PFALZGRAF AND M.-C. MASSIANI, Inorg. Chim. Acta, **165**, 1 (1989).
95. S. BOULMAAZ, R. PAPERNIK. L.G. HUBERT-PFALZGRAF, et. al. (in press).

96. S.I. KUTSHEIKO, N.YA. TUROVA AND N.I. KOZLOVA, Sov. J. Coord. Chem., **11**, N12, 945 (1985).
97. E.P. TUREVSKAYA, N.YA. TUROVA, et. al., Coord. Khimiya, **14**, 926 (1988); **15**, (1989).
98. E. COSTA, A. CAMUS AND N. MARSICH, J. Inorg. Nucl. Chem., **27**, 281 (1965).
99. H.E. PODALL, J.H. DUNN AND H. SHAPIRO, J. Amer. Chem. Soc., **82**, 1325 (1960).
100. G.A. RAZUVAEV, V.N. LATYAEVA, et. al., J. Organometal. Chem., **134**, 43 (1977).
101. M.H. CHISHOLM, F.A. COTTON, M.W. EXTINE, et. al., Inorg. Chem., **18**, 120 (1979).
102. G.E. COATES, et. al., J. Chem. Soc., **1965**, 1870; **1966A**, 1064; **1967A**, 1233.
103. H. FUNK, et. al., Z. Anorg. Chem., **296**, 36 (1958); **302**, 199 (1959).
104. E.P. TUREVSKAYA, A.I. YANOVSKIY, N.YA. TUROVA, et. al., Sov. J. Coord. Chem., **15**, 122 (1989).
105. R. FELD AND P.L. COWE, The Organic Chemistry of Titanium, (Butterworths, London, 1965).
106. N.YA. TUROVA, T.V. ROGOVA, et. al., Pat. USSR 1.310.381 (1987) Chem. Abstr., **107**, 153915 (1987); Sov. J. Coord. Chem., **9**, N9, 685 (1983).
107. E.P. TUREVSKAYA, N.YA. TUROVA, et. al., Sov. J. Coord. Chem., **12**, 619 (1986).
108. W.J. EVANS, M.S. SOLLBERGER AND J.R. HANUSA, J. Amer. Chem. Soc., **110**, 1841 (1988).
109. R.A. ANDERSEN, D.H. TEMPLETON AND A. ZALKIN, Inorg. Chem., **17**, 1962 (1978).
110. G.W. HALSTEAD, P.G. ELLER, et. al., Inorg. Chem., **17**, 2967 (1978); Inorg. Synth., **21**, 162 (1982).
111. H. ALBERS, M. DEUTSCH, et. al., Ber., **85**, 267 (1952).
112. R.W. ADAMS, R.L. MARTIN AND G. WINTER, Austral. J. Chem., **19**, 207 (1966).
113. B.P. BARANWAL AND R.C. MEHROTRA, Austral. J. Chem., **33**, 37 (1980).

114. M. BOCHMANN, G. WILKINSON AND G.B. YOUNG, J. Chem. Soc. Dalton Trans., **1980**, 1863.
115. A.L. SUVOROV AND S.S. SPASSKIY, Usp. Khim., **28**, 1267 (1959).
116. P. EDWARDS, G. WILKINSON, et al., J. Chem. Soc. Chem. Commun., **1979**, 1158.
117. Jap. Pat. 60-258.132 (1985); Chem. Abstr., **105**, 26611 (1986).
118. L. HANDY, F. BRINCKMAN, et. al., Chem. Commun., **1970**, 214; Inorg. Chem., **11**, 523 (1972); J. Fluor. Chem., **8**, 55 (1976).
119. D.W. WALKER AND J.M. WINFIELD, J. Fluor. Chem., **1**, 376 (1972); J. Inorg. Nucl. Chem., **34**, 759 (1972).
120. E.P. TUREVSKAYA, N.YA. TUROVA AND A.V. NOVOSELOVA, Bull. Acad. Sci. USSR, Div. Chem. Sci., **1968**, 1577; 2590.
121. F. UCHIDA AND T. YAZAKI, Jpn. Kokai, 63-22038-9 (1988).
122. P.S. GRADEFF, F.G. SCHREIBER, et. al., Inorg. Chem., **24**, 1110 (1985); J. Less-Comm. Met., **126**, 335 (1986); Eur. Appl. Pat. 183.603 (1986); 316.213; 324.671 (1989).
123. N.YA. TUROVA, A.V. NOVOSELOVA AND K.N. SEMENENKO, Zh. Neorg. Khim., **4**, 997 (1959).
124. M.N. CHISHOLM, K. FOLTING, et. al., J. Amer. Chem. Soc., **103**, 1305 (1981); Chem. Commun., **1982**, 189; Inorg. Chem., **23**, 1021 (1984); Polyhedron, **2**, 681 (1983).
125. N. HAGIHARA, et. al., J. Amer. Chem. Soc., **81**, 3160 (1959); Bull. Soc. Chem. Japan, **34**, 892 (1961).
126. H.L. KRAUSS AND G. MUNSTER, Z. Anorg. Chem., **352**, 24 (1967).
127. D.A. BROWN, D. CUNNINGHAM AND W.K. GLASS, J. Chem. Soc., A, **1968**, 1563.
128. G. HORNUFF AND K. KAPPLER, J. Prakt. Chem., **23**, 54 (1964).
129. C.J. WILLIS, J. Chem. Soc., Chem. Commun., **1972**, 944.
130. H. FUNK, H. MATSEHINER AND H. NEUMANN, Z. Anorg. Chem., **340**, 75 (1965).
131. K. YUNLU, P.S. GRADEFF, K.G. CAULTON, et. al., Inorg. Chem., **30**, 2317 (1991).

132. V.G. KESSLER, N.YA. TUROVA, A.I. YANOVSKY, et. al., Zh. Neorg. Khim., **36**, 1681 (1991).
133. T.G. BRIKLINA, V.A. SHUSHONOV, Reactions of Organometallic Compounds with Oxygen and Peroxides, "Nauka," Moscow (1966).
134. S.N. MISRA, T.N. MISRA, et. al., Chem. and Ind., **1963**, No. 3, 120.
135. D.C. BRADLEY, R.C. MEHROTRA AND W. WANDLAW, J. Chem. Soc., **1952**, 5020.
136. E.P. TUREVSKAYA AND N.YA. TUROVA, J. Coord. Chem., **9**, N8, 1084 (1983).
137. D.C. BRADLEY, B.N. CHAKRAVARTI, et. al., J. Chem. Soc., **1958**, 99; **1955**, 726.
138. V.G. KESLER, N.YA. TUROVA, A.V. KOROLEV, et. al., Mendeleev. Commun., **1991**, 89.
139. S.I. KUCHEIKO, N.YA. TUROVA, et. al., Sov. J. Coord. Chem., **11**, 867 (1985).
140. E.P. TUREVSKAYA, N.I. KOZLOVA, N.YA. TUROVA, et. al., Sverkhprovodimost, **2**, N9, 30 (1989).
141. J.V. SINGH, B.P. BORANWAL AND R.C. MEHROTRA, Z. Anorg. Chem., **477**, 235 (1981).
142. J. BARTHEL AND G. SCHWITZEGEBEL, Z. Phys. Chem. (BRD), **54**, 181 (1967); **55**, 33 (1967); **72**, 222 (1970).
143. M.S. BAINS, Canad. J. Chem., **42**, 945 (1964).
144. T. GREISER, E. WEISS, et. al., Ber., **110**, 3388 (1977); Acta Cryst., **B38**, 2251 (1982).
145. P.G. WILLIARD AND G.B. CARPENTER, J. Amer. Chem. Soc., **107**, 3345 (1985).
146. O.H. JONSON AND H.E. FRITZ, J. Amer. Chem. Soc., **75**, 718 (1953).
147. D.C. BRADLEY, L.J. KAY AND W. WARDLAW, J. Chem. Soc., **1956**, 4916.
148. D.A. WRIGHT AND D.A. WILLIAMS, Acta Cryst., **B24**, 1107 (1968).
149. D.C. BRADLEY, et. al., J. Chem. Soc., **1952**, 2027; **1956**, 1139, 4439; **1963**, 1023.

150. R.C. MEHROTRA, J. Indian Chem. Soc., **31**, 85 (1954).
151. M.S. BAINS, Canad. J. Chem., **40**, 381 (1962).
152. YU. T. STRUCHKOV AND G.M. LOBANOVA, Vestnik MGU **1959**, N2, 169.
153. J. SCHERLE AND F.A. SCHRODER, Acta Cryst., **B30**, 2772 (1974).
154. M.J. HAMPDEN-SMITH, T.A. WARK, et. al., Can. J. Chem., **69**, 121 (1991).
155. B.A. VAARSTRA, J.C. HUFFMAN, P.S. GRADEFF, L.G. HUBERT-PFALZGRAF, K.G. CAULTON, et. al., Inorg. Chem., **29**, 3126 (1990).
156. P. TOLEDANO, F. RIBOT AND C. SANCHEZ, Acta. Cryst., **46C**, 1419 (1990).
157. A.A. PINKERTON, D. SCHWARZENBACH, L.G. HUBERT-PFALZGRAF, et. al., Inorg. Chem., **15**, 1196 (1976).
158. V.G. KESSLER, A.I. YANOVSKY, N.YA. TUROVA AND YU. T. STRUCHKOV, Chem. Commun., (in press).
159. M. VEITH, J. HANS, L. STAHL, et. al., Z. Naturforsch., **46B**, 403 (1991).
160. J.A. IBERS, Nature, **197**, 686 (1963).
161. R.D. WITTERS AND C.N. CAUGHLAN, Nature, **205**, 1312 (1965).
162. O. PONCELET AND L.G. HUBERT-PFALZGRAF, Polyhedron, **8**, 2183 (1989).
163. E.A. SANDUZSKI, W.H. JESLEY AND R.D. THOMAS, J. Amer. Chem. Soc., **100**, 7761 (1978).
164. M.H. CHISHOLM, J.C. HUFFMAN, et. al., J. Amer. Chem. Soc., **103**, 6093 (1981).
165. L.F. DAHL, G.L. DAVIS, et. al., J. Inorg. Nucl. Chem., **24**, 357 (1962).
166. S. HARVEY, M.F. LAPPERT, et. al., J. Chem. Soc. Chem. Commun., **1988**, 1216.
167. E. WEISS, K. HOFFMAN AND H.F. GRUTZMACHER, Ber., **103**, 1190 (1970).
168. T. GREISER AND E. WEISS, Ber., **109**, 3142 (1976).
169. S.C. GOEL, M.A. MATCHETT, M.Y. CHIANG AND W.E. BUHRO, J. Amer. Chem. Soc., **113**, 1844 (1991).

170. H. DUNKEN AND J. KRAUSE, Z. Chem., **1**, 27 (1961).
171. E. WEISS, et. al., Helv. Chim. Acta., **46**, 2051 (1963); Z. Anorg. Chem., **372**, 206 (1970); **332**, 197 (1964).
172. H. STAGLICH AND E. WEISS, Ber., **111**, 901 (1978).
173. T.V. ROGOVA, N.YA. TUROVA AND A.V. NOVOSELOVA, Doklady Chem. Proc. Acad. Sci. USSR, **285**, 896 (1985).
174. T. GREISER AND E. WEISS, Ber., **112**, 844 (1979).
175. J.G. OLIVER AND I.J. WORRALL, J. Inorg. Nucl. Chem., **33**, 1281 (1971).
176. R.W. ADAMS AND G. WINTER, Austral. J. Chem., **20**, 171 (1967).
177. N.YA. TUROVA, V.A. KOZUNOV, A.I. YANOVSKY, et. al., J. Inorg. Nucl. Chem., **41**, 5 (1979); Doklady Chem. Proc. Acad. Sci. USSR, **244**, 26 (1979).
178. E.P. TUREVSKAYA AND N.YA. TUROVA, Bull. Acad. Sci. USSR Div. Chem. Sci., **1977**, N7, 1380.
179. S.I. KUCHEIKO, N.YA. TUROVA AND O.M. SOLOVEITCHIK, Zh. Obshch. Khim., **55**, 2353 (Russia) (1985).
180. N.I. KOZLOVA, V.A. SIPACHEV AND N.YA. TUROVA, Sov. J. Coord. Chem., **7**, 512 (1981).
181. N.YA. TUROVA, et. al., Sov. J. Coord. Chem., **11**, 366 (1985); **6**, 261 (1980); **13**, 149 (1987).
182. K.G. CAULTON AND L.G. HUBERT-PFALZGRAF, Chem. Rev., **90**, 969 (1990); L.G. HUBERT-PFALZGRAF, Ecol d 'ete Sol-Gel, **2**, (CNRS, Oleron, 1991), p. 429.
183. M.I. KARPOVSKAYA (YANOVSKAYA), N.YA. TUROVA, A.V. NOVOSELOVA, et. al., Russ. J. Inorg. Chem., **23**, 766 (1978); **24**, 107, 1005 (1979); Doklady Chem. Proc. Acad. Sci. USSR, **239**, 111 (1978); Sov. J. Coord. Chem., **8**, N2, 148 (1982).
184. E.P. TUREVSKAYA, N.YA. TUROVA AND A.V. NOVOSELOVA, Russ. J. Inorg. Chem., **23**, 355 (1978); Doklady Chem. Proc. Acad. Sci. USSR, **242**, N4, 271 (1978); Inorg. Mat. Izv. AN SSSR Neorg. Mater., **16**, N10, 950 (1980).
185. N.YA. TUROVA, N.I. KOZLOVA, et. al., Russ. J. Inorg. Chem., **25**, 1788 (1980); Sov. J. Coord. Chem., **8**, 346 (1982); **9**, N7, 540 (1983).

186. E.P. TUREVSKAYA, N.YA. TUROVA AND A.V. NOVOSELOVA, Sov. J. Coord. Chem., **9**, 755 (1983); **10**, N10, 770 (1984).
187. T.V. ROGOVA, N.YA. TUROVA, et. al., Sov. J. Coord. Chem., **11**, 710 (1985); **12**, 440 (1986).
188. N.YA. TUROVA, V.G. KESSLER AND S.I. KUCHEIKO, Polyhedron, **10**, 2617 (1991).
189. N.YA. TUROVA, Koord. Khim., **10**, 1165 (1984).
190. B.A. VAARSTRA, J.C. HUFFMAN, W.E. STREIB AND K.G. CAULTON, Inorg. Chem., **30**, 3068 (1991).
191. M.I. YANOVSKAYA, E.P. TUREVSKAYA, et. al., Ferroelectrics, (in press).
192. D. EICHORST, D.A. PAYNE, et. al., Inorg. Chem., **29**, 1458 (1990).
193. A.I. YANOVSKY, E.P. TUREVSKAYA, et. al., (in press).
194. M. VEITH, et. al., Angew. Chem., **103**, 92 (1991); Z. Naturforsch, **46B**, 403 (1991).
195. H.R. FISHER, J. GLERUP, et. al., Inorg. Chem., **21**, 3063 (1982).
196. C.D. CHANDLER, G.D. FALLON, et. al., Austral. J. Chem., **40**, 1427 (1987).
197. M.F. BERBAUSKAS, J.H. WENGROVIUS, et. al., Acta. Cryst., **C40**, 1536 (1984).
198. YU.A. SIMONOV, G.S. MATUSENKO, et. al., Dokl. AN SSSR, **250**, 99 (Russia) (1980).
199. K. TAFT AND S.J. LIPPARD, J. Amer. Chem. Soc., **112**, 9629 (1990).
200. N.N. SAUER, E. GARCIA, K.V. SALAZAR, et. al., J. Amer. Chem. Soc., **112**, 1524 (1990).
201. R. SCHMIDT, A. MOSSET AND J. GALY, Inorg. Chim. Acta., **179**, 167 (1991).
202. L.F. FRANCIS, D.A. PAYNE, et. al., Chem. Mater., **2**, 645 (1990).
203. J. LIVAGE, Solid State Chemistry, **18**, (1988), p. 259 (in progress).
204. D.C. BRADLEY, Inorganic Polymers, edited by F.G.A. Stong and W.A.G. Graham (Academic Press, N.Y., 1962), p. 300.

205. K.A. ANDRIANOV, Polymers with Inorganic Main Chains (Publishing house of the Academy of Sciences of the USSR, Moscow, 1962).
206. K. WATENPAUGH AND C.N. CAUGHLAN, Chem. Commun., **1967**, 76.
207. N.YA. TUROVA AND E.P. TUREVSKAYA, Sov. J. Coord. Chem., **3**, 679 (1977).
208. J. GALY, et. al., Compt. Rend., Ser. 2, **307**, 1747 (1988); XV Congr. Intern. Union of Cryst., Bordeux, 1990, PS-06-02-40.
209. B. MOROSIN, Acta. Cryst., **B33**, 303 (1977).
210. D.C. BRADLEY, B.N. CHAKRAVARTI AND WARDLAW, J. Chem. Soc., **1956**, 4439.
211. M.H. CHISHOLM, K. FOLTING, et. al., Inorg. Chem., **23**, 1021 (1984); **24**, 214 (1985).
212. C. GAFFNEY, P.G. HARRISON AND T.J. KING, J. Chem. Soc. Chem. Commun., **1980**, 1251.
213. K.G. CAULTON, M.H. CHISHOLM, et. al., ibid, **1990**, 1349; 1498;
214. A.I. YANOVSKY, N.YA. TUROVA, et. al., Sov. J. Coord. Chem., **8**, 76 (1982).
215. P. TOLEDANO, C. SANCHEZ, et. al., Compt. Rend., **311**, 1161, 1315 (1990).
216. I. LUNEAU-GAUTIER, A. MOSSET AND J. GALY, Z. Krist., **180**, 83 (1987).
217. S. DOEUFF, Y. DROMZEE AND C. SANCHEZ, ibid, **308**, 1409 (1989); Inorg. Chem., **28**, 4439 (1989).
218. I. LAAZIZ, A. LARBOT, et. al., Acta. Cryst., **46**, 2332 (1990).
219. A.I. YANOVSKY, E.P. TUREVSKAYA, N.YA. TUROVA AND YU.T. STRUCHKOV, Sov. J. Coord. Chem., **11**, 63 (1985).
220. A.I. YANOVSKY, M.I. YANOVSKAYA, et. al., J. Chem. Soc. Chem. Commun., 1605.
221. R. PAPERNICK, L.G. HUBERT-PFALZGRAF, et. al., J. Chem. Soc. Chem. Commun., **1990**, 695.
222. G. DATTA, H.S. MAITI AND A. PAUL, J. Mater. Sci. Lett., **6**, 787 (1987).
223. J.C. BERNIER, P.J.L. POIX AND J.L. REHSPRINGER, PCT Int. App., WO 86.02.345.

224. G. TOMANDI, H. ROSCH AND A. STEIGELSCHMITT, Better Ceram. through Chem. III: 3rd Mater. Res. Soc. Symp., (Pittsburgh, PA, 1988), p. 281.
225. P.P. PHULE AND S.H. RISBUD, Adv. Ceram. Mater., **3**, No. 2, 183 (1988); Mater. Sci. and Eng., **B3**, 241 (1989).
226. A. MOSSET AND I. GAUTIER=LUNEAU, J. Non-Cryst. Solids, **100**, 339 (1988).
227. S.S. FLASCHEN, J. Amer. Chem. Soc., **77**, 6194 (1955).
228. K. KISS, J. MAGDER, M.S. VUKASOVICH AND R.Y. LOCKHART, J. Amer. Ceram. Soc., **49**, 291 (1966); Pat. USA 3.292.994 (1966).
229. F. CHAPUT AND J.P. BOILOT, High Tech. Ceram.: Proc. World Congr. 6th Int. Meet. Mod. Ceram. Technol. (Amsterdam e.a., 1987), p. 1459; Silicat. Ind., **55**, 317 (1990); J. Amer. Ceram. Soc., **73**, 942 (1990).
230. S.A. BRUNO, Eur. Pat. 0.324.619 (1989).
231. R.C. PASTOR, Pat. USA 4.636.378 (1987).
232. A. LARBOT, F. GARCIA AND CH. GUIZARD, Eur. J. Solid State and Inorg. Chem., **26**, 327 (1989).
233. M.I. DIAZ-GUEMES, C.I. GONZALEZ, C.J. SERNA AND J.M. PALACIOS, J. Mater. Sci., **24**, 1011 (1989).
234. E. LILLEY AND R.R. WUSIRIKA, Pat. USA 4.764.493 (1988); Eur. Pat. Appl. EP 250.085 (1986).
235. M. KLEE, J. Mater. Sci. Lett., **B**, No. 8, 985 (1989); Eur. Pat. Appl. 0.324.619 (1989).
236. F. CHAPUT AND J.P. BOILOT, Electroceram.: Bacis Sci. Sec. Inst. Ceram. (Shelton, 1988), p. 21; Silicat. Ind., **51**, 187 (1989); Sci. Ceram. 14: Proc. 14th Int. Conf. (Stoke-on-Trent, 1988), p. 133; J. Mater. Sci. Lett., **6**, 1110 (1987).
237. M. OCANA AND C.J. SERNA, J. Mater. Sci. Lett., **9**, 772 (1990).
238. K.S. MAZDIYASNI, R.T. DOLLOFF AND J.S. SMITH, J. Amer. Ceram. Soc., **52**, 523 (1969); Pat. USA 3.647.364 (1972).
239. E. WAINER, Pat. USA 2.948.628 (1960).
240. S. HIRANO AND S. NAKA, Jpn. Kokai, 62-21.758 (1987).
241. A.D. HARLEY, L.G. DUQUETTE, J.A. KHOURY AND I. KOHATSU, Pat. USA 4.710.227 (1987).

242. Y. OZAKI, Y. AKUTSU, S. YAMOMOTO AND Y. SHINOKARA, Jpn. Kokai, 52.265.167; 62-265.118 (1987).
243. R. KASAI AND Y. OZAKI, J. Ceram. Soc. Jap., **95**, 912, 1000 (1987).
244. K. WAKAMOTO AND M. SUMIDA, Pat. Jap. 01.87.504 (1989).
245. K.W. KIRBY, Mater. Res. Bull., **23**, 881 (1988).
246. M. KLEE AND W. BRANDT, Klei/Glas/Keram., **8**, 60 (1987).
247. J.L. REHSPRINGER, P. POIX AND J.C. BERNIER, J. Non-Cryst. Solids, **82**, 286 (1986).
248. J.S. SMITH, R.T. DOLLOFF AND K.S. MAZDIYASNI, J. Amer. Ceram. Soc., **53**, 91 (1970).
249. M. KATO, et. al., Jpn. Kokai, 57-196.722 (1982).
250. K.S. MAZDIYASNI AND L.M. BROWN, J. Amer. Ceram. Soc, **54**, 539 (1971).
251. K.S. MAZDIYASNI AND L.M. BROWN, J. Amer. Ceram. Soc, **55**, 633 (1972).
252. M. KAHN AND M.P. PECHINI, Pat. USA 3.490.927 (1970).
253. M. SHARON AND A. SINHA, Sol. Energy Mater., **9**, 391 (1984).
254. R.E. RIMAN, R.R. LANDHAM AND H.K. BOWEN, J. Amer. Ceram. Soc., **72**, 821 (1989).
255. Y. SUWA, Y. SUGIMOTO AND S. NAKA, J. Jap. Soc. Powder and Powder Met., **25**, 164 (1978).
256. J.J. RITTER, R.S. ROTH AND J.E. BLENDELL, J. Amer. Ceram. Soc., **69**, 155 (1986).
257. P.P. PHULE, S.H. RISBUD AND J.E. PEMBERTON, 1st Int. Ceram. Sci. and Technol. Congr., Anaheim, CA: Meet. Abstr., (Columbus, OH, 1989), p. 95.
258. A. OGA, A. ITANI AND T. MIYADZAKI, Jpn. Kokai, 62-267.540 (1989).
259. S. HIRANO, et. al., Adv. Ceram., **19**, 139 (1986); Ceram. Dielec.: Compos., Process. and Prop.: Proc. Symp. (Westerville, OH, 199), p. 203; Ceram. Powder Sci. III: Proc. 3rd Int. Conf. Powder Process. Sci., (Westerville, OH, 1990), p. 733.
260. L. SUCHOFF, Pat. USA 3.002.861 (1961).
261. J.J. MITCHELL, Pat. USA 3.066.048 (1962).
262. M.I. YANOVSKAYA, N.YA. TUROVA, E.P. TUREVSKAYA, et. al., Izv. Akad. Nauk SSSR. Neorg. Mater., **17**, 307 (1981).

263. K.A. VOROTILOV, M.I. YANOVSKAYA, E.P. TUREVSKAYA, et. al., Ferroelectrics, (in press).

264. J. FUKUSHIMA, K. KADAIRA AND T. MATSUSHITA, Amer. Ceram. Soc. Bull., **55**, 1064 (1976).

265. K. SAKASHITO, Jpn. Kokai, 63-02.850 (1988).

266. T. YOKO, K. KAMIYA AND S. SAKKA, Res. Repts. Fac. Eng. Mie Univ., **12**, 49 (1987).

267. D.M.S. NELGY AND M.A. AEGERTER, Better Ceram. through Chem. III: 3rd Mater. Res. Soc. Symp., (Pittsburgh, PA, 1988), p. 515.

268. T. TUCHIYA, T. KAWANO, T. SEI AND J. HATANO, J. Ceram. Soc. Jap., **98**, 742 (1990).

269. G. DAENDLIKER AND H. MORAWETZ, Proc. Int. Symp. Decompos. Organometallic Compounds Refract. Ceram. Metals, Metal Alloys, **1968**, 157; Pat. USA 3.366.510 (1968).

270. M.I. YANOVSKAYA, A.P. LEONOV, S.YU. STEFANOVICH, et. al., J. Mater. Sci., **23**, 395 (1988).

271. S.I. HIRANO AND K. KATO, Adv. Ceram. Mater., **2**, 142 (1987); **3**, 503 (1988); J. Non. Cryst. Solids, **100**, 538 (1988); Jpn. Kokai, 63-238; 63-270.397 (1988); Jpn. Pat. 01.93.425 (1989); Bull. Chem. Soc. Jap., **62**, 429 (1989).

272. A.A. BALLMAN AND M. ROBBINS, Pat. USA 4.103.970 (1978).

273. T. SUDZUKI, Jpn. Pat. 1.290.533 (1989).

274. D.Y. EICHORST AND D.A. PAYNE, Better Ceram. through Chem. III: 3rd Mater. Res. Soc. Symp., (Pittsburgh, PA, 1988), p. 773.

275. A.X. KUANG, E.T. WU, A. NAZERI-ESHGHI, AND J.D. MACHENZIE, 1st Eur. Conf. Appl. Polar Dielec. and Int. Symp. Appl. Ferroelec.: Abstr. Booklet, (Zurich, 1988), p. 204; ISAF '86: Proc. 6th IEE Int. Symp. Appl. Ferroelec., (New York, NY, 1986), p. 388, p. 391.

276. N. PUYOO-CASTAINGS, F. DUBOUDIN, J. RAVEZ AND P. HAGENMULLER, Mater. Res. Bull., **22**, 261 (1987); Calorim. Et Therm., **18**, 35 (1987); J. Mater. Res., **3**, 557 (1988).

277. J.S. BALBAA AND G. GOWDA, J. Mater. Sci., **5**, 751 (1986).
278. S. HIRANO, T. HAGASHI, K. NOSAKI AND K. KATO, J. Amer. Ceram. Soc., **72**, 707 (1989).
279. Y. SUWA, K. KATO, S. HIRANO AND S. NAKA, J. Jap. Soc. Powder and Powder Met., **28**, 28 (1981).
280. Y. HIRAI, T. FUKUDA, H. KAWAHARA AND Y. KOBEYASHI, Solid State Commun., **61**, 267 (1987).
281. C.J. CHEN, Y.H. XU, R. XU, J.D. MACKENZIE, 1st Int. Ceram. Sci. and Technol. Congr.: Meet. Abstr. (Columbus, OH, 1989), p. 91.
282. O. INOUE, S. KAWASHIMA AND H. OCHI, Jpn. Kokai, 62-143.825 (1987).
283. O. INOUE, S. KAWASHIMA AND H. OCHI, Jpn. Kokai, 62-143.826 (1987).
284. O. INOUE AND S. KAWASHIMA, Jpn. Kokai, 63-170.216 (1988); Pat. USA 4.853.199 (1989).
285. O. RENOULT, J.P. BOILOT, F. CHAPUT, et. al., Ceramic Today - Tomorrow's Ceramics, edited by P. Vincenzini (El Sevier Science Publishers B.V., 1991), p. 1991.
286. Jpn. Kokai, 58-199.716 (1983).
287. O. INOUE AND S. KAWASHIMA, Pat. USA 4.918.035 (1990); Eur. Pat. 237.081 (1987).
288. A.D. HARLEY, L.G. DOUGUETTE, I.A. KHOURY AND T. KOHATSU, Pat. USA 4.816.072 (1989); Eur. Pat. 030.297.2 (1989).
289. F. CHAPUT, J.P. BOILOT, L. LEJEUNE, et. al., J. Amer. Ceram. Soc., **72**; 1355 (1989).
290. P. RAVINDRANATHAN, S. KOMARNENI, R. ROY, et. al., Ferroelectrics, **87**, 133 (1988); Mater. Lett., **8**, 161 (1989); J. Amer. Ceram. Soc., **73**, 1024 (1990).
291. L.F. FRANCIS, Y.J. OH AND D.B. PAYNE, J. Mater. Sci., **25**, 5007 (1990).
292. S. TSUKAMURA, T. YOSHIMOTO AND J. KATO, Jpn. Kokai, 61-198.518; 61-191.519 (1986).
293. I. JAWED, A.E. BAILEY, A.P. RITTER AND S.R. WINZER, ISAF '86: Proc. 6th IEEE Int. Symp. Appl. Ferroelec., (New York, NY, 1986), p. 384.

294. M. YASUDA AND Y. UEDA, J. Phys. Condens. Mater., **1**, 497 (1989).
295. P. GRIESMAR, G. PAPIN, C. SANCHEZ AND J. LIVAGE, J. Mater. Sci. Lett., **9**, 1288 (1990).
296. G.A. POPKOVICH, V.V. SVIRIDOV AND O.A. STENINA, Zh. Neorg. Khim., **21**, 1526 (1976); **25**, 3199 (1980).
297. S. YAMAKAWA AND M. TSUZAKI, Jpn. Kokai, 62-70.261; 62-72560 (1987).
298. T. OGIHARA, H. KANEKO AND M. KATO, J. Mater. Sci. Lett, **7**, 867 (1988).
299. J. KATO, Jpn. Kokai, 62-108.729 (1987).
300. Y. OZAKI, et. al., J. Ceram. Soc. Jap., **96**, 140 (1988); Jpn. Kokai, 57-821.20; 57-821.21 (1982); 59-419.30 (1984.
301. H. HIRASHIMA, E. ONISHI AND M. NAKAGAWA, J. Non-Cryst. Solids, **121**, 404 (1990).
302. M. SAITO, Jpn. Pat. 01.145.384 (1989).
303. Y. TAKAHASHI, et. al., Jpn. Kokai, 63-324497 (1990); J. Ceram. Soc. Jap., **98**, 754 (1990).
304. L.M. BROWN AND K.S. MAZDIYASNI, J. Amer. Ceram. Soc, **55**, 541 (1972).
305. K. TAKAHASHI, M. ODA AND H. SHIBATANI, Jpn. Kokai, 62-167217 (1987).
306. T. IINO, Jpn. Kokai, 61-26514 (1986).
307. Y. OGATA, H. ONO AND S. YAMANEKA, Jpn. Kokai, 62-216.965 (1987); Eur. Pat. 237.640 (1987).
308. P. DURAN AND C. MOURE, Amer. Ceram. Soc. Bull., **64**, 575 (1985).
309. K. KAKEGAWA, J. MOHRI, M. IMAI, et. al., J. Chem. Soc. Jap., Chem. and Ind. Chem., **1985**, No. 4, 692.
310. R.C. BUCHMAN AND D.E. WITTMER, Pat. USA 4.283.228 (1980).
311. T. KAVAYI, Jpn. Kokai, 60-103031 (1985).
312. J.V. BIGGERS AND S. VENKATARAMANI, Mater. Res. Bull, **13**, 717 (1978).
313. G.H. HAERTLING AND C.E. LAND, Ferroelectrics, **3**, 269 (1972).

314. R.H. DUNGAN AND G. SNOW, Amer. Ceram. Soc. Bull, **56**, 781 (1977).

315. R. BROOKS AND D.K. MCCARTHY, Ferroelectrics **27**, 179 (1980).

316. M.I. YANOVSKAYA, E.P. TUREVSKAYA AND N.YA. TUROVA, Izv. Akad. Nauk SSSR, Neorg. Mater., **23**, 658 (1987).

317. P. DURAN, Silicat. Ind., **49**, 211 (1984).

318. E.R. LA SERRA, Y. CHARBOUILLOT, P. BAUDRY AND M.A. AEGERTER, J. Non-Cryst. Solids, **121**, 323 (1990).

319. F. MELLOY, Pat. USA 3.331.659 (1967).

320. J. KATO, et. al., Jpn. Kokai, 58-41723 (1983).

321. R.A. LIPELES AND D.J. COLEMAN, Mater. Res. Soc. Symp., **73**, 666 (1986).

322. GURKOVICH AND J.B. BLUM, Ultrastruct. Process. Ceram., Glasses and Composites, (New York, NY, 1984), p. 152.

323. Y. HAYASHI AND J.B. BLUM, J. Mater. Sci., **22**, 2655 (1987).

324. K.O. BUDD, S.K. DEY AND D.A. PAYNE, Better Ceram. through Chem. II, (Pittsburgh, PA, 1986), p. 711; Electron Packag. Mater. Sci., **2**, (Pittsburgh, PA, 1986), p. 317; J. Amer. Ceram. Soc., **70**, 295 (1987).

325. CH. CHEN, D.F. RYDER AND W.A. SPURGEON, J. Amer. Ceram. Soc., **72**, 1495 (1989).

326. K. KAMIYA, H. HONDA AND H. NASU, J. Ceram. Soc. Jap, **98**, 759 (1990).

327. S.K. DEY AND R. ZULEEG, Ferroelectrics **108**, 37 (1990).

328. C.K. KWEK AND S.B. DESU, 1st Int. Ceram. Sci. and Technol. Congr.: Meet. Abstr., (Columbus, OH, 1989), p. 132.

329. S.L. SWARTZ, ibid, p. 90.

330. G. YI, ZH. WU AND M. SAYER, J. Appl. Phys., **64**, 2717 (1988).

331. L.E. SANCHEZ, S.-Y. WU AND I.K. NAIK, Appl. Phys. Lett., **56**, 2399 (1990).

332. S.D. RAMAMURTHI AND D.A. PAYNE, J. Amer. Ceram. Soc, **73**, 2547 (1990).

333. K.D. BUDD, S.K. DEY AND D.A. PAYNE, Brit. Ceram. Proc, **1985**, No. 36, 107 (1986).

334. S. VENKATARAMANI AND J.V. BIGGERS, Amer. Ceram. Bull., **59**, 462 (1980).

335. S.D. RAMAMURTHI AND D.A. PAYNE, J. Amer. Ceram. Soc, **78**, 2547 (1990).

CHAPTER 9

SOL-GEL SCIENCE AND PE-MOCVD OF DIELECTRIC PEROVSKITE FILMS

Sandwip K. Dey
Arizona State University, Tempe, Arizona 85287-6006

1. INTRODUCTION

Perovskite materials of the oxygen octahedra class (e.g., $BaTiO_3$, $SrTiO_3$, $PbTiO_3$, $(Ba,Sr)TiO_3$, and $(Pb,La)(Nb,Zr,Sn,Ti)O_3$) exhibit high permittivities, large electromechanical coupling coefficients, and pyroelectric, electro-optic, and ferroelectric effects.[1-3] In the last decade, an unprecedented interest has developed in the processing of polycrystalline, ferroelectric (point group: 4 mm), and paraelectric (point group: m3m) perovskite thin-film materials, due to growing demand for compatibility with integrated circuit technology. In particular, the multiaxial $Pb(Zr,Ti)O_3$ (or PZT) thin-film system are being evaluated for a variety of applications including non-volatile computer memories, ULSI DRAMs, decoupling capacitors, and neural network components.[4-5] The common thin-film processing techniques for these multicomponent systems are sputtering[6] (ion-beam,[6a] magnetron,[6b] rf-diode,[6c] and electron cyclotron resonance (ECR) plasma-assisted[6d]), excimer laser ablation,[7] chemical vapor deposition[8] (plasma-enhanced[8a] and thermal[8b]), evaporation[9] (thermal,[9a] e-beam[9b], and flash[9c]), mist deposition,[10] metal organic decomposition,[11] and sol-gel.[12] Due to the proliferation of conferences and publications devoted to the subject, the preceding list and references are by no means complete.

To date, research efforts of our group have focused on chemical processing techniques, namely polymeric solution gelation (sol-gel) and plasma-enhanced metal organic chemical vapor deposition (PE-MOCVD) of lead-based ferroelectric and paraelectric perovskite thin films on conducting and insulating substrates. Although, there are numerous, potentially useful barrier layers on Si substrates that are conducting (e.g., RuO_2, $(La,Sr)CoO_3$, $LaAlO_3$, $YBa_2Cu_3O_7$, Pt/TiN, and Ni alloys) or substrates that are non-conducting (e.g., SiO_2 glass, sapphire, MgO, and $SrTiO_3$),[13] our efforts, and those of most research groups have been primarily focused on depositing perovskite thin films on Pt/Ti/SiO_2 passivated Si substrates with various combinations and thicknesses of the passivation layers. A common goal has been to determine the interrelationships between structure, processing, properties, and performance. An obvious resource towards achieving this goal is the vast literature on bulk perovskite dielectrics, which indicate how numerous microstructural parameters, including stoichiometry, grain size, second phase, and dopant and its compensation mechanism, influence the properties.[1-3]

Admittedly, several research groups working on chemical processing of perovskite thin films, using a similar set of precursors, have reported a wide array of microstructures which exhibit a wide range of electrical properties. The lack of knowledge of the specimen history, and its implications, has lead to difficulties in interpretation of reported data and in building a consensus regarding the relationships. Additionally, the understanding of the influences of substrate-film interfacial effects on the time-dependent properties, such as fatigue, aging, retention, and imprint, has been elusive. Thus, a systematic, integrated approach that follows the evolution of the structure, from chemical precursors to the final perovskite film, should eventually lead to general agreement among laboratories about relationships between structure, processing, properties, and performance. This common knowledge base can then be used as a guide for optimizing the processing parameters. Such an approach will enhance our ability to tailor various dielectric thin-film properties. This, coupled with resolving current and projected problems in device integration should enable the transfer of technology from the laboratory to commercial scale production.

This chapter focuses on the sol-gel processing science and PE-MOCVD of perovskite thin films with particular emphasis on the structure-processing relations in the $Pb(Zr,Ti)O_3$ system. It is not intended to be a comprehensive review. The organization of this chapter is given in Table 1.

Table 1. Outline

SOL-GEL SCIENCE AND PE-MOCVD OF DIELECTRIC PEROVSKITE FILMS		
1.	INTRODUCTION	
2.	POLYMERIC SOL-GEL PROCESSING OF FERROELECTRIC LEAD-BASED PEROVSKITE THIN FILMS	
	2.1	General Processing Issues and Recommendations
	2.2	Precursor Synthesis, Empirical Formulas, and Structure
	2.3	Thin-Film Fabrication
	2.4	Film-Substrate Interfacial Reactions and Phase Evolution
	2.5	Microstructural Evolution in Systems with Excess-Lead Oxide
	2.6	Dopant Compensation Mechanisms and Their Effect on Leakage Current
3.	PLASMA-ENHANCED METAL ORGANIC CHEMICAL VAPOR DEPOSITION OF PEROVSKITE DIELECTRIC THIN FILMS	
	3.1	Phase Thermochemistry for the Ferroelectric Perovskite $Pb(Zr,Ti)O_3$ System
	3.2	Precursors and Limitations
	3.3.	Effect of Crystal Chemistry and Reaction Kinetics on $Pb(Zr,Ti)O_3$ Phase Evolution in a RF Reactor
	3.4.	Advantages of ECR Plasma Driven Deposition and Recent Results on Deposition of TiO_2 and $SrTiO_3$ Thin Films
4.	CONCLUDING REMARKS	
	Acknowledgements	
	References	

In sol-gel processing of Pb-based perovskite thin films, poor and variable microstructures, and lack of control of the perovskite phase stoichiometry can stem from a variety of sources. It will be shown that these sources include solution synthetic methods, hydrolysis and heat treatment parameters, and degree of dielectric film/substrate interactions. Thus, microstructural and electrical data must be interpreted cautiously, taking into account the influences of the aforementioned parameters. Outlined also will be; a) how one determines the composition and structure of a prehydrolyzed precursor, b) a model, based on electron microscopy and precusor chemistry, of microstructural evolution of PZT thin films with excess-PbO, and c) dopant compensation mechanisms and their effects on electrical properties.

The emphasis on forming Sr, Ba, and Pb perovskite thin-film structures below 500 °C, with conformality over device topography, makes PE-MOCVD an appropriate choice among numerous deposition methods. Currently, a limiting step appears to be the availability of precursors that are volatile, inexpensive, and non-toxic. The important characteristics in solid state chemistry and processing parameters which control the formation of the perovskite PZT phase below 500 °C are discussed on the basis of calculated, thermodynamic phase stability diagrams. Based on preliminary experimental data on rf-PECVD of PZT, simple, mechanistic explanations are introduced using concepts of crystal chemistry and deposition kinetics. Additionally, the numerous advantages of ECR driven MOCVD are discussed and recent results of ECR-MOCVD of TiO_2 and $SrTiO_3$ thin films below 450 °C are reported.

Currently, advanced work in processing and integration of perovskite dielectric thin-film materials is ongoing in various industries and academic institutions; both domestic and abroad. The source of all illustrations here are, however, from our research group at Arizona State University. Since electrical behavior and interpretation are outside the scope of this chapter, the reader is referred to the additional topical areas that have been published by our group[14]; a) non-volatile programmable random access memories on GaAs,[14a] b) improved measurement of current transients and observation of sub-nanosecond switching,[14b] c) measurement and simulation of partial switching,[14c] d) programmable ferroelectric thin films for IC neural networks,[14c] e) photovoltaic effects and interpretations in ferroelectric thin films,[14d] f) metal-dependent formation of ohmic and Schottky contacts, and the relative importance of work function and surface states,[14e] g) composition dependence of fatigue, retention, and imprint,[5,14b,14f] and h) the origin and magnitude of spatially dependent electric fields in thc surface regions and interpretation of dielectric and leakage current data, in paraelectric thin films.[14g]

2. POLYMERIC SOL-GEL PROCESSING OF FERROELECTRIC LEAD-BASED PEROVSKITE THIN FILMS

Sol-gel processing is one of a number of chemical processing techniques currently under investigation for the fabrication of inorganic non-metallic thin films. These are reviewed by Sakka[15] and Dislich.[16] A sol is a suspension or dispersion of discrete colloidal particles or polymeric species in a solution while a gel is a colloidal or polymeric solid which has an internal network structure with mixed or highly dispersed solid and fluid components. Thus, there are two distinct types of sol-gel systems, namely polymeric and colloidal.[17] In colloidal gels, electrolytic and steric effects in the precursor solution, or sol, determine gel formation, while in polymeric gels the relative rates and extents of chemical reactions (including polymerization) are important.[18] Polymeric gels are formed from the three-dimensional linkages (initiated by hydrolysis and polycondensation reactions) of complexed solutions containing the desired cations.

Polymeric sol-gel processing offers unique forming advantages. In particular, thin films can be deposited by spin-casting, dip-coating, or spraying appropriately hydrolyzed, alkoxide solutions containing the desired cations onto various substrates, including those of large or complex shapes. The amorphous gel films are then pyrolyzed at relatively low temperatures (300-700°C) to form amorphous or crystalline oxides.

To fabricate multicomponent perovskite thin films, numerous metal organic precursors including alkoxides, carboxylates (ethyl hexanoates, propionates, and neodecanates), diketonates, and various organic salts are used to form initially metal-complexed solutions.[19] From a commerical viewpoint, alkoxides of most metals can be synthesized and are convenient starting materials with respect to availability and cost. Metal alkoxide (general formula of $M(OR)_n$, where M is a metal ion and R is an alkyl group) complexes are partially hydrolyzed by addition of less-than-stoichiometric quantities of water to produce a spin-casting solution. This hydrolysis step must be controlled so that precipitation is avoided while polymerization proceeds. The hydrolysis reaction can be simply represented as:

$$M(OR)_n + x\,H_2O \longrightarrow M(OH)_x(OR)_{n-x} + x\,ROH \qquad (1)$$

Acid and base catalysts can influence both the hydrolysis and condensation reaction rates and the structure of the condensed product.[18] The partial hydrolysis products condense (by dehydration and dealcoholation reactions) to form metal-oxygen-metal linkages which form the basis for the final crystalline structure. The structure of the condensed product depends on the relative rates of the hydrolysis, dealcoholation, and dehydration reactions. The contribution from each of these reactions in turn depends on internal parameters (such as nature of metal atom, alkoxy group, and molecular complexity) as well as external parameters (such as choice of catalyst, concentration, solvent, and temperature). By judicious control of physical and chemical parameters of a particular process coupled with the control of gelation rate during spin-casting, one may obtain thin-film products with desirable properties.[14f,20,21]

2.1. General Processing Issues and Recommendations

Prehydrolyzed $Pb(Zr,Ti)O_3$ or PZT solutions were synthesized according to the flow scheme in Figure 1, using standard Schlenk techniques.[22,23a] This was a modification of a procedure originally reported by Blum and Gurkovich.[24] The titanium and zirconium alkoxides were refluxed separately, deliberately avoiding the formation of the double alkoxide. The individual alkoxides in the complexation reaction with a lead precursor promotes the replacement of the acetate ligands to alkoxide ligands. This strategy minimizes the amount of acetate groups (to ~7%), which do not undergo hydrolysis, and thereby gives a cleaner pyrolysis.

Note, in Figure 1, that the synthesis of lead-based prehydrolyzed precursors may be carried out by method I or method II. In method I, lead acetate trihydrate, $Pb(OOCCH_3)_2{\bullet}3\ H_2O$ (or $Pb(Ac)_2{\bullet}3\ H_2O$), was refluxed in methoxyethanol, $CH_3OCH_2CH_2OH$ (or ROH), at 124.5°C for 2.0 hours to remove the water of hydration. The complexed prehydrolyzed PZT solutions were synthesized by mixing this lead precursor with Zr and Ti methoxyethoxides followed by refluxing for 2.0 hours at atmospheric pressure. Next, the solutions were vacuum distilled at 45-50°C at 20 Torr for 2.0 hours, then cooled and diluted to concentrations of 1.0 M. In method II, following the initial refluxing of $Pb(Ac)_2{\bullet}3\ H_2O$, the lead precursor solution was further heated at 100°C for 24 hours, in order to increase the condensation reactions within $Pb(OR)_x(Ac)_y{\bullet}z\ H_2O$. This caused more Pb-O-Pb bonds to form in the Pb precursor prior to mixing and refluxing with the Zr and Ti methoxyethoxides. Fourier Transform Infrared Spectroscopy (FTIR) indicated that relatively lower amounts of alkoxy groups were present in the Pb-(Zr,Ti) alkoxide precursor (oil) derived by Method II. This also led to a reduction in the total weight loss of the corresponding gel.[23b]

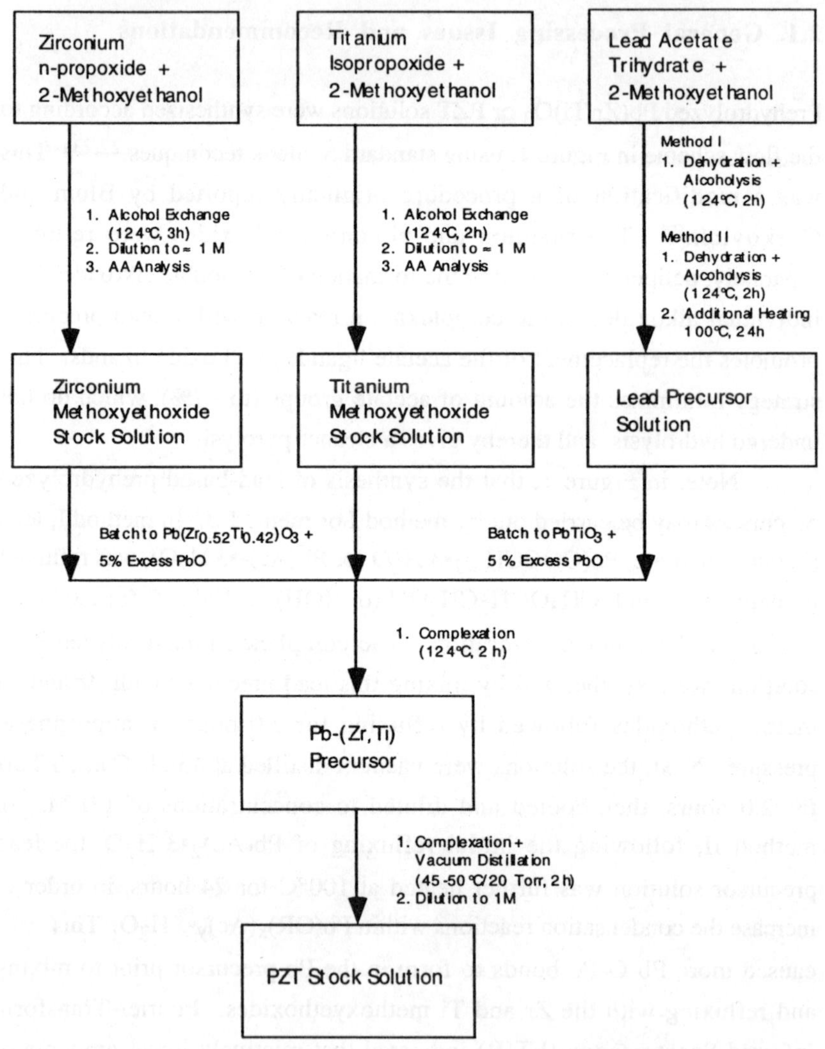

FIGURE1. Flow diagram for the synthesis of Ti, Zr, and Pb(Zr,Ti) alkoxides.

In addition to the differences in the precursor synthetic methods, the structural differences between acid and base-catalyzed gel, influence the pyrolytic behavior and the consequent structural evolution. Dynamic Mass Spectrometric Analysis (DMSA) showed that, during pyrolysis of gels in mildly reducing conditions, 2-methoxyethanol, acetone, and carbon dioxide were evolved primarily for both acid- and base-catalyzed PZT gels. The presence of the latter two gases was a result of the decomposition of residual acetate groups via the carbonate route. It is of practical interest to note that a high decomposition temperature (650°C) of lead cabonate is observed in a nonoxidizing atmosphere as compared to 350°C in pure oxygen. Even during heat treatments in air, any existing carbonate groups below the surface of the gels or films will be subjected to a reducing atmosphere created by cabonaceous residue. Thus, an oxygen-rich atmosphere is recommended. In an oxygen ambient, acid and base-catalyzed PZT gels showed exothermic decomposition patterns, corresponding to the combustion of residual organic materials in the gel. The total weight loss in acid-catalyzed gels was higher than that in base-catalyzed gels in both oxygen and mildly reducing ambients.[23b] This was a direct result of base-catalyzed gels having a more cross-linked structure and exhibiting phase separation through syneresis, and thus, evolving a lower amount of trapped organics during the sol to gel conversion than their acid-catalyzed counterparts.[25] These results are in agreement with the sol-gel silica system.[18]

It was observed by Thermogravimetric Analysis (TGA) and Differential Thermal Analysis (DTA) that acid-catalyzed PZT gels underwent an exothermic decomposition under inert and reducing atmospheres. Based on the differences in the pyrolysis behavior of acid-catalyzed and base-catalyzed PZT gels, coupled with results of DMSA observations, the predominant exothermic decomposition mechanisms of acid-catalyzed gels in a specific temperature range are:[23b]

1) 220-290°C: condensation polymerization of alkoxy groups,

2) 290-380°C: decomposition of acetate groups along with some condensation polymerization, and
3) 380-600°C: polymeric particle rearrangement (increase in packing efficiency).

Note that particle rearrangement and/or skeletal densification (i.e., condensation polymerization and structural relaxation) are expected to occur throughout the temperature range of 220-600°C. A sharp exothermic peak (328°C), corresponding to the melting of metallic lead, was observed in all PZT gels which were pyrolyzed under an inert or reducing atmosphere. The origin of the elemental lead stems from the reduction of Pb-O-Pb species. These species exist in either PbO-rich regions within condensed PZT units or excess-PbO. The source of the reducing conditions to form metallic Pb is carbonaceous residue created from the decomposition of methoxyethoxide and/or acetate ligands. This reduction occurs over a wide temperature range but is masked during a raw run. Lead melts (328°C) endothermally during the background run but when subtracted from the raw data, an exotherm is observed. Note that the inhomogeneity (i.e., the presence of Pb-O-Pb rich regions) of stoichiometric, base-catalyzed PZT gels leads to higher amounts of elemental Pb during pyrolysis under reducing conditions.

Homogeneous and compliant acid-catalyzed PZT gels, capable of polymeric rearrangement, attain a very low energy amorphous structure (to the extent of being nano-crystalline at 300-400°C). This structure requires minimal energy to undergo a reconstructive transformation to the crystalline state at about 615°C. In contrast, inhomogeneous, but cross-linked, base-catalyzed gels require higher thermal energy to overcome the activation energy barrier of transformation to the crystalline state at about 635°C.[23b,25] Typically, microstructures of PZT thin films obtained from acid-catalyzed sols exhibit a final grain size that is smaller than those prepared from base-catalyzed sols. This result may be explained by the high initial nucleation density achievable in acid-catalyzed gels. Here lies an independent avenue, in addition to specialized heat treatments, of tailoring the grain size of thin films.

From a processing point of view, base-catalyzed, high-water sols are preferred for two reasons: 1) Since the crystallization is sluggish in highly cross-linked gels, complete gel (amorphous) densification may be achieved prior to crystallization. This is necessary to obtain dense, crystalline films; 2) Due to its structure, evolution of trapped organics is facilitated during drying and, therefore, the observed weight loss during the pyrolysis is small. Consequently, the cracking tendency of the thin films is reduced. Note that the removal of residual organics, followed by the densification of the amorphous gel-matrix in thin films, prior to crystallization, is necessary by using a rapid rate of heating (250°c/s) in a rapid thermal annealing (RTA) furnace.[26] The use of a RTA furnace allows free organics to vaporize instantaneously due to the open structure created during rapid thermal expansion. Also, volatiles from thermolytic or combustion processes are easily removed from this open structure before time-dependent shrinkage sets in. Thus, a decrease in the carbon content in thin films is observed when rapid heating rates are utilized.

In addition to the decomposition of acetate groups via the carbonate route, the source of carbonate groups appears to be absorbed carbon dioxide from the air. Since these carbonate groups do not decompose until a high temperature is reached, it is recommended that sols be stored in air-tight containers.

2.2. Precursor Synthesis, Empirical Formulas, and Structure

It is clear that differences in the precursor synthetic routes and hydrolysis conditions affect the amount of residual acetate groups, gel structure, pyrolysis behavior, and, therefore, the resulting microstructures of sol-gel derived PZT gels and thin films. Thus, the preferred precursor synthetic route (Method II, Figure 1) and the fingerprint of the prehydrolyzed PZT precursor are described below in detail. Note that, by achieving predominantly a single type of organic functional group (alkoxy ligand), decomposition occurs within a low and narrow temperature range.

The standard Schlenk technique was used to synthesize the Pb-(Nb, Zr, Sn, Ti, Fe) precursor complexes. The methoxyethoxides of Ti and Zr were made via the alcohol exchange reactions of methoxyethanol with titanium isopropoxide and zirconium n-butoxide. The details of the experimental setups for atmospheric and vacuum distillation were reported previously.[14a,14f,23a] During distillation, the vapor temperature was monitored by a thermometer to indicate completion of alcoholysis reactions for formation of the Ti and Zr methoxyethoxides. A constant vapor temperature (122-124°C) indicates distillation of only methoxyethanol and that further exchange of alkoxide groups with methoxyethanol is not possible. During preparation of the PZT alkoxide precursor, lead acetate trihydrate (99.9% pure) was first dissolved in 2-methoxyethanol. The complete dehydration of lead acetate trihydrate ($Pb(OOCCH_3)_2 \cdot 3\,H_2O$ or $Pb(Ac)_2 \cdot 3\,H_2O$) can be represented by:

$$Pb(Ac)_2 \cdot 3\,H_2O \longrightarrow Pb(Ac)_2 + 3\,H_2O \tag{2}$$

The anhydrous lead acetate, $Pb(Ac)_2$, product can undergo an alcoholysis reaction with methoxyethanol or ROH to form lead methoxyethoxide (Pb $(OCH_2CH_2OCH_3)_2$ or Pb $(OR)_2$), and acetic acid ($HOOCCH_3$ or HAc) according to:

$$Pb(Ac)_2 + 2\,ROH \longrightarrow Pb(OR)_2 + 2\,HAc \tag{3}$$

Additionally, water and methoxyethylacetate, $CH_3OCH_2CH_2COOCH_3$ or RAc, can then be generated by an esterification reaction between HAc and ROH as follows:

$$HAc + ROH \longrightarrow RAc + H_2O \quad (4)$$

Thus, the overall atmospheric distillation dehydration of a mole of hydrated lead acetate, considering only a fractional completion of reaction 3, may be represented as follows:

$$Pb(Ac)_2{\bullet}3H_2O + ROH \longrightarrow P(Ac)_{2\text{-}2x}\,(OR)_{2x}{\bullet}[(3{+}2x){-}z]\;H_2O + z\;H_2O + 2x\;RAc \quad (5)$$

where z and 2x are the moles of H_2O and RAc per mole of lead acetate, as determined by Karl Fisher titration and Gas Chromatography, respectively. Note that x is the fractional conversion of reaction 3. Using the data of row 1, Table 2, the composition of the lead precursor that underwent only atmospheric distillation was determined to be $Pb(OOCCH_3)_{1.61}(OCH_2CH_2OCH3)_{0.39}{\bullet}0.48\;H_2O$. For the lead precursor which underwent vacuum distillation, the value of x was found to be 0.573 (2x = 1.15, row 2, Table 2). Thus, the composition of the lead precursor was computed as $Pb(OOCCH_3)_{0.85}(OCH_2CH_2OCH_3)_{1.15}{\bullet}0.63\;H_2O$. This specie should have the same molecular complexity of six; a number identical to the lead precursor which was vacuum distilled to dryness.[23a] Therefore, the molecular formula is $Pb_6(OOCCH_3)_{5.10}(OCH_2CH_2OCH_3)_{6.90}{\bullet}3.78\;H_2O$, which can be approximated as $Pb_6(OOCCH_3)_5(OCH_2CH_2OCH_3)_7 \bullet 4\;H_2O$. This is the molecule that exists in the prehydrolyzed lead precursor solution, if lead is in excess of the stoichiometric composition.

Stoichiometric amounts of Zr and Ti methoxyethoxides (with the required Zr/Ti ratios) were mixed with the lead precursor (that underwent only atmospheric distillation) and the resulting solution was refluxed for two hours (for a 50 cc batch), followed by redilution.[23a] Note that excess-Pb precursor was added at this stage, as necessary. Iron (III) ethoxide was prepared and then converted to its methoxyethoxide by a separate

TABLE 2. Amount of RAc and water recovered by Gas Chromatography and Karl Fisher analysis, respectively, during dehydration, alcoholation, esterification and PZT complexation reactions.

Reaction # (atmospheric and/ or vacuum)	H_2O Recovered		RAc Recovered	
	moles	%	moles	%
5 (atm.)	2.91 (i.e., z)	59.8	0.39 (i.e., 2x)	19.5
5 (atm. + vac.)	3.52	72.3	1.15 (i.e., 2x)	57.3
6 (atm.)	0.44 (i.e., w)	9	1.29 (i.e., y)	64.5
6 (atm. + vac.)	0.83 (i.e., w)	17	1.48 (i.e., y)	73

atmospheric distillation.[27] At this stage, Fe or Nb methoxyethoxide was also added, along with Zr and Ti methoxyethoxide, as required. This allowed for complexation of the Pb, Nb or Fe, Zr and Ti components. Representing the term (3 + 2x) - z in equation 5 as u, the overall complexation reaction of lead precursor with Zr and Ti methoxyethoxides can be represented by:

$$Pb(Ac)_{2-2x}(OR)_{2x} \cdot u\ H_2O + v\ Zr\ (OR)_4 + (1-v)\ Ti\ (OR)_4 + 2y\ ROH$$
$$\longrightarrow PbZr_vTi_{(1-v)}\ (OR)_{4+2x+y}\ (Ac)_{2-2x-y} \cdot (u+y-w)\ H_2O + y\ RAc \uparrow + w\ H_2O\uparrow \quad (6)$$

The variables y and w are the experimentally recovered RAc and H_2O per mole of lead acetate, respectively, during the combined atmospheric and vacuum distillations to 1M concentration. Note that y moles of H_2O are generated along with y moles of RAc for 2y moles of

ROH consumed in reaction 6. The quantity of water and RAc recovered during reactions 5 and 6 are listed in Table 2. The solution was rediluted to about 0.5 molarity and vacuum distilled at 18-20 torr pressure at 45-50°C for 2 hours. Typically, the vacuum distillation was carried out until the vapor temperature reached 80°C. The final complexed solution was adjusted to a 1.0 molar concentration.

Starting with the atmospherically-distilled lead precursor, and by using equation 6 and the data in the Table 2 (i.e., u = 0.48, v = 0.5, 2x = 0.39, y = 1.48, and w = 0.83), the composition of the Pb-(Zr,Ti) alkoxide precursor (Zr/Ti = 1) vacuum distilled to 1 molar was determined to be $PbZr_{0.5}Ti_{0.5}(OR)_{5.87}(Ac)_{0.13}{\bullet}1.13H_2O$ or simply $PbZr_{0.5}Ti_{0.5}(OR)_6{\bullet}H_20$. Based on the number average molecular weight (M_n), and Nuclear Magnetic Resonance (NMR) and IR results, the consumption of this associated water resulted in oxo-bridges by the condensation reaction according to:[23a,c]

$$Pb_2ZrTi(OR)_{12}{\bullet}2\,H_2O \longrightarrow Pb_2ZrTiO_2\,(OR)_8\,(ROH)_2 + 2\,ROH \qquad (7)$$

After accumulation of all of the data, two structures containing oxo ligands were proposed[23a] for the prehydrolyzed Pb-(Zr,Ti) alkoxide precursor, or stock solution. They are the geometric isomers having a molecular formula of $Pb_2M_2O_2\,(OR)_8\,(ROH)_2$, and a molecular weight of 1336 where M is Zr and/or Ti.

2.3. Thin-Film Fabrication

The steps involved in thin-film fabrication are indicated in Figure 2. A 50cc hydrolysis solution, containing 0.2 moles of distilled and deionized water and 0.03 moles of ammonium hydroxide, was prepared in 2-methoxyethanol by weighing appropriate amounts. The stock solution was then hydrolyzed by mixing one molar stock solution with the hydrolysis solution in the volume ratio of 2:1. The hydrolyzed solution, thus, contained 2.0 moles of H_2O per mole of PZT alkoxide and 0.2 molar basic catalyst (NH_4OH). This hydrolyzed solution was then aged for 24 hours. Methoxyethanol and a drying control chemical additive (formamide) were mixed before obtaining bulk gels or thin films. The volumetric proportion of stock solution, hydrolysis solution, methoxyethanol, and formamide in

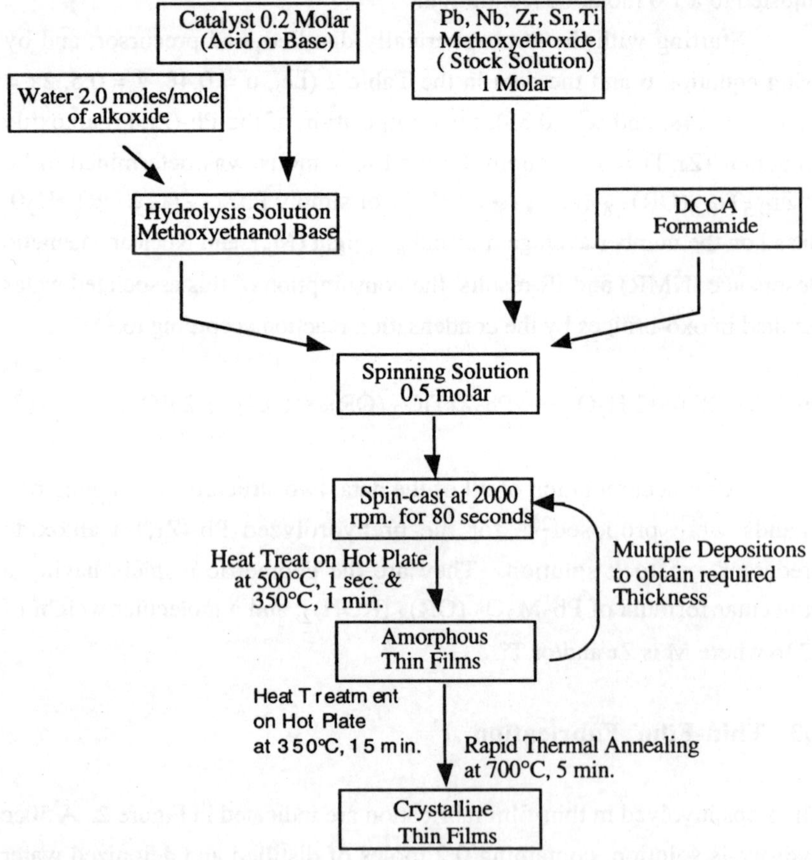

Figure 2 A flow diagram for the fabrication of thin films in the Pb(Nb,Zr,Sn,Ti)O_3 system.

the spin-casting solution was 2:1:0.5:0.5, respectively. Thin films were obtained by spin-casting onto three-inch diameter $Pt/Ti/SiO_2/Si$ wafers, at 2000 rpm for 80 seconds. A typical thickness of 1000 Å per layer was obtained. Each layer was heat treated at 500°C for 1 second and 350°C for 60 seconds per layer, followed by viscous sintering at 350°C for 15 minutes after the final layer was deposited. The final crystallization and densification of the amorphous PZT and doped PZT thin films were carried out at 700°C for 5 minutes using an AG Associates Heat Pulse 610 rapid thermal annealing (RTA) furnace.

2.4. Film-Substrate Interfacial Reactions and Phase Evolution

A survey of the literature on Pt/Ti bottom electrodes for integration of Pb-based perovskites with existing Si technology indicates that the understanding of interfacial reactions is limited.[28-31] In practice, significant differences in ferroelectric properties are observed for films of the same composition, deposited on substrates supplies by different vendors. This is a direct consequence of the interdiffusion of various species within the substrate, reactions at the ferroelectric/substrate interface at densification temperatures, poor and variable microstructures, and a lack of control of the perovskite phase stoichiometry. A few critical questions related to these issues are:

1) Why is there a PbO deficiency in stoichiometrically-batched PZT films? Is the consequent pyrochlore phase formation associated with any film-substrate interfacial reactions?

2) When excess-PbO is added into PZT batch compositions, where is it being incorporated in the multilayer $PZT/Ti/Pt/SiO_2/Si$ structure? and

3) What is the role of excess-PbO in the formation of the perovskite phase and in the grain growth mechanism?

In order to answer these questions, microchemical and cross-sectional (TEM) Transmission Electron Microscopic (TEM) analyses were carried out on stoichiometric niobium-and-tin modified PZT thin films on two different types of substrates.[14f] These substrates were identified as Type I (Pt(1000Å)/Ti(1000Å)/SiO_2 (200Å)/Si) and Type II (Pt(1500Å)/Ti(200Å)/SiO_2 (2000Å/Si). The Pb(Nb,Zr,Sn, Ti)O_3 or PNZST films were RTA heat treated at 700°C for 5 minutes in air. Figures 3, 4, and 5 represent Secondary Ion Mass Spectroscopy (SIMS), Auger Electron Spectroscopy (AES), and TEM results, respectively, of PNZST films on a Type I substrate. The general interfacial reaction features from SIMS, AES and TEM studies were:

1) a large fraction of the original Ti adhesion layer diffuses through the Pt layer;

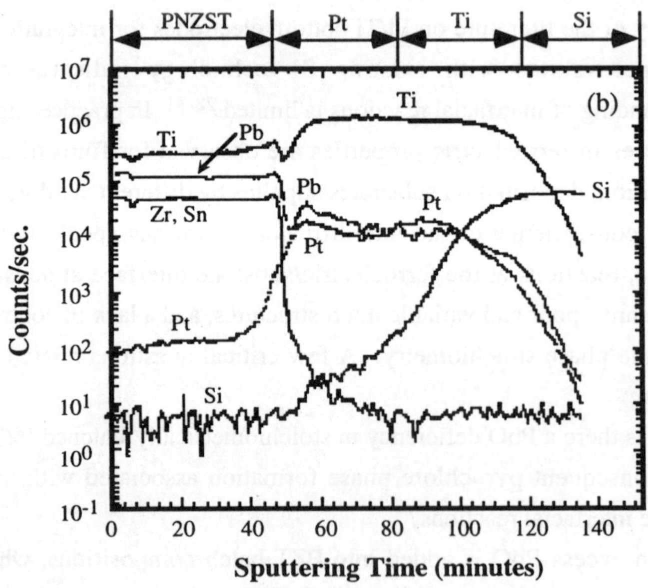

Figure 3. Secondary ion signal density as a function of depth in 0.5 μm thick PNZST/Type I multilayer structure.

2) most of the Ti oxidizes to TiO_2 and the amount of residual Ti film on the SiO_2 layer is a function of the original Ti thickness;
3) the lower section of the continuous Pt layer breaks away and forms dense ellipsoidal islands surrounded by the TiO_2 matrix; and
4) a substantial amount of Pb from the PNZST film diffuses into the substrate and the Pb is mainly associated with the two regions of Pt.

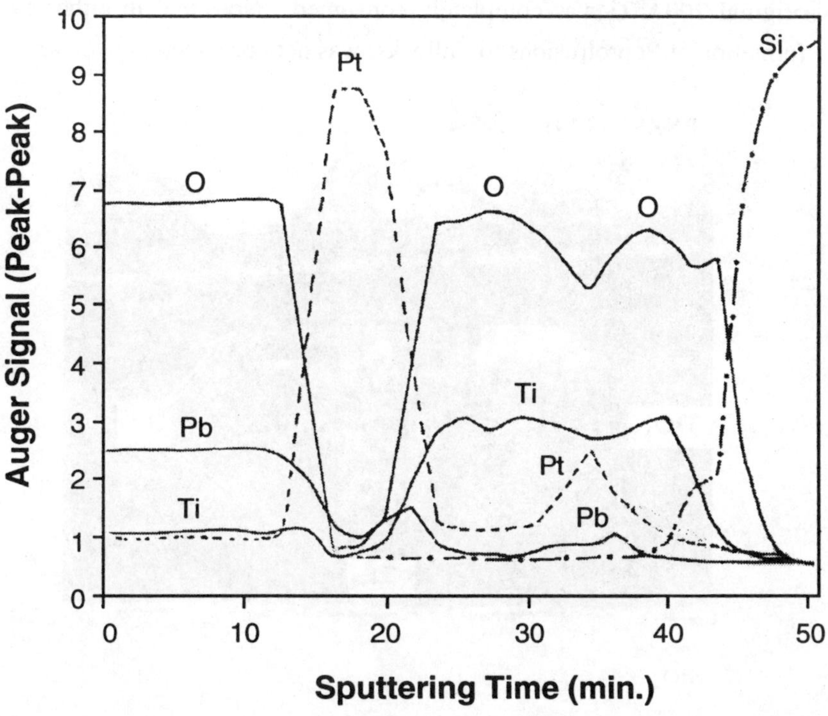

Figure 4. Auger electron spectroscopic depth profile of the same (Figure 3) specimen.

The nature of the interfacial reactions in the PNZST/Type I and PNZST/Type II multilayer structures were similar; the only difference was the degree of interactions in (1) through (4). For example, the original

1000Å continuous-and-dense Pt layer (Type I substrate) supporting the PNZST film reduced to an approximately 500-700Å thick continuous layer and also 200-500Å thick islands. Of the 1000Å Ti adhesion layer that was originally present in the as-received Type I substrate, about 100Å of it remained, whereas the rest oxidized to TiO_2. In contrast, the 1500Å Pt layer in Type II substrate remained intact with only small quantities of isolated islands (100-200Å wide) at the TiO_2/SiO_2 interface. In this case the original 200Å Ti was completely consumed. Note that in either case, formation of Pt protrusions (or hillocks) was not observed.

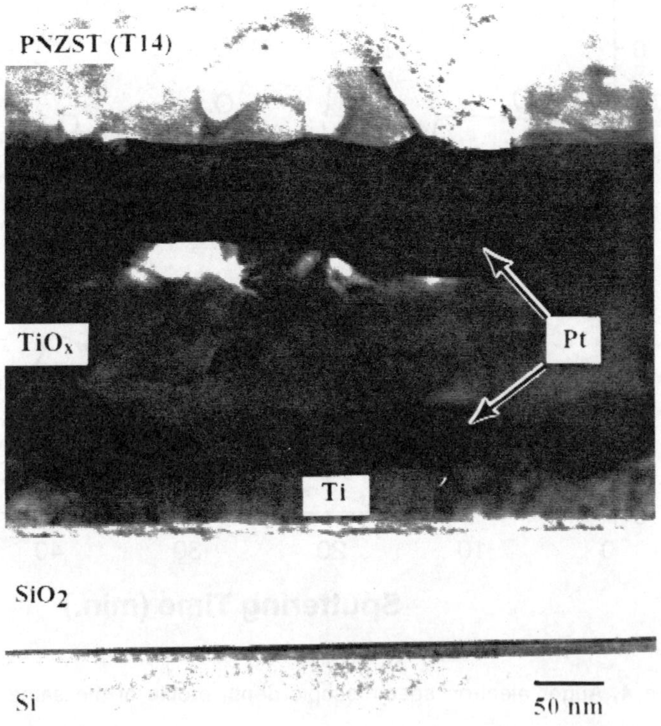

Figure 5. Cross section transmission electron micrograph of the same (Figure 3) specimen.

The above results indicate that in order to interpret the phase and microstructure evolution during heat treatments of PZT in air, the effects of Pb-initiated disintegration of Pt, oxidation of Ti to TiO_2, stress relaxation in the Pt/TiO_2/Ti substructures, and interfacial reactions during heat treatments must be considered. The reduction of PbO (probably caused by the reducing environment created from carbonaceous residue (Section 2.1)) to Pb initiates the formation of low-melting point phases with Pt. Note that PbO could also be reduced by out-diffused Ti from the substrate; PbO having a higher formation energy ($\Delta G^\circ \approx$ -188 kJ/mol) than TiO_2 ($\Delta G^\circ \approx$ -889 kJ/mol).[21] It is well known that Pt forms various intermetallic compounds of varying stoichiometry with Pb (e.g., PbPt, Pb_4Pt and $PbPt_x$). Since Pb and Pb_4Pt melt at a T_m = 327°C and 350°C, respectively, and PbPt (T_m = 795°C) and $PbPt_x$ (T_m = 915°C) phases are sufficiently ductile at the processing temperature of 700°C (i.e., 0.7 T_m; where T_m is melting temperature), the stresses developed due to conversion of Ti to TiO_2 are released by the separation of the lower section of the Pt layer into ellipsoidal islands, as opposed to the formation of protrusions on the Pt surface. The ellipsoidal shape of low curvature and, therefore, of low surface energy, is also consistent with the liquid-like nature of various Pt-Pb phases. The overlap of ^{208}Pb and ^{196}Pt signals observed in SIMS analysis is probably a direct consequence of this type of alloying. The formation of PtO and PtO_2 with melting temperatures (T_m) of 500°C and 450°C, respectively, may be discounted since the oxygen AES signal within the substrate is not associated with Pt. Note that, during the vapor phase deposition of Pb-based perovskites with a heated substrate and high oxygen activity, the formation of protrusions on Pt are observed since Pb (in the form of PbO only) cannot exert its effect. Indeed, the tendency of formation of protrusions is drastically reduced in a reducing atmosphere.[31]

One of the questions raised earlier, i.e., whether the PbO deficiency in stoichiometrically-batched PNZST films is related to second phase formation in Pb-based perovskite films can now be answered. Figures 6a and b illustrate SEM images of heat treated PNZST thin films from stoichiometrically batched solutions on Type I and Type II substrates, respectively. The microstructure on Type I substrate consists of circular perovskite grains with rough edges that are surrounded by a continuous amorphous or nanocrystalline second phase. This interphase was undetectable by x-ray diffraction. The circular perovskite regions did not grow with longer (~ 15-30 min.) heat treatments. In contrast, the microstructure on the Type II substrate is characterized by well defined perovskite grains with a small amount of second phase. The electron diffraction analyses indicated that the nanocrystalline second phase was a Pb-deficient pyrochlore $Pb_{2-\delta}(Zr,Ti)_2O_{7-x}$, with a lattice parameter of 10.4Å.[14f] It is evident that the formation of a lead-deficient pyrochlore phase in PNZST films can be explained by the loss of Pb to the substrate and the uptake of Ti from the substrate; the greater the loss is to the substrate, the greater is the quantity of pyrochlore phase formation in the film. The extent of Pb loss from the film is probably dictated by both the grain size and density of Pt. The amount of Ti loss from the substrate is controlled by the Ti and Pt layers thicknesses. In comparison to a Type II substrate, a Type I substrate has a thinner (1000Å) Pt layer of smaller initial grain size (200Å as compared to 300Å) and a thicker (1000Å) Ti layer. Thus, a larger amount of continuous Pb-deficient pyrochlore (Py) interphase (Figures 6a and b) in perovskite (Pe) PNZST films on the Type I substrate can be explained by a higher Pb loss from the dielectric film to the substrate and Ti loss from the substrate. The type II configuration is, therefore, favorable but to reproduce the nature of the Ti, Pt, and SiO_2 layers will require identical processing conditions.

Finally, it should be noted that the amount of perovskite phase increases with the thickness of the film. This can be explained on the basis of a fixed amount of Pb loss to the substrate for a given time and temperature of heat treatment. The PbO deficiency in a thinner film is so pronounced as to promote the formation of pyrochlore only. Interestingly,

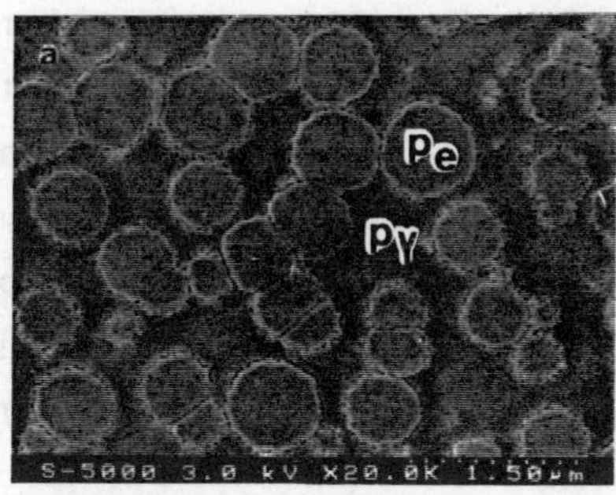

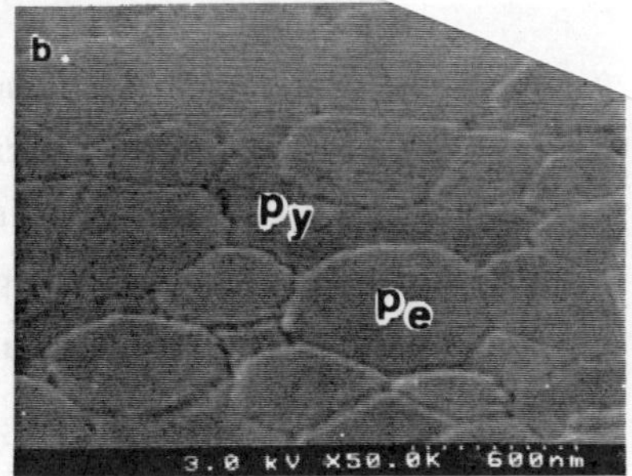

Figure 6. Scanning electron micrographs of stoichiometrically batched PNZST thin films on a) Type I and b) Type II substrates.

a high fraction of perovskites has been obtained in very thin films or films (derived from stoichiometrically batched PNZST compositions) of varying thickness when the substrate was coated with PbO prior to spin-casting PNZST solutions. To compensate for the PbO loss to the substrate at 700°C in 5 minutes and to achieve dense perovskite films (without a PbO coating on the substrate), an addition of at least 5-10% excess PbO with respect to the stoichiometric composition is necessary.

2.5. Microstructural Evolution in Systems with Excess-Lead Oxide

The microstructures of heat treated PNZST thin films (derived from solutions with 10 mole % excess lead over the stoichiometric composition) on Type-I and Type-II substrates are illustrated in Figures 7a and b. Note that the PNZST films are dense, have a thickness of a single grain, and exhibit a columnar growth habit on the underlying Pt grain (1000Å). The PNZST grains (0.2-0.5 μm) in Figure 7a are, on the average, smaller compared to grains (≈ 1 μm) in Figure 7b. Since the perovskite grains are incoherently aligned with the substrate, elastic energy constraints appear to have played a minor role on the growth process. Moreover, since both samples were heat treated simultaneously and the initial Pt grain size in both substrates was similar, the difference in the PNZST grain sizes cannot be explained on the basis of nucleation-and-growth-rate differences. Therefore, the difference in the grain sizes in Figures 7a and b may be attributed to the difference in the amount of residual excess-PbO within the PNZST films after heat treatment. It is clear that a higher (lower) amount (due to interfacial reaction and diffusion) of Pb within Type-I (Type-II) substrate translates into a lower (higher) amount of residual excess PbO within the PNZST films; the consequence of which is the smaller (larger) grain size as illustrated in Figure 7a (7b).

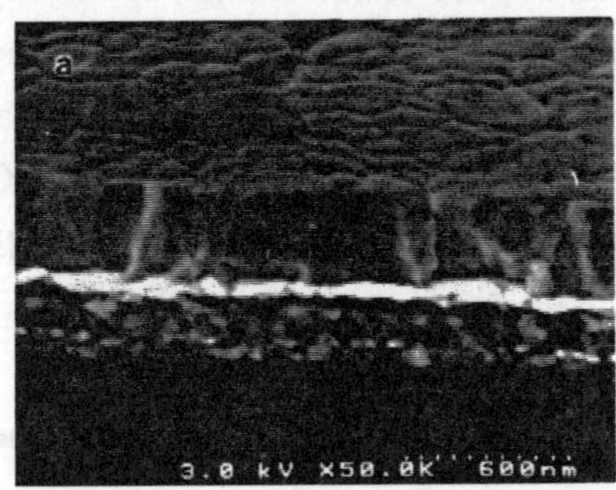

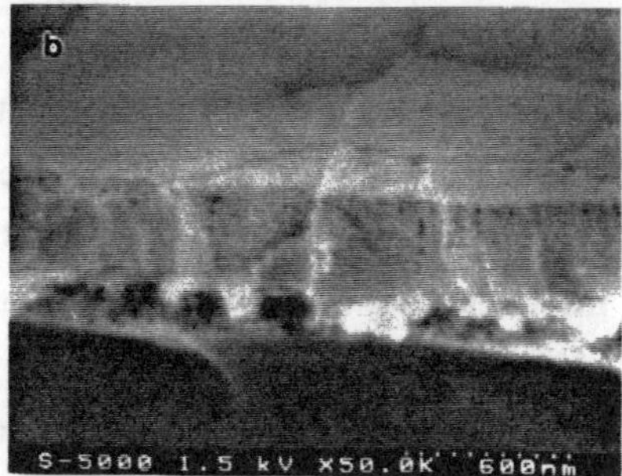

Figure 7. Cross-sectional SEM of PNZST thin films derived from solutions that were batched with 10% excess lead on a)Type I and b) Type II substrates.

In concurrence with our observations, it has been reported in independent studies that PZT compositions containing higher than ≈ 10 mole % excess-PbO exhibit an increase in the grain size.[32,33] Since phase equilibrium data in the PbO-TiO_2 system indicate that the first liquid appears at 838 °C,[34] the increased grain size and dense microstructure in PZT thin films with 10 mole% excess PbO, heat treated at 700 °C, cannot be explained by liquid phase sintering. Moreover, since a narrow range of cation stoichiometry has been known to exist in $PbO-ZrO_2-TiO_2$ system,[35] it was surprising that x-ray diffraction patterns did not reveal excess-PbO peaks in PZT thin films.[36]

High Resolution TEM (HREM), however, detected 100-200Å PbO precipitates within PZT (52/48) grain (Figure 8a).[37] The distribution of these precipitates was nonuniform and different grains had different volume percentages of PbO. Since the precipitates were embedded in the PZT matrix, moiré fringes were visible in the region of overlap between the precipitate and the matrix. A higher magnification view of a single precipitate is shown in Figure 8(b). The precipitates exhibit inclined (101) lattice fringes (3.11Å) in a PZT matrix with (101) lattice fringes of 2.89Å width. A consistent orientational relationship between the precipitate and

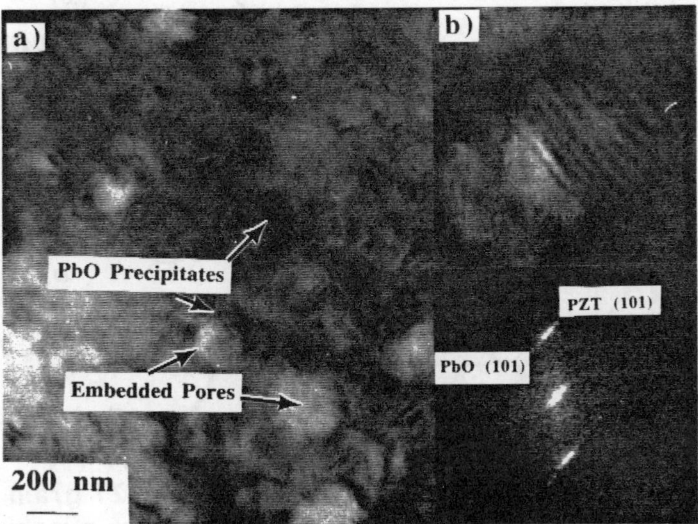

Figure 8. HRTEM image showing a) distribution of PbO precipitates within PZT grains, and b) high magnification of a single PbO precipitate along with optical diffractogram. Note that the films were derived from solutions that were batched with 15% excess lead.

matrix was absent. Careful observation of the PbO precipitate/PZT matrix interface revealed the presence of dislocations, indicating the incoherent nature of the interfaces. From these results, the aforementioned one-to-one relationship between residual excess-PbO in the PNZST film and grain size can only be explained on the basis of precursor chemistry (Figures 9 & 10).

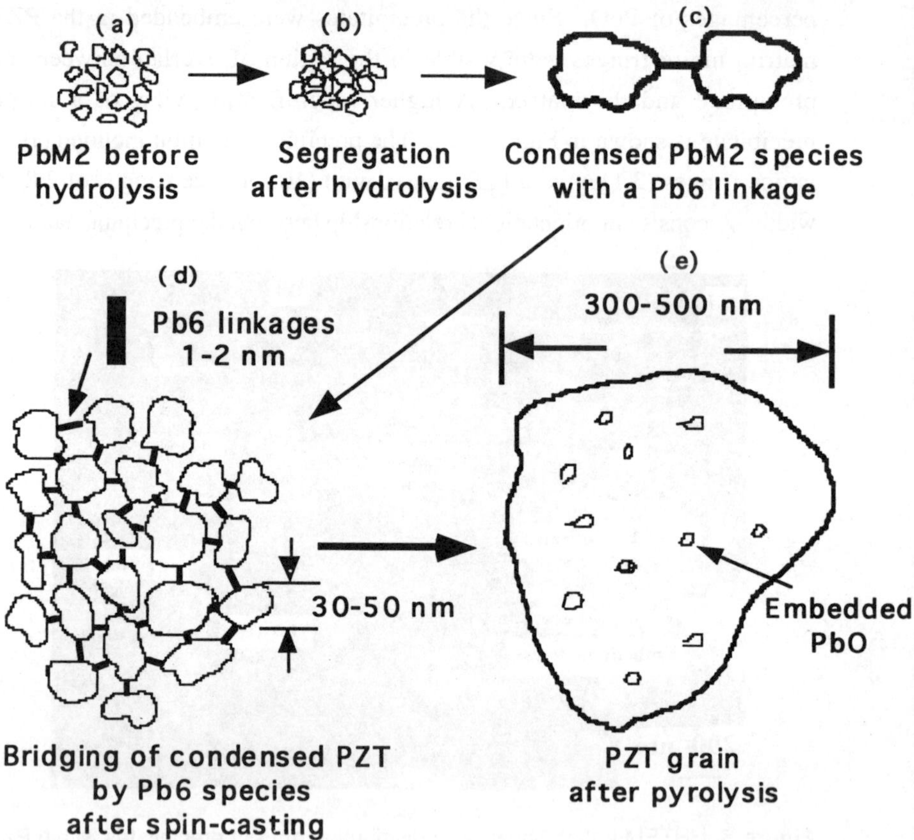

Figure 9. A schematic of microstructure evolution (a-e) in PZT thin films.

The effect of excess-PbO on the grain growth of PZT thin films, along with the proposed model for microstructure evolution, is illustrated schematically in Figure 9. Condensed PZT species (30-50 nm) are connected by excess Pb-O linkages (1-2 nm). As described below, these oligomeric PbO species eventually condense and cluster to form crystalline PbO that remains embedded within 300-500 nm PZT grains after heat treatment.

From the study of the prehydrolyzed precursor structure, the molecular formulas of the lead and Pb-(Zr,Ti) alkoxide precursors are $Pb_6(OOCCH_3)_5(OCH_2CH_2OCH_3)_7 \cdot 4H_2O$ and $Pb_2M_2O_2(OR)_8(ROH)_2$ (where M is Ti and/or Zr), respectively. The former and the latter are referred to as Pb6 and PbM2, respectively. The addition of hydrolysis water to a solution containing PbM2 (Figure 9a) with 15% excess Pb6 results in preferential hydrolysis and condensation of the former. This is a direct result of the highly electropositive nature of Zr and Ti in PbM2. Consequently, there is segregation (Figure 9b) of PbM2 oligomers to form larger (30-50 nm) species. Additionally, during condensation, water and alkoxy groups associated with Pb6 can form some Pb-O-M oxy bridges (Figure 9c). The role of the Pb6 oligomeric precursor, therefore, is to bridge the segregated and condensed PbM2 species. The details (Figure 10) of the polymeric species in the spin-casting solution show condensed PbM2 (shown horizontally), which is linked by a bridging Pb6 oligomeric species (shown vertically). The structure contains alkoxy, hydroxy, and oxy species which bridge the metals. Note that a) the terminal alkoxy groups have been hydrolyzed, with some bridging groups remaining; b) the acetate and carbonate ligands, which are bonded to Pb, remain in a bridging configuration; c) some ROH may be bonded to Pb atoms that initially have a low coordination number.[23a]

After spin-casting the solution, further hydrolysis and condensation can occur as the oligomeric species, represented in 2-dimensions in Figure 10, come into close proximity. In reality, the 3-dimensional thin-film gel structure is spherical, since the chains would collapse in order to lower the free volume. The gel structure (Figure 9d) can also be viewed as a 3-D network of condensed PbM2 polymers which are linked by bridging Pb6 species. This allows the metals to achieve higher coordination numbers through coordinate covalent bonding.

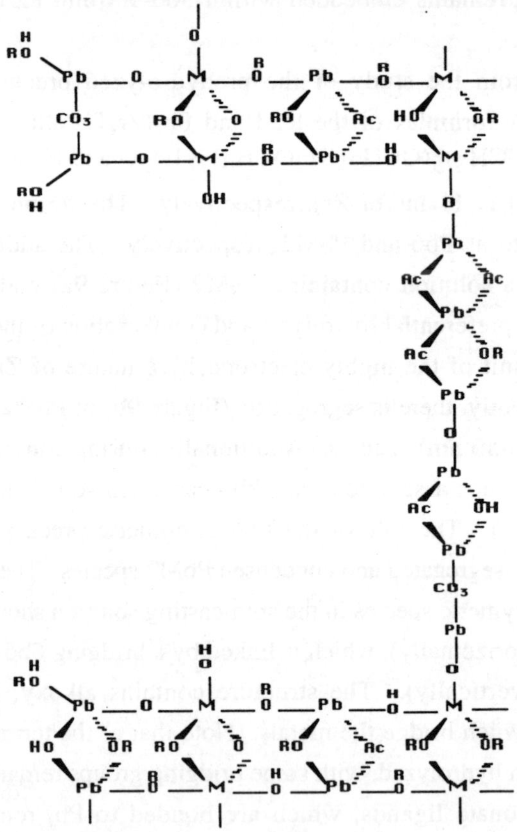

Figure 10. Atomic level details of a $Pb_6(OOCCH_3)_5(OCH_2CH_2OCH_3)_7 \cdot 4H_2O$ or Pb6 bridge between $Pb_2M_2O_2(OR)_8(ROH)_2$ (where M is Ti and/or Zr) or PbM2 species.

The structure of the condensed gel in as-deposited PZT thin films with excess-PbO is the basis for the observation of enhanced grain growth during the heat treatment procedure. The mechanism involves trapping of the bridging Pb6 precursor groups which eventually crystallize and coalesce to form PbO precipitates. These precipitates then remain embedded within the PZT grains (Figure 9e). Thus, the residual excess-PbO is present in the form of embedded precipitates, as opposed to being located at grain boundaries and triple junctions. Pyrolysis of the thin-film gel results in thin films of dense perovskite. The observed weight loss of between 13-19% during pyrolysis[23b] is in reasonable agreement with the theoretical (from Figure 10) weight loss of 22%.

From the above discussion, it is clear that the grain growth mechanism in sol-gel processing of PZT thin films is different from that of conventional ceramics where very high temperatures (i.e., above the PbO melting point) are often used. The addition of excess-PbO not only aids in perovskite phase formation (i.e., by maintaining stoichiometry), but also in the grain growth of PZT thin films. Indeed, PZT films derived from solutions with excess-PbO show an increase in the grain size if the loss to the substrate is over compensated. The proposed model explains how large amounts of residual excess-PbO in the film translates to the formation of large grains.

One final point regarding densification of PZT thin films in the presence of excess-PbO merits clarification. In principle, a high fraction (≈ 8-10%) of excess-PbO in the form of metallic Pb is lost to the substrate from a 0.5 μm thin film during heat treatment in air at 700°C for 5 minutes. Therefore, initial compositions with 8-10% excess-PbO maintain stoichiometry of the perovskite film after heat treatment. Without this excess-PbO, various phases, including pyrochlore, coexist with the perovskite phase and the thin films appear porous. In essence, stoichiometric PZT films can achieve full densification, only if Pb is not lost to the substrate at 700°C, or to the atmosphere at much higher temperatures

(>800 °C). This point was clearly illustrated in PNZT thin films derived from stoichiometrically-batched solution, which formed epitaxially on inert sapphire substrates and exhibited perovskite phase purity.[21]

2.6. Dopant Compensation Mechanism and Their Effect on Leakage Current in Thin Films

It is well known that both microstructure and solid state chemistry influence the electrical properties of bulk perovskite materials.[38-40] To date, distinct effects of aliovalent acceptor or donor dopants have not been reported[41,42] in the Pb-based perovskite thin-films. Based on the above discussions, this was probably due to the lack of phase stoichiometry and microstructure control.

In the event of Pb loss to the substrate, a high donor concentration (within the solubility limit) would be required in order to obtain phase pure insulating films. In essence, donor doping in perovskite PZT thin films is not successful if the loss of PbO is greater than the donor concentration. Indeed, Smyth et.al.,[43] have reported that a higher fraction of donor is required to suppress the $V_O^{\cdot\cdot}$ concentration in the presence of lead loss. When loss of PbO is so significant (≈ 10 mole %)as to form several phases, studies to determine the effects of donor doping are meaningless.

Recently,[23c] the defect compensation mechanism in donor-doped (2-5 mole %) $Pb(Zr_{0.52},Ti_{0.48})O_3$ thin films, with deliberately added 10 mole % excess-PbO, was determined by transmission electron microscopy (TEM). Note that the excess-PbO, was necessary to compensate for the aforementioned loss to the substrate during processing. The effects of donors and acceptors on the microstructure and leakage current in the PZT(52/48) thin-film system were reported for the first time. The compositions with aliovalent dopants were batched, assuming A, B, and O site vacancy models as follows:

1) A-site vacancy model: $Pb_{1-x/2}\,[Zr_{0.52}Ti_{0.48}]_{1-x}\left[Nb_{Ti}^{\bullet}\right]_{x}\,O_3$, where X = 2, & 5 mole %.

2) B-site vacancy model: $Pb[Zr_{0.52}Ti_{0.48}]_{1-5x/4}\left[Nb_{Ti}^{\bullet}\right]_{x}\,O_3$, where X = 5%

3) Oxygen vacancy model: $Pb[Zr_{0.52}Ti_{0.48}]_{1-x}\left[Fe_{Ti}^{'}\right]_{x}\,O_{3-x/2}$, where X= 2%

All compositions were batched such that they would be single phase, if the assumed defect model was operative. The A-to-B-site ratio and TEM results are summarized in Table 3, along with the designation of compositions. The results indicate that donors in sol-gel derived, PZT thin films are compensated by $V_{Ti}^{'''}$, rather than $V_{Pb}^{''}$, since a single phase microstructure is exhibited only by B5 (i.e., non-stoichiometric and Ti-deficient). Additionally, the high B/A ratio, determined by energy dispersive spectroscopy (EDS), in the grain boundary (GB)phase of A2 and A5, can be explained by the exsolved B-site cations if compensation occurs according to the B-site vacancy-model. Moreover, the amount of second phase is higher in A5 than A2, as expected, since the Nb content in the former composition is higher. The microstructure of the O2 (acceptor-doped) thin film, exhibited some embedded porosity within the perovskite grains with clean grain boundaries.

The results of this study were in concurrence with an earlier report of $V_{Ti}^{''''}$ as the compensating defect in Nb-doped (> 5 mole %) $BaTiO_3$.[38] Contrary to the above observations, $V_{Pb}^{''}$ has been reported to be the compensating defect in Nb-doped, bulk PZT ceramics.[33] This discrepancy may be explained by the increased activity of PbO (since excess-PbO was added to compensate for the loss during processing) and, therefore, the observed $V_{Ti}^{''''}$ compensating defect model. Indeed, Atkin and Fulrath[35] have shown that, at a constant oxygen partial pressure, the concentration of V_{Pb}'' decreases and V_{Ti}'''' increases with an increase in the activity of PbO. The observed V_{Ti}'''' model may also be inherent for highly donor-doped PZT, but simulation studies, as carried out by Catlow et al.[44] for the $BaTiO_3$ system, must be done for verification.

Table 3. Vacancy compensation model, designation, batched composition, along with A/B ratio and observed phases in the TEM for acceptor and donor doped PZT thin films.

Vacancy model / Designation	Batched Composition	Batched A/B Ratio	TEM Results
A-site/A2	$Pb_{0.99}[Zr_{0.52}Ti_{0.48}]_{0.98}[Nb^{\bullet}_{Ti}]_{0.02}O_3$	0.99	Lead deficient, GB phase
A-site/A5	$Pb_{0.975}[Zr_{0.52}Ti_{0.48}]_{0.95}[Nb^{\bullet}_{Ti}]_{0.05}O_3$	0.975	Lead Deficient, GB phase
B-site/B5	$Pb[Zr_{0.52}Ti_{0.48}]_{0.9375}[Nb^{\bullet}_{Ti}]_{0.05}O_3$	1.013	no GB phase
Oxygen/O2	$Pb[Zr_{0.52}Ti_{0.48}]_{0.98}[Fe'_{Ti}]_{0.02}O_3$	1	no GB phase, embedded porosity

The effects of both microstructure (especially in A5 and A2 with second phases) and solid state chemistry (in B5 and O2) were reflected on the leakage currents of PZT(52/48) thin films.[23c] The leakage current characteristics of phase pure (disregarding porosity) perovskite films, where the solid state chemistry plays an important role, are illustrated in Figure 11. The O2 film exhibited a higher leakage current density ($J_L \approx 10^{-7}$ A/cm^2) at low applied fields (< 200 kV/cm) when compared to the B5 film. Additionally, the breakdown strength (400 kV/cm) of O2 was approximately an order of magnitude lower than the single phase B5 film. The observed results for the O2 film are consistent with the oxygen vacancy, $V_O^{\cdot\cdot}$, model for bulk ceramics. Thus, a high, initial $V_O^{\cdot\cdot}$ concentration leads to a field-induced concentration polarization of $V_O^{\cdot\cdot}$, with a pronounced increase in the electrical conductivity.[40] If the initial concentration of $V_O^{\cdot\cdot}$ is low, the leakage current should be low. Indeed, the low J_L (10^{-7} A/cm^2 at 1.2 MV/cm) of the donor-doped (5% Nb), B-site compensated thin film (B5) was as expected, since the $V_O^{\cdot\cdot}$ concentration was suppressed in this composition.

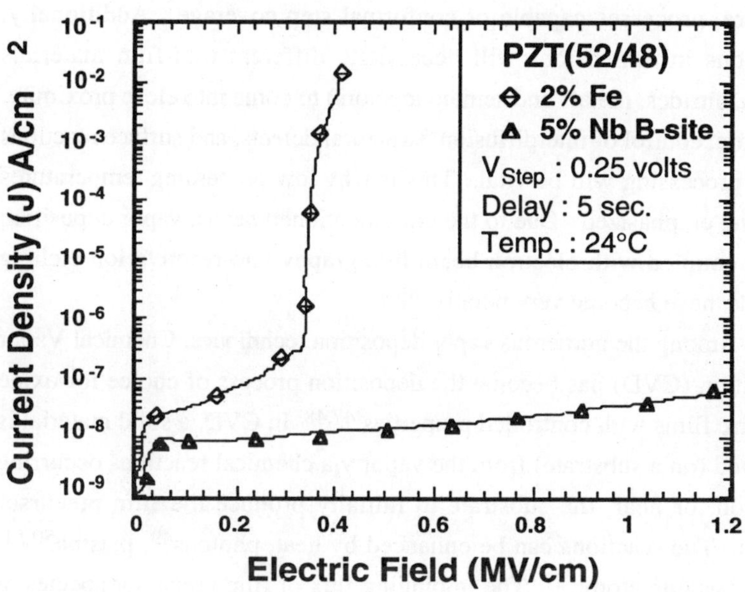

Figure 11. Leakage current density versus electric field for acceptor and donor doped PZT thin films (0.46μm) with capacitor area of 90 x90 μm^2.

3. PLASMA-ENHANCED METAL ORGANIC CHEMICAL VAPOR DEPOSITION OF PEROVSKITE DIELECTRIC THIN FILMS

In the past thirty years, as integrated circuit design dimensions have decreased, device density has advanced from large scale integration (LSI, 10^3-10^5 devices per circuit), via very large scale integration (VLSI, 10^5-10^6 devices per circuit), to ultra large scale integration (ULSI, > 10^6 devices per circuit). This, in turn, has placed stringent demands on device fabrication technology, particularly with respect to control of chemistry over both lateral and vertical dimensions of features. Further decreases of lateral dimensions (below sub half-micron) will require deeper side walls, and therefore, processes capable of conformal step coverage. Additionally, reductions in dimensions will necessitate different thin-film materials (oxides, nitrides, metals and semiconductors) to come into close proximity. Therefore, control of interdiffusion, structural defects, and surface reactions during processing will be vital. This is why low processing temperatures are being emphasized. Due to the aforementioned needs, vapor deposition routes, coupled with electron beam lithography and reactive ion etching methods, have become very popular.[45-47]

Among the numerous vapor deposition techniques, Chemical Vapor Deposition (CVD) has become the deposition process of choice for oxide dielectric films with controlled properties.[47,48] In CVD, a solid material is deposited (on a substrate) from the vapor via chemical reactions occurring either on, or near, the substrate to initially produce the film precursor species. The reactions can be enhanced by heat, photons[49], plasma[50,51], and metastable atoms.[52] The impinging flux of film precursor species or adatoms are thermally accomodated with the substrate. Next, the mobile adatoms interact with each other to form clusters or reevaporate. Cluster growth results in islands which eventually coalesce to form a continuous film. There are numerous deposition parameters in CVD which determine the nucleation and growth kinetics, microstructural evolution, and thin-film properties. These parameters include: the chemical state, kinetic energies, and fluxes of film precursor species; growth temperature; and the substrate material's composition, surface cleanliness, orientation, and crystallinity.

3.1. Phase Thermochemistry for the Ferroelectric Pervoskite $Pb(Zr,Ti)O_3$ System

Chemical vapor deposition is a non-equilibrium process dictated by chemical reaction kinetics and transport.[47] However, in approaching a new, multicomponent chemical system for deposition by CVD, it may be useful to theoretically examine the solid-state equilibria in terms of parameters that can be controlled or measured.[54] Thermodynamic calculations allow the most stable state to be specified in terms of temperature, composition, and partial pressures.[55] If experimental results comply with thermodynamic predictions, then the calculations provide a useful tool for predicting the outcome of experiments. On the other hand, if there is a deviation, comparisons between results and calculations allow the nature of reaction mechanisms to be studied. Phase-stability diagrams, which plot the ranges of stability of solid phases as a function oxygen partial pressure and cation ratios (parameters adjustable within a CVD reactor), are derived from equilibrium compositions computed by the method of direct minimization of the system free energy.[54]

A phase stability diagram was computed in the Pb-Zr-Ti-O system at 480°C (with a fixed Zr/Ti ratio of 53/47) using a modified version of the program SOLGASMIX-PV,[56] available thermodynamic data, [34,57-58] and certain assumptions.[56,59,60] Although numerous reports indicate the formation of "pyrochlore" phase[61] prior to perovskite PZT phase formation, it was assumed to be metastable and, therefore, omitted from the calculation. Since the pyrochlore phase does not transform to the perovskite phase for Pb-poor compositions (Sections 2.4 and 2.5), its omission from the calculations seemed reasonable. In general, the calculations indicated that perovskite PZT was stable at high oxygen partial pressures above about 10^{-25} atm of oxygen. It coexists with PbO for high concentrations of lead and with $PbTi_3O_7$, Ti_2ZrO_6, and ZrO_2 for successively smaller amounts of lead. At low oxygen partial pressures, it equilibrates with liquid lead.

The Zr and Ti fractions in the perovskite phase were also calculated as a function of oxygen partial pressure for three different regimes of Pb content (Figure 12).[60] Note that the perovskite composition does not necessarily reflect the cation ratios as that for the entire system (i.e., Zr/Ti = 53/47), since other phases may co-exist. For a high content of Pb, the perovskite is slightly Zr-rich, relative to PZT(53/47), because PbO has the capacity to dissolve small amounts of titanium oxide. This implies that perovskites with precisely the morphotropic (53/47) composition cannot be attained at thermal equilibrium. It is only at low oxygen partial pressures in the range of about 10^{-20} to 3×10^{-22} atm of oxygen that the perovskite composition of $Pb(Zr_{0.53}Ti_{0.47})O_3$ would appear to be stable by itself with no other phase. At successively lower oxygen partial pressures, the composition of the perovskite shifts increasingly to the Zr-rich side, where $PbZrO_3$ is the most stable perovskite composition at the lowest oxygen pressure. Since most CVD reactions on this system will typically be carried out between 10^{-4} to 1 atm O_2, minor quantities of a Pb-rich phase in contact with the perovskite will result if the process is entirely thermally driven. Therefore, plasma enhanced CVD (PECVD) may be an alternate processing route of choice since the injection of rf energy will upset the system chemistry from approaching a quasi, thermal equilibrium state.

3.2. Precursors and Limitations

The precursors for CVD of Ba, Sr, and Pb-based perovskite dielectrics are predominantly metal alkyls, cyclopentadienyls, alkoxides, and β-diketonates. Major drawbacks to date have been limited studies on volatile

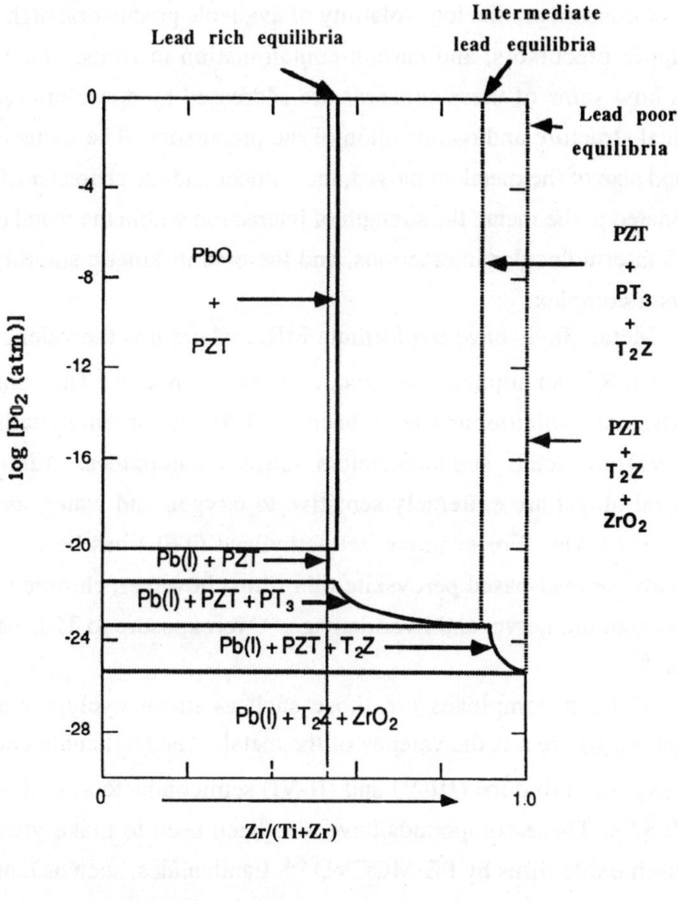

Figure 12. Zirconium fraction (x) of perovskite $Pb(Zr_xTi_{1-x})O_3$, in equilibrium at 480°C with various phases as a function of oxygen partial pressure for three different regimes of lead.

alkaline-earth and lanthanide compounds for CVD, high cost of starting chemicals, and negligence of environmental safety concerns. Additional issues of concern are the low volatility of available precursors, high toxicity of volatile precursors, and carbon contamination in films. This section shows how some of these concerns are addressed by consideration of the chemical structure and composition of the precursors. The focus is on the type and size of the metal employed, the number and the character of ligands coordinated to the metal, the strength of interaction within the metal complex and its intermolecular interactions, and the overall kinetic stability of the precursor complex.

Metal alkyls have the formula MR_n, where n is the valency of the metal, and R is an aliphatic, olefinic, or aromatic moeity. The compounds are extremely volatile, and have been used for the organometallic CVD (OMCVD) of metals, semiconductors, nitrides, and oxides. Additionally, the metal alkyls are extremely sensitive to oxygen and water, as well as being very toxic. For example, tetraethyllead (TEL) has been used as a precursor for lead-based perovskite thin films; however, chronic exposure can cause brain, nerve, and liver damage. Overexposure to TEL vapor can be fatal.[62]

Metal π-complexes are those such as metal cyclopentadienyls, $M(C_5H_5)_n$, where n is the valency of the metal. The lanthanide complexes have been used to dope (III-V) and (II-VI) semiconductors, such as GaAs and ZnS.[63] These compounds have also been used to make yttrium and rare-earth oxide films by PE-MOCVD.[64] Lanthanides, such as lanthanum, can have large coordination numbers (typically 6 to 9 in organolanthanide compounds) due to their large size and large number of orbitals available for bonding. The π-complexing ligands (such as cyclopentadienyls) donate electrons to the central metal atoms. This electron donation promotes volatility of the complexes by satisfying the metals, electronically and coordinately, and preventing intermolecular associations with adjacent ligands. The vapor pressures of the lanthanide cyclopentadienyls, $M(C_5H_5)_3$, increase as the metals vary from lanthanum to lutetium across the lanthanide series. This is a direct result of decreasing size of the central metal atom. At 227°C, the vapor pressure of the lanthanum compound is 0.16 Torr, while the value of the lutetium compound is 7.7 Torr.[63] A couple of advantages of these compounds over the trivalent β-diketonates are that there is less than one-half as much carbon and volatilities are higher. An external source of oxygen, however, must be used, as there is none in the complex. Since they are extremely air-sensitive, handling is difficult.

Metal alkoxides are the predominant precursors in the MOCVD of metal oxides. Titanium and zirconium alkoxides have shown sufficient volatilities and ease of decomposition, but Pb, Ba, Sr, and La alkoxides behave differently. Due to their large atomic radius (1.75 Å) and tendency for higher coordination numbers (4,5, and 6 in organolead compounds), strong intermolecular associations occur with Pb alkoxides even with bulky ligands such as the tert-butoxide, $OC(CH_3)_3$. For example, $Pb(^tO\,C(CH_3)_3)_2$ has been found to be trimeric.[65-67] This compound has the advantage of having as much as one-third less carbon than Pb β-diketonate, but its volatility is much lower and is much more air-sensitive. The alkoxides of the alkaline earth group metals are generally nonvolatile and not useful as CVD precursors. Recently, liquid barium alkoxides, having a formula: $Ba[O(CH_2CH_2O)_nR]_2$, where n = 2,3 and R = CH_3, C_2H_5, have been synthesized.[68] When n = 2, barium has a coordination number of 6, and when n = 3, the coordination number increases to 8. These molecules are very stable with respect to hydrolysis, but unfortunately decompose in the gas phase.[69] The alkoxides of the lanthanides are somewhat volatile. Like the π-complexes, the volatility increases with atomic number, with lanthanum being the most difficult to volatilize. By utilizing bulky alkoxide groups, the oligomeric character can be reduced to two, as is observed with $[La(OCMeEtPr^i)_3]_2$.[70] This alkoxide can be sublimed at 135°C at a 10^{-3} Torr pressure.

The β-diketonates are quite useful ligands, especially in metals which do not form any other type of volatile compounds. For example, there is the barium derivative, barium bis(2,2,6,6-tetramethyl-3,5-

heptanedionate), $Ba(O_2C_{11}H_{19})_2$ or $Ba(thd)_2$[71]. A major problem associated with these solid compounds is that the volatilities are often low, and fairly high temperatures must be used to cause sublimation. The fluoride derivative, i.e., Pb or Ba bis-heptafluorodimethyloctadione, Pb or $Ba(O_2C_{10}H_{10}F_7)_2$, can be vaporized at 30-50°C lower than the parent β-diketonates.[72] The substitution of fluorine for hydrogen increases volatility, but can lead to the deposition of metal fluorides.[73,74] Even though β-diketonates are not as reactive towards water as alkoxides are, they often form hydrates and are, thus, hygroscopic. Heating the hydrates at elevated temperatures causes ligand loss to form nonvolatile β-diketonato hydroxo oligomers.[75] This could lead to CVD reproducibility problems. Thus, even though the β-diketonates are not air-sensitive, they should be treated as such and handled in an inert atmosphere.[76,77] Recently, there have been a number of advances in metal β-diketonates chemistry. The first is the use of donor-acceptor bonding between neutral ligands and the Lewis-acid metal acceptor. As stated previously, by increasing the coordination number of the metal atoms, intermolecular interactions are reduced, giving rise to increased volatility. For example, $Ba(thd)_2$ is an oligomer in the gas phase. By the use of complexes such as $Ba(thd)_2$•nD where D is ammonia,[78] methanol,[79] and ethers,[80] the molecular complexity can be reduced to two. These compounds offer an advantage over the homoleptic β-diketonates compounds only if sublimation occurs before fission of the donor-acceptor bond. The second advance has been the synthesis of the first metal β–diketonates that is a liquid, i.e., bis-2,2dimethyl-8-methoxyoctane-3,5-dionato barium (11).[69] This novel ligand is basically the thd-ligand with an n-propyl methoxide group attached at the 5th carbon. Since the coordination number of the barium atom is six, a high volatility is achieved (100 Torr at 85°C).

To modulate volatility, decomposition temperature and/or reactivity of precursors, it may be desirable to use a combination of ligands. For example, alkyllead alkoxides having the formula $Pb(C_2H_5)_3(OR)$ where R

= $C(CH_3)_3$ or $CH_2C(CH_3)_3$,[81] have been used as the Pb source in CVD of PZT. There are several advantages of these compounds over the homoleptic, highly toxic and air-stable parent compound, $Pb(C_2H_5)_4$. First, the alkoxide group renders the molecule sensitive to atmospheric moisture and this would make it nonvolatile if a leak was to occur. Secondly, its vapor pressure is reduced to the vapor pressures of Zr and Ti alkoxide sources. Another mixed ligand compound used for CVD of $BaTiO_3$ is alkoxy β-diketonates such as titanium bis(isopropoxide)bis(2,2,6-tertramethyl-3,5-heptanedionate), $Ti(^iOC_3H_7)_2(O_2C_{11}H_{19})_2$.[82] This compound was synthesized in order to avoid the gas phase reaction between $Ti(^iOC_3H_7)_4$ and $Ba(thd)_2$, which gave rise to nonvolatile, barium alkoxy β-diketonate.

From the brief discussions above, it is obvious that, in order to obtain reproducible, high permittivity perovskite dielectric thin films at low temperatures, precursors must be handled under moisture free conditions. Some times they may require modifications or development, especially for the alkaline earths, lanthanides, and heavy metals (Pb and Bi). One must synthesize and characterize volatile precursors that leave minimal carbon residue in thin-film products. It is desirable to use component precursors with similar volatilities, and those that have adequate mass transport characteristics at the lowest possible temperature.

3.3. Effect of Crystal Chemistry and Reaction Kinetics on $Pb(Zr,Ti)O_3$ Phase Evolution in an RF Reactor

Metal organic CVD of perovskite $BaTiO_3$,[82] $SrTiO_3$,[83,84] $PbTiO_3$,[85-88] $(Ba,Sr)TiO_3$,[89] $Pb(Sc_xTa_{1-x})O_3$,[73,74] $Pb(Zr_xTi_{1-x})O_3$,[60,81,90] and $Pb,La(Zr,Ti)O_3$[91] have been reported using various precursor systems. Although such films have good crystallinity, acceptable electrical properties are obtained at temperatures between 600°-700°C. For development of ULSI circuits, low deposition temperatures and conformal step coverage are indispensable requirements. Although the latter is achievable by thermal MOCVD, the former requirement must be met by non-thermal modes of enhancement, e.g., by plasma or photons. The advantages of low temperature PE-MOCVD are: a) previously deposited layers, interfaces or doping profiles are not perturbed, b) solid-state diffusion and defect formation is minimized, c) compatibility with low-melting-point substrates/films is maintained, and d) complex and metastable compounds may be synthesized. Although the chemistry and physics of a plasma glow discharge are complex, the plasma performs a couple of key functions: (i) reactive chemical species generated by electron-impact overcome kinetic (activation) barriers for a particular reaction, and (ii) activated gas atoms and molecules, energetic neutrals and ions, metastable species, electrons, and ions bombard and stimulate surface processes. These effects are equivalent to raising the temepratute of the reaction without raising the temperature of the substrate substantially. The combination of physical processes with chemical reactions results in material properties, and etching profiles and rates unachievable by either process acting alone.

A preliminary study on a traditional radio frequency (rf) PE-MOCVD process was undertaken to determine whether it is possible to produce PZT thin films (on platinum passivated silicon substrates), which are phase-pure perovskite, below 500°C, in the as-deposited condition.[60, 92] The source materials were selected on the basis of their volatility at 25°C in order to avoid heating the bubblers and lines. The tertiary butoxides of Ti and Zr were synthesized[93] and TEL was purified. The purity of all precursors was verified by NMR. The vapor pressure of the two tert-butoxides at room temperature and TEL at 0°C were similar, at about 0.06 Torr. The reactor was of the cold-wall, horizontal-tube variety and relied on radiant heating of a graphite susceptor directly behind the substrate platform. This graphite disc also served as the driven electrode for rf (13.56 MHz) plasma generation. Potentially hazardous exhaust gases were neutralized by oxidation in the thermal decomposition unit. The introduction of the oxidant (99.997% O_2) and carrier gases (99.999% Ar) was regulated by mass flow controllers.

The typical operating conditions for rf PE-MOCVD experiments are outlined in Table 4. A series of experiments were conducted for three different pressures, i.e., 1000 mTorr, 310 mTorr and 160 mTorr. At a given temperature, the perovskite content of the films was dependent on deposition rate, plasma power, and reactor pressure. As expected, a decrease in pressure was found to suppress the deposition rate due to lower reactant concentrations, which is controlled through the mass-flow controllers and hydrostatic bubbler pressure. Interestingly, the percentage of the perovskite phase was found to increase correspondingly. The thermal CVD (without plasma) gave significantly lower deposition rates. The plasma-enhanced deposition times ranged from 50 to 140 minutes which, depending on the operating conditions, yielded films anywhere from 0.5 to 12.5 μm thick. The phase content of the films was determined from x-ray diffraction peak intensities. The composition of the films was determined by electron probe microanalysis, since they were sufficiently thick. It was found that films existing initially as a pyrochlore phase could be converted completely to perovskite below 500°C for lead-rich compositions. One example concerned a film with the nominal composition $Pb_{1.03}Zr_{0.50}Ti_{0.50})0_3$. After annealing for 24 hours at 480°C in air, the pyrochlore converted completely to the perovskite phase with a (100) orientation.

An important advantage for using PE-MOCVD is illustrated in figure 13. There is a direct, one-to-one correspondence between gas and solid compositions. This is clearly shown for the compositions ranging from 10 to 42% zirconium and from about 33 to 75% lead. In contrast, the results from a single thermal CVD experiment at 480°C shows a large deviation from this linear relationship. Evidently the deposition efficiency of each metal during a thermal process is not the same where Zr deposits most efficiently and Pb deposits least efficiently. Since the molecular distributions in the gas phase are usually very sensitive to temperature, the introduction of an rf-plasma apparently overcomes this undesirable thermal effect.

TABLE 4. General process parameters for PE-MOCVD of PZT

Parameters	Conditions
Substrate temperatures	425-495°C
Pressures	160, 310, 1000 mTorr
Deposition times	50-140 min
RF Power (forward)	20 or 30 W
(reflected)	1 W
Gas flow rates: Ti carrier	4-15 sccm
Zr carrier	1-10 sccm
Pb carrier	2-10 sccm
O_2	10-300 sccm
Source temperatures: Ti	24 °C
Zr	24 °C
Pb	0 °C
O_2	24 °C

The generation of reactive chemical species, alluded to above, translates to an increase in the deposition rate with plasma power. For example, the deposition rates for thermal and plasma enhanced (20W) CVD, at 450°C, 100 mTorr, and 22 scfm total carrier flow rate, were 0.34 and 12.5 μm/hr, respectively. If the deposition rates were too high, the structural reorganization at solid surface were unable to keep pace with the rate at which film species arrived. This led to the formation of high entropy, metastable pyrochlore phase. For example, with a plasma power of 20W, the deposition rates increased with increasing pressure as was observed in proceeding from 160 to 300 to 1000 mTorr. Deposition rates on the order of 0.5μm/hr at 160 mTorr were lower than rates typically found at 300 mTorr (1.0-2.0 μm/hr) and 1000 mTorr (8.3-12.5 μm/hr). At the lowest pressure (160 mTorr), a significant fraction of perovskite phase was observed, but only the pyrochlore phase were observed at the higher pressures. However, at 300 mTorr, a higher plasma power of 30 W served to stimulate 22 vol % perovskite formation in the as-deposited condition. Therefore, the perovskite formation could be initiated at high deposition rates by increasing the plasma power, up to some limit. It was observed that plasma etching, due to physical processes, occurred on PZT thin-film surfaces at plasma powers greater than 40 to 45W.

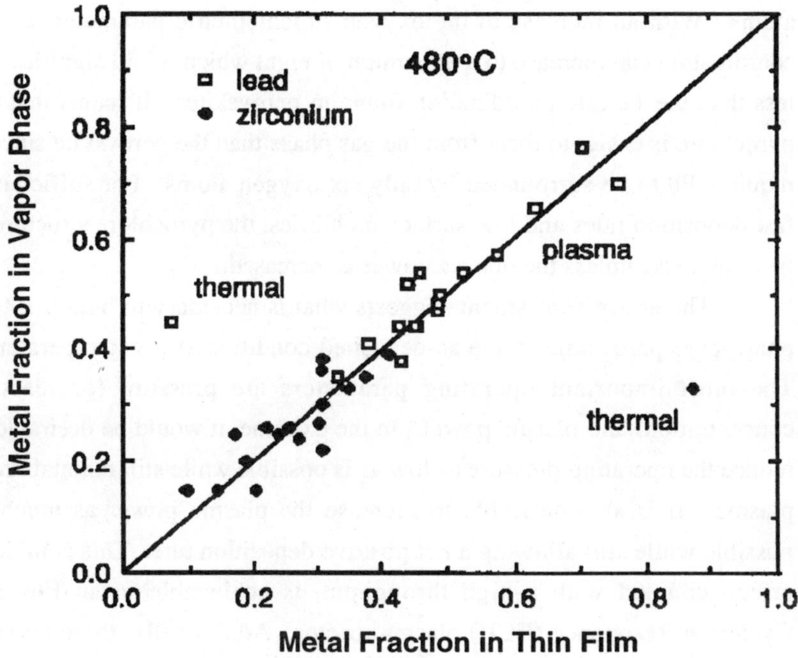

Figure 13. The relationship between metal fraction in the vapor phase (determined from gravimetric losses of the precursor bubblers) and metal fraction in the deposit for thermal and plasma-enhanced CVD.

To understand the above results, it is worthwhile reviewing the subtle differences between the crystal chemistries of pyrochlore and perovskite structure.[60] The perovskite structure has a relatively dense packing of ions. Lead is coordinated by twelve oxygen atoms, whereas the Zr and Ti ions are coordinated by six oxygens. These two polyhedra share trigonal faces such that each Pb ion is surrounded by eight nearest neighboring cations (Zr and Ti), and similarly Zr and Ti are surrounded by eight Pb ions. The general formula for most Pb-based pyrochlore structure is $Pb_{2-\delta}(Zr, Ti)\ O_{7-x}$ where x is very near unity.[94] Like the perovskites, both Ti and Zr are 6-fold coordinated but with significantly distorted octahedra. On the other hand, Pb is coordinated by only about six oxygen atoms. With an increase in the oxygen stoichiometric parameter, x, this coordination can increase to a maximum of eight which is still significantly less than the 12-fold coordination found in perovskites. It seems that the pyrochlore is easier to form from the gas phase than the perovskite since it requires Pb to be surrounded by only six oxygen atoms. For sufficiently fast deposition rates and low surface mobilities, the pyrochlore structure is fixed in place, unless the plasma power is increased.

The above assessment suggests what is necessary to obtain 100%, phase-pure perovskite in the as-deposited condition at low temperatures. The most important operating parameters are pressure (or reactant concentration) and plasma power. In the extreme, it would be desirable to reduce the operating pressure as low as is possible while still maintaining a plasma. It is also desirable to increase the plasma power as much as possible while still allowing a net positive deposition rate. This combined effect, coupled with a high throughput, is achievable in an Electron Cyclotron Resonance (ECR) plasma reactor. Additionally, there remains considerable lattitude in the adjustment of these experimental parameters to realize the goal of depositing phase-pure perovskites at temperatures below 450°C.

3.4. Advantages of ECR Plasma Driven Deposition and Recent Results on Deposition of TiO_2 and $SrTiO_3$ Thin Films

In the field of microelectronics, plasma-enhanced chemical vapor deposition (PECVD) has become a very diverse, commercial processing technique.[50, 95] Deposition and etching of thin films over large wafers and extremely small selective areas has made it production worthy. It must be noted that the merit of a deposition or etching process is measured by controllability of the process over selective regions for the isotropy or anisotropy. For instance, conditions that are extremely important in the case of deposition are to keep the ion energies low (so that damage to the deposited films are minimized) and electron energies high, along with a high ion flux. This can be achieved by either choosing proper gases that are fed into the plasma or by changing the pressure. The gas type can be controlled independently of the plasma source, whereas the latter has to be chosen with caution, because most plasma sources are ineffective when operating at lower vacuum pressures (typically not lower than 100 mTorr). The ECR plasma, with its ability to operate over a wide range of pressures (as low as 1 to 10 mTorr), is an obvious choice.

The ECR plasma process involves the coupling of microwave energy (typically of frequency 2.45 GHz) to the natural resonant frequency of the electron gas under a static magnetic field (875 Gauss for a frequency of 2.45 GHz). At resonance, the magnetic-field-induced confinement of energized electrons allows efficient energy transfer to the neutrals and ions. Due to this inherent confinement process, the ion flux density that can be generated in an ECR plasma is very high ($5x10^{11}$ to 10^{13} ions/cm^3). Traditional, non-magnetic plasma sources cannot generate a high density plasma at low pressures because of the lack of an effective mechanism to couple the energy to the ions and neutrals. An additional method of ion flux density control in an ECR source is through the intensity of the magnetic field; a feature absent in traditional sources.

In traditional sources the ion energies can be changed by changing the power and bias potential, but the flux densities are insufficient at low pressures (at about 100 to 1000 times less than the ECR source). Moreover, precise ion energy distributions are possible in an ECR, without any energy tails. Typically, at low pressures, an RF source generates ion energies of several hundreds of eV while an ECR source generates ion energies of 10 to 20 eV. This makes more amenable the control of chemical reactions on substrate surfaces. An added advantage of an ECR plasma reactor is its flexibility to use an independent RF source to bias the wafer. This provides an independent control to change the energy of the ions. In summary, an ECR source, with its ability to generate high ion flux densities at low pressures, and with good control over the species energy distributions, is a versatile technique for both etching and deposition processes of thin films. Consequently, control over the microstructure and microchemistry of the thin films may be feasible, a factor important for tailoring electrical properties.

Figure 14 is an illustration of a Plasma Quest unit, dedicated to high permittivity dielectric thin-film research, housed at Arizona State University. It has dual ECR and RF plasma enhanced chemical vapor deposition chambers, each capable of handling 8-inch wafers for deposition and reactive ion etching. The system is fully automated for R&D and pilot line use. The two chambers are connected through a load lock and the whole system is interfaced to a 386 PC on an electronics rack. Around the ECR chamber, is a lower magnet that allows collimation of ions, and therefore, a fine control of the ion flux density at the wafer level. Additionally, the wafer chuck can be rf-biased. To control the stoichiometry of the films, the precursors can be delivered individually through heated lines. Initial experiments, using this ECR-PEMOCVD system, have given encouraging results. For example, uniform TiO_2 films (on 6-inch Si wafers) were deposited at 300°C. The deposition parameters are tabulated in Table 5.

Figure 14. Plasma Quest reactor, model 357-D95, housed at ASU.

By far the most impressive data reported, to date, on $SrTiO_3$ and $(Ba,Sr)TiO_3$ perovskite thin films is by a research group at NEC.[96, 97] Their work has focused on the prospects of using ECR-PEMOCVD for the development of gigabit dynamic random access memory (DRAM) storage capacitors. Since the pressure in the reactor is lower than typical pressures in the bubblers, reproducibility of the stoichiometry has been maintained by control of the bubbler pressure, as well as the carrier gas flow. They have been successful in depositing conformal, $SrTiO_3$ films at 450°C on 0.5 μm high, stacked storage electrodes in a 0.125 μm^2 area (thickness: 1000Å on top and 800Å on sidewalls), that provides a cell capacitance of 25 fF. This is a number that exceeds the minimun capacitance requirements of ULSI DRAMs. A quick inspection of their process parameters and microstructure results reveals that conformal step coverage occurs predominantly, not by adatom migration, but by deposition under conditions such that growth rate is limited by surface chemical reactions. Mechanistically, reaction-limited growth conditions are easier to achieve if the sticking coefficients of reactants or reactive intermediate species are low. The arriving species, instead of being thermally accomodated instantaneously on exposed surfaces, proceed through multiple surface collisions towards less accessible sites. In general, low sticking coefficients result in good step coverage.[98-100] The above results indicate that conformal growth of $SrTiO_3$ thin films occurs in the presence of species with low sticking coefficients.

Table 5. Process parameters for ECR-PEMOCVD of TiO_2 on 6-inch Si wafers

Parameters	Conditions
Substrate temperature	300 °C
Pressure	5 mTorr
Deposition time	10 min
Microwave power	1000 W
Gas flow rates: Ti-isopropoxide	3.1 sccm
O_2	10 sccm
Ar	100 sccm
Ti-source temperature	70 °C
Line temperature	90 °C
Upper magnet	185 Amps

4. CONCLUDING REMARKS

In the next few paragraphs, some thoughts regarding sol-gel processing of Pb-based perovskite thin films are assessed. Additionally, some fundamental issues that one needs to address before a reliable CVD (thermal or plasma-enhanced) technology can be transferred from the laboratory to commercial scale production are highlighted.

An issue that is frequently debated is the viability of the sol-gel process for commercial scale production of planar, low-density microelectronic devices. The sol-gel process has a number of advantages over vacuum deposition methods, including low capital investment, excellent dopant incorporation and multicomponent composition control, and quick turn-around time for fabricating thin-film devices. It appears to be a good developmental method for novel systems that require speedy compositional and microstructural optimization. For commercial scale reliability and reproducibility of a specific thin-film composition, however, one must characterize systematically all the steps in the processing cycle. This requires the use of numerous microchemical, microstructural, analytical, and spectroscopic techniques, which can be extremely tedious. Only upon completion can one gain the understanding, and possibly the control, of the prehydrolyzed solution->sol->gel->perovskite ceramic thin-film conversion. The tasks are further compounded in the Pb-based perovskite system due to the high reactivity of Pb with various electrode material systems. Low temperature processing, therefore, is vital if Pb-based systems are to be manufactured by the sol-gel process.

Currently, the heat treatment temperatures of commercially useful, stoichiometric Pb,Sr,Ba-based perovskite compositions are relatively high. A recent report suggests the potential for significant reductions in the processing temperatures via the alkoxide-solution grown single-crystal route.[101] Such a suggestion must be evaluated with caution. It is important to bear in mind that prehydrolyzed alkoxide precursor solutions often convert incongruently into single crystals. These crystals, which have an improper stoichiometry, when redissolved in a suitable solvent, spin-cast, and heat treated, invariably form mixed phases. In the PZT system, it is difficult to grow crystals (with cation stoichiometries that are useful) from prehydrolyzed solutions used to fabricate sol-gel thin films. This is due to a) mixture of isomers, b) polydispersity, and c) the possibility of variable cation stiochiometry.[23a] Thus, growth of single crystals from solutions, therefore, remains to be a field of fundamental scientific interest unless electrical data and analyses of films derived from such crystals prove otherwise.

It is well known that microstructural parameters influence the properties of dielectrics. The questions that remain open in sol-gel processing of crystalline thin films are: a) can grain size be controlled? b) is it possible to obtain pore-free microstructures? c) can carbonaceous residues be completely eleminated by special techniques acceptable in IC processing laboratories? and d) how will the particle problems in processing large area wafers be resolved?

As for CVD of multicomponent perovskite materials, there are numerous issues that need urgent attention. These include precursor development, understanding of the thermodynamics and kinetics of a specific system, mechanisms of microstructural evolution, etchability and integratability of specific electrode/perovskite oxide combinations, and scaleability of a given process that yield high density devices with acceptable properties. Some thoughts on these issues are expressed below.

For precursors, it is desirable to have vapor pressures in excess of 50 mTorr at room temperature. As for currently available metal organic precursors, the rate of decomposition is significant at temperatures required for adequate mass transport. Therefoie, precise control of stoichiometry may be difficult in the event that transport properties vary (due to decomposition) during the course of a CVD run. Recent advances in using a liquid delivery system, where individual precursors (with[80,102] or without adducts[102]) are diluted in a common solvent such as tetrahydrofuran, could be an improved alternative to a system using individual solid components.

A measure of merit of a CVD process is the controllability and/or predictability of its conformality over device topography. Simulation of feature evolution as a function of deposition parameters would be extremely useful in design of experiments and optimization of thin-film properties.

The accuracy of simulations[99] would, of course, be dictated by our knowledge of fundamental information such as transport behavior, flux distribution of the reacting species, the intrinsic heterogeneous kinetics of the reactions, and surface mobilities. These types of information are currently unavailable for multicomponent species that comprise the perovskite phase. Thus, a synergistic approach to the problem, coupled with data from various diagnostic[47] techniques, should lead to some success in the future.

Another area that needs understanding is microstructural evolution and it's relationship with precursor type, plasma chemistry, and process parameters. The importance of low-energy ions/surface interactions, in a CVD reactor, at various stages of perovskite thin-film growth has not been recognized. Additionally, the interrelationships between atomic-level processes, effect of stress, and generation of point and extended defects on growth mechanisms, using experimental and computer-simulated results, must be determined. This fundamental approach will lead eventually to a device-worthy CVD process that can be readily modified for fully automated commercial scale production.

As reiterated by Fledderman,[103] the etch rate, anisotropy, and selectivity are critical issues that must be addressed in patterning (by plasma-assisted reactive ion etching with gases that are environmentally safe) perovskite thin films for device fabrication. Lower etching temperatures and control of damage should facilitate the fabrication of reliable devices. Recent results of ECR etching[104] indicate that $Pb(Zr,Ti)O_3$, $(Ba,Sr)TiO_3$, and Pt can be etched at 275-1075, 105-473, and 205-800 Å/min, respectively. Additionally, the etching selectivity of PZT to Pt can be controlled in the range of 1:1 to 5:1.

The above demanding tasks will be time consuming and difficult, but it is important for interdisciplinary research to be directed to this area so that perovskite oxide materials may be designed with properties and performance prescribed for the purpose at hand. Based on the current world-wide research efforts focused in the area of integrated perovskite dielectrics, including the layered perovskites,[1,2] it is felt that novel and useful fundamental concepts for tailoring thin-film ferroelectric and paraelectric material properties will be introduced. This surely will impact the electronic and photonic industries significantly in the 21st century.

ACKNOWLEDGEMENTS

I would like to thank Advanced Projects Research Agency (MTO and ESTO), National Science Foundation, California Institute of Technology (President's Fund), Physics Division of Army Research Office, McDonnell Douglas Electronic Systems Company, Motorola, Symetrix, and Matsushita Electric in providing the support for dielectric thin-film research at Arizona State University. I appreciate the significant contributions of my colleagues, L. Eyring, J.B. Wagner, Jr., W.T. Petuskey, R.O. Grondin, D.K. Schroder, I. Grozdanov, D.L. Jacobson, and S. Krause, in fulfilling the synergistic nature of this program. I acknowledge all my graduate students, in particular, P. Alluri, C.K. Barlingay, L. Clark, P.R. Coffman, and T. Gloerstad for their unique contributions to this field. My thanks are extended to my friends, L.E. Cross, J.F. Scott, G.W. Taylor, and R. Zuleeg for their support and guidance. Finally, I am indebted to my wife, Asha S. Dey, for her patience and moral support throughout the years.

REFERENCES

1. M.E. LINES and A.M. GLASS, Principles and Applications of Ferroelectric and Related Materials (Clarendon Press, Oxford, 1977).
2. B. JAFFE, W.R. COOK and H. JAFFE, Piezoelectric Ceramics (Academic Press, New York, 1970).
3. J.C. BURFOOT and G.W. TAYLOR, Polar Dielectrics and Their Applications (University of California, Los Angeles, 1979).
4. Integrated Ferroelectrics, edited by G.W. Taylor, **1-4** (Gordon and Breach, New York, 1992-1994).
5. S.K. DEY, Ferroelectrics: Arthur R. Von Hippel Special Issue, **135**, 117-130 (1992).
6. a) S.B. KRUPANIDHI, H. HU and V. KUMAR, J. Appl. Phys., **71**, 376 (1992); R.N. CASTELLANO and L.G. Feinstein, ibid, **50**, 4406 (1979); D. XIAO, Z. XIAO, J. ZHU, Y. LI and H. GUO, Ferroelectrics, **108**, 59 (1990); A.I. KINGON, T.M. GRAETINGER, S.H. ROU, H.N. AL-SHAREEF, K.D. GIFFORD, P.D. HREN and O. AUCIELLO, Proc. 5th US-Japan Seminar on Dielectric and Piezoelectric Ceramics (Kyoto, Japan,

1990) 324; A.R. KRAUSS and O. AUCIELLO, U.S. Patent no. 4,923,585 (1990).
b) S.B. KRUPANIDHI, N. MAFFEI, M. SAYER and K. EL-ASSAL, J. Appl. Phys., **54**, 6601 (1983); M. OKUYAMA and Y. HAMAKAWA, Ferroelectrics, **63**, 243 (1985); K. SRINIVAS and M. SAYER, J. Appl. Phys., **64**, 1484 (1988); H. ADACHI, T. KAWAGUCHI, K. SETSUNE, K. OHJI and K. WASA, Appl. Phys. Lett., **42**, 867 (1983).
c) A. Okada, J. Appl. Phys., **49**, 4494 (1978).
d) M. OKUYAMA, Y. TOGANI and Y. HAMAKAWA, Appl. Surf. Sci., **33/34**, 625 (1988); Y. MASUDA, A. BABA, H. MASOMOTO, T. GOTO, M. MINAKATA and T. HIRAI, Proc. 5th US-Japan Seminar on Dielectric and Piezoelectric Ceramics (Kyoto, Japan, 1990) 337.

7. H. TABATA, T. KAWAI, S. KAWAI, O. MURATA, J. FUJIOKA and S.I. MINAKATA, Appl. Phys. Lett., **59**, 2354 (1991); S. OTSUBO, T. MAEDA, T. MINAMIKAWA, Y. YONEZAWA, A. MORIMOTO and T. SHIMUZU, Jpn. J. Appl. Phys., **29**, L133 (1990); K.L. SAENGER, R.A. ROY, K.F. ETZOLD and J.J. CUOMO, Mater. Res. Symp. Proc., **200**, 115 (1990); O. AUCIELLO, A.R. KRAUSS, J. SANTIAGO-AVILES, A.F. SCHREINER and D.M. GRUEN, Appl. Phys. Lett., **52**, 239 (1988); D. ROY, S.B. KRUPANIDHI and J.P. DOUGHERTY, J. Appl. Phys., **69**, 7930 (1991); J.S. HORWITZ, K.S. GRABOWSKI, D.B. CHRIZEY and R.E. LEUCHTNER, Appl. Phys. Lett., **59**, 1565 (1991).

8. a) P-Y. LESAICHERRE, H. YAMAGUCHI, T. SAKUMA, Y. MIYASAKA, M. YOSHIDA, A. ISHITANI, MRS Proceedings, **310**, 487 (1993).
b) M. OKADA, S. TAKAI, M. AMEEMYA and K. TOMINAGA, Jpn. J. Appl. Phys., **28**, 1030 (1989); P.J. VAN VELDHOVEN and G.J.M. DORMANS, MRS. Proceedings, **310**, 223 (1993).

9. a) A. MANSINGH and S.B. KRUPANIDHI, J. Appl. Phys., **51**, 5408 (1980).
b) M. OIKAWA and K. TODA, Appl. Phys. Lett., **29**, 491 (1976).
c) J.R. SLACK and J.C. BURFOOT, Thin Solid Films, **6**, 233 (1970); A. MANSINGH and S.B. KRUPANIDHI, Thin Solid Films, **80**, 359 (1981).

10. L.D. McMILLAN, M. HUFFMAN, T.L. ROBERTS, M.C. SCOTT, C.A.P. ARAUJO, Integrated Ferroelectrics, **4**, 319 (1994).

11. R.W. VEST, Ferroelectrics, **102**, 53 (1990); R.W. VEST and J. XU, ibid, **93**, 21 (1984); G.H. HAERTLING, J. Vac. Sci. and Tech., **A9**, 414 (1991); C.T. LIN, L. LI, J.S. WEBB, R. LIPELES and S. LEUNG, ibid, **3**, 333 (1993).

12. S.R. GURKOVICH and J.R. BLUM, Ultrastructure Processing of Ceramics, Glasses and Composites, edited by L.L. Hench and D.R.

Ulrich 152 (John Wiley and Sons, 1984); Y. XU and J.D. MACKENZIE, Integrated Ferroelectrics, **1**, 17 (1992); P. PHULE, ibid, **4**, 315 (1994); G. TEOWEE, J.M. BOULTON, C.D. BAERTLEIN, R.K. WADE and D.R. UHLMANN, ibid, **4**, 231 (1994); R.W. SCHWARTZ, D. DIMOS, S.J. LOCKWOOD and V.M. TORRES, ibid, **4**, 165 (1994); M.I. YANOVSKAYA, L.I. SOLOVEVA, E.P. KOVSMAN, I.E. OBVINTSEVA, K.A. VOROTILOV and N.V. TUROVA, ibid, **4**, 315 (1994); J. FUKUSHIMA, K. KODAIRA, A. TSUNASHIMA and T. MATSUSHITA, Yogyo Kyoaishi, **83**, 204 (1975); G. YI, Z. WU and M. SAYER, J. Appl. Phys., **64**, 2717 (1988); K.D. BUDD and D.A. PAYNE, Inst. Phys. Conf. Ser., U.K., **103**[1], 13 (1989) M. KLEE, R. EUSEMANN, R. WASER and W. BRAND, ibid, **72**, 1566 (1992); N. TOHGE, S. TAKAHASHI and T. MINAMI, J. Am. Ceram. Soc., **74**, 67 (1991); U. SELVARAJ, K. BROOKS, A.V. PRASADARAO, S. ROY and L.E. CROSS, ibid, **76**, 1441 (1993); C.C. HSUEH and M.L. MECARTNEY, J. Mater. Res., **6**, 2208 (1991);

13. T. Ogawa, Integrated Ferroelectrics, **1**, 99 (1992); R. Ramesh, W.K. Chan, B. Wilkens, T. Sands, J.M. Tarascon, V.G. Keramidas and J.T. Evans, ibid, **1**, 1 (1992).

14. a) S.K. DEY and R. ZULEEG, Ferroelectrics, **108**, 37 (1990).
b) T.K. GLOERSTAD M.S. Thesis, Arizona State University (1993); T.K. GLOERSTAD, L.T. CLARK, R.O. GRONDIN, and S.K. DEY, IEEE Trans. on Instru. and Meas., **41(5)**, 716 (1992).
c) L.T. CLARK, Ph.D. Dissertation, Arizona State University (1992); L.T. CLARK, S.K. DEY and R.O. GRONDIN, Ferroelectrics, 116, 205 (1991); L.T. CLARK, T.J. GLOERSTAD, R.O. GRONDIN and S.K. DEY, Integrated Ferroelectrics, 3, 309 (1993).
d) P. ALLURI, M.S. Thesis, Arizona State University (Dec., 1993); P. ALLURI and S.K. DEY Thin Solid Films, accepted publication; S.K. DEY and P. ALLURI, Proc.6th U.S.-Japan Seminar on Dielectrics and Piezoelectric Ceramics (Maui, USA, 1993).
e) J.J. LEE, P. ALLURI and S.K. DEY, Appl. Phys. Lett., **65(16)**, 2027 (1994); J.J. LEE, Ph.D. Dissertation, Arizona State University (May 1994); S.K. DEY and J.J. LEE, IEEE Trans. on Electron Devices, 39, 1607 (1992).
f) S.K. DEY, Ferroelectrics, **135**, 117 (1992); C.K. BARLINGAY, Ph.D. dissertation, Arizona State University (Dec., 1994).
g) S.K. DEY, P. ALLURI, J.J. LEE and R. ZULEEG, Integrated Ferroelectrics, in press

15. S. SAKKA, Treatise on Materials Science and Technology, **22**, (Academic Press, New York, 1982).

16. H. DISLICH and P. HINZ, J. Non Cryst. Sol., **48**, 11-16 (1982).

17. D.P. PARTLOW and B.E. YOLDAS, J. Non Cryst. Sol., **46**, 153-161 (1981).
18. C.J. BRINKER and G.W. SCHERER, The Physics and Chemistry of Sol-Gel Processing, (Academic Press, New York, 1990).
19. Better Ceramics Through Chemistry IV, Materials Research Society Symp. Proc., edited by B.J. Zelinski, C.J. Brinker, D.E. Clark, and D.R. Ulrich, **180** (1990).
20. S.K. DEY and R. ZULEEG, Ferroelectrics, **112**, 309-319 (1990).
21. C. K. BARLINGAY and S.K. DEY, Appl. Physics Lett., **61** [11] 1278-1280 (1992).
22. D.F. SHRIVER and M.A. DREZDZON, The Manipulation of Air Sensitive Compounds, (Wiley-Interscience, New York, 1986).
23. a) P.R. COFFMAN and S.K. DEY, J. Sol-Gel Science and Technology., Part 1: **1**, 251-265 (1994); b) P.R. COFFMAN, C.K. BARLINGAY, A. GUPTA and S.K. DEY, ibid, Part II: submitted Nov. (1994); c) C.K. BARLINGAY, P.R. COFFMAN, and S.K. DEY, ibid, Part III: submitted Nov. (1994).
24. J.B. BLUM and S.R. GURKOVICH, J. Mater Sci., **20,** 4479 (1985).
25. Z.C. KANG, A. GUPTA, M.J. McKELVY, L. EYRING and S.K. DEY, MRS Proceedings, **230**, 301 (1992); **183**, 291 (1990).
26. C.K. BARLINGAY and S.K. DEY, Rapid Thermal and Integrated Processing : MRS Symp. Proc., (Mat. Res. Soc., Anaheim, 1991) **224**, 311-316.
27. D.C. BRADLEY, R.K. MULTANI and W. WARDLAW, J. Chem. Soc., **126**, 1958 (1958).
28. H.N. AL-SHAREEF, K.D. GIFFORD, P.D. HREN, S.H. ROU, O. AUCIELLO, A.I. KINGON, in Proc. 4th International Symposium on Integrated Ferroelectrics, edited by C.A. Araujo and G.W. Taylor Gordon and Breach, New York, 1992) 181.
29. A. TANAKA, S.C. CHOI, M. MIYAYAMA, A.S. BHALLA, and H. YANAGIDA, Integrated Ferroelectgrics, **3**, 51 (1993).
30 G.A.C.M. SPIERINGS, J.B.A. VAN ZON, P.K. LARSEN and M. KLEE, Integrated Ferroelectrics, **3**, 283 (1993).
31. R. BRUCHHAUS, D. PITZER, O EIBL, U. SCHEITHAUER and W. HOESLER, in Ferroelectric Thin Films II, : MRS Symp. Proc., **243,** edited by E.R. Myers, A.I. Kingon and B.A. Tuttle (Materials Rsearch Society, Pittsburgh, 1992) 123.
32. M. KLEE, A. DE VEIRMAN, D.J. TAYLOR and P.K. LARSEN, Integrated Ferroelectrics, **4**, 197-206 (1994).
33. T. IKEDA, Y. TANAKA, T. AYAKAWA, and N.H. NAOKE, Jpn. J. Appl. Phys., **3** [10], 581-587 (1964).
34. M.A. AISA, M.F. ABADIR and A.M. GADALLA, Trans. Brit. Cer. Soc., **79,** 100-104 (1980).
35. R.B. ATKIN and R.B. FULRATH, J. Am. Ceram. Soc., **54** [5], 265-270 (1971).

36. A.R. MODAK, Effect of Chemical Additives on Structure and Electrical Properties of Sol Gel Derived Ferroelectric $PbZrO_3$-$PbTiO_3$ Thin Films, M.S. Thesis, Arizona State University (1991).
37. A.R. MODAK and S.K. DEY , Integrated Ferroelectrics, **5**, in press (1994).
38. H.M. CHAN, M.P. HARMER and D.M. SMYTH, J. Am Ceram. Soc., **69** [6], 507-510 (1986).
39. K.H. HÄRDTL and D. HENNINGS, J. Am. Ceram. Soc., **55** [11], (1972).
40. R. WASER, T. BAIATU and K.H. HÄRDTL, J. Am. Ceram. Soc., **73** [6], 1645 (1990).
41. D.F. RYDER and N.K. RAMAN, J. Electronic Mat., **21**, 971-5 (1992).
42. A.H. CARIM, B.A. TUTTLE, A. DOUGHTY and S.L. MARTINEZ, J. Am. Ceram. Soc., **74** [6], 1455 (1991).
43. M. RAYMOND, J. CHEN and D.M. SMYTH, Integrated Ferroelectrics, **5** [1], 73-78 (1994).
44. G.V. LEWIS and C.R.A. CATLOW, J. Phys. Chem. Solids, **47** [1], 89-97 (1986).
45. D.L. FLAMM and G.K. HERB, in Plasma Etching, An Introduction, edited by D.M. Manos and D.L. Flamm (Academic Press, New York, 1989) **1**, 1-89.
46. P. SINGER, Semiconductor International, **52-57**, July (1992); P. SINGER, ibid, 52-56, November (1994); P. BURGGRAAF, ibid, 191-194, July (1994); P. BURGGRAAF, ibid, 46-50, August (1994).
47. K. JENSEN and W. KERN, in Thin Film Processes II, edited by J.L. Vossen and W. Kern (Academic Press, New York, 1991) **III-1**, 283-367.
48. J.O. CARLSSON, Acta Chemica Scandinavica, **45**, 864-869 (1991).
49. V.C. McCRARY and V.M. DONNELLY, "Photo-Assisted Vapor Deposition," in Chemical Vapor Deposition, edited by M.L. Hitchman and K.F. Jensen (Academic Press, London, 1993), 8, 437-513.
50. D.W. HESS and D.B. GRAVES, in Advances in Chemistry Series 221, edited by D.W. Hess and K.F. Jensen (American Chemical Society, Washington DC, 1989) **8**, 377-440.
51. C. DESHPANDEY and R.F. BUNSHAH, J. Mater. Educ., **10**, 439-482 (1988).
52. Y. TOYOSHIMA, K. KUMATA, U. ITOH, K. ARAI, A. MATSUDA, N. WASHIDA, G. INOUE and K. KATSUMI, Appl. Phys. Lett., **46**, 584 (1985).
53. J.E. GREEN, in Multicomponent and Multilayered Thin Films for Advanced Microtechnologies: Techniques, Fundamentals and Devices, edited by O. Auciello and J. Engemann (Kluwer Academic Publishers, Boston, 1993) 39-85.

54. G. ERICKSSON, Chemica Scripta, **8**, 100 (1975).
55. T.M. BESMANN, "SOLGASMIX-PV, A Computer Program to Calculate Equilibrium Relationships in Complex Chemical Systems", Report ORNL/TM-5775 (Oak Ridge National Laboratories, 1977).
56. R.A. ROBIE, B.S. HEMINGWAY and J.R. FISHER, Thermodynamic Properties of Minerals and Related Substances at 298.15 K and 1 Bar Pressure and at Higher Temperatures (U.S. Geological Survey Bulletin, 1979) 1452.
57. K.H. HÄRDTL and H. RAU, Solid State Comm., **7**, 4 (1969).
58. A.E. McHALE and R.S. ROTH, J. Am. Ceram. Soc., **69**, 827 (1986).
59. B.A. TUTTLE, R.W. SCHWARTZ, D.H. DOUGHTY and J.A. VOIGHT, Mat. Res. Soc. Symp. Proc., **200**, 159-166 (1990).
60. W.T. PETUSKEY, D.A. RICHARDSON and S.K. DEY, Integrated Ferroelectrics, **2**, 269-295 (1992).
61. K. IIJIMA, Y. TOMITA, R. TAKAYAMA and I. UEDA, J. Appl Phys., **60**, 361 (1986).
62. H. SHAPIRO and F.W. FREY, The Organic Compounds of Lead (Wiley-Interscience, New York, 1968).
63. G.B. DEACON, P.K. MACKINNON, R.S. DICKSON, G.N. PAIN and B.O. WEST, Appl. Organomet. Chem., **4**, 439-449 (1990).
64. A. WEBER, H. SUHR, H. SCHUMANN and R.-D. KOHN, Appl. Phys A., **51**, 520-525 (1990).
65. R. PAPIERNIK, L.G. HUBERT-PFALZGRAF and M.-C. MASSIANI, Inorg. Chim. Acta., **165**, 1-2 (1989).
66. R. PAPIERNIK, L.G. HUBERT-PFALZGRAF and M.-C. MASSIANI, Polyhedron, **10**, [14] 1657-62 (1991).
67. S.C. GOEL, M.Y. CHANG and W.E. BUHRO, Inorg. Chem., **29**, 4640-46 (1990).
68. W.S. REES, Jr., and D.A. MORENO, J. Chem. Soc., Chem. Comm., 1759 (1991).
69. W.S. REES, Jr., C.R. CABALLERO and W. HESSE, Agnew. Chem. Int Ed. Engl., **31** [6], 375 (1992).
70. D.C. BRADLEY, H. CHUDZYNSKA, M.B. HURSTHOUSE and M. MOTEVALLI, Polyhedron, **10** [10] 1049 (1991).
71. R.C. MEHROTRA, R. BOHRA and D.P. GAUR, Metal β-diketonates and Allied Derivatives (Academic Press, New York, 1978).
72. H. HOLZSCHUH, S. OEHR, H. SUHR and A. WEBER, Mod. Phys. Lett., **B2**, 1253 (1988).
73. F.W. AINGER, C.J. BRIERLEY, M.D. HUDSON, C. TRUNDLE and R.W. WHATMORE, Mat. Res. Soc. Symp. Proc., **200**, 37-47 (1990).

74. C.J. BRIERLEY, C. TRUNDEL, L. CONSIDINE, R.W. WHATMORE and F.W. AINGER, Ferroelectrics, **91**, 181-92 (1989).
75. S.B. TURNIPSEED, R.M. BARKLEY and R.E. SIEVERS, Inorg. Chem., **30**, 1164 (1991).
76. T. HASHIMOTO, K. KITAZAWA, Y. SUEMUNE, T. YAMAMOTO and H. KOINUMA, Jap. J. Appl. Phys., **29**, [12], L2215 (1990).
77. T. HASHIMOTO, H. KOINUMA, M. NAKABAYASHI, T. SHIRAISHI, Y. SUEMUNE and T. YAMAMOTO, J. Mater Res., **7**, [6], 1336 (1992).
78. W.S. REES, JR., M.W. CARRIS and W. HESSE, Inorg. Chem., **30**, 4481 (1991).
79. A. GLEIZES, S. SANS-LENAINE, D. MEDUS and R. MORANCHO, C.R. Acad. Sci. Paris, **312**, II, 983 (1991).
80. R. GARDINER, D.W. BROWN, S.P. KIRLIN and A.L. RHEINGOLD, Chem. Mater., **3**, 1053-59 (1991).
81. G.J.M. DORMANS, M. de KEIJSER, P.J. VAN VELDHOVEN, D.M. FRIGO, J.E. HOLEWIJN, G.P.M. VAN MIER and C.J. SMIT, Chem. Mater., **5**, 448 (1993).
82. P.C. VAN BUSKIRK, R. GARDINER, P.S. KIRLIN and S. KRUPANIDHI, J. Vac. Sci. Tech., submitted (1992).
83. W.A. FEIL, B.W. WESSELS, L.M. TONGE, and T.J. MARKS, J. Appl Phys., **67**, 3858 (1990).
84. H. YAMAGUCHI, P.Y. LESAICHERRE, T. SAKUMA, Y. MIYASAKA, A. ISHITANI and M. YOSHIDA, Jpn. J. Appl. Phys., **32**, 4069-4073 (1993).
85. M. KOJIMA, M. OKUYAMA, T. NAGAKAWA and Y. HAMAKAWA, Jpn. J. Appl. Phys., **22**, 14 (1983).
86. B.S. KWAK, E.P. BOYD and A. ERBIL, Appl. Phys. Lett., **53**, 1702 (1988).
87. S.G. YOON and H.G. KIM, J. Electrochem. Soc., **135**, 3137 (1988).
88. S.L. SWARTZ, D.A. SEIFERT, G.T. NOEL and T.R. SHROUT, Ferroelectrics, **93**, 37 (1989).
89. P.C. VAN BUSKIRK, R. GARDINER, P.S. KIRLIN, and S. KRUPANIDHI, in Proc. 8th Intl. Symp. Applications Ferroelectrics, Greenville, edited by G. Haertling (IEEE Piscataway, 1992) 340.
90. T. SHIOSAKI, M. FUJIMOTO, M. SHIMIZU, M. FUKAGAWA, K. NAKAYA, and E. TANIKAWA, Integrated Ferroelectrics, **5**, 39-45 (1994).
91. M. OKADA and K. TOMINAGA, J. Appl. Phys., **71** [4], 1955 (1992).
92. D.A. RICHARDSON, C.J. KUO, S.K. DEY and W.T. PETUSKEY, in Ferroelectric Films, edited by A. Bhalla and K.M.

Nair (Ceramic Transactions, The American Ceramic Society, Westerville, 1992) **52**, 195-202.
93. D.C. BRADLEY, R.C. MEHROTRA and D.P. GAUR, Metal Alkoxides (Academic Press, London, 1978).
94. O. MULLER and R. ROY, Crystal Chemistry of Nonmetallic Materials, vol. 4: Major Ternary Structural Families (Springer, New York, 1974).
95. R.D. COMPTON, in Semiconductor International, July, 60-65 (1992).
96. P-Y. LESAICHERRE, S. YAMAMICHI, H. YAMAGUCHI, K. TAKEMURA, H. WATANABE, K. TOKASHIKI, K. SATOH, T. SAKUMA, M. YOSHIDA, S. OHNISHI, K. NAKAJIMA, K. SHIBAHARA, Y. MIYASAKA, and H. ONO, Proceedings of the IEDM, (1994).
97. M. YOSHIDA, H. YAMAGUCHI, T. SAKUMA, Y. MIYASAKA, P.Y. LESAICHERRE and A. ISHITANI, Symposium on High Permittivity Materials, 184th Electrochemical Society Meeting, New Orleans, Oct. 1993.
98. T.S. CALE and G.B. RAUPP, J. Vac. Sci. Tech., **B8(6)**, 1242 (1990).
99. T.S. CALE, G.B. RAUPP and T.H. GANDY, ibid, **A10(4)**, 1128 (1992).
100. T.E. ZIRKLE, S.R. WILSON, S. SUNDARAM, T.S. CALE and G.B. RAUPP, ibid, **A11(4)**, 905 (1993).
101. C.D. CHANDLER, C. ROGER and M.J. HAMPDEN-SMITH, Chem. Rev., **93**, 1205-1241 (1993).
102. P-Y. LESAICHERRE, in conference report of 41st spring Oyo Butsuri Gakkai, Ferroelectricity Newsletter, **2(3)**, 15 (1994).
103. C. FLEDDERMANN, Integrated Ferroelectrics, **5**, 29-37 (1994).
104. J.E. SPENCER, Plasma Quest, private communication.

PART III

PHYSICAL AND CHEMICAL VAPOR DEPOSITION

CHAPTER 10

SPUTTER-DEPOSITION OF FERROELECTRIC FILMS AND HETEROSTRUCTURES: ANALYSIS OF ORIENTATION - MICROSTRUCTURE - PROPERTY RELATIONSHIPS AND *IN SITU*, REAL-TIME CHARACTERIZATION OF GROWTH PROCESSES

Orlando Auciello,[1,2] Alan R. Krauss,[3] and Ken D. Gifford[2]

1. MCNC, Electronics Technology Division, 3021 Cornwallis Rd., Research Triangle Park, NC 27709-2889

2. N.C. State University, Department of Materials Science and Engineering, Raleigh, NC 27695-7919

3. Argonne National Laboratory, Materials Science and Chemistry Divisions , Argonne, Illinois 60439

1. INTRODUCTION

Intensive research on the synthesis, characterization, and determination of processing-microstructure-property relationships of ferroelectric thin films and their integration into layered heterostructures has been performed during the last five years because of the potential application to the fabrication of electronic memory devices. The high dielectric permittivities of perovskite-type materials can be advantageously used in dynamic random access memories (DRAM),[1] while the large values of switchable remanent polarization of ferroelectric materials are suitable for non-volatile ferroelectric random access memories (FRAM).[2-7] Research is being performed to develop both the scientific basis for the synthesis and characterization of ferroelectric films and layered heterostructures and the applications to devices.[1-7]

The technological uses of oxide thin films are not limited to ferroelectric memories, but extend to a wide range of applications in discrete devices, utilizing the full range of bulk oxide properties, including dielectric, ferroelectric, piezoelectric, electrostrictive, pyroelectric, optical, electro-optic and magnetic properties, as well as electronic conduction, ionic conduction and superconduction. Applications include multilayer capacitors for memories, boundary layer capacitors, varistors, gas sensors, radiation detectors, temperature sensors, transducers, switches, shutters, MHD electrodes, fuel cell electrolytes, concentration cell electrolytes, and electrolytes for high energy density batteries. These ceramics represent an important world market, which has been experiencing steady growth in recent years.[8]

In reviewing the device applications mentioned above, it is worth noting that the use of oxide films integrated with semiconductors has been limited to the simple oxides. The wide range of properties which are available with multicomponent oxides (e.g., ferroelectricity and superconductivity) are not being fully exploited currently in microelectronic devices.

Various vapor phase deposition techniques [plasma and ion beam sputter-deposition (PSD and IBSD respectively), pulsed laser ablation deposition (PLAD), electron beam or oven-induced evaporation for molecular beam epitaxy (MBE), and chemical vapor deposition (CVD)] have been applied to produce ferroelectric films and layered heterostructures (see Refs. 9-10 for recent reviews). However, the wide range of properties which are available via the multicomponent oxides are currently not being exploited in microelectronic devices. The reason is primarily because of the difficulty in reliably producing device-quality films directly on large semiconductor substrates, in a way that is fully compatible with existing semiconductor process technology. As pointed out in recent reviews,[9,10] deposition methods suitable for these applications have not yet been optimized, despite intensive efforts based upon a variety of different deposition technologies. Therefore, the focus of research efforts in multicomponent oxide films must be the optimization of suitable process methods and the investigation of processing-composition–microstructure property relationships. This is the focus of the present review.

The emphasis in this review is on sputter-deposition of ferroelectric thin films as implemented in two main variants, i.e., plasma and ion beam sputter-deposition. However, we first present a brief discussion of alternative vapor-deposition techniques to put plasma and ion beam sputter-deposition in a proper perspective. Second, we discuss the plasma and ion beam sputter-deposition methods, including an analysis of deposition processes and their influence on composition-microstructure-property relationships of ferroelectric films and integration into heterostructures. Finally, we present first results on *in situ*, real-time characterization of growth processes relevant to the synthesis of ferroelectric and ferroelectric/oxide electrode heterostructures.

2. BRIEF REVIEW OF ALTERNATIVE VAPOR-DEPOSITION TECHNIQUES

A brief review of alternative deposition techniques, other than sputter-deposition methods, is presented in this section to provide a context for characterizing the capabilities of the sputtering methods used in manufacturing ferroelectric thin film-based devices and to make comparisons with the techniques of interest to this review.

A manufacturing process for producing ferroelectric thin film-based devices should at least include the following characteristics: (1) applicability of the processes to deposition of ferroelectric films and integration with metallic or conductive oxide electrode layers with different physical and chemical properties, (2) compatibility with integrated device processing, including production of as-deposited films with specific microstructures (perovskite, for example) on substrates at the lowest possible temperature, (3) production of device-compatible, highly oriented or polycrystalline films and heterostructures with specific properties (e.g., fatigue-free for a large number of polarization switching cycles, long polarization retention times, and no polarization imprint effects), (4) ability to produce patterned structures, superlattices and layered hetero-structures, (5) reproducibility of the deposition process, and (6) simple and low cost deposition processes with capacity for high deposition rates.

Electron Beam Evaporation. This technique has been used mainly for high temperature superconducting oxide thin films,[11,12] where it has typically been implemented by using individual thermionically produced electron beams, which are accelerated through 5-10 keV potentials, magnetically deflected, and then focused onto spatially separated elemental targets located in water-cooled holders symmetrically positioned in front of the substrate.[13] Process parameters, particularly precise film stoichiometry and abrupt interfaces, are difficult to control, since direct shuttering of individual beams near each source is difficult without disturbing the focused electron beam and thus quenching the target material evaporation. The evaporation of the elemental materials from different spatial locations involves complex hardware and tends to produce non-uniform films across large area substrates.

Pulsed Laser Ablation-Deposition (PLAD). This method has been extensively used by many groups to produce multicomponent oxide thin films since the first demonstration of the technique for the production of superconducting $YBa_2Cu_3O_{7-x}$ films by "ablation" of material from sintered $YBa_2Cu_3O_{7-x}$ targets.[14,15] The impact of a laser beam (generally an excimer laser with a wavelength of 248 or 193 nm) on the target produces localized melting and resolidification that for some target materials results in topographical and compositional changes on the impacted area of the target, which affects the reproducibility of the film characteristics after several hundred pulses.[16] This problem can be solved by continuous rotation of the target and/or laser beam scanning on the surface to permanently expose a fresh area to the laser beam.

For the synthesis of ferroelectric $Pb(Zr_x Ti_{1-x})O_3$ (PZT), initial work on laser ablation from continuously exposed fresh areas of a PZT target demonstrated the feasibility of producing deposited films at substrate temperatures 500-600 °C, and in oxygen atmospheres, needed for nucleation of the perovskite PZT structure.[17] However, even in the presence of a seemingly appropriate oxygen atmosphere, early as-deposited PZT films contained a pyrochlore phase,[18] possibly due to Pb deficiency in the film. More recently, problems related to PLAD synthesis of ferroelectric films and integration in heterostructure capacitors were solved, resulteing in the production of fatigue-free (up to the maximum 10^{12} polarization switching cycles currently tested) $La_{0.5}Sr_{0.5}CoO_3/PZT/La_{0.5}Sr_{0.5}CoO_3$ heterostructure capacitors, with good polarization retention.[19-21]

Comprehensive discussions of the PLAD technique can be found elsewhere.[22-24]

Molecular Beam Epitaxy (MBE). This technique is capable of producing highly strained (metastable) epitaxial films with the necessary properties to satisfy device requirements. This technique has been successfully used at the laboratory scale to produce a variety of compound semiconductors, HTSC, and other multicomponent thin films. However, in the case of semiconductor thin films, MBE techniques have a long history of difficulty with dopant incorporation. This may or may not be a problem if doping of oxide thin films becomes necessary. Other problems to be considered in relation to MBE, which make it difficult to use in manufacturing processes, include relatively low deposition rates, a questionable capability for covering large-area substrates with uniform films, difficulty of integration with other processing steps, and the relatively high cost of equipment. This technique has not been extensively used to investigate the synthesis of ferroelectric films and heterostructures.

Chemical Vapor Deposition (CVD). This technique is widely used for synthesizing multi- and single-element thin films, and has been extensively used to produce ferroelectric and electro-optic materials.[25-27] Major difficulties associated with this technique include the high substrate temperatures ($\geq$ 700 °C) required for producing most perovskite-based oxide films with device-compatible characteristics. The availability of precursors was initially a problem, although recent advances in CVD resulted in the development of new precursors which will advance the applicability of CVD to ferroelectric films.

3. PLASMA SPUTTER - DEPOSITION OF FERROELECTRIC THIN FILMS AND HETEROSTRUCTURES: SYSTEMS, DEPOSITION PROCESSES, AND PROCESSING - MICROSTRUCTURE-PROPERTY RELATIONSHIPS

Several variations of the plasma sputter-deposition technique have been developed and are still under investigation for the production of multicomponent oxide thin films, including HTSC, ferroelectric, and electro-optic materials. The plasma sputter-deposition techniques described in this review are being developed both for use in research laboratories and for commercial production of thin film-based devices. In the latter case, the development of automated systems for the production of oxide films in an integrated deposition/processing cycle will help meet the requirements for large scale integration of oxide materials in the fabrication of oxide/semiconductor thin film-based devices.

Basic phenomena occurring during the interaction of plasmas with the targets and substrates, during film synthesis, are important in that they determine to a large extent the composition, microstructure, and properties of the films. Therefore, a discussion of basic plasma-material interaction and transport of species through the plasma will be presented when discussing plasma sputter-deposition of ferroelectric and electro-optic thin films in this section.

Extensive work has been performed in recent years on plasma sputter-deposition of ferroelectric and electro-optic thin films. Both single multicomponent oxide and multiple elemental metallic targets have been used to synthesize a wide range of ferroelectric materials in thin film form, including $BaTiO_3$ (BTO),[28-30] $PbTiO_3$ (PT),[31] $Pb_{1-x}La_xTiO_3$ (PLT),[32] $PbZr_{1-x}Ti_xO_3$ (PZT),[33] and $Pb_{1-x}La_x(Zr_yTi_{1-y})O_3$ (PLZT).[34] Due to space limitations, only the work on the Pb-based family of materials is reviewed in some detail in this paper.

Iijima et al.[31] demonstrated that highly c-axis oriented PT thin films (98% of the film had the c-axis oriented perpendicular to the substrate surface) can be produced by r.f. magnetron sputtering only when using low deposition rates (< 20 Å/min), gas pressure of about 7×10^{-3} Torr, and PbO-rich targets (see Table 1a). According to Iijima et al.,[31] high gas pressures in the magnetron system increase the sputtering rate because of the production of a larger amount of ions in the plasma, but at the same time decrease the mean kinetic energy of the sputtered species arriving at the substrate, due to collisions with the plasma species. A reduced kinetic energy of the depositing species results in a lower mobility on the substrate surface at the deposition temperature. The combination of high sputtering rate (hence high deposition rate) and low mobility of the depositing species tend to inhibit the epitaxial growth of the film. The work of Iijima et al.[31] demonstrates the importance of controlling the deposition parameters to optimize certain film properties for particular device applications. In the case of the PT films discussed above, the main objective was to minimize the relative dielectric constant ε_r and maximize the pyroelectric coefficient g, to maximize the figure of merit $F_V = g / \varepsilon_r c_V$ (c_V: volume specific heat) for application of PT films as pyroelectric infrared sensors. The parameter ε_r was minimized and g maximized by growing highly c-axis oriented films.[31] Similar work on PLT [32] demonstrated that an accurate control of La dopant introduction in the films is also necessary to achieve optimized film properties tailored for a particular application.

Sreenivas et al. demonstrated that magnetron sputtering can be used to produce PZT films with controlled stoichiometry and properties, utilizing metallic elemental targets.[33] They designed a disk-shaped target consisting of several sectors of the Pb, Zr, and Ti,[35] in which they could change the size of the targets to control the amount of each material deposited on the substrate at a particular temperature. They performed extensive studies of the basic plasma-target-substrate interaction phenomena and their effects on the deposition processes. These investigations[35] revealed that the growth of stoichiometric highly oriented films with good electrical properties, by their particular magnetron sputter-deposition method, is governed by three main processes: (a) formation of a reproducible oxide layer on the target surface, (b) the stability of oxide species formed during transport through the plasma towards the substrate, and (c) the nucleation and growth of the film on the substrate surface. Sreenivas et al. observed that in order to obtain PZT films with good stoichiometry, microstructure, and electrical properties they needed to use pure oxygen at a pressure of 3×10^{-3} Torr and a target-substrate distance of approximately 10 cm. Computer simulations of the transport of sputtered species in the plasma and experimental measurements of deposition rate vs. gas pressure and film composition vs. substrate-target distance indicated that complex plasma-surface interaction and material transport processes control many of the film characteristics. For example, Sreenivas et al. argued that the metallic species sputtered from the Pb, Zr, and Ti targets undergo multiple collisions with the plasma species and among themselves during transport to the substrate. The collisions serve to thermalize and mix the sputtered species. Once the sputtered species are thermalized, they reach the substrate by diffusion through the plasma. However, the diffusional fronts of Pb, Zr, and Ti or their oxide species are formed at different distances from the target,[33] directly influencing the lateral deposition profiles and composition of the resulting films.

Sreenivas et al.[33] were unable to produce as-deposited PZT films at the temperature needed (550-600 °C) to produce the perovskite structure, mainly because of poor control of the sticking coefficient of Pb. Therefore, films were synthesized at 200-250 °C and subsequently annealed in air at 550 °C for 10-20 hrs. X-ray diffraction of as-deposited films at 200-250 °C revealed a (100) peak characteristic of the massicot or litharge phases of PbO, which indicated that the as-deposited film contained crystalline PbO grains, while the ZrO_2 and TiO_2 appeared to be deposited in an amorphous form. Annealing of the as-deposited films resulted in structural transformations at about 450 °C, which lead to a tetragonal or rhombohedral structure at 550 °C depending on the Zr : Ti ratio in the film (rhombohedral for Zr:Ti = 58/42, and tetragonal for Zr:Ti = 45/55). The surface topography of the films depended strongly on the annealing temperature and cooling time, such that rapid cooling resulted in film cracking and hillock formation that may originate from the bottom Pt electrode, depending strongly on the state of stress of the films.[33] On the other hand, long annealing times (20 hrs.) and very slow cooling resulted in smooth films with practically no hillock formation. Experiments involving annealing at different temperatures revealed that the lattice strain of the films was reduced from -0.01 to -0.001 as the annealing temperature increased from 450 °C to 700 °C in an oxygen atmosphere. In contrast to Sreenivas et al., other groups have more recently produced as-deposited PZT films without post-deposition annealing.[5,6] The electrical properties of the PZT films produced by magnetron sputter-deposition of metallic targets[33] are presented in Table 1b. Generally, good electrical properties were obtained for the films deposited under the optimized conditions described above. It is interesting to notice that other researchers have also produced good quality PZT films by magnetron sputter-deposition of multi-element metal targets.[36]

Adachi et al.[34] were able to grow epitaxial PLZT films on c-plane sapphire with epitaxial relations (111)PLZT // (0001) sapphire and [1$\underline{1}$0]PLZT // [10$\underline{1}$0] sapphire, by planar magnetron sputter-deposition using pressed powder PLZT targets of various compositions (Table 1a). They also determined that the perovskite phase is largely controlled by Pb incorporation into the films, which in their particular experiments appeared to be independent of the deposition rate. Adachi et al. also observed, in agreement with other researchers, that films deposited at temperatures ≤ 550 °C presented a pyrochlore structure, although in Adachi et al.'s case the films appeared to be remarkably rich in Pb. Stoichiometric PLZT films were obtained at substrate temperatures above 550 °C.[34] On the other hand, stoichiometric epitaxial PLZT films were produced with the same magnetron sputtering system at about 450 °C by using a PLZT (28/0/100) target with a reduced Pb content.[34] This latter result of Adachi et al.[34] seems to disagree with the work of Ishida et al.[37] and Okada,[38] who needed a Pb-rich PLZT target to produce stoichiometric PLZT films with a conventional diode sputtering system. In view of Sreenivas et al. work [33] related to the control of Pb incorporation in the films due to transport processes in the plasma, the seemingly contradictory results of Adachi et al.[34] and Ishida et al.[37] may be explained in terms of different transport processes in the plasma. In the diode sputtering system,[37] the higher gas pressure may reduce the transport of Pb towards the substrate, therefore requiring a larger supply of Pb from the target. On the other hand, the magnetron sputtering system used by Adachi et al.[34] requires lower gas pressure to sustain the plasma, resulting in a more effective transport of Pb, and therefore requiring a less enriched Pb target to incorporate the required Pb content in the film.

The limited discussion presented above indicates that the geometry and deposition conditions play fundamental roles in the synthesis of ferroelectric thin films by plasma sputter-deposition. Unfortunately, there is not enough systematic data, particularly related to the sputtering, transport and

TABLE 1a. Deposition / processing conditions and compositional and microstructural characterization of ferroelectric thin films produced by plasma sputter-deposition

MATER	TECHN	TARGET	SUBSTR	DEP TP (°C)	ANNEAL (°C)	GAS	PRESS (Torr)	POWER (w)	COMP	MICROSTR	MORPHOL	REF
PTO	RF MAG SPT	PTO/PbO	(100)MgO	≤550		Ar90:O(10)	7.00E-02	170	A2 B2 O7	PYROCL		IIJIMA [31]
		PTO/PbO(0)		550-600		" "	" "	"		NO C-PEROVSK		" "
		PTO/PbO(0.3)		550-600		" "	1.00E-02	"	PbTiO(3)	C-PEROVSK	MOSAIC	" "
PLTO	RF MAG SPT	PLTO/PbO10	(100)MgO	550-600		Ar90:O(10)	1.00E-02	170				
		PL(0.05)	" "	" "		" "	" "	"		C-PEROVSK (.86)		IIJIMA [32]
		PL(0.10)	" "	" "		" "	" "	"		C-PEROVSK (.71)		" "
		PL(0.15)	" "	" "		" "	" "	"		C-PEROVSK (.67)		" "
PZT	RF DIOD SPT	PbO10/PZT	Pt, FQ, InO/FQ	≤ 300		Ar20:O(80)	1.00E-03		Pb(-7%)	AMORPH		OKADA [38]
		PZO52/PTO48	" " " "	350-450		" "	" "		Zr50 Ti50	PYROCL		" "
		PZO42/PTO58	" " " "	350-450		" "	" "		Zr47 Ti53	" "		" "
									Pb (stoich)	PEROVSK		OKADA [38]
		PZO52 PTO48	" " " "	≥550		" "	" "		Zr60 Ti40	" "		" "
		PZO42 PTO58	" " " "	≥550		" "	" "		Zr47 Ti53	" "		" "
PZT	DC MAG SPT	Pb, Zr, Ti	Si,Gl,FQ	200		O(100)	4.00E-02	50	PbO,ZrO2,TiO2	(100)PbO		SAYER [33]
"	" "	" "	" " " "	"	550	"	" "	"	Zr(58):Ti(42)	(101) PZT/RHOM	SMO+HILL	" "
	" "	" "	" " " "	"	750	"	" "	"	Zr(45):Ti(55)	(101) PZT/TETR	SMO+HILL	" "
	" "	" "	ALUMINA	"	550	"	" "	"	Zr(58):Ti(42)	(101) PZT/RHOM	ROUGH	" "
PLZT	RF MAG SPT	PLZT(28/0/100)	C-SAPPHIRE	≤550		Ar(60)O(40)	3.00E-02	200	A2 B2 O7	PYROCL		ADACHI [34]
	" "	" "	" "	580		" "	3.00E-03	200	PLZT(STOICH)	(111)PLZT	SMOOTH	" "

RF MAG SPT: RF Magnetron Sputtering; DC MAG SPT: DC Magnetron Sputtering; RF DIOD SPT: RF Diode Sputtering; PYROCL: Pyrochlore; C-PEROVSK: Film with **c**-axis perpendicular to the substrate; AMORPH: Amorphous; RHOM: Rhombohedral; TETR: Tetragonal; SMO: Smooth; HILL: Film with Hillocks.

TABLE 1b. Deposition / processing conditions and electrical characterization of ferroelectric thin films produced by plasma sputter-deposition,

MATER	TECHNIQUE	TARGET	SUBSTR	DEP TP (°C)	ANNEAL (°C)	GAS	PRESS (Torr)	POWER (w)	Ps (μC/cm2)	Pr (μC/cm2)	Ec (KV/cm)	K	tan ∂	REF
PTO	RF MAG SPT	PTO/PbO	(100)PT/MgO	≤550		Ar90:O(10)	7.00E-02	170						
		PTO/PbO(0)		550-600		" "	" "	"	20	16	60	190	0.15	IIJIMA [31]
		PTO/PbO(0.3)		550-600		" "	1.00E-02	"	40	35	160	97	0.15	" "
PLTO	RF MAG SPT	PLTO/PbO10	Ni-Cr(100)MgO	550-600		Ar90:O(10)	1.00E-02	170						
		PL(0.05)	" "	" "		" "	" "	"	30	19	52	100	0.01	IIJIMA [32]
		PL(0.10)	" "	" "		" "	" "	"	45	30	36	200	"	" "
		PL(0.15)	" "	" "		" "	" "	"	52	39	32	280	"	" "
PZT	RF DIOD SPT	PbO10/PZT	Pt, FQ, InO/FQ			Ar20:O(80)	1.00E-03							
		PZO52/PTO48	" " " "	≥550		" "	" "		26.1	20.4	23.3	751	0.03	OKADA (38]
		PZO42/PTO58	" " " "	≥550		" "	" "		7.33	6.37	12.2	654	0.04	" "
PZT	DC MAG SPT	Pb, Zr, Ti	Si,Gl,FQ	200	550	O(100)	4.00E-02	50	50	30	25	800	0.1-0.3	SAYER [33]

PTO: Lead Titanate, PLTO: Lead Lantanum Titanate, PZT: Lead Zirconium Titanate, P_S :Saturation Polarization, P_r :Remanent Polarization, E_C : Coercive Field, K: dielectric constant, tan d: loss deposition processes, for many of the ferroelectric materials being investigated, to synthesize films under well controlled conditions. Therefore, further work is necessary to determine the optimum conditions needed to produce films of different ferroelectric materials with the best composition, microstructure, and properties. An overview of relevant results from studies of composition-microstructure-property relationships, for plasma sputter-deposited ferroelectric thin films and heterostructures, is presented in Tables 1a and 1b above.

3.1. Scaling of the Plasma Sputter-Deposition Technique to Cover Large Area Substrates.

Most of the work on scaling of the plasma sputter-deposition technique has been performed for high temperature superconductig (HTSC) thin films. Although this review is focused on the synthesis of ferroelectric thin films and heterostructures, we briefly discuss here the scaling of this technique, because the work done for HTSC applies equally to ferroelectric films. Although plasma sputter-deposition is a method well adapted to covering large area substrates, problems of thickness and compositional uniformity were observed in early work, directed at producing YBCO films by sputtering of bulk $YBa_2Cu_3O_{7-x}$ targets.[39,40] In this work, it was found that growing films were bombarded by O^- ions, produced during sputtering of multicomponent oxide HTSC targets. This bombardment produced uneven resputtering and compositional changes across the film surface. Various techniques were developed to minimize or eliminate the O^- ion bombardment problem; e.g. the off-axis method where the substrates are positioned perpendicular to the target on one side,[9] and more recently the shield method. In the latter approach, the target and substrate are parallel to each other and the O- ions travel in a direction along the perpendicular to the two plates. A plate positioned between the target and the substrate permit the sputtered target material to deposit on the substrate while preventing the more energetic O- ions from striking the substrate.[41] These techniques result in improvements of the thickness and compositional uniformity of YBCO films deposited over relatively large areas.[41-43] However, films with transition temperatures in the range 88-90 K still exhibit thickness variations of about 35% across a 38 x 38 mm substrate,[42] presumably because they were not continuously rotated during deposition (the importance of substrate rotation will be discussed in Section 2.3, in relation to ion beam sputter-deposition of large area YBCO films). On the other hand, Migliuolo et al.[43] were able to produce YBCO films with ± 10% thickness variation across 40 mm diameter substrates, using an "on-axis" magnetron sputtering geometry, with an enhanced magnetic field in the vicinity of the target surface, which permits the plasma to be sustained at a lower sputtering voltage (110 V) than normally required in other magnetron sputtering systems. The reduced voltage may have resulted in a reduced energy of the species bombarding the growing film, thus in a reduced resputtering effect.

On the other hand, other groups have been able to produce *in situ*, large area (50 mm diameter) superconducting films by using the plasma sputter-deposition technique.[44] Although most of the work on scaling the plasma sputter-deposition techniques has been done for HTSC films, the same approaches can be used for depositing ferroelectric thin films on large area substrates.

4. ION BEAM SPUTTER - DEPOSITION OF FERRO - ELECTRIC THIN FILMS AND HETEROSTRUCTURES

4.1. Systems and Sputter-Deposition Phenomena.

The IBSD technique was used for synthesizing multicomponent oxide ferroelectric thin films as early as 1979, when Castellano and Feinstein used the method to produce $Pb(Zr_x Ti_{1-x})O_3$ (PZT) thin films.[45] A major problem with the IBSD method as applied by Castellano and Feinstein was that they used PZT hot pressed oxide targets, which were exposed to an Ar^+ ion beam directed at ~ 45 ° with respect to the target surface normal. Problems associated with this early implementation of IBSD were: (a) preferential sputtering of multicomponent oxide targets, that results in modification of the target surface composition and therefore the composition of the film; and (b) extensive scattering of low mass ions (Ar) on heavier mass targets (Pb, Zr), which leads to resputtering of the films as well as ion bombardment-induced damage and incorporation of gas into the films, as indicated by studies on HTSC film systems.[46]

An alternative IBSD technique recently implemented by various groups[47,48] involves the use of multiple ion beams, generally produced by Kaufman-type ion sources,[49] where each beam is directed at elemental target materials (e.g., Pb, Zr, and Ti) (see Fig. 1 for example[47]). The multiple ion beam sputter-deposition method is described in detail elsewhere,[47,48] including a chapter in this book; therefore we will not discuss it further here. Instead, we will concentrate on describing the SIBMT deposition technique and the work on thin film synthesis performed using this method.

Our group developed an automated single ion beam / multitarget (SIBMT) sputter-deposition technique,[50,51] where the single ion beam is produced by a Kaufman-type ion source.[49] Our technique was initially demonstrated by producing $YBa_2Cu_3O_{7-x}$ thin films with very good orientation and electrical properties,[52] and thickness and composition uniformity over 4" substrates.[53] Subsequently, we extensively used the SIBMT deposition method to synthesize PZT thin films and integrate them into heterostructure capacitors with metal[54] and metal-oxide electrodes[55,56] in studies focused on controlling polarization fatigue, retention and imprint, as discussed below. We also used the SIBMT method to produce $KNbO_3$ thin films.[57,58] PZT and $KNbO_3$ materials were selected because of their technological importance, and to demonstrate the capability of the automated SIBMT deposition technique.

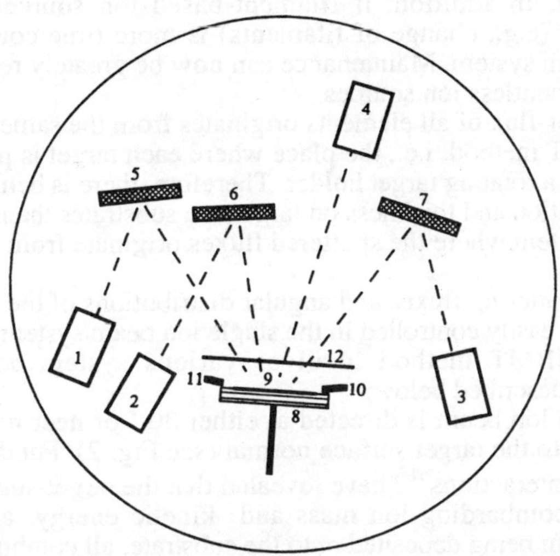

FIGURE 1. Schematic of a multiple ion beam-target system as implemented by one group. (1 - 3) Primary ion sources; (4) low - energy ion source for ion - assisted deposition; (5 - 7) metal targets (e.g., Pb, Zr, and Ti); (8) heated substrate holder; (9) substrates; (10) thickness/rate probe; (11) bombarding ion flux probe; (12) shutter; (reproduced from Ref. 47).

The implementation of the IBSD technique in any of the variations described above (i.e., multi-beam and target approach and SIBMT) present several advantages over other sputter-deposition methods such as plasma sputtering, namely:

1) Independent control of ion current density and energy, which can be used to better control the sputtering process and consequent film characteristics.

2) Lower operating pressure in the deposition chamber and minimization of the plasma-substrate interaction, which can produce undesirable effects such as resputtering of the film, damage, and gas incorporation during film deposition.

3) The ability of controlling the composition of the films by tailoring the deposition of each elemental material.

4) The ability to produce smoother films because of better control of the deposition rate.

Although there are general advantages common to the multiple ion beam-targets technique[47,48] and our SIBMT deposition method,[50,51] there are some differences that makes the SIBMT technique a more versatile and controllable approach, namely:

1) The use of a single ion beam eliminates the complications of simultaneous control of various sources, makes the hardware assembly much simpler, improves reliability, and reduces equipment cost substantially. In addition, if filament-based ion sources are used, the maintenance (e.g., change of filaments) is more time consuming for the multiple beam system. Maintenance can now be greately reduced using the new r.f. filamentless ion sources.
2) The sputter-flux of all elements originates from the same spatial location in the SIBMT method, i.e., the place where each target is positioned under the beam via a rotating target holder. Therefore, there is better control of the film composition and thickness on large area substrates than for the multiple ion beam system where the sputtered fluxes originate from different spatial locations.
3) The ion scattering fluxes and angular distributions of the sputtered fluxes may be more easily controlled in the single ion beam system.

The SIBMT method involves various system components and functions as described below:
(1) A single ion beam is directed at either 30 ° or near normal incidence with respect to the target surface normal (see Fig. 2). Fundamental studies of ion-solid interactions [46] have revealed that the target-substrate-ion beam geometry, bombarding ion mass and kinetic energy, and mass of the sputtered atom being deposited onto the substrate, all combine to control the degree of beam-induced secondary effects, as discussed below.
(2) A rotating target holder driven by a computer-controlled stepper motor, serves to sequentially position elemental (or simple oxide) targets in front of the sputter-beam (Fig. 2).
(3) A quartz crystal resonator (QCR) (Fig. 2), measures the amount of each elemental material deposited and sends a feedback signal to the computer. When a pre-programmed amount of an element, has been deposited, the computer shuts off the ion beam, rotates the target holder to position the next target under the beam, resets the QCR reading, and turns the ion beam back on. Alternatively, the the deposition rate for each elemental target can be utilized as a user imput parameter to control the stoichiometry of the multicomponent films.
(4) The entire process, including the elemental layer thicknesses, gas introduction in the deposition chamber, ion beam energy and current, and thermal treatment is under computer control, allowing reproducible execution of complex processing sequences..

A specially designed computer program[50] is a fundamental component of the automated ion beam sputter-deposition system.

A very important part in the development of the SIBMT deposition technique has been the studies, including experiments[59] and computer modeling,[46] that were performed to understand sputtering and primary ion scattering processes which have major influence in deposition phenomena and the resulting composition, microstructure, and properties of the film. Although the work performed to understand the processes mentioned above was done for the $YBa_2Cu_3O_{7-x}$ system,[47] the results can also be applied to the PZT system for optimization of the deposition geometry and conditions.

In addition, our studies are relevant to other ion beam sputter-deposition concepts such as those involving multiple ion beams and targets.

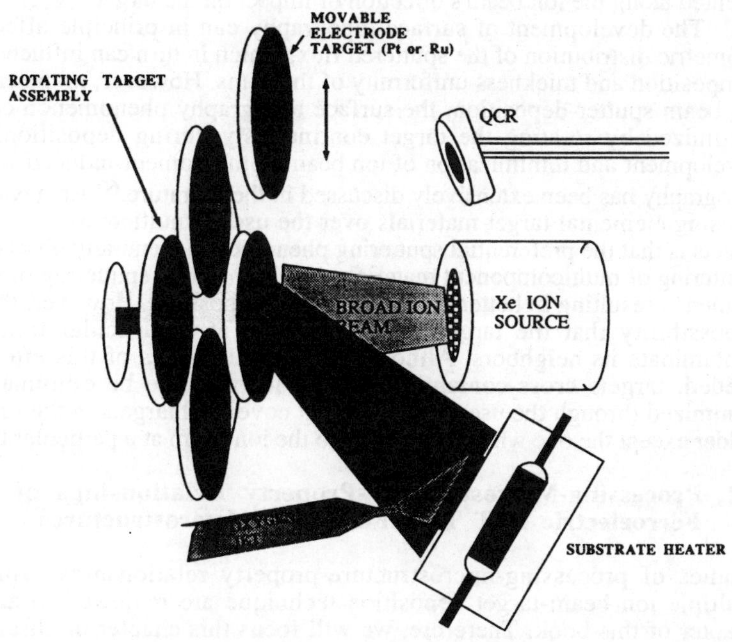

FIGURE 2. Schematic of the SIBMT thin film deposition system (from Ref. 9).

Computer simulations[46] have demonstrated that relatively light ions such as Ar^+ impacting the target with kinetic energies in the range of 1000-1400 eV, at 45° with respect to the surface normal, result in an undesirably high flux of primary ions which scatter from the target to the film with kinetic energies of up to 1000 eV, resulting in film damage, preferential sputtering, and consequent compositional changes, and gas implantation into the film. This effect is particularly deleterious when Ar^+ ions are used to sputter heavy atoms such as Ba and Pb. It was found that by using Kr^+ or Xe^+ ions and carefully selecting the system geometry, the scattered ion effect can be minimized or eliminated.[46,59] Additionally, a mapping of the sputtered flux angular distribution suggests that the QCR and the substrate should be positioned symmetrically with respect to the target surface normal in order to receive similar fluxes of sputtered species.

Sputtering-induced changes in the surface topography of the target represent another phenomenon which is frequently ignored but nevertheless affect the properties of the thin film. These topography changes generally consist of the formation of conical and ridge-type features with axes oriented along the ion beam's direction of impact on the target [60,61].

The development of surface topography can in principle affect the geometric distribution of the sputtered flux, which in turn can influence the composition and thickness uniformity of the films. However, in the case of ion beam sputter-deposition, the surface topography phenomenon can be minimized by rotating the target continuously during deposition. The development and minimization of ion beam bombardment-induced surface topography has been extensively discussed in the literature.[60] An advantage of using elemental target materials over the use of multicomponent oxide targets is that the preferential sputtering phenomenon, generally observed in sputtering of multicomponent materials, is eliminated in sputtering of single elements, resulting in better control of film composition. However, there is a possibility that the target being sputtered at a particular time will contaminate its neighbors. Although additional studies of this effect are needed, targets cross-contamination can in principle be eliminated or minimized through the use of a shield that covers all targets in the rotating holder except the one which is exposed to the ion beam at a particular time.

4.2. Processing-Microstructure-Property Relationships of Ferroelectric PZT Thin Films and Heterostructures

Studies of processing-microstructure-property relationships using the multiple ion beam-target deposition technique are reviewed in another chapter of this book. Therefore, we will focus this chapter on discussing our own work on this subject, using the SIBMT deposition method. In previous work, we presented preliminary data that indicated that the formation of the desirable perovskite phase in PZT thin films can be achieved through strict control of A:B-site cation ratio, via the substrate temperature, oxygen partial pressure, and composition of the fluxes.[9] From our previous work, it became clear that the lead oxide sticking coefficient depends strongly on the partial pressure of oxygen and substrate temperature. In addition, we determined, in agreement with other groups, that PZT-based capacitors with Pt electrodes present a strong polarization fatigue that make these capacitors unsuitable for non-volatile memories. Our previous work has been extensively reviewed in a recent book chapter,[9] so it will not be repeated here. Instead, we will concentrate this chapter on reviewing our work of the last two years, which resulted in the production of PZT-based capacitors with negligible fatigue and very good retention properties that make current PZT-based heterostructure capacitors compatible with non-volatile memory requirements.

It became clear in the last two years that the two most promising ferroelectric materials for application to non-volatile memories are $Pb(Zr_xTi_{1-x})O_3$ (PZT) [5,6,9] and a proprietary material referred to as Y1.[62]

Although not much has been published on the compositional, crystallographic orientation, and microstructural characteristics of Y1, a substantial amount of information exists in the literature about PZT.[5,6,9] Both materials, however, have been used to fabricate capacitors with practically no polarization fatigue up to about 10^{12} polarization switching cycles, which is the maximum number of cycles used in reported tests performed by different groups until now.[5,6,9] Fatigue-free Y1-based capacitors involve the use of Pt electrodes. On the other hand, PZT-based capacitors exhibit large polarization fatigue, after about 10^6-10^7 switching cycles, and substantial polarization retention loss when used in conjunction with Pt electrodes.[5,6,9] However, recent work on integration of metal-oxide electrodes [e.g., RuO_2[55,56,63] and $La_{0.5}Sr_{0.5}CoO_3$ (LSCO)[19-21,64]] with PZT films into heterostructure capacitors has shown that fatigue and retention effects are no longer a problem for PZT-based capacitors, within the range of tests performed until now. The work recently performed by various groups indicates that electrode material and orientation, and/or the PZT/electrode interface (including intermediate template layers)[19-21,55,56,63-67] are key factors involved in the control of orientation of the ferroelectric layer and electrical properties of PZT-based heterostructure capacitors.

The synthesis of PZT, Pt, RuO_2, and LSCO layers and their integration into heterostructures discussed in this section was done using the SIBMT deposition technique (described in Section 4.1), utilizing a Xe^+ ion beam of 1400 eV and 9-20 mA total current. Substrate temperatures were in the range 550-600 °C. Platinum and RuO_2 layers were produced by IBSD in vacuum and 5×10^{-4} Torr of oxygen, respectively, at temperatures in the range 400-500 °C, which resulted in epitaxial, highly oriented, or polycrystalline layers depending on substrate material.

The orientation and phases present in the layered heterostructures were characterized by X-ray diffraction (XRD) using a Rigaku system (with a Cu-K_a source). The microstructures were determined using transmission electron microscopy (TEM). Electrical characterization was performed using an RT66A ferroelectric test system (from Radiant Technologies) in the virtual ground mode to measure polarization fatigue and retention losses (using square waves within a frequency range of 500 kHz-1 MHz supplied by a function generator).

The work reviewed here has been focused on studying the RuO_2 / PZT / RuO_2 system, including three different PZT-based heterostructure capacitors as described below. The layers are identified from the top electrode on the left to the substrate on the right, namely:
(a) RuO_2/PZT/RuO_2/(100)MgO, is hereafter referred to as (R / PZT / R / M); (b) RuO_2/PZT/$PbTiO_3$/RuO_2/(100)MgO as (R / PZT / PT / R / M); and (c) RuO_2/$PbTiO_3$/PZT/$PbTiO_3$/RuO_2/(100)MgO as (R / PT / PZT / PT / R / M) .

4.2.1. X-Ray Diffraction (XRD) and Transmission Electron Microscopy Studies. The layered structures were produced using general deposition conditions described above. The synthesis of the PT and PZT layers was accomplished by depositing sequential layers of Pb and Ti (for PT) and Pb, Zr, and Ti (for PZT). The thickness of each individual Pb, Ti, and Zr layer was intentionally kept very thin (<10Å) in order to ensure complete interdiffusion and form a homogeneous film with the desired crystal structure. X-ray diffraction analysis of the systems mentioned above revealed that all three of them have PZT layers with a preferred (101) crystallographic orientation (see Fig. 3a for example). The preferred (101) orientation of the PZT layer on (110) oriented RuO_2 may result from the combined effect that this orientation minimizes the lattice mismatch between film and substrate, plus a possible favorable electrostatic interaction of ionic species with opposite signs at the film-substrate interface.[68] The XRD patterns also show a weak peak corresponding to the (001) orientation. It should also be noted that XRD of the R/PZT/R/M heterostructure shows some additional non-ferroelectric second phase in the PZT layer (see the shoulder at $2\Theta \sim 30.4°$ in Fig. 3(a)). This second phase is eliminated by the inclusion of the PT template layer (Fig. 3b), which must be thicker than ~ 100 Å to be effective in promoting the nucleation of the perovskite PZT structure and inhibiting nucleation of the non-ferroelectric phase. The $PbTiO_3$ phase grows quite easily with the perovskite structure, without the presence of pyrochlore phases. Although the PT film is a well identified layer when deposited independently, as confirmed by XRD of PT / (110) RuO_2 / (100) MgO (Fig. 3c), the PT layer does not survive as a separate layer when depositing the whole PZT/PT structure, due to interdiffusion produced at the deposition temperature of 550-600 °C. This has been confirmed by AES sputter-depth profiling.

To further demonstrate the importance of controlling the initial stages of PZT growth, particularly avoiding the nucleation of non-ferroelectric regions, growth of a PZT film was attempted on both a PZ (~ 200 Å thick) layer deposited on (110) RuO_2 / (100) MgO, to compare the results against the PZT film grown on the PT layer. The film grown on the PZ layer shows an XRD reflection at about 34° (Fig. 3d), which has been tentatively identified as cubic zirconium oxide. In addition, a second noticeable peak at 40° was tentatively identified as corresponding to a Ti-rich phase. The small peak at 30.3° could be related to the formation of a $Pb_2Ru_2O_{6.5}$ or a zirconia phase, but further work is needed to confirm this hypothesis. The XRD analysis of the PZT-based heterostructures described above is relevant to underscore the effect of the first layers in contact with the substrate in controlling the nucleation and growth of PZT films. It is relevant to point out that our work is in agreement with previous investigations,[67] which demonstrated the effectiveness of a "template" PT layer in controlling the nucleation and growth of the perovskite PZT phase. It has been proposed that the PT layer contributes to lowering the pyrochlore-to-perovskite transformation temperature by about 100° C, due to the PT layer presenting a lower energy barrier for PZT perovskite nucleation[66] than Zr-rich regions that may form when starting the PZT growth directly on the substrate. A detailed discussion of possible explanations for the appearance of the second phases described above can be found in ref. 56.

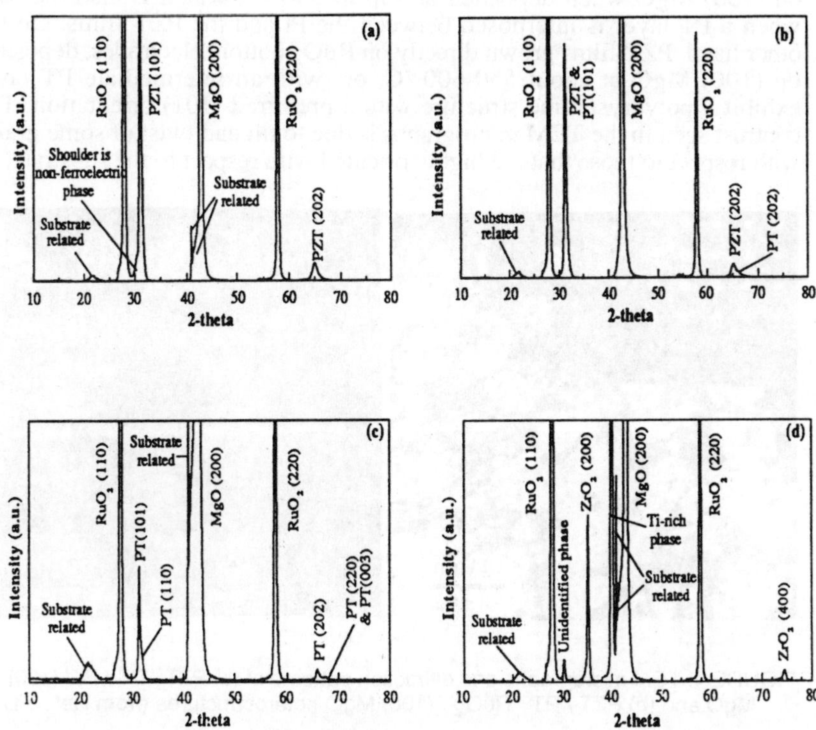

FIGURE 3. XRD spectra for: (a) PZT / R / (100) MgO , (b) PZT / PT / R / (100) MgO, (c) PT / (110) RuO_2 / (100) MgO, (d) a film grown on a PZ / (100) MgO heterostructure. The features marked "from substrate" originate from the MgO substrate (from Refs. 55 and 56)

Finally, in addition to showing that a PT layer is effective in controlling the nucleation and growth of IBSD PZT films on RuO_2 electrodes, we also demonstrated that further substantial improvement in the control of the PZT layer orientation produced by IBSD can be achieved by producing a hybrid Metal - RuO_2 bottom electrode (not disclosed at this time because of patent considerations), which results in highly (001) oriented PZT layers that yield the largest saturation and remanent polarization observed in the RuO_2/PZT/RuO_2 system.

TEM studies of PZT layers produced on Pt and RuO_2 bottom electrodes using the IBSD technique revealed that in general these films present a microstructure characterized by small grains (about 500-1000 Å), as shown in Fig. 4. Epitaxial (001) PZT films grow on (100) epitaxial Pt on (100) MgO when deposited at 550 to 600 °C, which is also the case when a PT layer is interposed between the Pt and the PZT films. On the other hand, PZT films grown directly on RuO_2 bottom electrodes, deposited on (100) MgO at about 550-600 °C, or with an intermediate PT layer exhibit a polycrystalline structure with a preferred (101) orientation. The contrast seen in the TEM micrographs is due to tilt and twist of some grains with respect to those that are highly oriented with respect to the substrate.

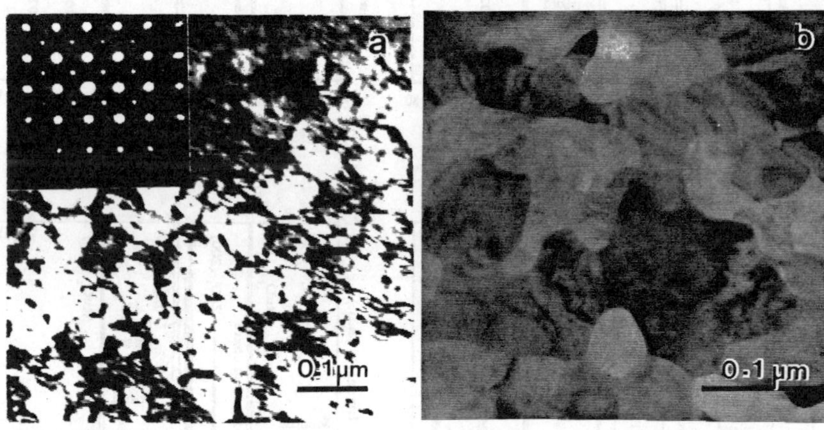

FIGURE 4. TEM micrographs and diffraction patterns of (a) PZT / PT / Pt / (100) MgO and (b) PZT / PT / RuO_2 / (100) MgO heterostructures (from Ref. 71)

4.2.2.Electrical Characterization. Electrical characterization of all three systems described above revealed substantial differences in the fatigue behavior of the PZT-based capacitors (tested to about 10^{10} polarization switching cycles in all cases) with and without the PT layers, as shown in Fig. 5. Capacitors without the PT layer present a substantial fatigue (about 71% reduction in remanent polarization, while those with one or two PT layers have a substantially smaller fatigue (about 33% and 29% decrease, in remanent polarization, for one and two PT layers, respectively). Furthermore, we have recently demonstrated that a proprietary modification of the RuO_2 film to produce a hybrid bottom electrode, prior to producing the PT layer, results in a highly oriented (001) PZT film and negligible fatigue (≤ 7% reduction in remanent polarization[55,56]). Polarization hysteresis loops were measured before and after fatigue to confirm the validity of the fatigue data presented in Fig. 5 (details of these measurements and a more specific analysis can be found in ref. 56).

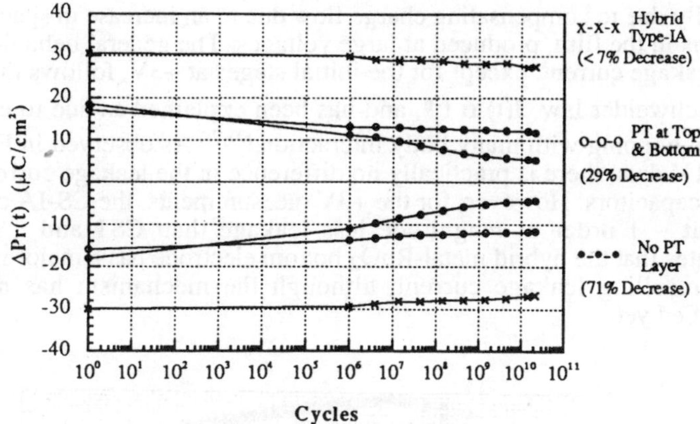

FIGURE 5. Change in remanent polarization vs. number of switching cycles (fatigue) of PZT - based capacitors with and without PT layers and of a PZT capacitor with a hybrid metal-RuO_2 bottom electrode (from Ref. 56).

The results discussed here and elsewhere[55,56] indicate that the PT layer may help eliminate the formation of undesirable second non-ferroelectric phases and/or charged defects (e.g., oxygen vacancies), which can produce internal bias fields in the PZT layer or at the PZT/electrode interface, all of which may play a role in the fatigue process. The beneficial effects of the PT layer(s) are enhanced by the hybrid metal-RuO_2 bottom electrode. In any case, it is relevant to point out that in spite of the use of optimized hybrid metal-RuO_2 and PT layers, fatigue is not totally eliminated in the IBSD PZT capacitors discussed in this section, since there is still a small decrease ($\leq 7\%$) in remanent polarization after ~ 10^{10} switching cycles (Fig. 5). This suggests that not only the electrode material, but also the structure of the PZT/electrode interface may play a role in the fatigue process. Further studies of this interface, preferably at the atomic scale level using analytical techniques such as that described in Section 5, may contribute to clarify this point.

In addition to controlling polarization fatigue, retention, and imprint (i.e., the reversal of dipoles orientation to a direction imprinted in the ferroelectric material previously to reorienting the dipoles into the opposite direction) phenomena in PZT-based capacitors, control of DC leakage current is very important. Recently, we have measured DC leakage currents for R / PZT / R / M (named CS1 in Fig. 6) and R / PZT / PT / R / M (named CS2 in Fig. 6) heterostructure capacitors defined above, and for a PZT capacitor involving a hybrid metal - RuO_2 bottom electrode processed with our proprietary method (named CS-IA in Fig. 6). Tests were performed at +3V and +1V (see Fig. 6). For the +3V test, there is an initial increase in DC leakage, which although not fully explained yet, may be partially due to compensating charge flow due to an increase in space charge regions in the film, produced at large voltages. The general behavior of the DC leakage current, except for the initial stages at +3V, follows the Curie-von Schweider law, $J(t) \alpha t^{-n}$, and has been explained as due to electrical charge hopping with many body interaction.[69,70] As observed in Fig. 6 for the +1V data, there is practically no difference in the leakage current of all three capacitors. However, for the +3V measurements, the CS-IA capacitor exhibit ~ 1 order of magnitude less leakage than CS1 and CS2. This indicates that the hybrid metal-RuO_2 bottom electrode has a major influence in controlling leakage current, although the mechanism has not been identified yet.

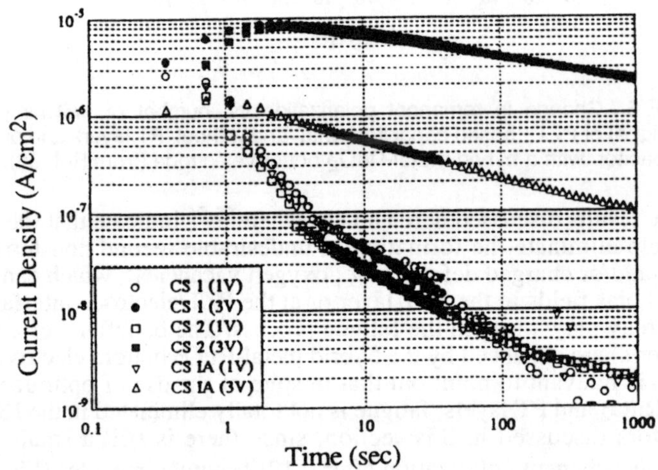

FIGURE 6. Leakage current measurements for PZT-based heterostructure capacitors CS1, CS2, and CS1A (from Ref. 12).

A comparison of leakage currents between IBSD and sol-gel deposited PZT-based heterostructure capacitors, involving RuO_2 electrodes, shows that the IBSD capacitors have ~ 1 order of magnitude lower leakage than those synthesized by sol-gel.[71] Preliminary data obtained recently indicates that the different leakage currents can be largely attributed to different microstructures.[71] The main difference observed between IBSD and sol-gel synthesized PZT-based heterostructure capacitors is the small grain (500-1000 Å) characteristic of the IBSD PZT layers vs. the larger grains (1-2 μm) observed in the sol-gel PZT films. An hypothesis to rationalize the higher leakage current in the sol-gel PZT capacitors is that the second non-ferroelectric phase is conductive. A candidate phase is $Pb_2Ru_2O_3$, although it has not been confirmed yet.

Further work is necessary to better understand the influence of microstructure and other parameters on leakage. This is an important subject, since leakage plays a fundamental role in the operation of ferroelectric memories.

5. *IN SITU*, REAL-TIME CHARACTERIZATION OF GROWTH PROCESSES OF Pb / Zr / Ti LAYERS AND HETERO-STRUCTURES INCLUDING OXIDE ELECTRODES

The work on metal-oxide/PZT/metal-oxide heterostructure capacitors discussed in Section 4 has shown that the electrode material and/or composition, orientation, and microstructure of the PZT layer and/or PZT / electrode interface have a strong influence on the electrical characteristics of the capacitor. For example, fatigue has practically been eliminated in the PZT-based heterostructure capacitors mentioned above, at least up to 10^{11}-10^{12} polarization switching cycles.

In addition to controlling the PZT/electrode interface, the work discussed in Section 4 has shown that the growth of the PZT layer needs to be controlled to avoid the formation of pyrochlore or other undesirable non-ferroelectric phases,[55,56] which contribute to degradation of the electrical properties of PZT-based capacitors.

The progress in producing PZT-based capacitors with controlled composition, microstructure, and electrical properties has been largely based on semiempirical considerations including thermodynamic and kinetic arguments. Missing from the work published in the literature are studies directed at obtaining valuable information on the film growth dynamics using *in situ*, *real-time* characterization techniques. For this purpose, Krauss and Auciello recently started a program to study growth processes of PZT films and heterostructures using a time-of-flight ion scattering and recoil spectrometry (TOF-ISARS) technique[72-75] integrated with an automated ion beam sputter-deposition (IBSD) system previously developed by the same authors.[50,51]

This Section is dedicated to reviewing our recent work directed at understanding the growth process of Pb/Zr/Ti layered structures on RuO_2, particularly the initial stages of nucleation and growth on the surface of RuO_2 layers used as bottom electrodes for PZT-based heterostructure capacitors.

5.1. Time-of-Flight Ion Scattering and Direct Recoil Spectroscopy

A schematic of the experimental system is shown in Fig. 7, while a detailed description of the IBSD and TOF-ISARS systems can be found in previous publications.[72,73] Briefly, the integrated system consist of an UHV surface analysis chamber containing the TOF-ISARS and an Auger electron spectrometer, coupled to a multitarget IBSD system through a 6mm differential pumping aperture. The base pressure of the analysis chamber is in the 10^{-9} Torr range. Oxygen can be introduced into the chamber through a leak valve for *in situ, real-time* characterization of adsorption of oxygen on surfaces and growth of ferroelectric and other oxide layers. A highly focused inert gas ion beam (normally 10 keV) is produced by an Atomika ion source and electronically processed into pulses 10-500 ns long by two sets of electrostatic deflection plates and four apertures (which can be adjusted over the range 0.125-4 mm diameter). The analysis beam impacts on the layer growing on the substrate, located on the axis of the analysis chamber, and is scattered at an angle of 165° into a channel electron multiplier detector located in a differentially pumped column (see Fig. 7). A second detector is positioned in the forward direction to detect direct recoil-ejected surface atoms when using the system in the direct recoil spectrometry mode.

The TOF-ISARS technique can be implemented either in the ion scattering spectroscopy (ISS) mode, where ions, impacting at normal or oblique incidence angle, are scattered from atoms on the surface of the sample and the time of flight to the detector is measured; or in the direct recoil spectroscopy (DRS) mode, where atoms are ejected from the surface of the sample by direct impact of the primary ions (directed at grazing incidence angle to the sample surface). The mass of the surface atoms is determined by pulsing the primary beam and measuring the sample-detector flight time, assuming classical two body collision dynamics. Both techniques are extraordinarily surface specific in that there is a "shadowed" region behind any surface atom exposed to the ion beam such that a second atom inside the shadow cone is invisible to the analysis beam.

Films are deposited using our automated SIBMT technique, which involves a single ion beam directed at a fixed region in space (Fig. 7), where different targets are positioned sequentially under the ion beam by means of a rotating target holder driven by a computer-controlled stepper motor.

Films with appropriate stoichiometry can be produced using a layer-by-layer deposition method, in which elemental (e.g. Pb, Zr, Ti) targets are sequentially sputtered and a film is deposited on a heated substrate to induce the layer interdiffusion necessary to produce the compound film. The amount of material needed for each layer is controlled by measuring with a quartz crystal resonator (QCR). A feedback signal from the QCR is provided to a computer driven by a custom made program that controls the deposition of each element, thus the film stoichiometry.

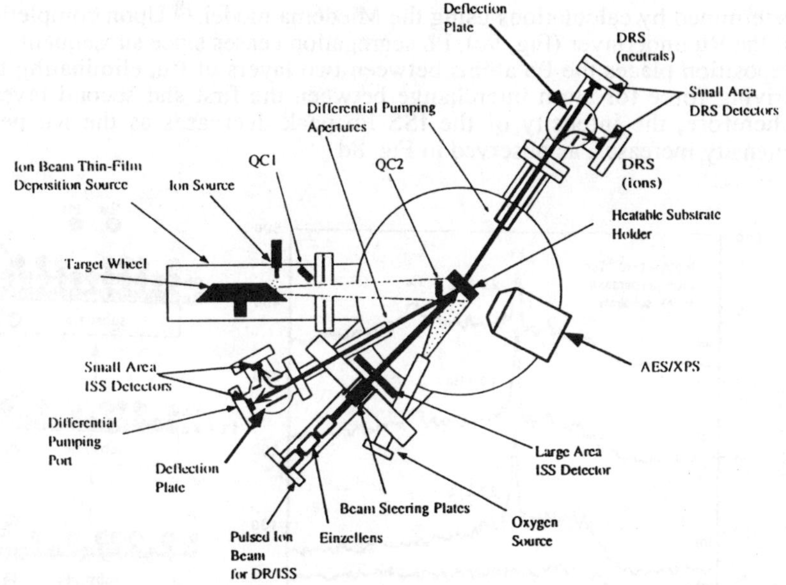

FIGURE 7. Schematic of the integrated IBSD / TOF-ISARS system used for the work reviewed here. (from Ref. 74).

5.2. *In Situ* ISS Analysis of Pb/Ru Heterostructures Growth

In order to characterize the capability of the TOF-ISARS technique, we first studied the growth of a Pb/Ru layered heterostructure under relatively simple conditions. A layer of Pb (about 3.7 Å thick) was deposited at room temperature on a stainless steel substrate while performing *in situ*, real-time analysis. The sputtered flux was directed at 45 ° to the substrate normal, while the ISS ion beam impacted normally to the growing layers. A peak appeared in the ISS spectrum at a time-of-flight of about 7.8 µsec (Fig. 8b), corresponding to 10keV Ar^+ ions scattered from Pb atoms deposited on the substrate surface. The sharp structure of the Pb peak is indicative of a single layer-type distribution of Pb atoms on the surface. Following the Pb deposition, a Ru layer was grown (Fig. 2c). Instead of observing the Ru peak growing and the Pb peak decreasing, the ISS spectrum revealed that after depositing about one half monolayer of Ru the Pb peak was unchanged, while only a small Ru peak appeared (Fig. 8c). Finally, after deposition of slightly more than one monolayer Ru, the Pb peak decreased to approximately one half of its initial value, while the Ru peak intensity increased (Fig. 8d). These results can be interpreted as indicative of a layered structure where the Pb atoms initially segregate to the surface (Fig. 9b), driven by a segregation energy of about 2.6 eV, leaving the deposited Ru in a salvage region between two layers of Pb atoms, as determined by calculations using the Miedema model.[18] Upon completion of the Ru underlayer (Fig. 9c), Pb segregation ceases since subsequent Ru deposition places the Pb atoms between two layers of Ru, eliminating the driving force for atom interchange between the first and second layers. Therefore, the intensity of the ISS Pb peak decreases as the Ru peak intensity increases, as observed in Fig. 8d.

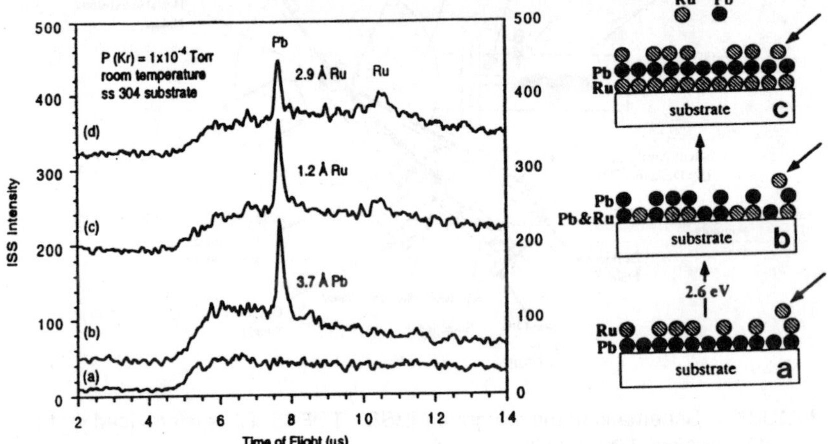

FIGURE 8. In situ, real - time ISS spectra obtained with scattering of Ar^+ ions from growing layers of Ru and Pb, (from Ref. 74).

FIGURE 9. Atomic scale model that correlates with the ISS spectra shown in Fig. 8, (from Ref. 74).

5.3. *Ex Situ* ISS Analysis of RuO_2 Layers

A highly oriented (110) layer of RuO_2 (as determined by XRD) was deposited on a (100) MgO substrate at 500 °C using a separate IBSD system with a higher deposition rate than that available with the SIBMT / TOF-ISARS integrated system. The layer was deposited by sputtering a Ru metal target while simultaneously introducing molecular oxygen into the film in the form of a jet that produced a 5×10^{-4} Torr background pressure in the deposition chamber. The RuO_2/MgO sample was subsequently transported in air to the TOF-ISARS system.

The surface of the RuO_2 layer was analyzed using the DRS mode of the TOF - ISARS instrument. This mode is used to detect the presence of light atoms such as H, C, and O on the surface, since the ISS mode is unable to detect atoms which are lighter than the primary ions. The DRS spectra shown in Fig. 10, taken at 5×10^{-5} Torr of O_2 background pressure indicate that H, C and O are present on the surface of the RuO_2 layer. The presence of H and C was most probably due to the exposure of the RuO_2 film to air during transport to the analysis chamber, since a spectrum similar to that at 5 x 10^{-4} Torr of O_2 was obtained in vacuum at 10^{-7} Torr. The introduction of oxygen into the analysis chamber, as a jet directed at the film, resulted in a substantial increase of the H and C signals above that corresponding to O, probably due to H and C impurities in the oxygen stream depositing on the surface of the RuO_2 layer to partially cover the O atoms of the RuO_2. These results demonstrate the high sensitivity of the technique.

In situ heating while acquiring DRS spectra as a function of substrate temperature in the range 30-400 °C, revealed that the H and C species are desorbed from the surface until they practically disappear (Fig. 11). The spectrum taken at 330 °C shows the presence of a Ru peak coexisting with the O peak, indicating that at this temperature some Ru atoms reside in the uppermost atomic layer. However, the Ru peak disappears for the spectrum taken at 400 °C, which can be interpreted as resulting from an O layer covering the Ru surface. However, the O atoms appear not to be positioned on top of Ru atoms, since an ISS analysis done at 545 °C in 5 x 10^{-4} Torr of oxygen, with a 10 keV Ar^+ ion beam incident normally on the surface, shows a spectrum characteristic of a thick clean Ru layer (i.e., a sharp rise on the short time-of-flight side of the peak, and a long tail on the long time-of-flight side). The combined results of the DRS and ISS analysis described above indicate that the RuO_2 surface may consist of a single layer of O atoms positioned between Ru atoms located in an underlying atomic row. This finding is relevant in the sense that it provides evidence for a possible mechanism for oxygen vacancy anhilation at the PZT/RuO_2 interface, which may contribute to the elimination of fatigue in PZT-based capacitors.

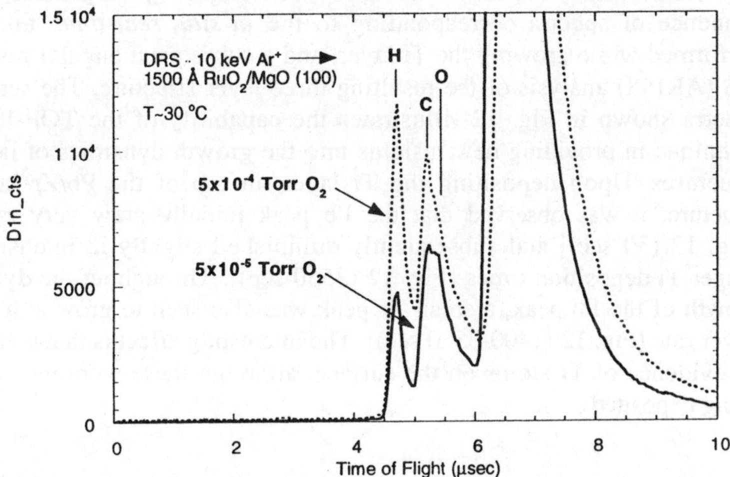

FIGURE 10. DRS spectra at two different O_2 background pressures for a 1500 Å thick RuO_2 layer deposited on (100) MgO (from Ref. 74).

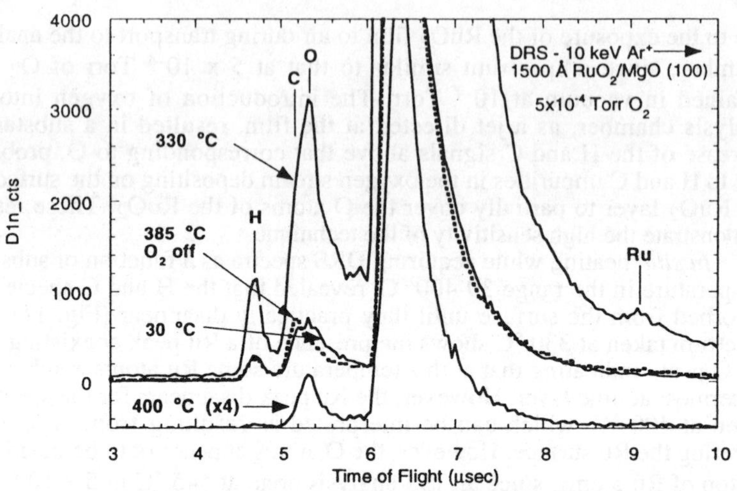

FIGURE 11. DRS spectra following those shown in Fig. 10. The spectra here were obtained in real - time as the sample was heated (from Ref. 74).

5.4. *In Situ,* Real - Time Analysis of Ti/Pb/Zr Layered Structures

A Pb/Zr layered structure was deposited at room temperature on (100) MgO, followed by deposition of a Ti layer. For brevity we present only a sequence of spectra corresponding to the *in situ, real-time* analysis performed while growing the Ti layer, and a subsequent angular resolved ISS (ARISS) analysis of the resulting three layer structure. The series of spectra shown in Fig. 12 illustrates the capability of the TOF-ISARS technique in providing new insights into the growth dynamics of layered structures. Upon depositing the Ti layer on top of the Pb/Zr layered structure, it was observed that the Pb peak initially grew very rapidly [Fig. 12 (50 sec)] and subsequently diminished slightly in intensity for longer Ti deposition times [Fig. 12 (3550 sec)]. Throughout the dynamic growth of the Pb peak, a small Zr peak was also seen to grow at a much lower rate (Fig. 12 (1400-3550 sec)]. The interesting effect is that there was no evidence of Ti atoms on the surface, although these were the species being deposited.

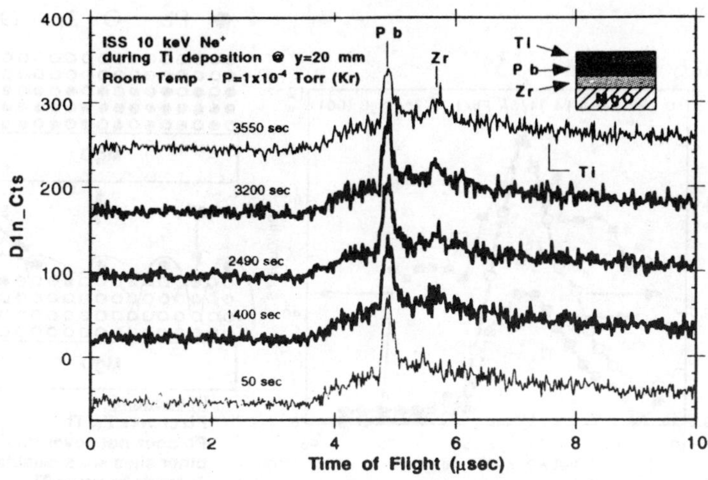

FIGURE 12. ISS sequential spectra taken when depositing a Ti layer on top of a Pb/Zr layered structure previously grown on (100) MgO at room temperature.

Following the *in situ, real-time* analysis shown in Fig. 12, an Angle-Resolved ARISS analysis was performed (Fig. 13 (a), which showed a rapid increase of the Pb peak at about 70° angle of incidence of the analyzing Ne^+ ion beam with respect to the surface normal, followed by a simultaneous increase of the Pb, Zr and Ti peak intensities at about 50°, to a maximum value at about 33°. Following the peak value of all three signals (at 33 ° polar angle) described above, the Pb intensity decreases to a value similar to that observed for polar angles in the range 50-70°, while the Ti and Zr peak intensities decrease to practically zero at normal incidence of the Ne^+ ion beam.

The data presented in Fig. 13 (a) can be interpreted in terms of the atomic scale model shown in Fig. 13 (b), where it is proposed that Pb atoms may be positioned on the top two surface layers in such a way that those on the top-most layer are located on top of Zr or Ti, but not on top of Pb atoms in the layer underneath. On the other hand, the larger intensity of the Zr peak with respect to the Ti indicates that Zr atoms tend also to cover Ti atoms. This interpretation is supported by calculations of (vacuum) surface segregation which show that ΔE_{seg} (Pb-Ti) = -1.11 eV > ΔE_{seg}(Pb-Zr) = - 0.83 eV > ΔE_{seg} (Zr-Ti) = - 0.02 eV.

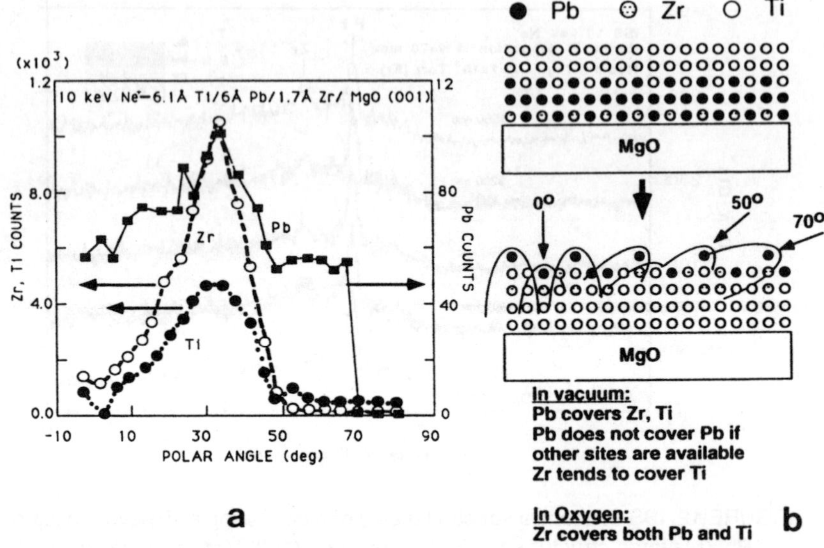

FIGURE 13. (a) ARISS analysis of the sequentially deposited Ti/Pb/Zr layers (Fig. 12) at room temperature on (100) MgO; (b) atomistic model that correlates with the ARISS data shown in (a).

The data presented in Fig. 13 can be interpreted in terms of the atomic scale model shown in Fig. 13 (b), where it is proposed that Pb atoms may be positioned on the top two surface layers in such a way that those on top-most layer are located on top of Zr or Ti, bu not on top of Pb atoms in the layer underneath. On the other hand, the larger intensity of the Zr peak with respect to Ti indicates that Zr atoms tend also to cover Ti atoms. This interpretation is supported by calculations of (vacuum) surface segregation energies, which show that ΔE_{seg} (Pb-Ti) = -1.11 eV > ΔE_{seg} (Pb-Zr) = -0.83 eV > ΔE_{seg} (Zr-Ti) = -0.02 eV.(may be the first evidence of the existence of Pb vacancies formed during deposition of PZT by the layer-by-layer SIBMT deposition technique. The Pb atoms in the first atomic layer have a very uniform but relatively large lateral spacing compared with the Pb atoms spacing in the second layer, as indicated by the sharpness of the Pb intensity increase at 70°. After deposition is terminated, the surface composition slowly changes, resulting in a decrease of the Pb signal, the appearance of Ti on the surface, and a very strong enhancement of the Zr signal. This evolution is generally in accord with behavior expected for oxygen-induced segregation. However, further work is necessary to confirm this hypothesis, particularly involving analysis of films grown at high temperature in an oxygen atmosphere, corresponding to the conditions used to synthesize PZT films by SIBMT.

6. CONCLUSIONS

The extensive basic and applied research performed in the field of synthesis and characterization of ferroelectric thin films, particularly PZT, has advanced the materials science and the technology basis of these materials for application in the manufacturing of advanced film-based devices such as non-volatile ferroelectric memories. Fatigue has been practically eliminated from PZT-based capacitors, using metal-oxide electrodes, while measurements of polarization retention have also yielded values compatible with a reliable operation of memories. It has been demonstrated that physical vapor-deposition techniques, particularly plasma and ion beam sputter-deposition, are at the forefront of the basic science related to the synthesis of multicomponent oxide thin films, and have a great potential for application in the technology of thin film-based device fabrication. The techniques described in this review will continue to provide the means for studying and producing films needed for advancing the basic knowledge of multicomponent oxide thin film science and thin film-based device technology.

A time-of-flight ion scattering and recoil spectroscopy (TOF-ISARS) technique has been used to perform the first *in situ*, *real-time* analysis of layered heterostructures relevant to the synthesis of ferroelectric PZT films, and conductive oxide layers, during growth. Initial results presented here show various major effects, namely:

(a) RuO_2 films on MgO substrates appear to be terminated in O atoms on the top layer, which are located in between Ru atoms lying in the layer underneath. This effect may have major implications for the explanation of the elimination of polarization fatigue demonstrated for RuO_2/PZT/RuO_2 heterostructure capacitors, indicating that fatigue may be eliminated because oxygen vacancies (considered as main contributors to fatigue through polarization domain pinning) at the PZT/metal electrode interface may be eliminated by the use of a conductive oxide electrode such as RuO_2.

(b) Deposition of a Ru monolayer on top of a Pb film results in surface segregation of Pb until a complete Pb layer develops over the Ru film.

(c) A Pb/Zr/Ti layered structure yields a top Pb layer with first evidence of the existence of Pb vacancies, suggesting major implications in relation to electrical characteristics of PZT-based capacitors. Further work involving deposition at high temperature in oxygen atmosphere, now in progress, will provide new insights into the growth dynamics of ferroelectric / metal oxide heterostructures under conditions used for the synthesis of capacitors by the IBSD deposition technique. Finally, it is relevant to point out that the TOF - ISARS technique can be applied to the *in situ, real-time* analysis of any film produced by vapor phase-deposition, and for *ex situ* analysis of films deposited with any other technique such as sol-gel or liquid phase epitaxy.

7. ACKNOWLEDGEMENTS

The authors would like to acknowledge the support of the Advanced Research Project Agency under contract N00014-93-1-0591, and the U.S. Department of Energy, BES Materials Science, under contract W-31-109-ENG-38. The authors also acknowledge the important contribution of all members of the NCSU, MCNC, and ANL groups.

8. REFERENCES

1. J. CARRANO, C. SUDHAMA, J. LEE, A. TASCH, W.H. SHEPHERD, AND N. ABT, IEEE Trans. Ultrasonics, Ferroelectrics and Freq. Control, **38**, 690 (1991).
2. J.T. EVANS AND R. WOMACK, IEEE J. Solid State Circuits, **23**, 1171 (1988).
3. J.F. SCOTT AND C.A. PAZ DE ARAUJO, Science, **246**, 1400 (1989).
4. S. DEY AND R. ZULEEG, Ferroelectrics, **108**, 37 (1990).
5. Proceedings of the 3th International Symposium on Integrated Ferroelectrics, edited by C.A. Paz de Araujo (University of Colorado Press, Colorado Springs, 1991). Also, Proceedings of the 4th International Symposium on Integrated Ferroelectrics, edited by C.A. Paz de Araujo (University of Colorado Press, Colorado Springs, 1992).
6. Ferroelectric Thin Films I : MRS Symposia Proc. **200**, (Materials, Research Society, Pittsburgh, 1990), edited by E.R. Myers and A.I. Kingon . Also, Ferroelectric Thin Films II: MRS Symposia Proc. **243**, (MRS, Pittsburgh, 1991) edited by A.I. Kingon, E.R. Myers, and B. Tuttle
7. Multicomponent and Multilayered Thin Films for Advanced Microtechnologies: Techniques, Fundamentals and Devices, NATO/ASI Book Series E, **234**, edited by O. Auciello and J. Engemann (Kluwer Academic Publishers, The Netherlands, 1993).
8. G. FISHER, Ceram. Bull., **64**, 545 (1985) .

9. O. AUCIELLO, A.I. KINGON, D.J. LICHTENWLANER, and A.R. KRAUSS, in Multicomponent and Multilayered Thin Films for Advanced Microtechnologies: Techniques, Fundamentals and Devices, NATO/ASI Book Series E, **234**, edited by O. Auciello and J. Engemann, (Kluwer Academic Publishers, The Netherlands, 1993) 151.
10. G.K. HUBLER (Ed.), various papers in Pulsed Laser Deposition: Special Issue of the MRS Bulletin (USA), **17**, (1992).
11. R.L. SANDSTROM, W.J. GALLAGHER, T.R. DINGER, R.H. KOCH, R.B. LAIBOWITZ, A.W. KLEINSASSER, R.J. GAMBINO, B. BUMBLE, AND M.F. CHISHOLM, Appl. Phys. Lett. **53**, 444 (1988) .
12. R.B. LAIBOWITZ, R.H. KOCH, P. CHAUDARI, AND R.J. GAMBINO, Phys. Rev., **B35**, 8821 (1987).
13. R.F. BUNSHAH ET AL. (Eds.), Deposition Technologies for Films and Coatings (Noyes Publications, New York, 1982).
14. D. DIJKKAMP, T. VENKATESAN, X.D. WU, S.A. SHAHEEN, N. JISRAWI, Y.H. MIN-LEE, W.L. MCLEAN, AND M. CROFT, Appl. Phys. Lett. **51**, 619 (1987).
15. J. NARAYAN, N. BIUNNO, R. SINGH, O. W. HOLLAND, AND O. AUCIELLO, Appl. Phys. Lett., **51**, 1845 (1988).
16. O. AUCIELLO, A.R. KRAUSS, J. SANTIAGO-AVILES, A. F. SCHREINER, AND D.M. GRUEN, Appl. Phys. Lett., **52**, 239 (1988) .
17. J.S. HORWITZ, K.S. GRABOWSKI, D.B. CHRISEY, AND R.E. LEUCHTNER, Appl. Phys. Lett., **59**, 1565 (1991) .
18. C. M. COTELL AND K.S. GRABOWSKI, MRS Bulletin, **17**, 44 (1992) .
19. R. RAMESH, H. GILCHRIST, T. SAND, V.G. KERAMIDAS, R. HAAKENAASEN, AND D.K. FORK, Appl. Phys. Lett., **63**, 3592 (1994) .
20. R. DAT, D.J. LICHTENWALNER, O. AUCIELLO, AND A.I. KINGON, Appl. Phys. Lett., **64**, 2673 (1994).
21. R. RAMESH, J. LEE, T. SANDS, V.G. KERAMIDAS, AND O. AUCIELLO, Appl. Phys. Lett. **64**, 2511 (1994) .
22. G.K. HUBLER (Ed.), various papers in Pulsed Laser Deposition: Special Issue of the MRS Bulletin (USA), **17**, (1992).
23. T. VENKATESAN, S. BHATTACHARYA, C. DOUGHTY, A. FINDIKOGLU, C. KWON, QI LI, S.N. MAO, A. WALKENHORST, AND X.X. XI, in Multicomponent and Multilayered Thin Films for Advanced Microtechnologies, edited by O. Auciello and J. Engemann, NATO/ASI Series E, **234**, (Kluwer Academic Publishers, The Netherlands, 1993) 209.
24. O. AUCIELLO, in Thin Films and Epitaxi: Handbook of Crystal Growth, vol. **3** (North Holland Materials-Science and Engineering, The Netherlands, in press, Fall 1994).

25. W.C. HENDRICKS, C.H. PENG, AND S.B. DESU, in Proc. 8th International Symposium on Applications of Ferroelectrics, IEEE Catalog No. 92CH3080-9, edited by M. Liu, A. Shafari, A.I. Kingon, and G. Haertling, (IEEE Service Center, Piscataway, 1992) 293.
26. T. KATAYAMA, M. SHIMIZU, AND T. SHIOSAKI, in Proc. 4th International Symposium on Integrated Ferroelectrics, edited by C.A. Paz de Araujo, (University of Colorado Press, Colorado, 1993) 382.
27 L.D. MCMILLAN, T.L. ROBERTS, M.C. SCOTT, AND C.A. PAZ DE ARAUJO, in Proc. 4th International Symposium on Integrated Ferroelectrics, edited by C.A. Paz de Araujo, (University of Colorado Press, Colorado, 1993) 666.
28. V.S. DHARMADHIKARI AND W.W. GRANNEMANN, J. Appl. Phys., **53**, 8988 (1982) .
29. I.H. PRATT, Proc . of the IEEE , **59**, 1440 (1971).
30. Y. SHINTANI AND O. TADA, Electron. Commun. Jpn., **56**, 74 (1973) .
31. K. IIJIMA, Y. TOMITA, R. TAKAYAMA, AND I. UEDA, J. Appl. Phys., **60**, 361 (1986).
32. K. IIJIMA, R. TAKAYAMA, Y. TOMITA, AND I. UEDA, J. Appl. Phys., **60**, 2914 (1986) .
33. K. SREENIVAS, M. SAYER, AND P. GARRETT, Thin Solid Films, **172**, 251 (1989) .
34. H. ADACHI, T. MITSUYU, O. YAMAZAKI, AND K. WASA, J. Appl. Phys., **60**, 736 (1986).
35. A. CROUTEAU AND M. SAYER, in Proc. of the 6th Intern. Symp. on Applications of Ferroelectrics, edited by V.E. Wood, IEEE (New York , 1986) 606.
36. T. FUKAMI AND M. NAOE, Trans. IEE Jpn., **103-A,** 278 (1983).
37. M. ISHIDA, S. TSUJI, Y. HAMAWAKA, AND T. NAKAGAWA, Appl. Phys., **21**, 339 (1978) .
38. A. OKADA, J. Appl. Phys., **49**, 4495 (1978).
39. S. I. SHAH, AND P. F. CARCIA, in Thin Film Processing and Characterization of High Temperature Superconductors: American Institute of Physics Conf. Proc. **165**, edited by J. M. E. Harper, R. J. Colton, and L. C. Feldman, (American Physical Society, New York, 1988), 50.
40. S. M. ROSSNAGEL, AND J. J. CUOMO, in Thin Film Processing and Characterization of High Temperature Superconductors: American Institute of Physics Conf. Proc. **165**, edited by J. M. E. Harper, R. J. Colton, and L.C. Feldman, (American Physical Society, New York, 1988), 106.
41. J. H. XU, G.G. ZHENG, A.M. GRISHIN, B.M. MOON, AND K.V. RAO, Appl. Phys. Lett., **64**, 1874 (1994) .
42. N. NEWMAN, K. CHAR, S.M. GARRISON, R.W. BARTON, R.C. TABER, C.B. EOM, T.H.GEBALLE, AND B. WILKENS, Appl. Phys. Lett., **57**, 520 (1990) .

43. M. MIGLIUOLO, R.M. BELAN, AND J.A. BREWER, Appl. Phys. Lett., **56**, 2572 (1990).
44. J. TALVACCHIO, M.C. FORRESTER, J.R. GAVALER, AND T.T. BRAGGINS, IEEE Trans. on Magnetics, **27**, 978 (1991).
45. R.N. CASTELLANO AND L.G. FEINSTEIN, J. Appl. Phys., **50**, 4406 (1979) .
46. A.R. KRAUSS AND O. AUCIELLO, in Materials Modification by Energetic Atoms and Ions: MRS Symp. Proc. **268**, edited by K.S. Grabowski, S. A. Barnett, S.M. Rossnagel, and K. Wasa , (Materials Research Society, Pittsburgh, 1992) 107.
47. S.B. KRUPANIDHI, H. HU, AND V. KUMAR, J. Appl. Phys. **71**, 376 (1992) .
48. I. KANNO, S. HAYASI, T. KAMADA, M. KITAGAWA, T. HIRAO, Jap. J. Appl. Phys., **32 (No. 9B)**, 4057 (1993) .
49. H.R. KAUFMAN, J.J. CUOMO, AND J.M.E. HARPER, J. Vac. Sci. Technol., **21**, 725 (1982) .
50. A. R .KRAUSS, AND O. AUCIELLO, U.S. Patent No 4,923,585 (1990).
51. A. R. KRAUSS, O. AUCIELLO, A.I. KINGON, M.S. AMEEN, Y.L. LIU, T. BARR, T.M. GRAETTINGER, S.H. ROU, C.S. SOBLE, AND D.M. GRUEN, Appl. Surface Sci., **46**, 67 (1990) .
52. D.J. LICHTENWALNER, O. AUCIELLO, R.R. WOOLCOTT, C.N. SOBLE, S.H. ROU, AND A.I. KINGON, J. Vac. Sci. and Technol., **A10**, 1537 (1992).
53. D.J. LICHTENWALNER, R.R. WOOLCOTT, C.N. SOBLE, O. AUCIELLO, AND A.I. KINGON, J. Appl. Phys., **70**, 6952 (1991).
54. K.D. GIFFORD, H.N. AL-SHAREEF, S.H. ROU, P.D HREN, O. AUCIELLO, AND A.I. KINGON, Ferroelectric Thin films II: MRS Symp. Proc. **243**, edited by A.I. Kingon, E.R. Myers, and B. Tuttle, (MRS, Pittsburgh , 1992), 191.
55. O. AUCIELLO, K.D. GIFFORD, AND A.I. KINGON, Appl. Phys. Lett , **64**, 2873 (1994).
56. K.D. GIFFORD, O. AUCIELLO, AND A. I. KINGON, Integrated Ferroelectrics (in press, 1994).
57. M.S. AMEEN, T.M. GRAETTINGER, S.H. ROU, H.N. AL-SHAREEF, K.D. GIFFORD, O. AUCIELLO, AND A.I. KINGON, Ferroelectric Thin Films: MRS Symp. Proc. **200** , edited by E.R. Myers and A.I. Kingon , (MRS, Pittsburgh, 1990), 65.
58. A.F. CHOW, D.J. LICHTENWALNER, R.R. WOOLCOTT, T.M. GRAETTINGER, L.A. BOATNER, N.R. PARIKH, O. AUCIELLO, AND A.I. KINGON, Appl. Phys. Lett. (in press, 1994).
59. O. AUCIELLO, M.S. AMEEN, A.R. KRAUSS, A.I. KINGON, AND M.A. RAY, Beam-Solids Interaction: Physical Phenomena: MRS Symp. Proc. **157**, J.M.E. Harper et al., (MRS, Pittsburgh, (1990) 287.

60. O. AUCIELLO AND R. KELLY (Eds.), Ion Bombardment Modification of Surfaces: Fundamentals and Applications, (Elsevier Science Publishers, The Netherlands, 1984).
61. O. AUCIELLO, AND A. R. KRAUSS, in Thin Film Processing and Characterization of High Temperature Superconductors: Amer. Inst. Conf. Proc. **165**, edited by J. M. E. Harper, R. J. Colton, L.C. Feldman, (American Physical Society, New York, 1989) 114.
62. T. MIHARA, H. WATANABE, C.A. PAZ DE ARAUJO, J. CUCHIARO, M. SCOTT, L.D. MCMILLAN, in Proc. 4th Intern. Symp. on Integrated Ferroelectrics, edited by C.A. Paz de Araujo, (University of Colorado Press, Colorado Springs, 1992) 137.
63. H.N. AL-SHAREEF, K.R. BELLUR, O. AUCIELLO, AND A.I. KINGON, in Proc. 5th Intern. Symp. on Integrated Ferroelectrics, edited by C.A. Paz de Araujo, (University of Colorado Press, in press, 1993).
64. D.J. LICHTENWALNER, R. DAT, O. AUCIELLO, AND A.I. KINGON, Ferroelectrics (in press, 1994).
65. S.L. SWARTZ, S.J. BRIGHT, P.J. MELLING, AND T.R. SHROUT, Ferroelectrics, **108**, 71 (1990).
66. C.K. KWOK AND S. DESU, J. Mater. Res., **8**, 339 (1993).
67. I. KANNO, S. HAYASHI, T. KAMADA, M. KITAGAWA, T. HIRAO, Jap. J. Appl. Phys, **32, Part 1, No. 9B**, 4057 (1993)
68. H. N. AL - SHAREEF, K. D. GIFFORD, P. D. HREN, S. H. ROU, O. AUCIELLO, AND A.I. KINGON, Integrated Ferroelectrics, **3**, 225 (1993).
69. A.K. JONSCHER, in Dielectric Relaxation in Solids, (Chelsea Dielectric Press, London, 1993) 117.
70. X. CHEN, A.I. KINGON, L. MANTESE, O. AUCIELLO, AND K.Y. HSIE, Integrated Ferroelectrics, **3**, 355 (1993).
71. O. AUCIELLO, K.D. GIFFORD, D.J. LICHTENWALNER, H.N. AL-SHAREEF, R. DAT, K.R. BELLUR, AND A.I. KINGON, in Thin Film Oxide Heterostructures: MRS Proc., edited by J.M. Phillips and R. Ramesh, (MRS, Pittsburgh, in press, 1994).
72. A.R. KRAUSS, Y. LIN, O. AUCIELLO, G.J. LAMICH, D.M. GRUEN, J.A. SCHULTZ, AND R.P. H. CHANG, J. Vac. Sci.Technol. **A**, (in press, 1994).
73. Y. LIN, A.R. KRAUSS, O. AUCIELLO, Y. NISHINO, D.M. GRUEN, R.P.H. CHANG, AND J.A. SCHULTZ, J. Vac. Sci.Technol. **A**, (in press, 1994).
74. O. AUCIELLO, A.R. KRAUSS, Y. LIN, R.P.H. CHANG, AND D.M. GRUEN, in Thin Film Oxide Heterostructures: MRS Proc., edited by J.M. Phillips and R. Ramesh, (MRS, Pittsburgh, in press, 1994).
75. A.R. KRAUSS, O. AUCIELLO, Y. LIN, R.P.H. CHANG, AND D.M. GRUEN, Integrated Ferroelectrics, (in press, 1994).

CHAPTER 11

MULTI-COMPONENT OXIDE FERROELECTRIC THIN FILMS BY MULTI-ION-BEAM REACTIVE SPUTTER (MIBERS) DEPOSITION

S.B. KRUPANIDHI*, H. HU, AND G.R. FOX

*Department of Engineering Science & Mechanics
Materials Research Laboratory
The Pennsylvania State University, University Park, PA,USA.

1.0 INTRODUCTION

Physical vapor deposition of ferroelectric thin films has been studied by many researchers. The techniques popularly used include rf diode sputtering[1,2] and magnetron sputtering,[3-7] and recently laser ablation.[8,9] Ion beam sputter deposition has also been investigated to some extent,[10-15] emphasizing the process control related to composition, structure and properties of the resultant films. Initially, compound ceramic targets were used to deposit such ferroelectric films as PZT, PLT and PLZT by rf diode sputtering and magnetron sputtering.[1-5] Evidences of success for these techniques, however, often masked by the limitations such as (a) limited control over the composition of the films due to the large differences in the sputter yield of different components and as a result the preferential deficiency of volatile elements, especially Pb, in the films, and (b) the presence of negative ions and reflected neutrals during sputtering of oxide targets, which tend to bombard the growing film uncontrollably and cause nonstoichiometry, surface damage and inferior morphology. Multiple target magnetron sputtering has been used as an alternate approach to allow better control over film composition.[6,7] However, unwanted bombardment of the films by the negative ions and reflected neutrals is still unavoidable.

Ion beam sputter deposition is believed to be more flexible approach due to its independent control over the energy and flux density of the ion beam and the localization of the discharge away from the substrates.[16] Furthermore, this technique can be easily combined with controllable low energy ion bombardment which was proven to offer great opportunities to modify growth process, microstructure and physical properties of the films.[16-19] Therefore, it has been extensively used for growing semiconducting thin films (such as Si and Ge) and optical coatings (such as TiO_2, ZrO_2 and Al_2O_3).[20,21] Initial studies on ion beam sputter deposition of ferroelectric thin films used the configuration of a single ceramic target with a single ion beam.[10] With this configuration, controlling composition and uniformity of the resultant films was still a problem. Recently, efforts have been directed towards using multiple targets, sputtered by either a single ion beam[12] or multiple ion beams.[13-15] The configuration of multiple targets with multiple ion beams seems noticeably attractive as it allows high flexibility and good control over stoichiometry, uniformity and reproducibility[14,15].

This chapter summarizes work on the deposition of ferroelectric thin films while considering Pb-based perovskites as specific examples by the quite flexible multi-ion-beam reactive sputtering (MIBERS) system. The following part of the chapter describes the general aspects of the technique, and its capabilities in depositing high quality multicomponent ferroelectric thin films. The flexibility and versatility of this technique were exploited to conduct a study on effects of minor changes in composition on structure, microstructure and properties of the ferroelectric thin films.[15] The results are presented, in the instance of La doped $PbTiO_3$ (PLT), in the second part, showing the importance of accurate control of composition in structure and property modifications. The availability of controlled low-energy ion bombardment is another important feature of this technique and has been employed to modify beneficially the growth process, microstructure and electrical properties of ferroelectric thin films.[14,22,23] The last part of the chapter compiles such bombardment-induced effects in ferroelectric PZT films observed up to date.

2.0 THE MIBERS SYSTEM AND THE DEPOSITION PROCESS

The basic idea of the MIBERS technique is that individually mounted single-component targets are sputtered independently yet simultaneously by different ion beam sources, so that a highly controllable deposition process can be

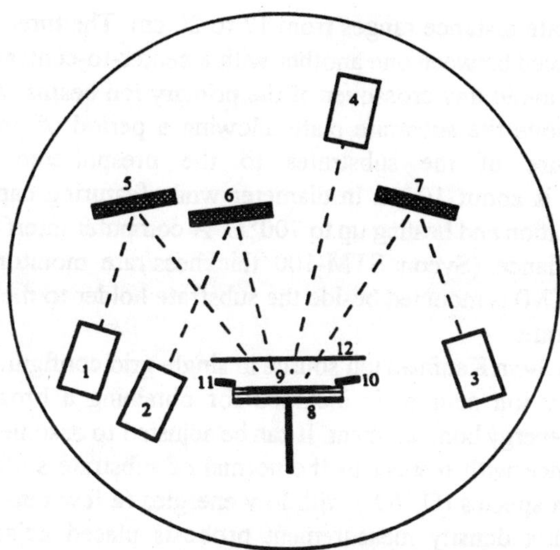

Figure 1. Schematic of MIBERS System

achieved. Such a technique, as can be expected, is rather complicated in instrumentation and requires a series of adjustments, modulations and calibrations. However, once the system is optimized and calibrated, deposition of any multicomponent films can be made with accurately controlled reproducibility. The following sections are intended to depict the flow chart of implementing a successful deposition with this technique.

2.1 Description of MIBERS

A schematic illustration of the MIBERS deposition system is shown in Fig. 1. The entire deposition assembly is placed in a water cooled vacuum chamber about 50 cm in diameter and 70 cm in length. The three water cooled target holders were designed to be 3-dimensionally movable and rotatable so that the target could be held at any position in the chamber and faced in any orientation. Three focused 3-cm Kaufman ion sources[24] (referred to as primary ion sources) were arranged facing the three targets, and were coordinately adjusted with the target holders so that optimum profiles of sputtered species were obtained on the plate of substrate holder. These primary ion sources use Ar^+ and are excited with high voltages (300-1500 volts). The source-to-target distance ranges from 11 to 16 cm, while the target-to-substrate distance ranges from 17 to 20 cm. The three targets are nearly evenly spaced between one another with a center-to-center distance of about 15 cm to avoid any crossover of the primary ion beams. A shutter is placed 2 cm above the substrate plate allowing a period of presputtering without exposure of the substrates to the presputtered flux. The substrateholder is about 10 cm in diameter while featuring capabilities of rotation or oscillation and heating up to 700°C. A computer interfaced quartz crystal microbalance (Sycon STM-100 thickness/rate monitor, hereafter referred to as QCM) is mounted beside the substrate holder to monitor the *in situ* deposition rate.

A fourth 3-cm Kaufman ion source in single grid configuration (often called secondary ion source) is included for obtaining a broad beam to accomplish low energy bombardment. It can be adjusted to assume a desirable angle of incidence with respect to the normal of substrate surface. It uses reactive oxygen species (O_2^+/O^+) with low energies (a few tens to hundred eV). An ion flux density measurement probe is placed adjacent to the substrates to measure the flux density of bombarding ions at the substrates. This measurement, in conjunction with the atomic flux density of depositing species determined from the QCM, establishes a parameter known as the ion/atom ratio, which is critical to achieve reproducible bombardment effects. To deposit the intended films, three 7.5 cm diameter metal targets of different kinds (e.g. Pb, Zr and Ti for PZT films, Pb, La and Ti for PLT, etc.) are mounted on the three water cooled target holders. By adjusting the tilt of the ion sources and the tilt of the targets, the sputtered fluxes, combined with substrate rotation of the substrate holder, a high degree of optimum uniformity of deposition, in thickness and composition, could be achieved.

Since different single-element targets are sputtered individually by independently operated ion sources, the MIBERS allows the utmost capability of varying and controlling the composition of depositing films. To make use of this capability and attain process reproducibility, the deposition rates of each component were calibrated in terms of deposition parameters such as primary beam voltage, beam current and oxygen partial pressure. Then specific deposition conditions were set to grow films with chosen compositions. Depositions can be made either with secondary ion bombardment or without the bombardment. In both cases the Ar gas is bled through the primary ion source as the precursor of Ar^+, while the O_2 gas is either bled through the secondary ion source as the precursor of O_2^+/O^+ ions if the secondary ion bombardment is to be applied or bled into the chamber

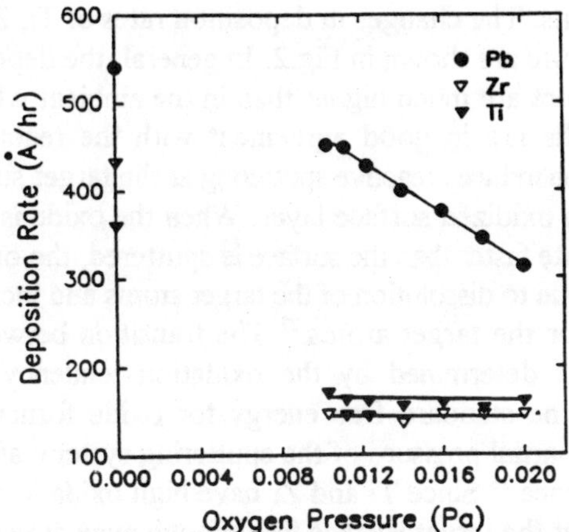

Figure 2. Variation of deposition rates of Pb, Zr & Ti with Oxygen partial pressure.

without ionization if the secondary ion source is not to be used.

2.2 Deposition Rate Calibration

To facilitate determining optimal combination of deposition parameters of any possible values, many calibration curves were obtained while the deposition parameters were scanned within a practical range. Only a few examples areshown in this section. All the deposition rate measurements were carried out by using the QCM, which had been previously calibrated.[25]

2.2.1 Deposition Rate Dependence on Oxygen Partial Pressure

It was found that, at constant oxygen partial pressure, varying total pressure by changing the argon partial pressure has no significant effect on thedeposition rates. This insensitivity is presumably due to the large mean free path of the gas species (on the order of 10 cm or larger) under the operating pressure range of the MIBERS system and the inertness of the argon. Since the mean free path is large, few collisions take place in the gas phase. The pressure changes will only influence interactions at the surfaces of targets or substrates. Since argon is an inert gas, it is expected to have little effect on these interactions.

Unlike argon, oxygen has a significant effect on the deposition rates of all the components. The changes in deposition rates of Ti, Zr and Pb with oxygen partial pressure are shown in Fig.2. In general, the deposition rates in a pure argon ambient are much higher than in the ambient where oxygen is present. The results are in good agreement with the reactive sputtering processes.[26] Oxygen induces reactive sputtering at the target surface, causing the formation of an oxidized surface layer. When the oxide is formed at the target surface at a rate faster than the surface is sputtered, the oxide decreases the sputtering rate due to dissolution of the target atoms and increased surface binding energies for the target atoms.[27] The transition between metal and oxide sputtering is determined by the oxidation tendency of the metal (characterized by the standard free energy for oxide formation from the metal), the oxygen partial pressure of the sputtering system, and the incident ion energy and fluence.[26] Since Ti and Zr have high oxidation tendency, the oxide layers forms at the target surface fast enough even at very low oxygen partial pressure under the indicated sputtering conditions.[26] Thus, the deposition rates of Ti and Zr drop dramatically upon introducing oxygen, and then remain nearly constant with increasing oxygen partial pressure indicating the saturation of oxide layer formation on the target surfaces. The deposition rate dependence of La on the oxygen partial pressure exhibits similar behavior to those of Ti and Zr. The Pb, by contrast, has less oxidation tendency and can assume different oxidation states,[28] and therefore the transition between metal and oxide sputtering at the target surface is partial over a range of oxygen partial pressure, resulting in a gradual decrease in the sputtering rate and thus the deposition rate.[25]

2.2.2 Deposition Rate Dependence on Incident Ion Energy and Beam Current

Figure 3 shows the deposition rates of Pb, Zr and Ti as functions of sputtering ion beam voltage, while the ion beam current and oxygen partial pressure were kept constant. The beam voltage determines the energy of the sputtering ions which must be kept above a threshold level necessary to initiate sputtering ofeach metal. However, often energies higher than the threshold are necessary to accomplish respectable deposition rates. It can be seen from the figure that while Zr and Ti show comparable deposition rate dependencies on the beam voltage and a common tendency of saturation at higher beam

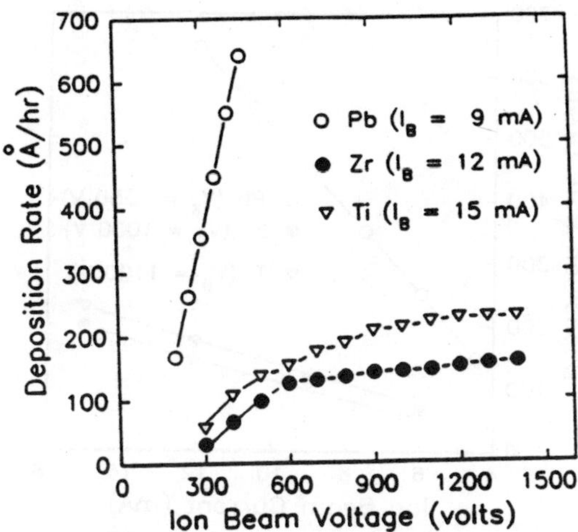

Figure 3. Deposition rates of Pb, Zr & Ti with ion beam voltage at constant ion beam current.

voltages, Pb shows much higher deposition rate than Zr and Ti and its saturation is not observed in the voltage range examined.

Figure 4 shows the dependencies of deposition rates of Pb, Zr and Ti on ion beam current, while the ion beam voltage and oxygen partial pressure were kept constant. The beam current determines the flux density of ionized species (Ar^+ in the present case) participating in the sputtering process. Again, Pb shows much higher deposition rate and faster increase with ion beam current, compared to Zr and Ti. It is noted that for Pb a beam voltage of 350 V is large enough to obtain comparable deposition rates to those of Zr and Ti at beam voltage of 1000 V or higher. Under the indicated conditions, nearly linear relations exist between deposition rate and beam current for all the three components, which is in good agreement with theory of sputtering yields[29] since the deposition rate is proportional to the sputtering rate.

At higher ion energies (i.e. higher beam voltages), however, the deposition rate of Pb versus the beam current exhibits a nonlinear behavior as shown in Fig.5. This can be related to the complex transitions between metal and oxide sputtering of the Pb target, as mentioned above. At fixed oxygen partial pressure, higher incident ion energies cause both higher sputtering rate and deeper penetration of the incident ions at the surface, so that more

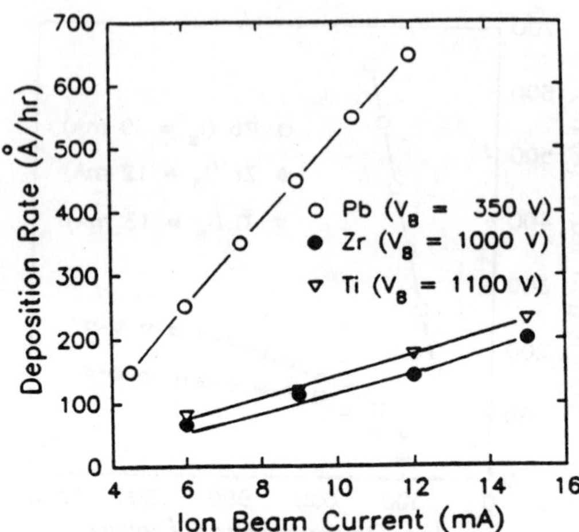

Figure 4. Deposition rates of Pb, Zr & Ti with ion beam current at constant ion beam voltage.

metallic species are sputtered than in the case of lower incident ion energies. The nonlinear behavior of the deposition rate begins when the rate of formation of the oxide surface becomes slower than the sputtering rate.[25]

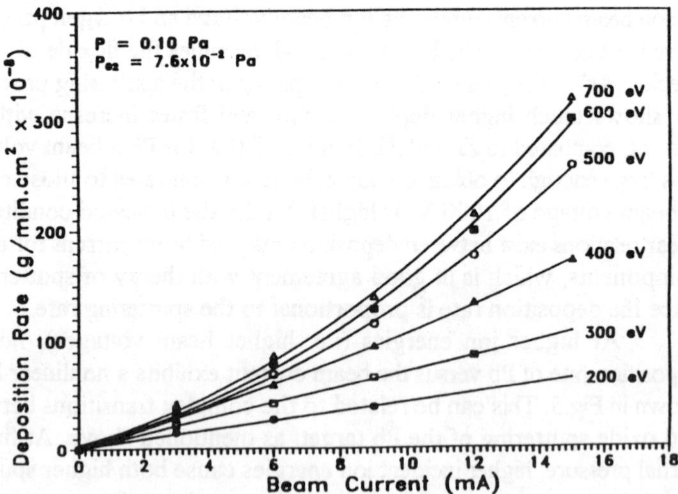

Figure 5. Pb depositiion rate with ion beam current for increasing ion energy & constant pressure.

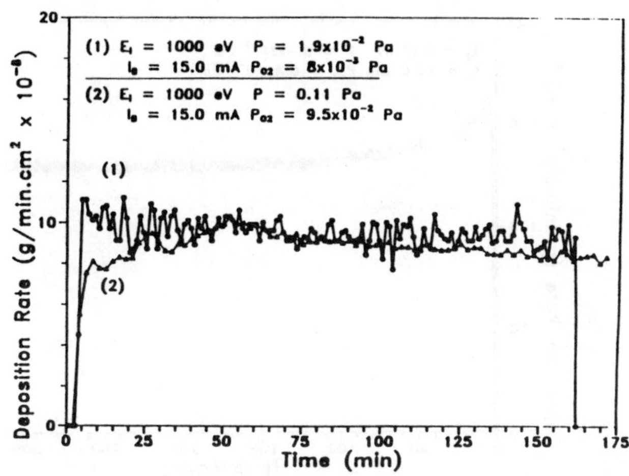

Figure 6. Time dependence of Ti deposition rate for two pressure conditions at constant ion beam current & energy.

2.2.3 Time Stability of Deposition Rates

The deposition rate as a function of time is shown in Fig.6 for Ti sputtered under two different conditions. Both conditions exhibit short term variations in deposition rate (approximately ±10%) which are probably due to variations in the ion source or noise from the QCM. These short term deposition rate changes, which occur over a few minutes, become insignificant when averaged over the length of the entire film deposition. Averaging over several data points eliminates the short term fluctuations in the deposition rate, and reveals a slight decay in the deposition rate observed over the entire length of the deposition. This decay in deposition rate generally remains within a few percent of the initial deposition rate setting. The deposition rates of Zr and La are also very stable over a long deposition time.

Unlike the Ti, Zr and La, the Pb exhibits a significant change in the deposition rate after long periods of sputtering as shown in Fig.7. The initial deposition rate for the Pb decays up to 30% within a 90 min period after which a constant deposition rate is achieved. This initial decay is repeated whenever the vacuum chamber is vented and necessitated for the Pb target to be presputtered for the same time (e.g. 80 min), to get a steady state deposition rate, for every run. Both decay in the initial deposition rate during

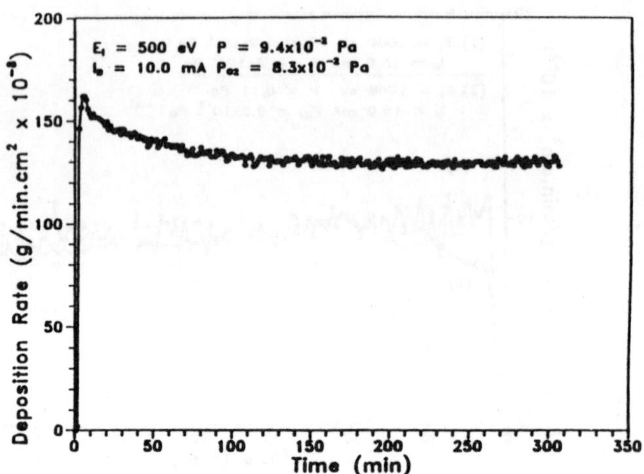

Figure 7. Time dependent Pb depositiion rate at constant sputtering conditions.

the run and the long term degradation of the deposition rates between runs of the Pb target are believed to be due to changes in the target surface morphology.

The most convincing observation on the changes in the target surface morphology is the cone formation, which can be seen after just a few hours of sputtering and becomes more severe with increasing sputtering time. Cones are formed when impurity atoms mask an area of the target and material is sputtered at a higher rate in areas surrounding the impurity atom.[30,31] Since oxygen increases the biding energy of the Pb, the cones may be a result of the reactive sputtering in oxygen.

The cone formation on the Pb target also changes the spatial distribution of sputtered species and thus deposition profile[25]. It was found that the change in the spatial distribution is insignificant only during the first 70 h of sputtering. Therefore, the Pb target needs to be periodically resurfaced (every 70 h or so) to minimize the influence of the cone formation on deposition rate and spatial distribution.

2.3 Film Deposition by MIBERS

Once the deposition rate calibration was done, growing ferroelectric thin films

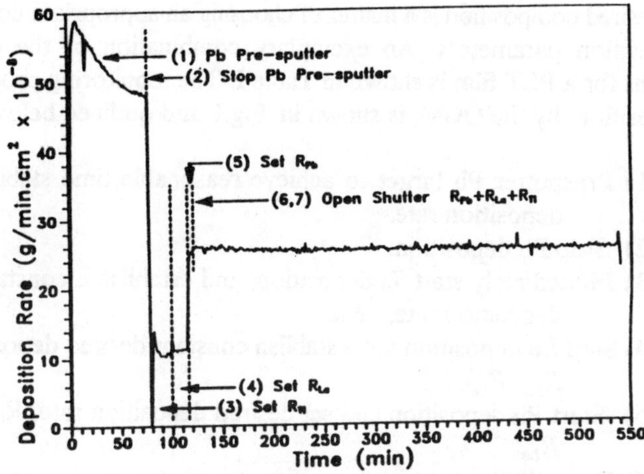

Figure 8. Typical deposition rate versus time for PLT growth.

Table I. Typical deposition conditions for PLT thin films with L/T = 0.05

	Pb target	La target	Ti target
Presputtering Time	90 min	15 min	15 min
Presputtering Ion Energy	500 eV	300 eV	1000 eV
Presputtering Ion Beam Current	5.0 mA	4.5 mA	15.0 mA
Sputtering Ion Energy	500 eV	300 eV	1000 eV
Sputtering Ion Beam Current	1.5 mA	1.8 mA	15.0 mA
Base Pressure	8×10^{-4} Pa		
Working Pressure	3.0×10^{-2} Pa		
Oxygen Partial Pressure	7×10^{-3} Pa		
Substrate Temperature	90-110°C		

with a desired composition is a matter of choosing an appropriate combination of deposition parameters. An exemplary combination of the deposition conditions for a PLT film is shown in Table I. The monitoring procedure for the deposition, by the QCM, is shown in Fig.8 and outlined below:

1) Presputter Pb target to achieve reasonable time stability of the deposition rate.
2) Stop Pb deposition.
3) Immediately start Ti deposition, and establish a constant desired deposition rate, R_{Ti}.
4) Start La deposition and establish constant desired deposition rate, $R_{La}=R_{tot}-R_{Ti}$.
5) Start Pb deposition and set desired deposition rate, $R_{Pb}=R_{tot}-R_{Ti}-R_{La}$.
6) Open shutter to start deposition on substrates.
7) Monitor $R_{tot}=R_{Ti}+R_{La}+R_{Pb}$ during deposition to confirm rate stability.

Since R_{Ti} and R_{La} are stable over time, slight adjustments in R_{Pb} can be made if necessary. Generally, once presputtering was completed and the deposition rates were established, R_{tot} remained nearly constant unless problems occurred with the ion sources, gas supply, or vacuum system. This method of composition monitoring can be used for any combination of target materials.

2.4 Evaluation of The MIBERS Technique

The success of the MIBERS technique was evaluated, in the instance of depositing PZT films, in terms of process reproducibility, deposition uniformity and basic characteristics of the resultant films such as crystallization behavior, morphology and electric properties. These are briefly outlined in this section.

2.4.1 Reproducibility and Uniformity

To evaluate the reproducibility of the system in composition control, Rutherford Backscattering Spectrometry was used to examine the composition of the as-grown films from several deposition runs made with the

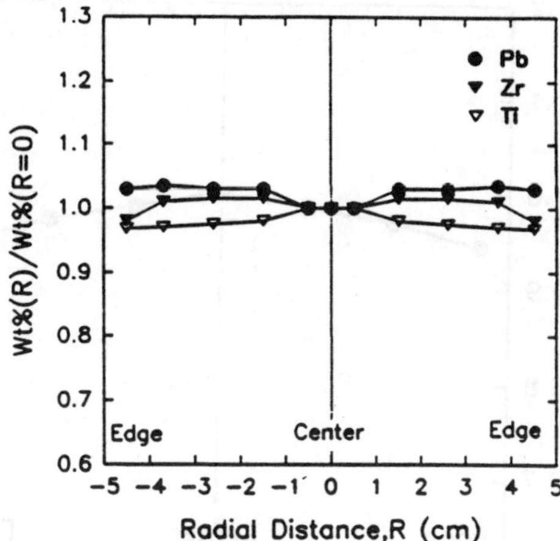

Figure 9. Radial variation of composition across substrate diameter with respect to center composition.

same deposition conditions. It was found that even for the most fluctuated component Pb, the deviation was less than 2%. To evaluate the uniformity of MIBERS grown films, measurements of both compositional variation and thickness variation across the whole area of the substrate holder were carried out, in the case of depositing PZT(50/50) films.

The results are shown in Figs.9 and 10 which indicate that for the composition (based on EPMA signal taken on a submillimetric average) the variation is within 5% across a diameter of 10 cm, and for the thickness, within 10%. Within a 7.5 cm diameter area, both variations are within 5%[14]. Such high uniformity has rarely been reported so far with other sputtering techniques. The capability of offering large area uniformity of the present MIBERS system may be attributed to its particular design, in addition to the rotation of substrates. First, the orientation of each target was independently adjusted such that each component might achieve an optimum spatial distribution of the sputtered species over the substrates. Second, unlike in other plasma based sputtering techniques, in this system both source-to-target and target-to-substrate distances were not restricted to any specific value so that they could be varied within some range to achieve the required deposition coverage (i.e. the coverage can be varied by varying either of the distances since either the sputtering ion beam or the sputtered flux is cone-like rather

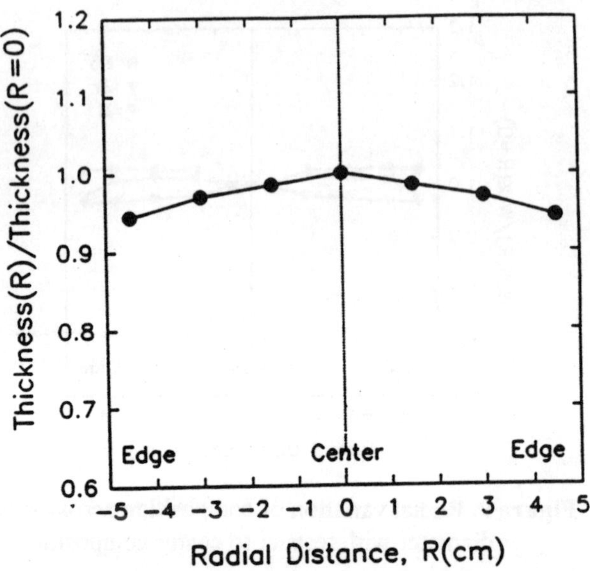

Figure 10. Radial variation of film thickness with respect to center thickness.

than collimated.).

2.4.2 Structural Characteristics of PZT films

The PZT films deposited without substrate heating are amorphous. *In-situ* crystallinity is achieved by heating the substrates to elevated temperatures. Two technique were employed to induce crystallization after deposition: conventional furnace annealing and rapid thermal annealing (RTA). It was found that crystallization behaviors of the films during the two annealing processes are very different.[32] Perovskite phase formation during furnace annealing is a slow and sluggish process, featuring a two stage transition. In heating the amorphous as-grown films by furnace (with a heating rate of 10°C/min), pyrochlore phase forms at about 470°C, exists over a wide temperature range, and starts to transform to perovskite phase after 600°C. At 600°C, it is necessary to soak the films for nearly 1 h to complete the perovskite phase formation. By contrast, during RTA crystallization starts at much lower temperatures and also proceeds much faster. The pyrochlore phase forms at about 340°C, and the perovskite phase forms above 420°C. If

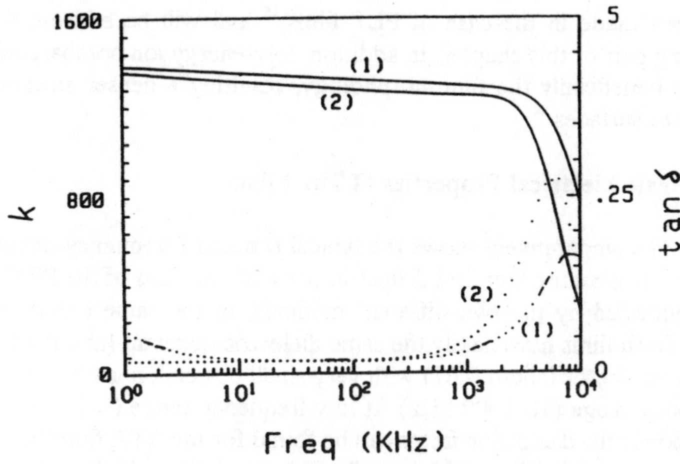

Figure 11. Frequency dependent k and tanδ for RTA and oven annealed PZT films.

the annealing temperature goes beyond 450°C, the perovskite phase formation is completed instantaneously (in 1 sec), without undergoing pyrochlore phase formation. Such a fast process appears much like an "explosive crystallization".[33] This marked difference in crystallization behavior between the two annealing methods is believed to be caused by the significant difference in heating rate as well as the highly amorphized structure of the as-grown films.[32] *In-situ* crystallization by substrate heating can reduce the perovskite formation temperature to about 520°C, which however is still not as low as that by RTA. Many other factors can also influence the crystallization behavior of Pb-based perovskite type ferroelectric thin films. These include, but are not limited to, amount of excess Pb,[14,15] oxygen partial pressure during the deposition,[15] kind of substrate[14] and low-energy ion bombardment.[14,22] The MIBERS deposited ferroelectric films are commonly of fine-grained and fairly dense structure (typically with average grain size ranging from 10 to 40 nm), which seems to be intrinsic to the ion beam sputter deposition technique.[16,20] However, low-temperature deposited and post-annealed films usually assume typical columnar or cluster-and-void morphology,[14,15] which is attributable to the low adatom mobility during the deposition.[34] The Pb content can significantly alter the morphology.[14,15] A systematic observation on the Pb effect on morphology and microstructure has been made in the case of PLT films,[15] and will be addressed in the following part of this chapter. In addition, low-energy ion bombardment can modify beneficially the film morphology, resulting a denser structure and smoother surfaces.[14]

2.4.3 Basic Electrical Properties of The Films

Figure 11 comparatively shows the typical results of frequency-dependence of room-temperature low-field dielectric constant and loss of the PZT(50/50) films annealed by the two different methods, at the same temperature of 650°C. Both films have nearly the same dielectric constant (about 1400) and dissipation factor (about 0.02) with very small dispersion over a fairly wide frequency range (10-1000 kHz). At low frequency range (< 10 kHz), small dispersion in the dissipation factor can be found for the RTA films but not for the furnace annealed films. This is attributable to relatively higher conductivity in the RTA films caused by a higher residual Pb content. In the high frequency range (> 1000 kHz), the furnace annealed films are more dispersive than the RTA films, which is characteristic of thicker interface barriers.[35] Therefore, it can be suggested that RTA is definitely better in the sense of minimizing interface reaction, while it should take a little longer to reduce the residual Pb content although 1 s looks enough for crystallization of perovskite phase.[32]

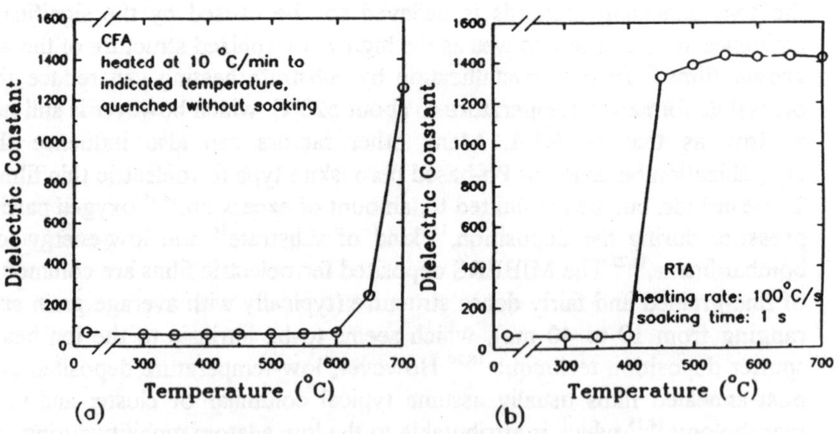

Figure 12. Dielectric constant with process temperature, by (a) Oven & (b) RTA annealing process.

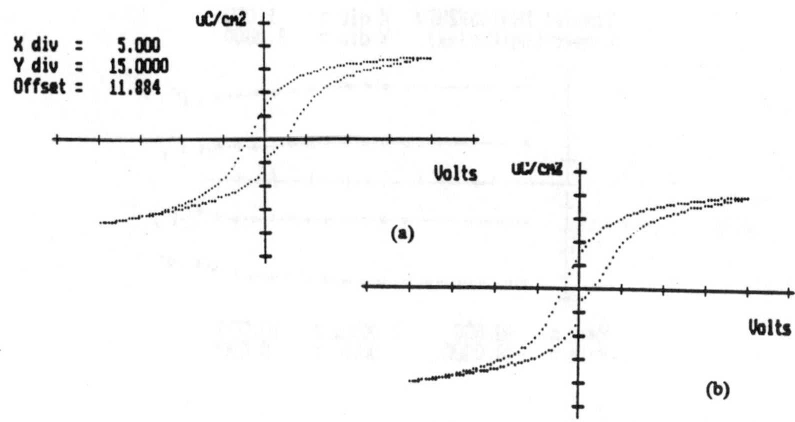

Figure 13. P-E hysteresis loops for PZT (50/50) films by RTA at 650 C for (a) 1 sec & (b) 30 secs.

The evolutions of dielectric constant of the films with annealing temperature are shown in Fig.12 for both annealing methods. As can be seen, while the evolution of dielectric constant during furnace annealing is rather slow, an abrupt increase to the saturation value in the dielectric constant is observed in the case of RTA within the temperature range from 400 to 450°C. These results are in good agreement with the crystallization behaviors of the films as seen above.

Figure 13 shows two P-E hysteresis loops of the films annealed by RTA at 650°C for different time. It can be noted that they have nearly the same remanent polarization of 20-25 $\mu C/cm^2$, although the coercive field seems decreasing slightly with increasing annealing time. This is consistent with the results of dielectric measurements described above, and again exhibits the nature of "explosive crystallization" by RTA. The lower coercive field in the case of longer annealing time is presumably due to relative higher crystallinity and "purer" structure of the films, as more defects (including residual Pb) and grain boundary phase would be annealed out as annealing time increased.

Development of ferroelectric thin films directed towards memory devices demands reliability over long time performance. Figure 14 shows fatigue behavior in polarization of the PZT(50/50) films under ±170 kV/cm

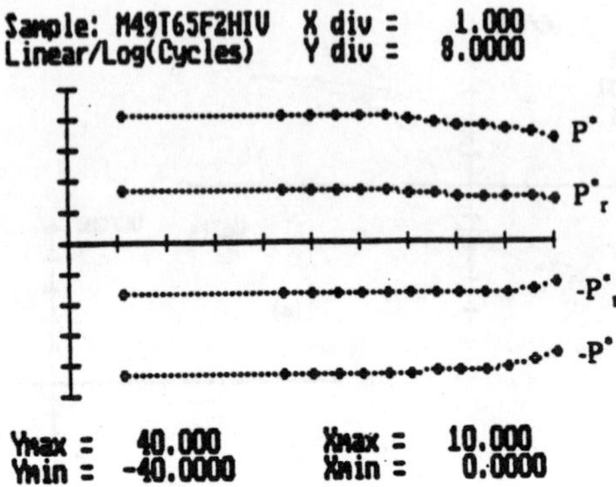

Figure 14. Fatigue behavior of PZT (50/50) films tested with 170 kV/cm bipolar pulse switching.

bipolar pulse cycling. It can be seen that the decay in polarization is less than 20% after 10^{10} switching cycles. Retention of the polarization of such films was measured by applying initially a -5 V write pulse and afterwards +5 V and -5 V read pulses. The loss in the stored charge is less than 10% after 10^5 s (Fig.15).

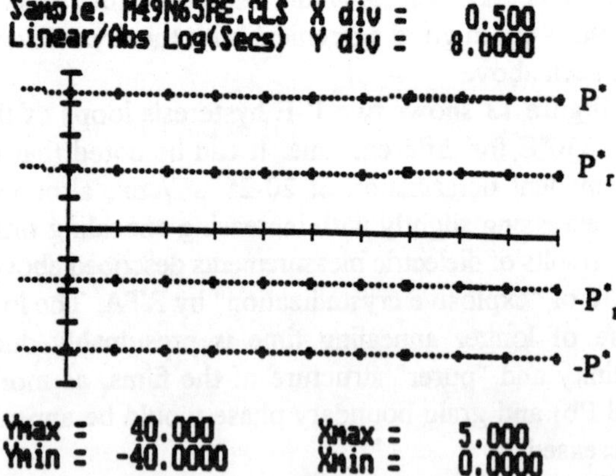

Figure 15. Polarization retention of PZT (50/50) films under zero bias, for a write -12 V followed by +12 V reading pulses.

3.0 COMPOSITION/STRUCTURE/PROPERTY RELATIONS OF MIBERS DEPOSITED La DOPED $PbTiO_3$ (PLT) THIN FILMS

The electrical properties of ferroelectric thin films are strongly tied to the microstructure and the crystal structure of the thin film material. Both the microstructure and the crystal structure are controlled by the thin film composition and the method by which the thin film is processed.[15] The relationship between composition, crystal structure, microstructure, and properties of MIBERS deposited ferroelectric thin films was explored in La doped $PbTiO_3$ (PLT) system, by observing the evolutionary development of these physical features during post-deposition annealing. Particular emphasis was placed on determining how small changes in composition affect the physical state of films annealed under constant conditions.

Because of the versatility of the MIBERS deposition technique, it was possible to study the effects of minor changes in anion and cation composition on crystal structure and microstructure development in both as-deposited and annealed films. The anion composition was adjusted by altering the reactive oxidation of the sputtered species, while the cation composition was modified by changing the relative sputtering rates of the Pb, La, and Ti metal targets. By studying these PLT films with small systematic changes in composition, the anion and cation composition-induced changes in the crystal structure and microstructure were differentiated.

The relationships between composition and microstructure form a link between alterations in composition and modification of thin film electrical properties. A microstructure model, developed from the observed relationship between composition and microstructure, was combined with a simple mixing rule model for a heterogenous material. Composition induced changes in the electrical properties are readily explained through the use of the combined microstructure and mixing rule model.

3.1 Characteristics of As-deposited PLT Films

The PLT films were deposited onto a Pt(200nm)/Ti(20nm)/SiO_2(1000nm)/Si substrate with a low substrate temperature of approximately 100°C.[25] Because the low substrate temperature did not allow significant adatom mobility, the sputtered species condensed on the substrate as an amorphous phase. According to X-ray and electron diffraction analysis, neither small changes in the anion composition nor small changes in the the cation

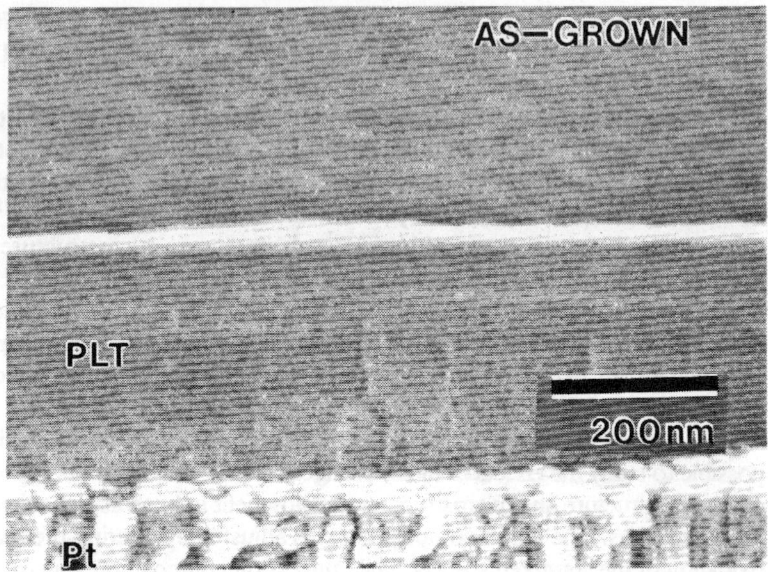

Figure 16. SEM fracture cross section of a typical as-deposited PLT thin film.

composition of the as-deposited films induce the formation of crystalline material. The amorphous nature of as-deposited films is insensitive to composition (for the range of compositions and deposition conditions studied) and is observed both inside and outside the PLT solid solution region.[41] The microstructure of the as-deposited films is also independent of the film composition. A typical fracture cross-section of an as-deposited film is shown in Figure 16. Approximately 10 nm particles are packed to form columns which are approximately 50 nm in diameter. Low density boundaries or void networks separate the tightly packed amorphous particles and columns which resemble the structures observed in the 1-T zone of the SZM.[36] The columnar structure of the PLT film mimicks the columnar structure of the Pt layer which consists of approximately 20 nm columns.

The consistency of the amorphous structure and the tightly packed columnar growth over a range of cation and anion compositions can be attributed to the basic features of the MIBERS technique. Since surface binding energy primarily determines the kinetic energy of the sputtered species, small changes in the sputtering rates of the Pb, La, and Ti targets (which are used to change the cation composition of the growing film) do not cause a significant change in the energy of the depositing species. Deposition pressure also does not significantly affect the energy of the depositing species since the pressures used for MIBERS deposition provide a large mean free path with minimal probability for collisions between the background gas and the depositing species. The modal energies of the Pb, La, and Ti species sputtered under oxidizing conditions are 4.5, 8.4, and 17.9 eV, respectively. Because the energies of the sputtered species remain constant over the range of sputtering conditions used for depositing films of various compositions, all the as-deposited films exhibit an amorphous and tightly packed columnar structure.[15,25]

Since the amorphous as-deposited PLT thin film microstructure is independent of composition, the influence of composition on crystallization and microstructure development during annealing can easily by studied. The effects of the anion and cation composition on this evolutionary process are presented in the following sections.

3.2 Effects of Anion Composition on Perovskite/Pyrochlore Phase Formation

Oxidation of the Pb in as-deposited PLT films controls the formation of perovskite and pyrochlore phases in annealed films. Understanding this behavior requires the knowledge of the steady state reactive sputtering process and the reactions occuring during annealing. Once the dependence of perovskite/pyrochlore formation upon Pb oxidation is clarified, perovskite films with a range of cation compositions can be prepared without the interference of pyrochlore formation.[25,35]

During PLT deposition, the MIBERS technique relies on an oxygen background gas to provide the oxidation of the depositing film. An oxide layer is formed at the surface of the Pb, La, and Ti metal targets by adsorption, which is subsequently followed by an oxidation reaction. If the sputtering rate is slower than the rate of oxide formation at the target surface, metal oxide species are sputtered from the target. For sputtering rates that are faster than the oxidation rate, both metal and metal oxide species are sputtered, and the relative amounts of each species is dependent on both the sputtering rate and the concentration of oxygen at the target surface.[25-27]

Observations of the positive ion energy distribution can be used to reveal changes in the speciation of sputtered material. The species sputtered from a target can be neutral, positive, or negatively charged (charged species are refered to as sputtered ions), and the relative quantities of each type of species is determined by the atmosphere and the material that is being sputtered.[37] When oxygen is adsorbed onto the surface of a target, the formation of a surface oxide layer causes metal oxide species (e.g., TiO_2 and PbO) to be sputtered along with the metal species (e.g., Ti, Ti+). The change in sputtered speciation modifies the relative quantities of neutral, positive, and negative species composing the sputtered material.[43-45] Because of the relationship between speciation and the ion fractions of sputtered material, comparison of sputtered ion energy distributions, measured for a constant sputtering rate, can be used to detect changes in sputtered speciation under various sputtering conditions.

An electrostatic ion probe was used to measure the energy distributions of the positively charged sputtered species for the Pb, La, and Ti metal targets. The constancy of the positive ion energy distributions for material sputtered from the La and Ti targets confirmed that there was no change in speciation over the entire range of oxygen partial pressure (P_{O2} = 8×10^{-3} to 7.6×10^{-2} Pa) used for PLT deposition. Both the La and Ti exhibit a high free energy for oxide formation which caused a rate of target surface oxidation faster than the sputtering rates utilized; therefore, oxidized species of La and Ti were sputtered from the targets and deposited in the PLT films.[25]

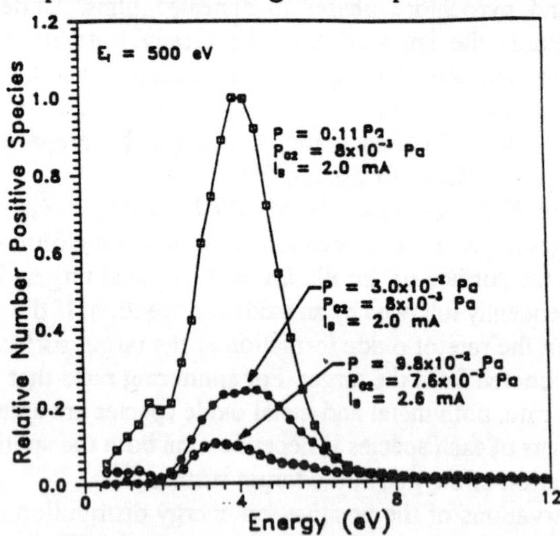

Figure 17. Energy distribution of secondary positive ion species from Pb target with increasing oxygen concentration.

Unlike La and Ti, the speciation of material sputtered from the Pb target is sensitive to the oxygen concentration (at the target surface) used for PLT deposition; this is demonstrated by the changes in the positive ion energy distributions (recorded for a constant deposition rate) shown in Fig. 17. The fraction of positive sputtered ions decreases as the oxygen concentration at the target surface is decreased. The oxygen concentration is decreased by either decreasing P_{O2} or increasing P_{Ar}. As the oxygen concentration at the target surface increases, the fraction of sputtered metal oxide species (e.g., PbO and PbO+) increases while the fraction of reduced metal species (e.g., Pb and Pb+) decreases; this results in a decrease of the total number of positive sputtered ions ejected from the target. The sensativity of the Pb speciation to the oxygen concentration is attributed to the low free energy of PbO formation (as compared with La and Ti oxides) which produces an oxidation rate that is similar to the sputtering rate of the target surface.[25]

These results, revealing the dependence of Pb speciation oxygen concentration, were used to deposit PLT films with two different oxygen contents. Films deposited at low P_{O2} = 8x10^{-3} Pa were oxygen deficient, containing $PbO_{1-\delta}$, and films deposited at high P_{O2} = 7.6x10^{-2} Pa contained PbO with no significant oxygen deficiency. Annealing the films with the two different oxygen contents (but with equal cation compositions) produces two different crystalline phases. Oxygen deficient films produce the perovskite phase, and the fully oxidized films produce the pyrochlore phase, as indicated by the X-ray diffraction patterns shown in Fig. 18. The phase formation was found to be independent of cation composition for constant oxygen content, and in oxygen deficient films, the oxidation of $PbO_{1-\delta}$, contributes an excess energy of 220 kJ/mol to the simultaneously occuring crystallization reaction, resulting in the formation of the ferroelectric perovskite phase. Without the excess energy contributed by the oxidation reaction, the metastable pyrochlore phase is crystallized.[35]

Measuring the sputtered ions ejected from the targets provides an *in situ* method for monitoring changes in the oxidation controlled speciation. Using this technique has made it possible to detect small changes in the oxygen content of Pb species that are deposited to form PLT films. By controlling the oxidation of Pb deposited in PLT it is possible to prepare pure pyrochlore and pure perovskite films over a wide range of cation compositions.[25] It is now possible to study microstructure evolution and cation composition effects in perovskite PLT films.

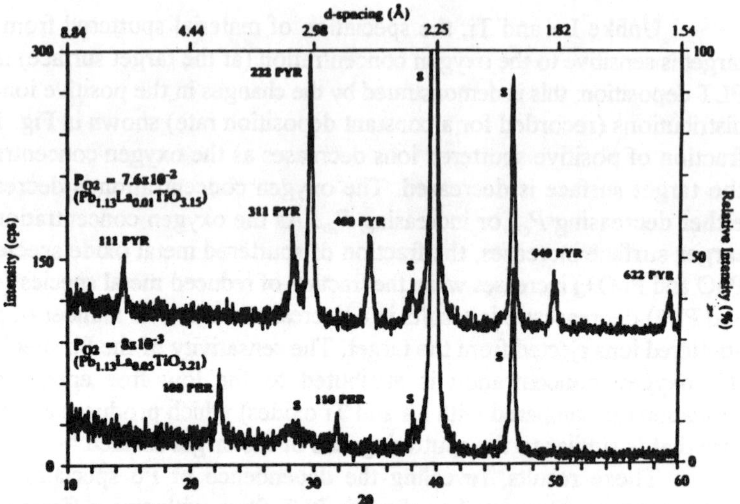

Figure 18. XRD patterns of annealed, x = 0.05, PLT films with same Pb content but deposited at different oxygen partial pressure conditions.

3.3 Effects of Cation Composition on Microstructure Development

Microstructure evolution in perovskite PLT films is greatly dependent upon the as-deposited cation composition and the post-deposition annealing process. Annealing initiates crystallization, densification, and PbO evaporation; all these evolutionary mechanisms are composition dependent and interdependent.[15] When comparing films annealed under different conditions it is necessary to characterize both the annealed composition and microstructure of the films for a proper comparison. If the annealing conditions are held constant, the relationships between as-deposited composition and annealing induced microstructure development can be uncovered.

Crystallization of PLT films begins at annealing temperatures as low as 400°C, but the crystallization process is not completed unless the films are annealed at 600°C or higher temperatures. Lead oxide evaporation occurs simultaneously with the perovskite crystallization; evaporation starts as low as 490 ± 50°C and the rate of evaporation increases with increasing annealing temperature. An example showing the loss of PbO due to evaporation is given

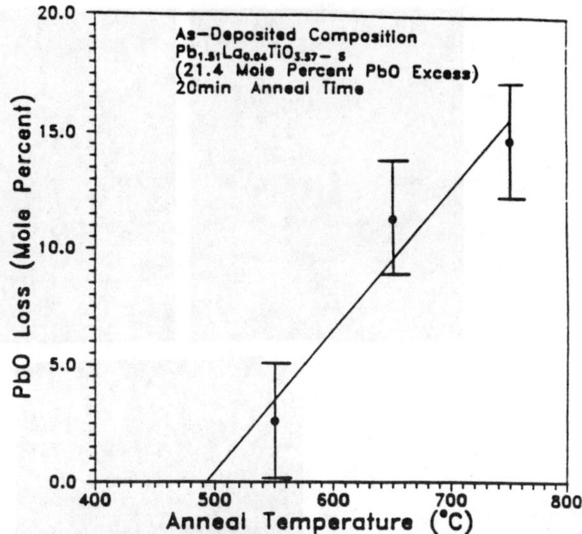

Figure 19. Pb loss with annealing temperature for PLT films.

in Fig. 19 for films with constant starting composition (21.4 mole percent excess $PbO_{1-\delta}$ in the as-deposited films) which were annealed at different temperatures for the same time (20 min) and heating rate (10°C/min).

Both crystallization and PbO evaporation during annealing cause the evolution of the as-deposited microstructure. The annealed microstructure consists of fine grains which assemble to form clusters that are separated by porous cluster boundaries. As crystallization and PbO evaporation proceeds during annealing, the cluster boundaries and fine grain structure become more pronounced (Fig. 20) due to a thermal etching effect, which results from grain growth and enhanced PbO loss at the grain and cluster boundaries. Annealing temperature does not change the cluster size, suggesting that the cluster boundaries observed after annealing develop from the low density boundaries present in as-deposited films. Because the film composition changes with increasing annealing temperature (due to PbO evaporation), it is difficult to differentiate between the effects of annealing temperature and the effects of composition on microstructure evolution.[15]

Films with systematic changes in the as-deposited cation composition (the anion composition was fixed by fixing the P_{O2} were studied under constant annealing conditions (650°C/20 min) to determine the relationship between as-deposited composition and microstructure formation. As

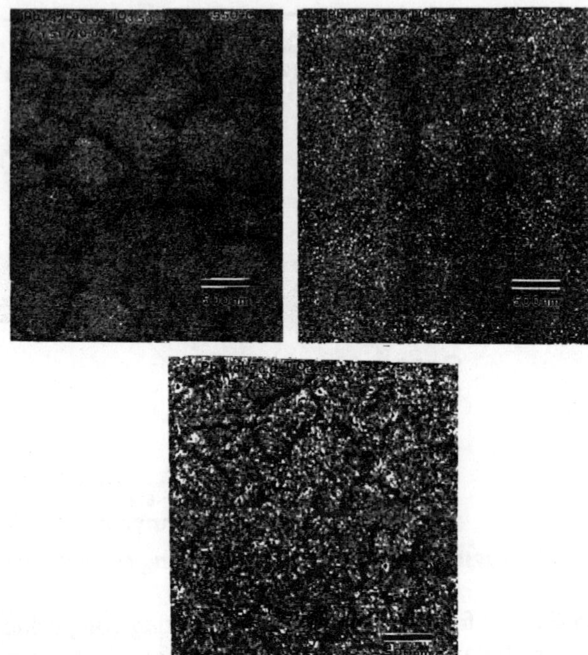

Figure 20. SEM surface images of films with a starting composition of $Pb_{1.51}La_{0.04}TiO_{3.57-\delta}$, annealed at 550, 650 and 750 C for 20 mins. Cluster boundaries are marked with arrows.

discussed previously, microstructure development is dependent on PbO evaporation and crystallization. Evaporation of PbO is not only temperature dependent but is also dependent on the starting composition of the film. As shown in Fig. 21, PbO loss increases with an increasing $PbO_{1-\delta}$ excess in the as-deposited film. The PbO loss with a small $PbO_{1-\delta}$ excess is controlled by the vapor pressure of PbO in PLT while the vapor pressure of pure PbO limits PbO loss in films with a high $PbO_{1-\delta}$ excess. An increase in La concentration also results in an increase in PbO evaporation due to the preference for an increased number of cation vacancies at the higher La content.[15,25]

All the films studied for the constant annealing conditions of 650°C/20 min were fully crystallized, but the crystallographic structure detected by X-ray diffraction (Fig. 22) changes dramatically with increasing $PbO_{1-\delta}$ content

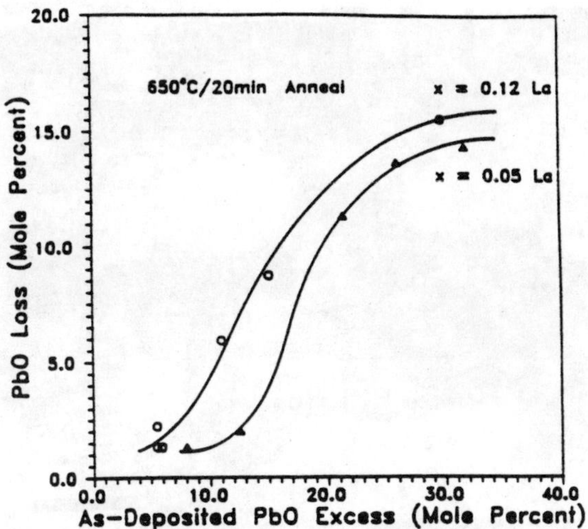

Figure 21. Pb loss with excess PbO content for PLT films of La=0.05 & 0.12 annealed at 650 C for 20 mins.

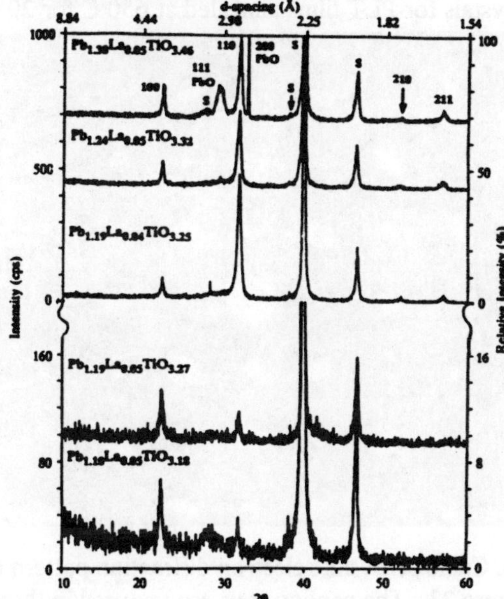

Figure 22. XRD patterns for x=0.05 PLT films with increasing excess PbO content and were annealed at 650 C for 20 mins.

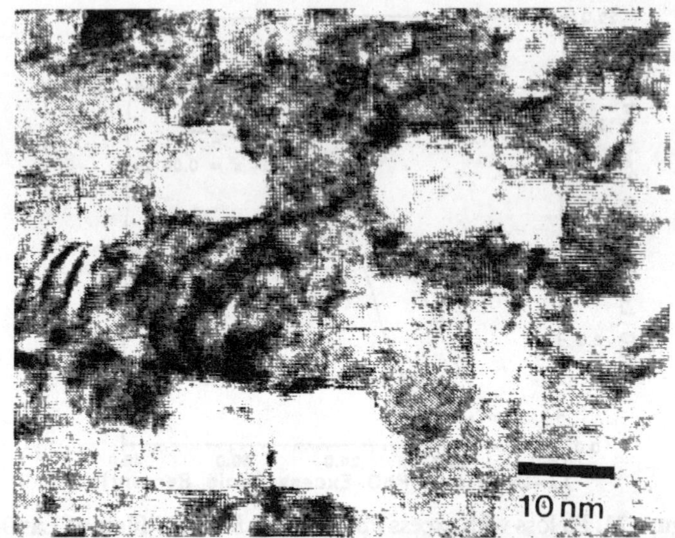

Figure 23. Plan-view TEM lattice image showing textured nanocrystals for PLT films annealed at 650 C for 20 mins.

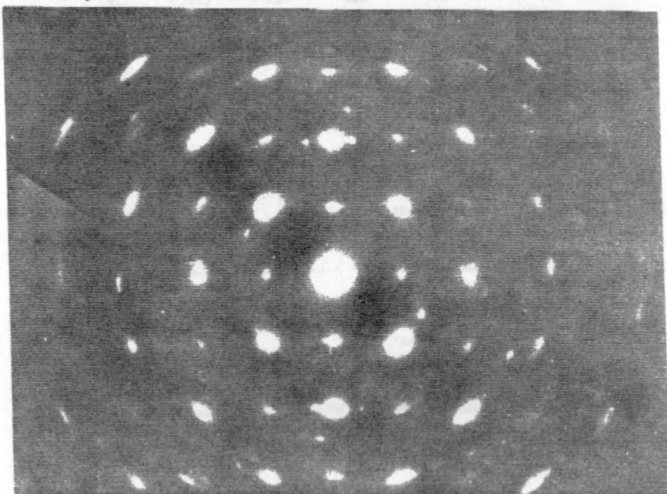

Figure 24. Cubic perovskite electron diffraction pattern of the area in figure 23. The nanocrystals are textured in three dimensions with boundaries along the {100} planes.

in the as-deposited films. Films with the lower PbO contents exhibit diffraction patterns indicative of a <100> textured structure, but at high PbO contents polycrystalline diffraction patterns are observed, suggesting that the <100> texture is destroyed. The composition dependence of film texture is similar for films with 5 and 12 mole percent La. To thoroughly understand the texture dependence on composition, the microstructure must be studied, in detail, over a range of size scales.

The smallest microstructural units composing the PLT films are 10 nm crystallites, as seen in Fig. 23. These nanocrystals, which are separated by low angle grain boundaries along the {100} planes, assemble to form clusters. Within a cluster, nanocrystals are textured in all three dimensions, as revealed by the electron diffraction pattern in Fig. 24. Cluster boundaries, which consist of approximately 30 nm pores, break the continuity of the textured nanocrystals, resulting in random orientation of clusters.

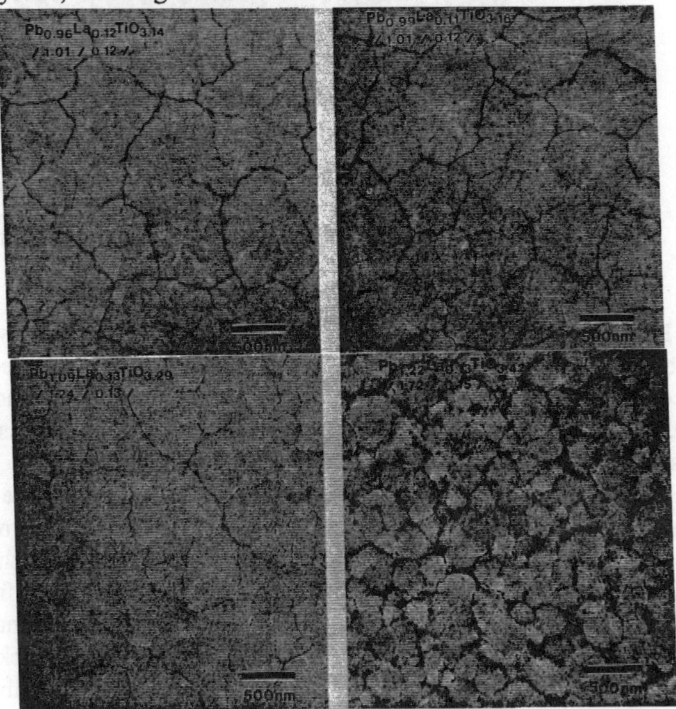

Figure 25. Cluster structure of PLT films with 12 mole % of La. Cluster size decreased with increasing PbO content.

Figure 26. Lattice image of a nanocrystal grain boundary in a film annealed at 650 C for 20 mins.

As the $PbO_{1-\delta}$ excess within an as-deposited film is increased, the cluster size observed after annealing decreases, as shown in Fig. 25. In the annealed films, the excess PbO and porosity formed by PbO evaporation disrupts nanocrystal texture, and the volume of continous texture (or cluster size) decreases with increasing quantities of second phase.

While the cluster size remains larger than the thickness of the film all the clusters are <100> oriented perpendicular to the film plane (i.e., perpendicular to the substrate surface) but randomly oriented within the plane of the film. The <100> texture perpendicular to the film plane results from the nucleation of nanocrystal growth at the interface between the PLT and the Pt substrate layer. When the cluster size becomes smaller than the thickness of the film, the clusters are randomly oriented through the film thickness (perpendicular to the film plane) and within the film plane. It is the loss of cluster orientation perpendicular to the plane of the film that causes the change from a <100> textured to a polycrystalline X-ray diffraction pattern.

Images of nanocrystal boundaries, such as the one shown in Fig. 26, do not show the presence of a PbO grain boundary phase. All evidence suggests that excess PbO in the <100> textured films (as determined by X-ray diffraction) is isolated in pockets within a continuous perovskite phase. With the addition of sufficient excess PbO (after annealing) the pockets of PbO form a continuous network which induces random cluster orientation, forming non-textured films (as determined by X-ray diffraction). Pores formed by PbO evaporation also contribute to the loss of <100> texture, and the pore distribution appears, in the textured and non-textured microstructure, in a fashion similar to the PbO second phase. Simplified models for textured and non-textured films are given in Fig. 27 (a) and (b), respectively. The model for textured films assumes a continuous PLT phase with isolated pores and pockets of PbO, and the model for non-textured films assumes a continuous mixed second phase of porosity and PbO which separates PLT clusters.[25,39,40]

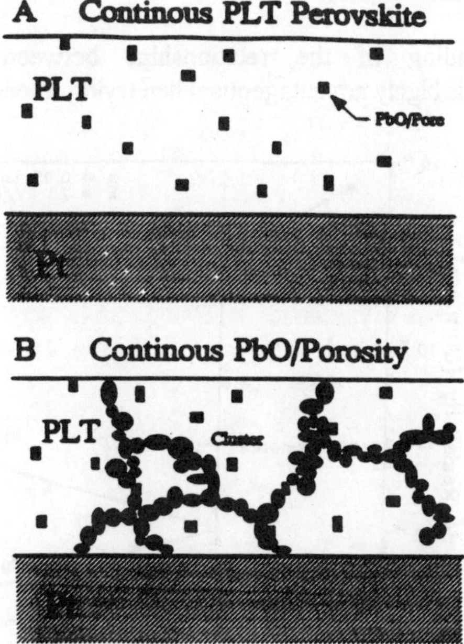

Figure 27. Microstructure models for (a) <100> textured & (b) non-textured PLT thin films with mixed phases of PLT perovskite, PbO and air. Both air and PbO are shown in shaded regions.

The detailed study of cation composition, crystallographic texture, and microstructure has made it possible to relate changes in as-deposited composition to changes in the texture and microstructure of annealed films. Since both the composition and the microstructure of an annealed film are dependent on the annealing procedure (i.e., annealing temperature, time, and atmosphere), the exact relationship between composition, texture and microstructure will change with modification of the annealing procedure. Nanocrystal size, cluster size and phase continuity are the primary factors controlling the annealed microstructure. The general relationship between composition and microstructure can be extended to annealing conditions other than those studied since the basic microstructural building blocks have been defined for MIBERS deposited PLT films.

3.4 Relationship Between Composition, Microstructure, and Electrical Properties

An understanding of the relationships between composition and microstructure is highly advantageous when trying to engineer the properties

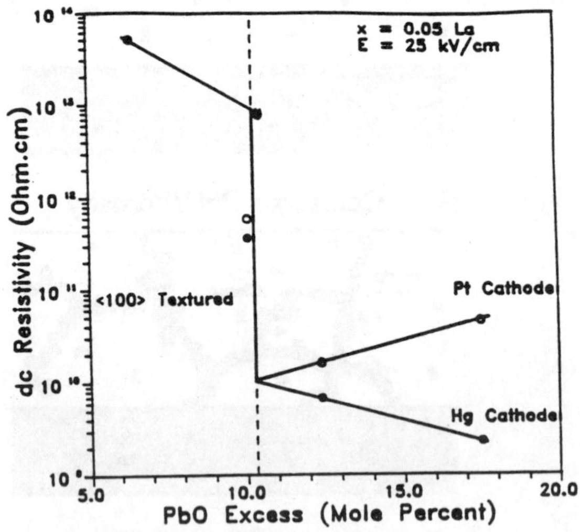

Figure 28. DC resistivity versus PbO for 5% La films. Dashed line shows different texture transitions.

of a thin film. By knowing the types of phases and connectivity of those phases, mixing rules can be used to predict the property dependence on composition. For the PLT films studied, the dependence of the electrical properties on composition can be understood by applying mixing rules to the simplified microstructure models for textured and non-textured films.[25,40]

The dc resistivity measured through the thickness of the film is given as an example of an electrical property which illustrates the relationship between composition, microstructure, and properties. As shown in Fig. 28, textured films exhibit a high dc resistivity (on the order of 10^{13} Ω·cm) which decreases slowly with increasing PbO excess. At the transition between <100> texture and non-textured, the resistivity drops discontinuously to a low resistivity (on the order of 10^{9} Ω cm). Because a mercury probe was used for the top electrode and the Pt substrate layer was used for the bottom electrode there are two curves, labeled Pt cathode and Hg cathode, which refer to measurements for opposite electric field polarities. The resistivity difference for the two polarties increases with increasing PbO content due to an increasing thin film surface roughness (resulting from PbO evaporation) which alters the contact between the Hg electrode and the PLT film.

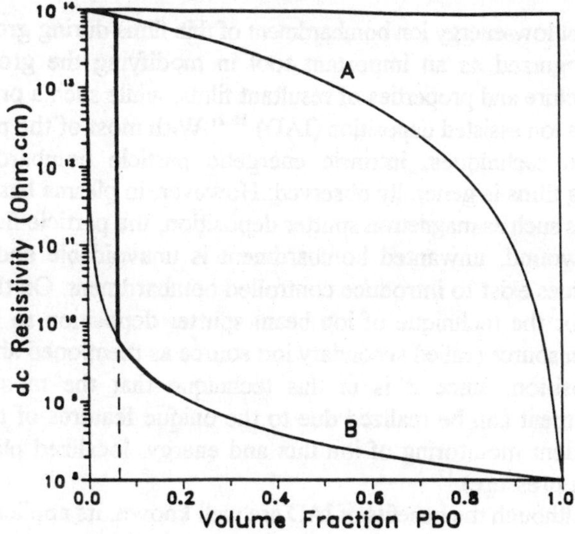

Figure 29. Calculated DC resistivity with PbO excess, for continuous (a) perovskite and (b) PbO phases.

A model of the resistivity dependence on excess PbO is presented in Fig. 29. This model applies the mixing rules developed by Hanai[46] to the simplified microstructure model shown in Fig. 27. In Fig. 29, curve A is calculated assuming a continuous perovskite phase with a resistivity of 10^{14} $\Omega\cdot$cm, and curve B is calculated assuming a continuous second phase of PbO with a resistivity of 10^{8} $\Omega\cdot$cm.[47] Curve A applies to textured films while curve B applies to non-textured films. The dashed verticle line marks the emperically determined transition between textured and non-textured microstructures. The heavy solid line indicates the portion of the calculated curve that corresponds to the experimental curves in Fig. 28. Such a close correlation between the experimental and calculated behavior of the resistivity not only supports the validity of the simplified microstructure model but also emphasizes the importance of understanding the relationships between composition and microstructure.

4.0 CONTROLLED LOW ENERGY OXYGEN ION BOMBARDMENT INDUCED MODIFICATIONS OF PZT THIN FILMS

Concurrent low-energy ion bombardment of thin films during growth has long been recognized as an important tool in modifying the growth process, microstructure and properties of resultant films, while such a process is often referred as ion assisted deposition (IAD).[16-21] With most of the physical vapor deposition techniques, intrinsic energetic particle bombardment of the depositing films is generally observed. However, in plasma based deposition techniques such as magnetron sputter deposition, the particle flux and energy are interwound, unwanted bombardment is unavoidable and thus limited opportunities exist to introduce controlled bombardment. On the contrary, it is ideal for the technique of ion beam sputter deposition to include a low energy ion source (called secondary ion source as mentioned above) to assist the deposition, since it is in this technique that the most controllable bombardment can be realized due to the unique features of the ion source (independent monitoring of ion flux and energy, localized plasma and low operating pressure).[17]

Although the benefits of IAD are well known, its application has been limited to the deposition of single-cation films such as metal, semiconductor and oxide optical ones.[16-21] This may be due to the complexity associated with multi-cation deposition process, especially the problem of preferential

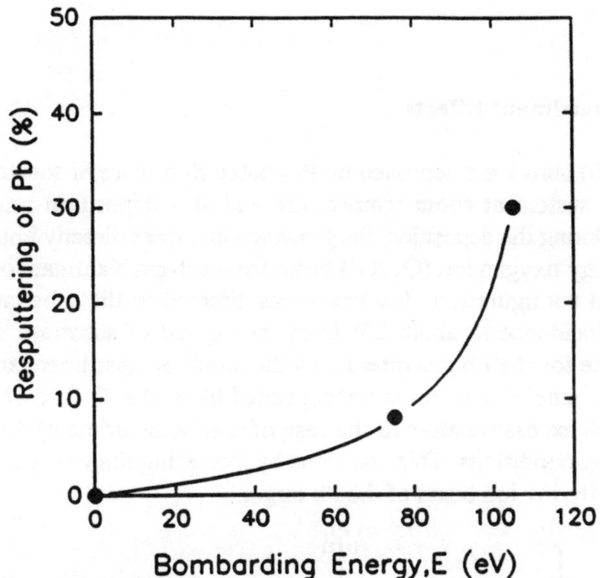

Figure 30. Resputtered Pb% due to bombardment on an unheated substrated.

resputtering. In the case of PZT, for instance, Pb is the volatile species and thus tends to be resputtered more easily than Zr and Ti when the growing films is directly bombarded. Figure 30 shows the percentage of resputtered Pb out of the total arrival Pb species onto the growing PZT films as a function of the bombardment energy of the secondary ion beam at constant ion flux and deposition rate. It is obvious from the figure that if the direct bombardment is applied, it is necessary to increase Pb flux onto the substrates to compensate for the ongoing resputtering. With a single ceramic target, there is no convenient way to make such a compensation. Therefore, it is very difficult, if not impossible, to introduce an effective ion bombarding assistance into the single ion beam / single target configuration for growing stoichiometric ferroelectric PZT films. However, with MIBERS technique compensating for such preferential resputtering is no longer a problem, since different components are sputtered independently from different targets by respective ion sources, so that the arrival flux of each component can be easily adjusted.

Low-energy oxygen ion bombardment was successfully used to modify and enhance the physical properties of MIBERS deposited PZT thin films. In the following sections, the bombardment induced effects are systematically described.

4.1 Bombardment Effects

PZT(50/50) films were deposited on Pt-coated Si or bare Si substrates by the MIBERS system at room temperature and at a deposition rate about 18 Å/min.[14] During the deposition, the growing films were directly bombarded by a low-energy oxygen ion (O_2^+/O^+) beam from a 3-cm Kaufman ion source in single grid configuration. The beam was directed to the substrates with an angle of incidence at about 25° from the normal of substrate surface. To compensate for the Pb resputtering by the direct bombardment and maintain nearly the same Pb content in the deposited films, the Pb flux was increased by 12-19% excess (relative to the case of non-bombardment) depending on bombarding conditions. This was done by increasing the voltage and current of the sputtering ion beam of the Pb target.

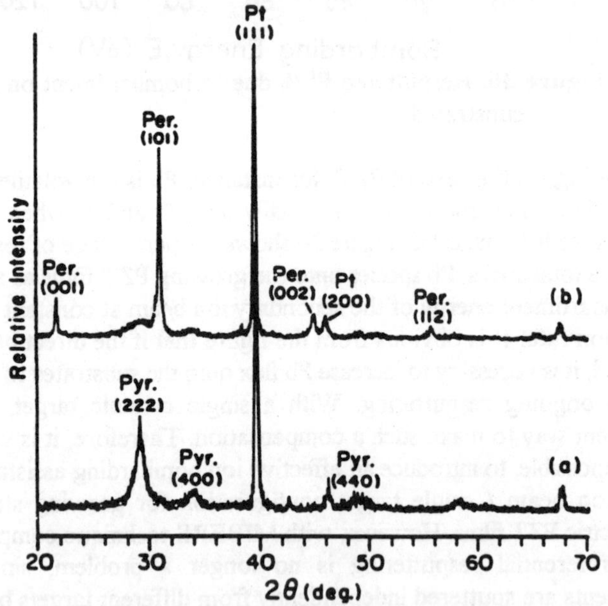

Figure 31. XRD patterns of PZT (50/50) films on Pt-coated Si substrates, annealed at 550 C & deposited (a) without & (b) with 75 eV oxygen ion bombardment.

To comparatively study the effects of bombardment, non-bombarded films were also deposited under otherwise the same conditions. It needs to be mentioned that here the word "non-bombarded" means without direct ion bombardment. The intrinsic bombardment effect of ion beam sputter deposition due to the backscattered ions and sputtered neutrals, which in the present case have energies of about 10 eV as measured,[15] is not specified, since it is common in both cases (with and without direct bombardment). The as-grown films were annealed at temperatures from 550 to 700°C in an oxidizing ambient to achieve crystallization, and were then characterized in terms of structure, morphology and electrical properties to examine the bombardment effects.

4.1.1 Crystallization Enhancement

The first noticeable effect of the bombardment is the reduction in crystallization temperature. As seen in Fig.31, films grown with the bombardment at ion energy of 75 eV and ion beam current of 3 mA show dominant perovskite phase after annealing at 550°C for 2 h, while films grown without the bombardment assume only the pyrochlore phase by the same annealing. This bombardment induced enhancement in crystallization was expected, since similar results have been widely reported for other materials.[17,18] The same effect was also observed for PZT films on bare Si substrates (Fig.32). It has been found that usually excess amount of Pb as high as 20-25% is needed for PZT(50/50) films on bare Si substrates to attain the perovskite phase by post-deposition annealing.[14] Figure 32(a) indicates that the perovskite phase is difficult to obtain in the non-bombarded PZT films of near-stoichiometric Pb content (about 3% excess Pb) on bare Si substrates. But the bombarded films of the same kind and by the same annealing assume dominant perovskite phase, as shown in Fig.32(b). Such crystallization enhancement may in part be attributed to enhanced adatom mobility by the bombardment. Muller visualized the possible mechanism of this effect in molecular dynamics simulations of crystal growth.[17] Low energy ion bombardment provides local atomic rearrangement allowing atoms to relax into lower energy sites. In the present case, although the as-grown films are mainly amorphous, it is quite possible that nucleation may have been initiated during deposition by these local atomic rearrangement and relaxation, so that some micro-crystallites have already existed in the as-grown films although they are too small to be detected by x-ray diffraction (XRD). In addition,

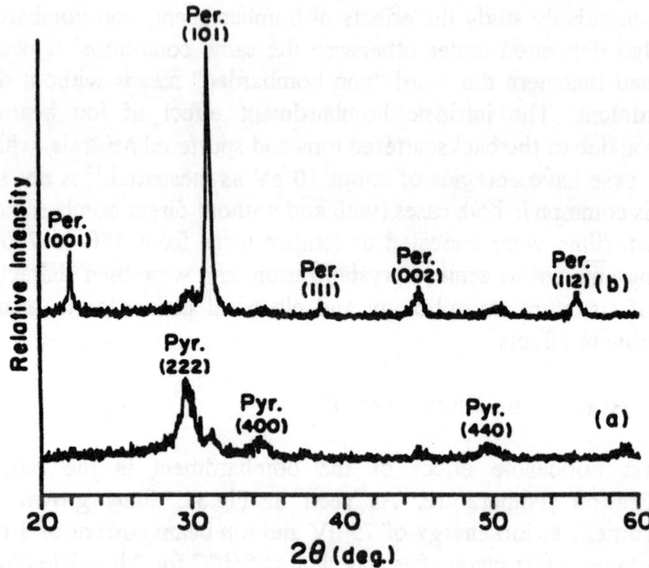

Figure 32. XRD patterns of PZT (50/50) near-stoichiometric Pb content films on bare Si, annealed at 650 C for 2 h; & deposited (a) without and (b) 75 eV oxygen ion bombardment.

enhanced incorporation of oxygen in the films by the reactive oxygen ion (O_2^+/O^+) bombardment, as similarly reported,[48] may be another reason for this crystallization enhancement. It has been noted that sufficient oxygen concentration is crucial for PZT films in forming and maintaining the perovskite structure.[10]

4.1.2 Inducing Preferred Orientation

Figure 33a indicates that while non-bombarded films usually exhibits a random orientation with the XRD pattern similar to those of randomly oriented PZT ceramics, bombarded films can assume preferred (100) orientation. The degree of such preferred orientation seems highly dependent on the bombardment conditions (Fig.33(b) and (c)). It was observed that the difference in Pb content between these two kinds of films is very small (nominally the same in the as-grown films, by Rutherford backscattering spectrometry). Also, it can be noted that the preferred orientation induced by

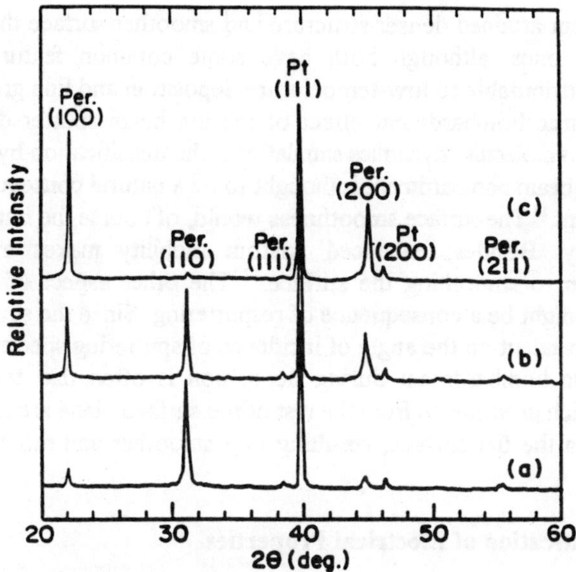

Figure 33. XRD patterns of PZT films on Pt/Si substrates, annealed at 580 C for 20 mins & deposited (a) without, (b) with 80 eV, 2.75 mA beam & (c) 80 eV, 3.5 mA beam bombardment.

the effect of Pb content (as seen in the previous part) may not assume so high degree as seen here, and usually is accompanied with second phase.[34] Therefore, the bombardment effect is thought to be the major reason for the preferred orientation in this case. The bombardment effect on preferred orientation has often been attributed to the occurrence of recrystallization associated with ion channelling effect.[49] Since there was no appreciable *in-situ* crystallization in the present case, it is thought that this preferred orientation might initiate from the possible *in-situ* nucleated micro-crystallites (as so mentioned above) which acted as seeds for the following crystallization. The other possible reasons are associated with bombardment induced anisotropic stress[50] and the bombarding energy stored in the as-grown amorphous films[33] which may tend to cause a preferred orientation of film upon instantaneous release during post-deposition annealing.[51]

4.1.3 Densification and Surface Smoothing

Scanning electron microscopy shows that the films grown with the bombardment attained denser structure and smoother surface than the non-bombarded ones, although both have some common features such as clustering attributable to low-temperature deposition and fine grain size due to the intrinsic bombardment effect of the ion beam sputter deposition.[14] According to molecular dynamics simulations, the densification by off-normal incident ion beam bombardment is thought to be a natural consequence of the bombardment.[17] The surface smoothness would, of course, be improved with film density. Besides, enhanced adatom mobility makes a significant contribution to smoothing the surface.[17] The other aspect of the surface smoothing might be a consequence of resputtering. Since the sputter yield is strongly dependent on the angle of incidence of sputtering species, the result of energetic bombardment during deposition is often that topographical features which protrude up from the rest of the surface plane are more rapidly etched than the flat surface, resulting in a smoother and more featureless films.[17]

4.1.4 Modification of Electrical Properties

4.1.4.1 Remanent Polarization and Coercive Field

As a natural consequence of structural modification, the impact of the low-energy oxygen ion bombardment on the electrical properties of the films is profound. Figure 34 shows the evolution of both remanent polarization (P_r) and coercive field (E_c) of both kinds of the films with annealing temperature (T_a).[22] A common trend is that, within the temperature range shown, P_r always increases with T_a while E_c initially drops and afterwards flatten out, which may be attributed to the development of the crystallinity of perovskite phase. More significant feature shown in the figure, however, is that for each T_a the bombarded films have higher P_r and lower E_c than the non-bombarded films. The following reasons can be noted responsible for this superiority of the bombarded films: a) better crystallinity for each corresponding T_a since the bombarded films started to form perovskite phase at lower temperature, b) higher degree of (100) orientation, and c) better electrode-film interfaces attributed to denser and smoother film surfaces.[14,22]

4.1.4.2 Dielectric Response

Figure 35 shows the change of room-temperature low-field dielectric constant

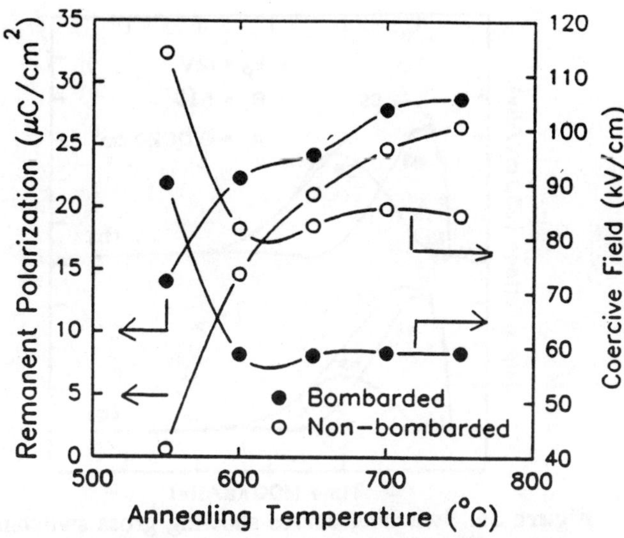

Figure 34. Annealing temperature dependence of P_R and E_C for films deposited with and without ion bombardment.

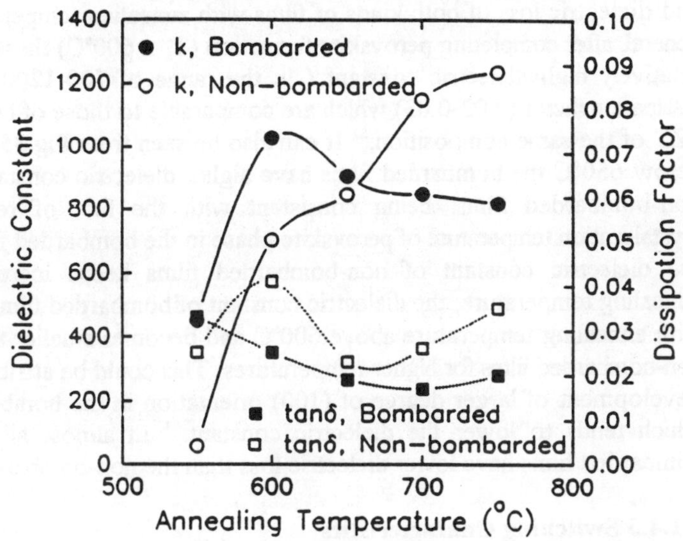

Figure 35. Annealing temperature versus dielectric behavior of bombarded and non-bombarded PZT films.

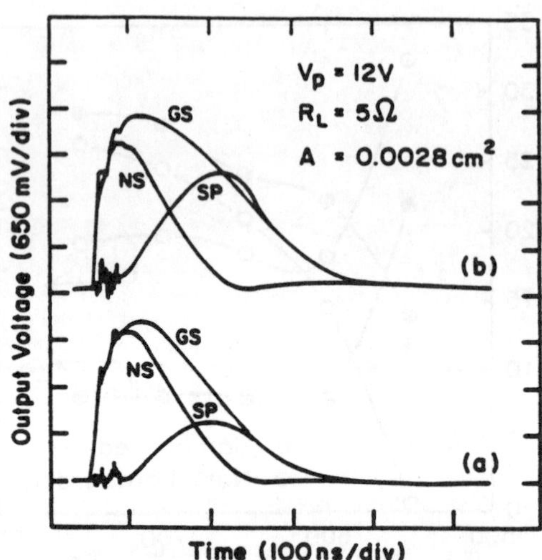

Figure 36. Switching curves showing gross switching (GS), non-switching (NS) & net switching (SP) for (a) non-bombarded and (b) bombarded films.

and dielectric loss of both kinds of films with annealing temperature T_a. In general, after completing perovskite formation ($T_a \geq 600°C$) these films have relatively high dielectric constant (in the range of 700-1200) and low dissipation factor (0.02-0.03) which are comparable to those of bulk ceramic PZT of the same composition,[52] It can also be seen from Fig.35 that for T_a below 650°C the bombarded films have higher dielectric constant than the non-bombarded films, being consistent with the fact of reduction in crystallization temperature of perovskite phase in the bombarded films. While the dielectric constant of non-bombarded films keeps increasing with annealing temperature, the dielectric constant of bombarded films decreases with annealing temperature above 600°C and becomes smaller than that of non-bombarded films for higher temperatures. This could be attributed to the development of larger degree of (100) orientation in the bombarded films which tends to lower the dielectric constant.[53] In almost all cases, the bombarded films have lower dielectric loss than the non-bombarded films.

4.1.4.3 Switching Characteristics

Typical switching characteristics of the films are shown in Fig.36, in which only V(t) transients resulting from the two positive pulses of double bipolar pulse input were recorded. In the measurement, the sample was initially polarized negatively with negative pulses. The first positive pulse resulted in a gross switching transient (GS) caused by the ferroelectric domain reversal (charge density of the capacitor changed by $\delta Q \sim 2P_r$) plus a normal capacitor charge transient ($\delta Q \sim P_s - P_r$), where P_s and P_r denote respectively the spontaneous polarization and remanent polarization of the films under the measurement condition. The second positive pulse gave rise to a non-switching transient (NS) merely due to the normal capacitor charge. The difference of these two curves then defines a net switching transient (SP) due to sole ferroelectric domain reversal. The switched charge density (Q_s) can be determined by the area under the SP curve divided by the load resistance R_L and the electrode area A and should be close to $2P_r$. It can be seen from Fig.36 that, under the same measurement conditions, the SP-covered area (and thus the Q_s) for the bombarded films is much larger than that for the non-bombarded films, and is consistent with the fact that the low-energy oxygen ion bombardment largely enhanced the remanent polarization of the films.

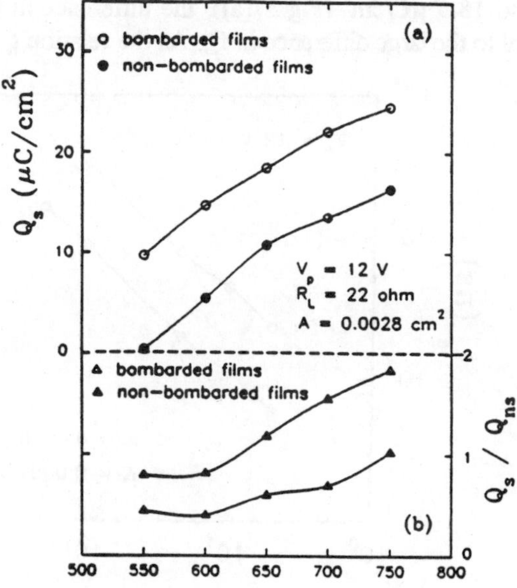

Figure 37. Annealing temperature versus Q_{SS} and the ratio of Q_S and Q_{SS} of PZT films.

Calculation of the switched charge density was extended to the films annealed at different temperatures and the results showing its dependence on annealing temperature are summarized in Fig.37a. It can be seen that for each annealing temperature the bombarded films exhibit constantly higher Q_s than the non-bombarded films. These results are in agreement with the data on P_r versus annealing temperature, and again demonstrates the superiority of the bombarded films over the non-bombarded films. Also shown in Fig.37b is the ratio Q_s/Q_{ns} as a function of annealing temperature, in which the Q_{ns} was calculated by using the area under the NS curve. This ratio is of significance in that it reflects the squareness of hysteresis loop under the applied electrical field of measurement and thus, to a large extent, the domain dynamics during switching.[55] It is of interest to note that the difference in this ratio between the bombarded films and the non-bombarded films is quite large (Fig.37(b)).

It can be found in Fig.36 that the switching time t_s, defined as the time from the onset to a point 90% down from the maximum value of the SP curve,[54] is about 230 ns for the non-bombarded films and about 260 ns for the bombarded films. Noting the values of Q_s in the respective cases are 10.8 $\mu C/cm^2$ and 18.5 $\mu C/cm^2$ (Fig.37(a)), the difference in t_s is small and not proportional to the large difference in Q_s. As the relation $t_s \propto Q_s$ holds in each

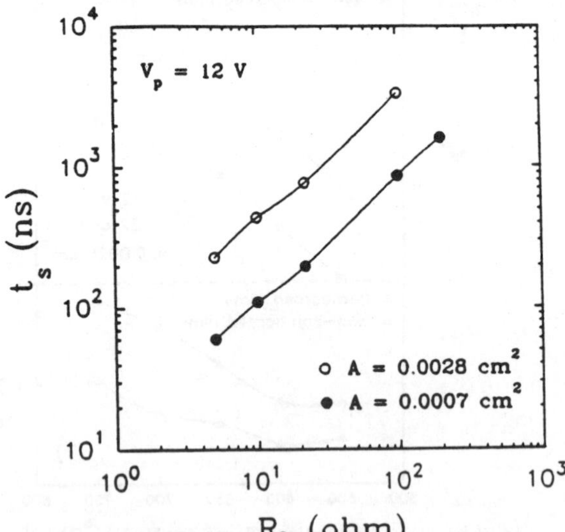

Figure 38. Switching time with load resistance & electrode area for non-bombarded films.

case and the coercive field comes into the proportional coefficient,[56] comparison can be made between these two kinds of the films by using the ratio t_s/Q_s. It is this ratio, rather than t_s, that reflects the easiness of a switching process. For the bombarded films this ratio is appreciably smaller than that for the non-bombarded films. This fact implies that the domain reversal in the bombarded films is dynamically more favorable and can be related to the lower coercive fields of these films compared with the non-bombarded films.

The switching time in the present case appears rather large, ranging from tens of nanosecond to a few microsecond. This is due to the relatively large dimension of the electrode areas used.[57-59] Figure 38 shows both load resistance (R_L) and electrode area (A) dependencies of the switching time. Assuming a linear dependence of switching time on the electrode area as reported by many authors,[57-59] the switching time is estimated to be about 6 ns for A=1000 μm^2 and less than 1 ns for A=1 μm^2 at V_p=12 V and R_L=50 □, being consistent with other reported results.[58,60]

To further quantify the dynamics of domain reversal, the relation[55] $t_s=t_o\exp(\alpha/E)$, i.e. $\log t_s=\log t_o+\alpha/(2.303E)$ was used to determine the activation field α for both bombarded and non-bombarded films.[23] Figure 39 shows data on field dependence of switching time and the fitting lines to the

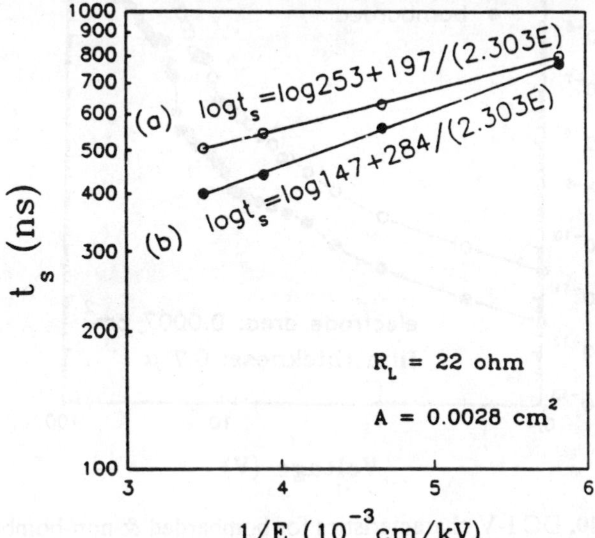

Figure 39. Switching time with applied electric field for (a) bombarded & (b) non-bombarded PZT films.

above relation, yielding an α of 284 kV/cm for the non-bombarded films and an α of 197 kV/cm for the bombarded films. As expected,[55] the α value in either case is about 4 times the respective coercive fields (E_c). As seen above, the E_c is about 55 kV/cm for the bombarded films and about 75 kV/cm for the non-bombarded films, obtained from the polarization hysteresis measurements.

4.1.4.4 Leakage Current and Time Dependent Dielectric Breakdown

Figure 40 shows the results of measurement on I-V response of both the bombarded and non-bombarded films, annealed at 600°C for 2 h.[24] Several interesting differences can be noted between the bombarded films and the non-bombarded films: (a) The ohmic resistivity of the bombarded films (~3×10^{11} □cm) is about one order of magnitude higher than that of the non-bombarded films (~3×10^{10} □cm). (b) The onset voltage of the space-charge-dominant conduction is much higher for the bombarded films (~12 V) than for the non-

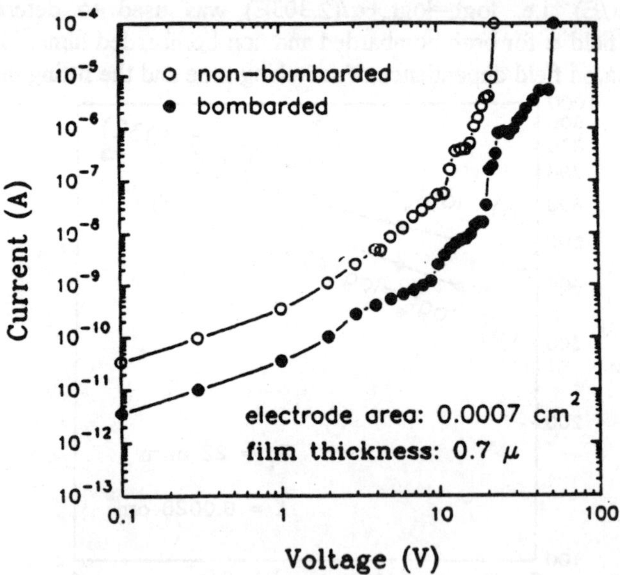

Figure 40. DC I-V characteristics for bombarded & non-bombarded PZT films, annealed at 650 C.

bombarded films (~3 V). (c) The bombarded films have much higher dielectric breakdown strength compared with the non-bombarded films (~770 kV/cm versus ~350 kV/cm, under the indicated measurement conditions). It was also found that for films annealed at higher temperatures, these differences become less significant while the values of the non-bombarded films approaches those of the bombarded films. These differences may be due to the difference in the microstructure and the degree of non-stoichiometry between these two kinds of films, and are being under further investigation.

The time dependent dielectric breakdown (TDDB) of the ferroelectric films is believed to have close relation to the I-V behavior since both of them are due to the motion of charge carriers.[62] Figure 42 shows the TDDB characteristics of the above mentioned films annealed at 680°C for 2 h. As expected, the bombardment induced effect on TDDB is amazingly large. Under the same dc field of 450 kV/cm, the non-bombarded films break down in about 50 min, while the bombarded films can last for 925 min.[24]

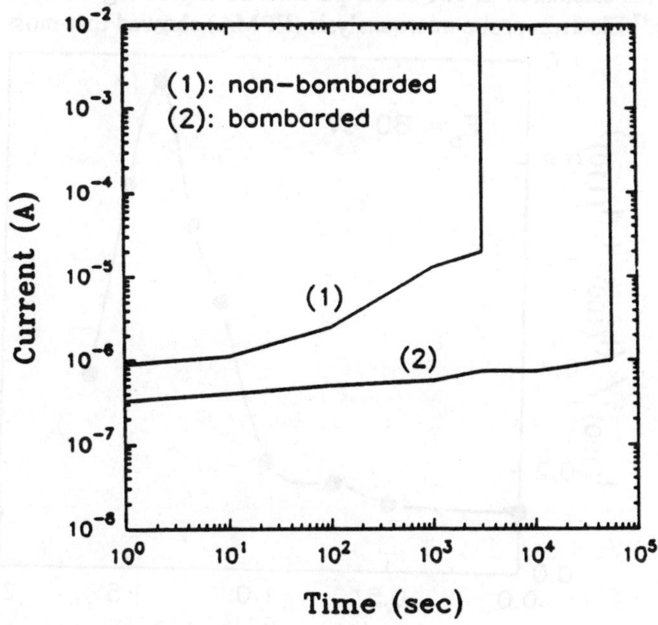

Figure 41. Time dependent dielectric breakdown behavior of bombarded and non-bombarded PZT films.

4.2 Optimization of Bombardment Conditions

After the effects of bombardment were qualitatively scanned, several bombardment-sensitive properties were chosen, and the dependencies of such properties on the bombardment parameters (ion beam flux and voltage) were investigated in order to define an optimal range of the bombardment conditions.[22]

For this purpose, the films were deposited with similar thickness ranging from 1700 to 2100 Å. While all the other deposition parameters were fixed the same, the bombarding ion energy (E_b) and beam flux (I_b) were changed, one at a time, from run to run to render different bombardment conditions. The I_b was changed by adjusting the current of cathode filament of the ion source only, without varying the O_2 flow, so that the oxygen partial pressure of the deposition chamber was kept the same at 0.01 Pa for all the runs. From the values of I_b and the measured deposition rates, ion/atom arrival ratio was calculated as one of the parameters expressing the bombardment effects.[17] Electron probe microanalysis (EPMA) showed that most of such

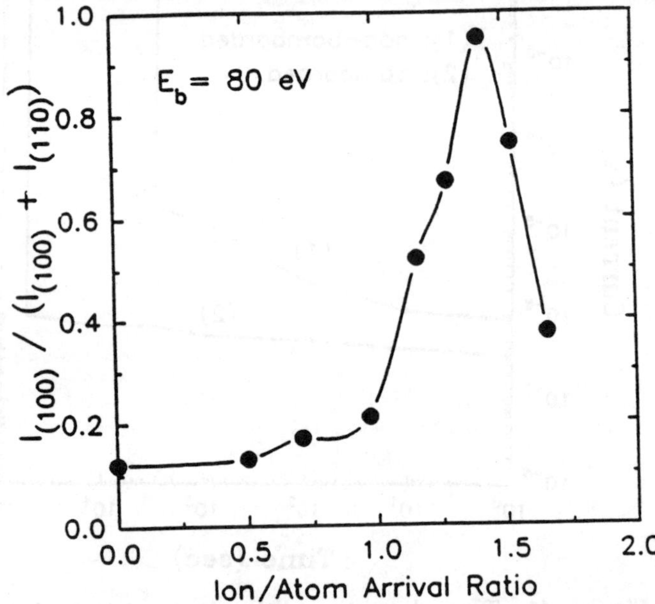

Figure 42. XRD peak intensity ratio of PZT films with Ion/Atom ratio.

deposited films have nearly the same Pb content (about 3% excess relative to the stoichiometry of PZT perovskite phase), except for a few deposited at too high E_b or I_b which are slight Pb deficient. The as grown films were annealed at the same condition (580°C for 20 min). All the crystallized films assumed pure perovskite phase in X-ray diffraction (XRD).

4.2.1 Ion flux

Figure 42 shows the XRD intensity ratio $I_{(100)}/(I_{(100)}+I_{(110)})$ of the crystallized films as a function of the ion/atom ratio. For clarity, it is worth mentioning that in the present case "ion" denotes both O_2^+ and O^+, and "atom" includes Pb, Zr and Ti. It is noted from Fig.42 that the degree of (100) orientation increases monotonically within the range of ion/atom ratio from 0 to about 1.4, gradually from 0 to 1.0, and sharply from 1.0 to 1.4. Afterwards, however, it steeply decreases with ion/atom ratio. The trend of the evolution of preferential orientation with ion/atom ratio in the region lower than 1.4 is consistent with the behaviors of low-energy ion bombardment modified

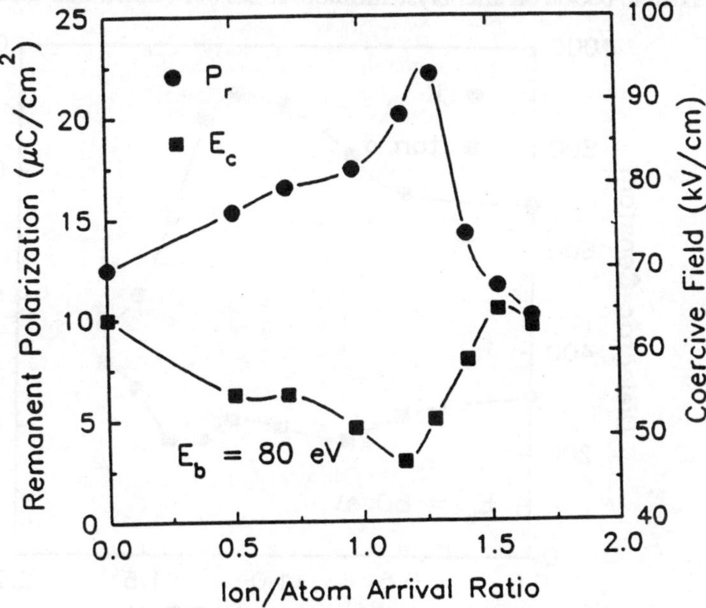

Figure 43. Remanent polarization and coercive field of PZT films with ion/atom ratio.

properties in general,[17] and confirms the strongly flux-dependent nature of the bombardment effect. It is of interest to note that there is a highly sensitive range between 1.0 and 1.4 of ion/atom ratio, which may be of particular importance in practice of property modifications. The abnormal steep decrease in the degree of (100) orientation in the higher ion/atom ratio region (>1.4) was presumably attributed to the deviation in the film stoichiometry, namely the resputtering-induced Pb deficiency by the high ion flux, as revealed by EPMA.

Figure 43 shows the P_r and E_c values as functions of the ion/atom ratio. Basically, the curve of P_r versus ion/atom ratio follows a trend similar to that of orientation in Fig.42, although it increases more steadily over the entire lower ion/atom region (from 0 to 1.3). It may be noted, however, that the peak position of ion/atom ratio is shifted to about 1.3, compared to 1.4 for the preferential orientation. This discrepancy may be stress related, since the films deposited at ion/atom ratio near 1.4 seemed highly stressed, as evidenced by the fact that thicker films (about 5000 Å) of this condition tended to buckle off after crystallization. It has been noted that the stress and

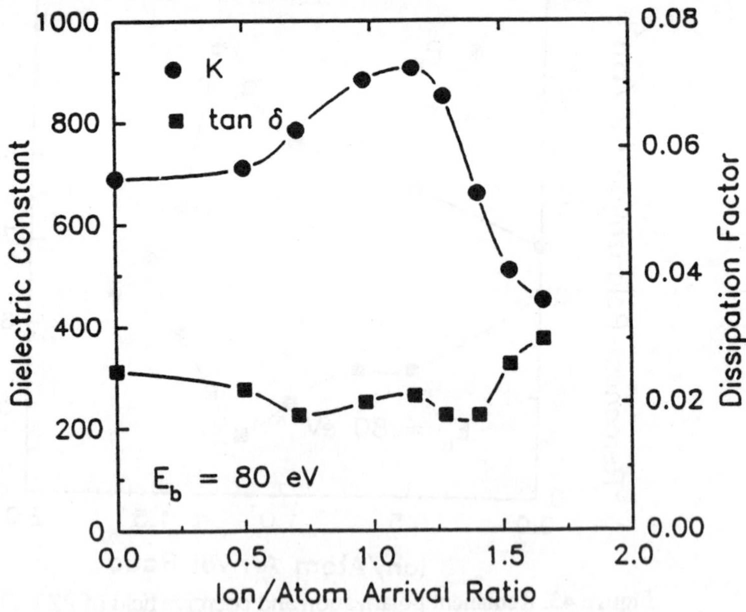

Figure 44. Dielectric behavior of PZT films with ion/atom ratio.

preferential orientation are often highly correlated.[17] The further decrease in P_r with ion/atom>1.4 was again thought to be the consequence of Pb deficiency. It can also be seen from Fig.43 that the E_c changes inversely with ion/atom ratio as compared to P_r. This fact is of great significance as it allows optimum conditioning to attain as high P_r and as low E_c as possible simultaneously.[22]

The dielectric constant (k) of the films at 100 kHz also showed a strong dependence on the ion/atom ratio, while the dissipation factor (tanδ) remained in the vicinity of 0.02 (Fig.44). Again, the k curve roughly mimics the P_r curve or orientation curve, except for a little further shift in the peak position towards the lower ion/atom side and some broadening of the peak. Comprehensively, the ion/atom ratio for optimum combination of P_r, E_c and k in the present conditions seems within the range from 1.0 to 1.3. Relative to the non-bombarding case (zero point of ion/atom ratio), the increments in P_r and k could be about 60% and 25% respectively, while the reduction in E_c could be about 20%.

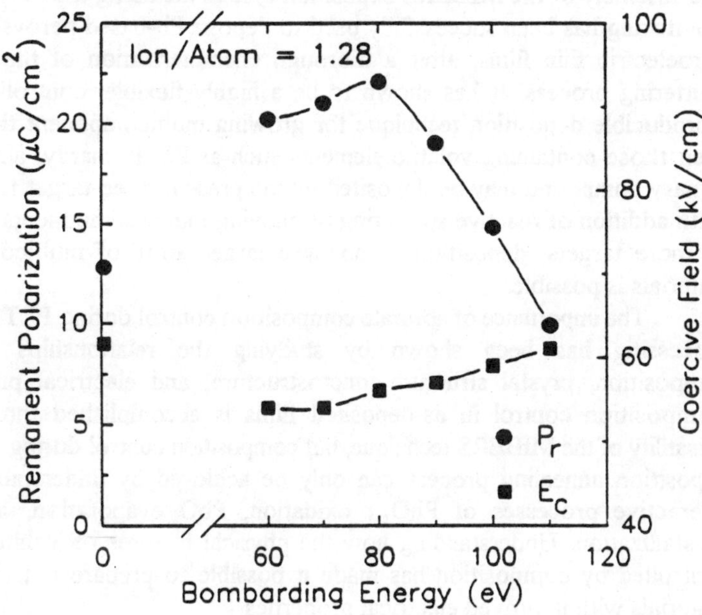

Figure 45. Remanent polarization and coercive field with bombarding energy.

4.2.2 Ion energy

While the ion/atom ratio was kept constant at 1.28, the bombardment effect was also studied in terms of P_r and E_c versus the bombarding ion energy E_b. The results were summarized in Fig.45. It is shown that for E_b from 60 eV to 80 eV, P_r had almost equally high values of about 20-22 μC/cm^2, while beyond 80 eV it continuously decreased as E_b increased. Therefore, it seems that, within the range of experimental condition explored in the present work, the bombarding energies between 60 and 80 eV are optimal to realize effective property modification. Higher E_b was not suitable because it led to significant Pb resputtering, which may have resulted in less effective bombardment for desirable modification[17] as well as Pb deficient in the films (observed by EPMA in the case of 110 eV).[17]

5.0 SUMMARY

The assembly of the MIBERS deposition system including the *in-situ* QCM monitoring has been successfully used to deposit Pb-based perovskite type ferroelectric thin films, after a thorough characterization of the reactive sputtering process. It has shown to be a highly flexible, controllable and reproducible deposition technique for growing multicomponent thin films, even those containing volatile elements such as Pb. Primarily, almost any ternary compound may be deposited by the present three-target technique. With addition of reactive sputtering or alloying minor components into one or more targets, deposition of an even larger array of multicomponent materials is possible.

The importance of accurate composition control during PLT thin film processing has been shown by studying the relationships between composition, crystal structure, microstructure, and electrical properties. Composition control in as-deposited films is accomplished through the versatility of the MIBERS technique, but composition control during the post-deposition annealing process can only be achieved by understanding the interactive processes of $PbO_{1-\delta}$ oxidation, PbO evaporation, and PLT crystallization. Understanding how the physical features of a thin film are controlled by composition has made it possible to prepare thin film PLT materials with improved electrical properties.

Low-energy oxygen ion bombardment during film growth was found to have a pronounced effect on the growth process, structure and electrical properties of PZT thin films, evidenced by reduced perovskite formation temperature, induced preferential orientation, improved film morphology and enhanced electrical properties. Some properties are strongly dependent on the bombarding in flux (characterized by ion/atom ratio) and energy, and therefore can be quantified to figure out optimal range of the bombardment conditions. The ion/atom ratios between 1.0 and 1.3 and the bombarding ion energies ranging from 60 to 80 eV were found to be optimal to realize desirable property modification. Relative to the non-bombarded case, the bombardment could increase the P_r and k by up to 60% and 25% respectively, and reduce the E_c by about 20%. The improved properties also include crystallization temperature, switching characteristics, I-V behavior and time dependent dielectric breakdown. The role of O_2 ion bombardment may be understood in terms of; (a) presence of activated oxygen species for a enhanced chemical reaction among the sputtered species and (b) additional low energy bombardment to affect the grain growth process. However, the observed effects are due to the combination of bombarding energy, atomic and ion flux densities, momentum transfer and the enhanced chemistry, which are yet to be resolved[63]. Most recently high density plasma sources, such as ECR (Electron cyclotron Resonance), were coupled during the physical vapor deposition of ferroelectric thin films[64], which improved physical and electrical properties. It will be essential to establish a much clearer understanding of low energetic ion-surface interactions, such that it offers a great potential of engineering the physical and electrical properties of multicomponent ferroelectric thin films for a variety of device applications.

REFERENCES

1. M. ISHIDA, H. MATSUNAMI, and T. TANAKA, J. Appl. Phys. **48**, 951 (1977).
2. A. OKADA, J. Appl. Phys. **48**, 2905 (1977).
3. S.B. KRUPANIDHI, N. MAFFEI, M. SAYER, and K. EL-ASSAL, J. Appl. Phys. **54**, 6601 (1983).
4. K. IIJIMA, Y. TOMITA, R. TAKAYAMA, and I. UEDA, J. Appl. Phys. **60**, 361 (1986).
5. H. ADACHI, T. MITSUYU, O. YAMAZAKI, and K. WASA, J. Appl. Phys. **60**, 736 (1986).
6. K. SREENIVAS and M. SAYER, J. Appl. Phys. **64**, 1484 (1988).
7. M. MATSUBARA, S. MIURA, Y. MIYASAKA, and N. SHOHATA,

J. Appl. Phys. **66**, 5826 (1989).

8. K.L. SAENGER, R.A. ROY, K.F. ETZOLD, and J.J. CUOMO, Mater. Res. Soc. Symp. Proc. **200**, 115 (1990).
9. D. ROY, S.B. KRUPANIDHI, and J.P. DOUGHERTY, J. Appl. Phys. **69**, 7930 (1991).
10. R.N. CASTELLANO and L.G. FEINSTEIN, J. Appl. Phys. **50**, 4406 (1979).
11. L.L. BOYER, A.Y. WU, G.W. METZGER, and J.R. MCNEIL, J. Vac. Sci. Technol. **A7**, 1199 (1989).
12. A.I. KINGON, T.M. GRAETTINGER, S.H. ROU, H.N. AL-SHAREEF, K.D. GIFFORD, P.D. HREN, M.S. AMEEN, and O.H. AUCIELLO, in Proceedings of the Fifth U.S.-Japan Seminar on Dielectric and Piezoelectric Ceramics, Kyoto, Japan, Dec. 1990, p.324.
13. D. XIAO, Z. XIAO, J. ZHU, Y. LI, and H. GUO, Ferroelectrics **108**, 59 (1990).
14. S.B. KRUPANIDHI, H. HU, and V. KUMAR, J. Appl. phys. **71**, 376 (1992).
15. G. FOX, S.B. KRUPANIDHI, K.L. MORE, and L.F. ALLARD, J. Mater. Res. (1992).
16. J.M.E. HARPER, J.J. CUOMO, and H.R. KAUFMAN, J. Vac. Sci. Technol. **21**, 737 (1982).
17. E. KAY, S.M. ROSSNAGEL, R.A. ROY, D.S. YEE, K-H.MULLER, J.M.CUOMO, S.M. ROSSNAGEL and H.R. KAUFMAN (Noyes, Park Ridge, NJ, 1989), p.170.
18. J.E. GREENE, S.A. BARNETT, J.-E. SUNDGREN, and A. ROCHETT, in *Ion Beam Assisted Film Growth,* edited by T. Itoh (Elsevier, Amsterdam, 1989), p.101.
19. D.M. MATTOX, J. Vac. Sci. Technol. **A7**, 1105 (1989).
20. U.J. GIBSON, in *Physics of Thin Films, Vol.13,* edited by M.H. Francombe and J.L. Vossen (Academic, New York, 1987), p.109.
21. F.A. SMIDT, International Mater. Rev. **35-2**, 61 (1990).
22. H. HU and S.B. KRUPANIDHI, Appl. Phys. Lett. **61,** 1246 (1992).
23. H. HU and S.B. KRUPANIDHI, Appl. Phys. Lett. (1992).
24. H.R. KAUFMAN, J.J. CUOMO, and J.M.E. HARPER, J. Vac. Sci. Technol. **21**, 725 (1982).
25. G.R. FOX, Doctoral Thesis, The Pennsylvania State University (1992).

26. R.N. CASTELLANO, Thin Solid Films **46**, 213 (1977).
27. J. ROTH, in *Sputtering by Particle Bombardment II*, edited by R. Behrish (Springer-Verlag, New York, 1983). p.91.
28. T. ROSENQVIST, *Principles of Extractive Metallurgy* (McGraw-Hill, New York, 1974), p.517.
29. R. BEHRISH, in *Sputtering by Particle Bombardment II*, edited by R. Behrish (Springer-Verlag, New York, 1983). p.3.
30. B. CHAPMAN, *Glow Discharge Processes* (Wiley-Interscience, New York, 1980). p.249.
31. R.H. CORNELY, in *Ion Beam Assisted Film Growth*, edited by T. Itoh (Elsevier, Amsterdam, 1989). p.13.
32. H. HU, C.J. PENG, and S.B. KRUPANIDHI, Thin Solid Films. (1992)
33. S.V. KRISHNASWAMY, R. MESSIER, P. SWAB, L.L. TONGSON and K. VEDAM, *J. Electron. Mater.*, **10**, 433 (1981).
34. H. HU, Multi-Ion-Beam Reactie sputter deposited Pb(Zr,Ti)O3 Thin Films for Memory Applications, Ph.D., doctoral thesis, The Pennsylvania State Univesity (1993).
35. G.R. FOX and S.B. KRUPANIDHI, J. Materials. Research, **9**,3, (1994)
36. J.A. THORNTON, J. Vac. Sci. Technol. **11**, 666 (1975).
37. H.HU and S.B. KRUPANIDHI, in Proceedings of 8th International Symposium on Applications of Ferroelectrics, (Greenville, SC, Aug, 1992).
38. M. SAYER, A. MANSINGH, A.K. ARORA and A. LO, Integrated Ferroelectrics **1**, 129 (1992).
39. G.R. FOX and S.B. KRUPANIDHI, J. Materials Research,**8**,9, 2191 (1993)
40. G.R. FOX and S.B. KRUPANIDHI, J. Mater. Research, **8**,9, 2203 (1993)
41. D. HENNINGS, Mat. Res. Bull., **6**, 329 (1971).
42. B. CHAPMAN, *Glow Discharge Processes*, Wiley-Interscience, New York, NY (1980), 177-184.
43. C.PLOG, L. WIEDEMAN, and A. BENNINGHOVEN, Surface Sci., **67**, 565 (1977).
44. D.H.W. CARSTENS, J.F. KOZLOWSKI, and D.M. GRUEN, High Temp. Sci., **4**, 301 (1972).
45. R.W. HEWITT & N. WINOGRAD, Surface Science, **78**, 1, (1978).

46. L.K.H. VAN BEEK, "Dielectric Behavior of Heterogeneous Systems", in *Progress in Dielectrics*, Vol.7, edited by J.B. Birks, Heywood Books, London, England (1967), 69-114.
47. H.E. BROWN, *Lead Oxide - Properties and Applications*, International Lead Zinc Research Organization, Inc., New York, NY 153-193 (1985).
48. J.M.E. HARPER, J.J. CUOMO, and H.T.G. HENTZELL, J. Appl. Phys. **58**, 550 (1985).
49. L.S. YU, J.M.E. HARPER, J.J. CUOMO, and D.A. SMITH, J. Vac. Sci. Technol. **A4**, 443 (1986).
50. D.W. HOFFMAN, Thin Solid Films **107**, 353 (1983).
51. R. MESSIER, T. TAKAMORI, and R. ROY, J. Non-crys. Solids **8-10**, 816 (1972).
52. B. JAFFE, W.R. COOK, and H. JAFFE, *Piezoelectric Ceramics* (Academic, New York, 1971), p.142.
53. J.C. BURFOOT and G.W. TAYLOR, *Polar Dielectrics and Their Application* (MacMillan, London, 1979), p.36.
54. J.F. SCOTT, L. KAMMERDINER, M. PARRIS, S. TRAYNOR, V. OTTENBACHER, A. SHAWABKEH, 4nd W.F. Oliver, J. Appl. Phys. **64**, 787 (1988).
55. J.C. BURFOOT and G.W. TAYLOR, *Polar Dielectrics and Their Application* (MacMillan, London, 1979), chap.12.
56. W.J. MERZ, J. Appl. Phys. **27**, 938 (1956).
57. B.P. MADERIC, L.E. SANCHEZ, and S.Y. WU, Ferroelectrics **116**, 65 (1991).
58. S.K. DEY, C.K. BARLINGAY, J.J. LEE, T.K. GLOERSTAD, and C.T.A. SUCHICITAL, in Proceedings of the Third Symposium on Integrated Ferroelectrics, Colorado Springs, Colorado, April 1991, p.30
59. J. CARRANO, C. SUDHAMA, V. CHIKARMANE, J. LEE, A. TASCH, W. SHEPHERD, and N. ABE, IEEE trans. Ultrason. Ferroelec. Freq. Contr. **38**, 690 (1991).
60. P.K. LARSEN, G.L.M. KAMPSCHOER, M.J.E. SPIERINGS, and R. CUPPENS, Appl. Phys. Lett. **59**, 611 (1991).
61. G.R. FOX and S.B. KRUPANIDHI, J. Appl. Phys, **74**, 3, 1949 (1993).
62. B.M. MELNICK, C.A. PAZ DE ARAUJO, L.D. MCMILLAN, D.A. CARVER, and J.F. SCOTT, Ferroelectrics **116**, 79 (1991).

63. S.B. KRUPANIDHI, J. Vac. Sci. & Tech., **A-10**, 1569 (1992).
64. J.R. BELSICK and S.B. KRUPANIDHI, J. Appl. Phys., **74**, 6851 (1993).

CHAPTER 12

SYNTHESIS OF FERROELECTRIC THIN FILMS BY ORGANOMETALLIC CHEMICAL VAPOUR DEPOSITION

M. de Keijser, G.J.M. Dormans, and P.K. Larsen

Philips Research Laboratories Eindhoven,
Prof. Holstlaan 4, 5656 AA Eindhoven, The Netherlands.

1. INTRODUCTION

Chemical vapour deposition (CVD) processes, where volatile compounds of the elements to be incorporated into the solid, the so-called precursors, are transported via the gas-phase to the region where deposition takes place, are well known. Historically, the most commonly used precursors are chlorides and hydrides. Such compounds are readily available, in most cases easy to handle and often have sufficiently high vapour pressures. Although in many cases good results are obtained, process control is not always optimal and the contamination of the film with halogen atoms is reported. To overcome these drawbacks, Manasevit in 1968 proposed the use of an alternative class of precursors; organometallic compounds.[1] Such molecules are characterised by the presence of direct bonds between the central metal atom and carbon atoms of the surrounding organic fragments, or ligands. If such compounds are used as precursors in a CVD process, the technique is referred to as organometallic chemical vapour deposition (OMCVD), or, in the case of epitaxial growth, organometallic vapour phase epitaxy (OMVPE). The use of organometallics as precursors has several advantages. In the first place they have a volatility that is usually, but not always, higher than that of inorganic compounds of the same metals. This enables higher deposition rates. In the second place, the decomposition temperature of such compounds is relatively low. This implies the possibility of the use of lower deposition temperatures. Another advantage is that physical and chemical properties can be tailored by introducing small changes in the ligands. In this way a number of useful compounds has become available. With proper choices of precursors and deposition conditions it turned out to be possible to deposit thin films of very high purity. Although the precursors contain a substantial amount of carbon, this is not incorporated in the films. Carbon incorporation is suppressed by the use of reactive gasses like hydrogen or oxygen. The decomposition of the organometallics yields volatile products, or fragments that react with the carrier gas to form volatile products, which then are exhausted. Today, organometallic chemical vapour deposition has established itself as a mature deposition technique in the world of compound semiconductor thin films. Impressive control of layer thickness and composition can now be achieved.[2] Ultimate deposition control is possible when using atomic layer deposition.[3]

Organometallic chemical vapour deposition has recently been applied successfully for the deposition of oxidic thin films such as high-T_c superconductors and ferroelectrics. For several reasons it is likely that this technique will play an important role in future developments of oxidic thin film deposition. OMCVD is a technique allowing direct formation of thin films in the proper crystalline phase and even epitaxial films for lattice matched substrates. It has the advantage over most other techniques that a high oxygen partial pressure can be applied during the deposition. OMCVD can be easily scaled-up, both in wafer throughput as well as in wafer diameter. This makes the technique a promising candidate for deposition on an industrial scale. A characteristic of OMCVD which is of technological importance, is that a very good step coverage can be achieved. With many other deposition techniques it is difficult to deposit films uniformly in, e.g., contact holes with large aspect ratios. A real breakthrough of the use of ferroelectric thin films in electronic devices has not occurred, yet. It is expected, however, that with the availability of new and improved deposition techniques allowing for higher quality films deposited at lower temperatures, such as OMCVD, this will change. Deposition techniques for ferroelectric thin films were reviewed by Roy *et al.*[4]

Among the ferroelectric materials the perovskite-type oxides attract the main attention and will be focused on in this chapter, with emphasis on the $PbZr_xTi_{1-x}O_3$ system. One should remember that also other types of ferroelectrics can be deposited using OMCVD. In fact, the range of materials that can be deposited using this technique is mainly limited by the availability of suitable precursors for the desired elements.

This chapter is organised as follows. The more general aspects of OMCVD will be summarised briefly since textbooks exist treating this subject in detail.[5] The choice of precursors is crucial and a section is dedicated to this topic. Finally results of ferroelectric thin film depositions are presented.

2. THE TECHNIQUE OF OMCVD

2.1 General

The process of (OM)CVD involves a number of physical and chemical sub-processes.

- evaporation and transport of precursors,
- chemical reactions in the gas phase,
- diffusion of reactants to the surface,
- adsorption and desorption,
- surface diffusion,
- chemical reactions at the surface,
- incorporation into the growing film,
- diffusion of reaction products from the surface to the gas phase.

In Figure 1 some of these processes are schematically shown. A complete

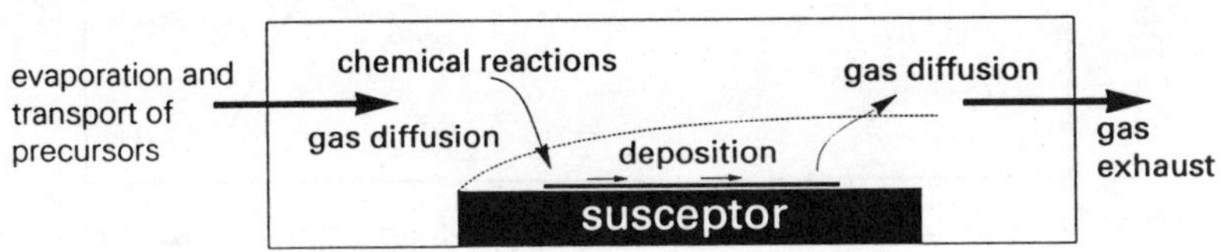

FIGURE 1. Schematic representation of sub-processes involved in an OMCVD process.

description of the process is very complicated since it strongly depends on the precursors involved, the reactor geometry and process conditions. Usually it turns out that one of the sub-processes determines the deposition rate. The overall process can then be adequately described by this rate-limiting process. A general approach to get an impression of the process is to study the deposition rate as a function of parameters such as the precursor partial pressures, the substrate temperature and the reactor pressure. The usual way to describe the deposition behaviour is the so-called Arrhenius plot of the deposition rate as a function of the reciprocal temperature. Such a plot is shown in Figure 2. Below, a qualitative description of this plot will be given to illustrate several of the aspects involved in the deposition process.

At low temperatures (regime I) the deposition is limited by chemical reactions (e.g., the decomposition of a precursor at the substrate). In this so-called kinetically-controlled regime, the deposition rate depends exponentially on the deposition temperature and is proportional to the precursor partial pressure. The temperature dependence is characterised by an activation energy which is typically in the order of 10-60 kcal/mole. This strong temperature dependence requires a very accurate control of the substrate temperature over the whole substrate area in order to get a uniform layer thickness and, in case of a multi-component layer, a uniform compo-

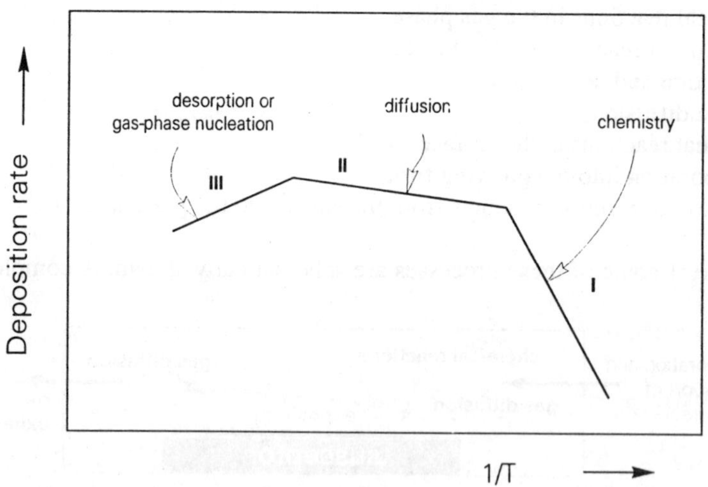

FIGURE 2. Schematic representation of an Arrhenius plot. Three different regimes can be pointed out in such a plot (see text).

sition. On the other hand, it might be advantageous to work in this regime in the case of deposition on non-planar substrates or on wafers with topography. Since the deposition rate only depends on the substrate temperature and not on the diffusion of species, a conformal step-coverage may be obtained. Also, as the deposition is kinetically limited it offers the possibility to deposit metastable phases of materials.

When the temperature is raised, chemical reactions become faster and the deposition rate increases. At a certain temperature, the deposition rate will no longer be limited by these reactions but by the diffusion of species to the substrate. In this case the growing film can reach its thermodynamical equilibrium. In the diffusion (or mass-transport) limited deposition regime (regime II) the deposition can be described by the diffusion of species through the gas phase to the substrate. The temperature dependence of the deposition rate is determined only by the temperature dependencies of the diffusion coefficients of the chemical species. Diffusion coefficients only slightly increase with temperature and, hence, the deposition rate will also depend only slightly on temperature. Therefore, it is advantageous to deposit layers in this regime since the temperature uniformity over the substrate is no longer critical. However, if the diffusion length is variable over the substrate (e.g., in the case of deposition in trenches) the deposition rate will also vary.

At very high temperatures often a decrease of the deposition rate is observed, regime III in Figure 2. Such a decrease can be attributed to the occurrence of parasitic side reactions. These reactions may, for example, cause the formation of clusters of molecules or even particles in the gas phase that do not contribute to the deposition. This generally results in a decrease of the growth efficiency.

From an OMCVD point of view, it is clear from these considerations that it might be advantageous to deposit layers in either regime I or II depending on the desired properties of the layer. For other reasons the deposition temperature may be restricted either to some maximum temperature (for example to suppress substrate degradation) or to a minimum temperature (for example to obtain a sufficiently crystallised film). The transition of regime I to II will not only depend on the temperature but also on the reactor pressure, gas-flow rate, precursors, reactor design, etcetera. This gives an extra degree of freedom in designing the process.

2.2 The OMCVD set-up

For a standard OMCVD process, no complicated hardware is necessary. But, as with many other techniques, also OMCVD systems can end-up as rather complicated pieces of equipment. In Figure 3 a simplified drawing of an OMCVD system is presented. Basically an OMCVD system consists of three parts; (i) a gas-mixing system (the gas system) connected to (ii) the reactor cell and (iii) a pumping system with gas exhaust.

The gas system Via the gas system, vapours of the precursors are transported from the evaporation zone to the reactor cell. This is done by directing an inert carrier gas, e.g., N_2, through the bubblers containing the precursors. The precursors are preferably liquids at convenient temperatures (i.e. -30-30°C) in order to ensure a reproducible pick-up of their vapours by the carrier gas. Gaseous precursors also are very useful. Since they are not available for the materials discussed here they will be ignored. In the case of solid precursors, the pick-up efficiency is influenced by the area of the solid which may vary in time. This will introduce a non-reproducibility in the deposition process. A solution to this problem might be to solve the solid in an inert solvent.[6] A more constant pick-up efficiency

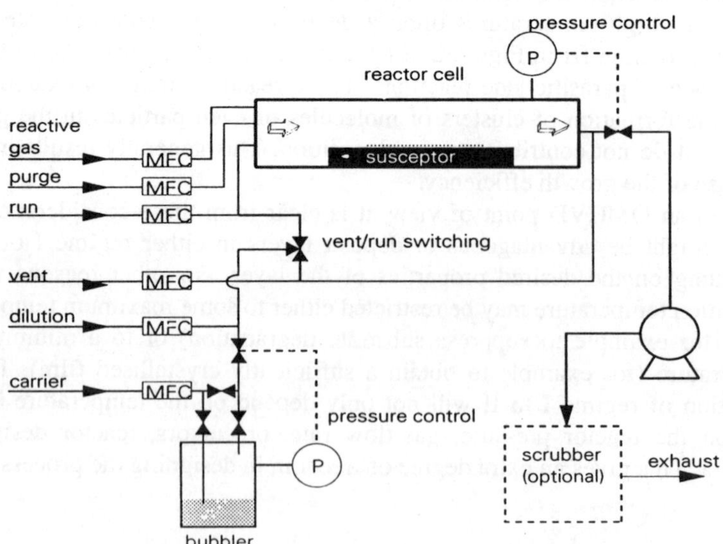

FIGURE 3. Schematic diagram of an OMCVD system.

may also be obtained by connecting two, or more, bubblers in series.

The main parameters controlling the composition of the film as well as the deposition rate are the partial pressures of the precursors in the reactor and the deposition temperature. Therefore, accurate control of these parameters is important.

The vapour pressure of a precursor, $p_i(T)$, is determined only by its temperature, T. The amount of transported material then is determined by the flow rate of the carrier gas through the bubbler, $\phi_{bubbler}$, and by the total gas pressure in the bubbler, $P_{bubbler}$. This results in a partial pressure of the precursor material in the reactor cell, $p_{reactor,i}$, according to Equation 1.

$$P_{reactor,i} = \frac{p_i(T)}{P_{bubbler}} \times \frac{\phi_{bubbler}}{\phi_{reactor}} \times P_{reactor} \qquad (1)$$

In this equation $\phi_{reactor}$ is the total flow rate through, and $P_{reactor}$ the total gas pressure in the reactor. Equation 1 assumes a complete saturation of the carrier gas with the precursor vapour, i.e. a pick-up efficiency of unity. This may not always be the case. Especially for solid precursors or liquid precursors used at high gas-flow rates and low bubbler pressures, this pick-up efficiency may be less than unity. Precursors with high vapour pressures can be transported without the use of a carrier gas. In this case the pressure in the bubbler, of course, is simply the vapour pressure of the precursor and is determined by the temperature only. For precursors with a low vapour pressure, the pressure in the bubblers, in practice, is determined by the pressure of the carrier gas. In an OMVCD system the bubbler pressure should be carefully monitored. Modern systems have a facility not only to measure but also to regulate this pressure. The flow rate through the bubbler, $\phi_{bubbler}$, is regulated by a (mass-)flow controller. So the amount of transported material can be controlled by manipulating the bubbler temperature, the bubbler pressure and the flow rate through the bubbler. With possible errors in monitoring the parameters as well as uncertainties in the vapour pressures of the precursors (literature values show considerable spread) and unknown pick-up efficiency, the calculation of the partial pressure in the reactor in most cases is not exact. In practice, an empirical relation between the deposition rate and precursor partial pressure (or the experimental parameters determining this pressure) is used. After such an empirical relationship has been established, a very accurate and reproducible process control is obtained. A more convenient way of controlling the partial pressures would be by direct measurement of the precursor concentrations. Devices are now available which, by making use of sound velocity measurements, are able to measure these.[7]

The gas mixture is fed into a gas-switching manifold. With this manifold it can be decided whether the gas mixture is switched to 'run' and enters the reactor cell to form the desired film, or is switched to 'vent' and is exhausted, contributing in a direct way to the environmental pollution (without a gas-scrubbing system) or in an indirect way (with a gas-scrubbing system).

In the case of OMCVD of oxidic materials, the hardware usually gets more complicated because a number of the available precursors of essential elements have a rather low vapour pressure at room temperature. This implies that the temperature at which these precursors must be evaporated has to be raised (typically to 50-150°C). Higher partial pressures of the precursors cannot be simply obtained by raising the temperature of the bubbler only, since in this case the precursor material would condense in cooler parts of the system that are in between the bubbler and the reactor cell. Therefore, all these parts have to be heated to the same temperature as the bubbler (unless the precursor vapour is diluted with additional gas but this reduces the effect of temperature increase of the bubbler). In practice, this appears to be rather difficult since heating not only concerns stainless steel tubing but also components like valves and in some cases sensitive electronic equipment. Also it is important to heat up the gas system in a very uniform way. If the temperature of the gas system is at some points significantly higher than the evaporation temperature, there is a risk of precursor decomposition. Since for a number of precursors the evaporation temperature is close to the decomposition temperature, this is a serious problem. Such a premature decomposition results in a contamination of the gas system, irreproducibilities and in a reduction of the amount of precursor material that reaches the reactor cell. In order to prevent these undesirable effects of cold and hot spots, the following solutions are used.

Ovens One approach is to put the relevant parts of the gas system entirely in an oven.[8] The main advantage is that all the parts included in the gas system are uniformly heated to the desired temperature. In practice, it is a rather complicated and expensive solution especially in the case that separate ovens are needed for each precursor. In this approach difficulties may arise at points where the gas system leaves the oven to enter another oven, or the reactor cell.

Electric heater tapes A simpler and less expensive solution is found in wrapping-up the gas system with heater tape. Using heater tapes, high temperatures can be reached. The main disadvantage in this approach is that it appears to be very difficult to obtain a uniform heating. Cold spots may occur since the gas system consists of several components all with different mass and shape. Hot spots develop when a part of the system gets slightly hotter than the rest of the system for example by a non-uniform wrapping of the heater tape. This higher local temperature results in an increased resistance of the heater tape which will, in its turn, cause locally an extra heat production, etcetera. The use of heater tapes with a negative resistance-temperature-coefficient will help to prevent this effect.

Liquid heating A third solution is to heat all the lines and components with a flow of some hot liquid, for example oil, which has at least the same temperature as the bubbler bath.[9] Heating of transport lines in this set-up is relatively easily accomplished. This can be done by making use of coaxial tubing; the inner tube is used for transport of the carrier gas containing the precursors. Through the outer tube flows the hot oil heating the inner tube. An advantage of a set-up like this is that the temperature distribution over the lines is very uniform. It is harder to heat components like valves. In practice this is done by mounting the components on metal blocks that are heated by the same oil that flows through the lines. It is difficult, however, to maintain the components at the desired temperature and the thermal isolation of the gas system is crucial. The use of oil, or another liquid, implies a certain risk; in case of a leakage in the gas system, the gas lines might get contaminated with the liquid.

Liquid-source CVD A way to circumvent the problem of heating the gas system is to make use of liquid-source CVD. In this technique the precursors are transported as liquids, with- or without a solvent, and the precursors are evaporated or injected by a nebulizer just before entering the deposition reactor. Liquid-source CVD requires dedicated equipment and the use of solvents may introduce difficulties. For processes using liquid precursors with very low vapour pressures or solid precursors this option may prove to be useful.[10]

Solid-source CVD Heating of the gas system, and in fact the necessity of the gas system itself, may be circumvented by making use of solid-source CVD. A mixture of solid state precursors containing the elements in the desired ratio is evaporated onto the heated substrate where the film forms.[11] With this technique thin films and heterostructures can be deposited uniformly over large areas.

The reactor cell The reactor cell is the central and most important part of an OMCVD system. The gas mixture, generated in the gas system and containing the vapours of the precursors in the desired ratio, enters the reactor cell where the actual decomposition and deposition take place. The precursor vapours decompose in a high temperature region and react to form the thin film. Chemical reactions, gas-flow dynamics and diffusion processes all play an important role here (see Figure 1).

The geometry of OMCVD reactor cells has been the subject of a number of studies, both experimental[2,12] and theoretical.[13] In general, it can be stated that no universal optimal cell design exists. It strongly depends on the type of process that is used. For example, the cell design for the deposition of III-V semiconductors may be quite different from the optimum cell design for the deposition of $PbZr_xTi_{1-x}O_3$ thin films.

The reactor cell contains one or more substrates on which the deposition takes place. In order to obtain a uniform temperature distribution over the substrate, it is common practice to make use of indirect heating. The substrate is placed on a block of material, a substrate-holder or susceptor, sometimes called a chuck having a large thermal mass in comparison with the substrate. In the case where only the susceptor is heated, the reactor is called a cold-wall reactor. The reactor walls in such systems may be cooled to suppress deposition on the walls. If the entire reactor is heated this is referred to as a hot-wall reactor. It depends very much on the process which type of reactor is preferred. Single-wafer cold-wall reactors allowing a high throughput are considered to fulfil modern demands best.

The material used for the construction of the susceptor is a point of concern. Graphite is the most commonly used susceptor material but in oxidising conditions this material will burn. Thus an alternative susceptor material has to be used or the graphite has to be provided with an inert coating. Alternative susceptor materials include silicon and ceramic materials, like for example aluminium oxide or aluminium nitride. Also, metal susceptors have been used successfully.[14]

Most modern OMCVD systems are operated at reduced reactor pressure. By lowering the reactor pressure, the role of gas-flow dynamics is reduced, due to an increased gas velocity. As a result of the reduced residence time at this increased gas velocity, homogeneous gas-phase reactions are suppressed. The diffusion rate of reactants will increase with decreasing reactor pressure. This effect can be used to change the deposition regime in which the OMCVD process is operated, see Figure 2. In general, no complicated pumping systems are required to achieve the moderate pressure ranges used in OMCVD. Unlike in ultra-high vacuum systems, the background pressure is not an important parameter in OMCVD systems. The leak tightness of the gas system and the purity of the carrier gasses are the important factors. A simple oil pump, operating with chemically inert low vapour pressure oil, will do. In order to pump large quantities of gas, a roots blower can be used in combination with an oil pump.

2.3 Precursors

An important characteristic of a precursor is its vapour pressure. This vapour pressure increases exponentially with the temperature according to the Clausius-Clapeyron equation:

$$\frac{d(\ln(p))}{d(T)} = \frac{\Delta H_v}{RT^2} \qquad (2)$$

In this equation, T is the temperature of the precursor, R the gas constant and ΔH_v the enthalpy of evaporation, respectively. Equation 2 is frequently rewritten into Equation 3:

$$\log(p) = A - \frac{B}{T} \qquad (3)$$

As a rule of thumb, the vapour pressure increases by a factor of 2 when the evaporation temperature increases by 10°C. Of course, the vapour pressure cannot be unlimitedly increased. An upper temperature is set by the onset of decomposition of the compound and, obviously, one has to keep the precursor temperature well below this.

Precursors that are most frequently used in (OM)CVD include the halides, alkyls, allyls, alkoxides and chelates. In fact, these are not all organometallic compounds since they do not have a direct bond between the metal and a carbon atom. For reasons of simplicity, the term 'organometallic compounds' is merely used here for compounds containing organic ligands and a metal. In the discussion to follow, the description of the precursors is restricted to compounds for the growth of the ferroelectric oxides (Pb,La)(Zr,Ti)O_3, $Bi_4Ti_3O_{12}$, $SrTiO_3$, and $BaTiO_3$.

Halides Halides can exhibit both extremely low as well as extremely high vapour pressures. For example, $PbCl_2$ is a stable salt with a very low vapour pressure, whereas $TiCl_4$ is a liquid with a high vapour pressure at room temperature. Disadvantages of the halide precursors may be the corrosivity of the reaction products (HCl) on hydrolysis and the frequently reported incorporation of halogen atoms in the layers, especially at low growth temperatures. Additionally, due to the strongly exothermic reaction upon oxidation or hydrolysis, the halides are very prone to homogeneous gas-phase reactions. This may cause depletion effects and give rise to poor surface morphologies.

Alkyls and allyls There are only few alkyls or allyls with high vapour pressures available for those metals of interest for ferroelectric thin films. Examples are tetraethyllead (TEL), tetraphenyllead, trimethylbismuth (TMB) and triphenylbismuth. Especially, TEL and TMB have convenient vapour pressures. Both compounds have been applied successfully for the growth of oxidic thin films. In contrast to the oxygen containing precursors to be dealt with next, the formation of oxides from the alkyls or allyls is by oxidation rather than by thermal decomposition. Therefore an additional oxidant such as oxygen is necessary to obtain the oxide as for example in the following reaction.

$$2Pb(C_2H_5)_4 + 3O_2 \rightarrow 2PbO + 8C_2H_4 + 4H_2O \tag{4}$$

The metal-carbon bond implies a risk of carbon incorporation into the solid. However, both for the lead and the bismuth precursors no such carbon contamination has been reported. On the other hand, it was found for the deposition of ZrO_2 from $Zr(C_5H_5)_2(OC_2H_5)_2$ that the films were heavily contaminated with carbon.[15] A comparable effect has been observed for the growth of ZrO_2 from the relatively unstable compound $Zr(CH_2C(CH_3)_3)_4$ by plasma enhanced OMCVD.[16] Under similar conditions, almost carbon free ZrO_2 layers could be obtained when using $Zr(C_5H_5)_2(CH_3)_2$ at temperatures as low as 100°C.[16]

Alkoxides Also for the metal alkoxides, compounds with a very low as well as compounds with a very high vapour pressure are known.[17] The volatility of the alkoxides depends on their degree of oligomerisation. If two or more alkoxides cluster together, the vapour pressure will decrease due to the increased molecular weight of the cluster. This tendency of clustering is correlated to the incomplete co-ordination of the central metal atom by the ligands and to the positive charge present on the metal which easily attracts negatively charged ligands of neighbouring alkoxides.

Clustering of the alkoxides can be suppressed by introducing bulky ligands causing steric hindrance around the metal atom. In this way the positive charge located on the central metal atom is shielded from the negative charge located on oxygen atoms in the ligands. This causes the phenomenon that alkoxides with larger molecular weights, which in general causes the vapour pressure to decrease, can have *higher* vapour pressures than the smaller alkoxides. For example, for tetravalent metal alkoxides the volatility increases going from the methoxides to the ethoxides and propoxides. If the ligand becomes too large, the volatility decreases again due to the increased molecular weight of the alkoxide.

Additionally, the branching of the alkoxide has a dramatic effect on the volatility. For alkoxides of zirconium it increases going from *n*-butoxide ($OCH_2CH_2CH_2CH_3$) via *i*-butoxide ($OCH_2CH(CH_3)_2$) to *t*-butoxide ($OC(CH_3)_3$) which is a monomeric liquid with a vapour pressure of ~0.1 mbar at 50°C. This effect, again, is due to steric hindrance by the ligands prohibiting the clustering by neighbouring compounds. Oxides can be formed from the alkoxide compounds by simple decomposition. In contrast with the alkyl and allyl compounds no additional oxidising species is necessary. For example the formation of ZrO_2 from zirconium-*t*-butoxide is given by the overall reaction:

$$Zr(^{t}OC_4H_9)_4 \rightarrow ZrO_2 + 4CH_2(CH_3)_3 + 2H_2O \qquad (5)$$

The alkoxides can easily be purified by distillation under reduced pressure and they can be stored in stainless steel bubblers without noticeable degradation. The main difficulty in handling the metal alkoxides is their ease of hydrolysis by even very small amounts of water or alcohols and stringent precautions are required to prevent this deterioration of the precursor. On the other hand this reactivity towards water may be used for a low temperature decomposition of these precursors.

Double alkoxides compounds (e.g. compounds containing two (different) metals) have been applied for the deposition of the corresponding double oxides. These precursors ensure a stoichiometric supply of metals to the deposition region.[17]

Chelates The chelates (or β-diketonates) are frequently used as precursors for the deposition of oxidic thin films, especially of high-T_c superconductors and ferroelectrics. In Table 1 some chelate precursors are listed and a schematic of its structure is presented in Figure 4. The chelate precursors are stable up to about 200°C, and they are non-hygroscopic and non-pyrophoric. The toxicity of these compounds is generally much smaller than that of alkyls. Similar to the alkoxides, the volatilities depend strongly on the ability of the ligands to prevent the metal atoms from increasing their co-ordination number by forming intermolecular bonds which shield off the positive charge on the central metal atom. Particularly, fluorinated groups are found to be effective to increase the vapour pressure. For instance the volatility of the Zr-diketonates is $Zr(hfa)_4 > Zr(tfa)_4 > Zr(thd)_4 \cong Zr(acac)_4$.[18] This effect is attributed to an increased concentration of negative charge on the ligands, thus providing an effective shielding of the positive nuclei. The deposition of oxidic thin films from chelates is by thermal decomposition and, theoretically, no oxidant is needed.[19]

In the chelates, the ligands form six-membered rings with the central metal atom, see Figure 4. Due to this keto-enol tautomerization, these chelates show aromatic character through the resonance of the π-electrons and they strongly absorb light at wavelengths exceeding 250 nm. Hence, these precursors might be used in photo-enhanced OMCVD.[20]

R_1	R_2	name
CH_3	CH_3	acetylacetonato(acac)
$C(CH_3)_3$	$C(CH_3)_3$	tetramethylheptadionato (tmd, or thd) or dipivaloyl-methadianato (dpm)
CCH_3	CCF_3	trifluoroacetylacetonato(tfa)
CCF_3	CCF_3	hexafluoroacetylacetonato (hfa)
$C(CH_3)_3$	$C(CF_2CF_2CF_3)$	heptafluorodimethyl-octadionato (fod)

TABLE 1. Examples of chelate precursors

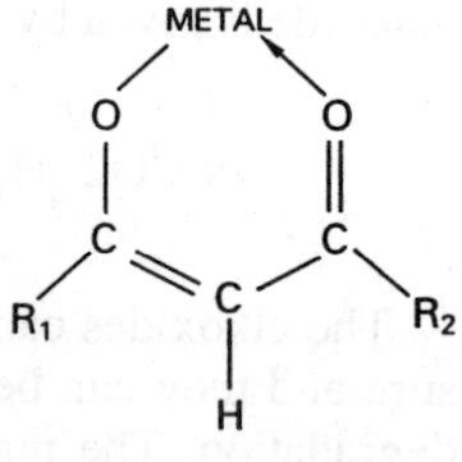

METAL, O, O, C, C, C, H, R_1, R_2

FIGURE 4. Structure

3. OMCVD OF FERROELECTRIC THIN FILMS

3.1 Introduction

The development of modern deposition techniques for oxidic thin films started in the 1980's. The first report on vapour deposition of $PbTiO_3$ dates back to 1982.[21] In following years only a few reports on (OM)CVD of ferroelectric thin films appeared. Around 1988 an increasing activity is observed in this field which was stimulated by two developments.

In the first place, the explosive interest in high-T_c superconductors, and in thin films of these materials, certainly has accelerated improvements in the oxidic thin-film techniques. The first reports on the deposition of high-T_c superconductor thin films by OMCVD appeared in 1988.[22-24] Since then, a large number of papers appeared in the literature.[25] OMCVD has produced layers of a quality comparable to or sometimes better than those obtained by other techniques.

In the second place, research in this field was pushed by the rapidly enhancing interest in the use of ferroelectric thin films for applications in (non-)volatile memories and sensor and actuator devices.

Before the deposition of ferroelectric thin films of the family based on $PbTiO_3$ will be dealt with in detail, a brief description is given of other ferroelectric thin films deposited by OMCVD. These include $Bi_4Ti_3O_{12}$, $BaTiO_3$, $SrTiO_3$, $Pb(Sc_{0.5}Ta_{0.5})O_3$ and $(Ba,Sr)Nb_2O_6$.

Barium titanate, $BaTiO_3$, has been grown by OMCVD using the precursors $Ba(thd)_2$ and $Ti(^iOC_3H_7)_4$.[26,27] The $Ba(thd)_2$ needs to be evaporated at temperatures around 250°C. As already mentioned this very high temperature causes technical problems because the gas system should be heated to above this temperature. Another serious problem is that this evaporation temperature is close to the region where $Ba(thd)_2$ starts to decompose. A third problem which has been encountered is the degradation of this compound in the bubbler with time leading to a decreasing vapour pressure. The evaporation temperature of the barium precursor could be reduced down to 105-125°C by using the $Ba(hfa)_2$-tetraglyme precursor.[28] Unfortunately, the layers then were found to be contaminated with fluorine. The precursor $Ti(^iOC_3H_7)_4$, which has a higher vapour pressure, was evaporated at temperatures between 25 and 60°C. Highly resistive dielectric layers were obtained on silicon substrates[26] at 600°C (ε=250, $\rho=10^9$ Ωcm) and on Pt/MgO stacks[27] at 800°C (ε=1040) with deposition rates of about 1 µm/hour. At 600°C barium carbonate is formed rather than barium titanate. In fact, the lower limit for the deposition temperature by thermal OMCVD is set by the formation of barium carbonate. Using a plasma-enhancement, films can be grown at 600°C and 680°C without any incorporation of carbon.[29,30]

The precursors used for the deposition of strontium titanate, $SrTiO_3$, were $Ti(^iOC_3H_7)_4$ and $Sr(dpm)_2$, evaporated at 50°C and 200-230°C, respectively.[31,32] Single-phase layers with a preferred orientation were deposited at 800°C on sapphire. The deposition rate was between 0.3 and 6 μm/hour. When using H_2O as oxidant, the layers were found to be contaminated with carbon. This could be suppressed effectively by the supply of additional oxygen.

Single-phase bismuth titanate, $Bi_4Ti_3O_{12}$, layers were deposited onto silicon substrates at 600-750°C at a deposition rate of 2.5-4.8 μm/hour. The precursors used were $Bi(C_6H_5)_3$ (evaporation temperature: 130-170°C) and $Ti(^iOC_3H_7)_4$ (40°C). It was found that the Bi/Ti-ratio decreased with increasing deposition temperature. Probably, this was due to the evaporation of the relatively volatile bismuth oxides from the growing film.[32,33] Strontium-barium niobate, $(Ba,Sr)Nb_2O_6$, thin films were grown by OMCVD with the precursors $Ba(thd)_2$ (250°C), $Sr(thd)_2$ (230°C) and $Nb(OC_2H_5)_5$ (130°C) on various substrates in a hot wall reactor at 550°C.[34] Attempts to deposit lead scandium tantalate, $Pb(Sc_{0.5}Ta_{0.5})O_3$, directly by OMCVD using the precursors $Ta(OC_2H_5)_5$ (95°C), $Sc(fod)_3$ (85°C) and $Pb(O^tC_4H_9)_2$ (120°C), were unsuccessful. In order to obtain this material, first $ScTaO_3$ layers were grown which were converted afterwards into $Pb(Sc_{0.5}Ta_{0.5})O_3$ by reaction with PbO.[35]

3.2 Lead titanate family

In the previous section, examples were given of ferroelectric thin films that have been synthesised using OMCVD. In this paragraph attention is focused on the lead lanthanum zirconate titanate, $(Pb,La)(Zr,Ti)O_3$, family. First, a short overview of the literature is given. Then, epitaxial films grown by OMCVD are treated in some detail followed by the deposition of $PbZr_xTi_{1-x}O_3$ on electroded substrates. Since 1982 a number of papers on the growth of $PbTiO_3$[35-55], $PbZr_xTi_{1-x}O_3$[9,56-63], $(Pb,La)TiO_3$[64,65], and $(Pb,La)(Zr,Ti)O_3$[65-68] have appeared.

The onset of CVD of $PbTiO_3$ thin films was undoubtedly marked by the work of Nakagawa *et al.* presented in a 1982 paper.[21] The use of chlorides as precursor materials is described. In a paper from the same research group by Kojima *et al.* the synthesis of $PbTiO_3$ thin films using

PbO and $Ti(OC_4H_9)_4$ (it was not specified which isomer was used) is described.[36] The use of $Pb(C_2H_5)_4$ as precursor for lead is suggested but for safety reasons this organometallic was not applied then. In a 1988 paper by Okada *et al.*[39] the deposition of $PbTiO_3$ using $Pb(C_2H_5)_4$ and $Ti(^iOC_3H_7)_4$ was described for the first time. An OMCVD system operating at atmospheric pressure was used and an initial study of the deposition of the individual oxides revealed that the deposition rate of TiO_2 was independent of the oxygen partial pressure. Obviously this is due to the fact that in this precursor oxygen is already bonded to the titanium. It was thus concluded that the formation of TiO_2 was via decomposition of $Ti(^iOC_3H_7)_4$ according to:

$$Ti(^iOC_3H_7)_4 \rightarrow TiO_2 + 4\,C_3H_6 + 2\,H_2O \tag{6}$$

The $Pb(C_2H_5)_4$ does not contain oxygen and therefore oxygen was necessary to form the lead oxide. It was proposed that PbO formation involves oxidation of $Pb(C_2H_5)_4$ according to Equation 4. The deposition rate of PbO saturated if the oxygen partial pressure exceeded a certain critical value.[37] The deposition rate of both PbO and TiO_2 first increased with substrate temperature, indicative of a kinetically controlled process (regime I in Figure 2) and then showed a decrease or a saturation, see Figure 5. This change in deposition rate was attributed to homogeneous gas-phase reactions (see Figure 2, regime III). Deposition of $PbTiO_3$ was achieved by matching the deposition rates of PbO and TiO_2:

$$PbO + TiO_2 \rightarrow PbTiO_3 \tag{7}$$

Single-phase perovskite-type $PbTiO_3$ was obtained on (001)Si, Al_2O_3 and (001)MgO. The deposition temperature was between 500 and 600°C. The films on (001)MgO showed a strong *c*-axis preferential orientation despite the relatively large lattice mismatch between the MgO and $PbTiO_3$. The Pb/Ti-ratio was found to be unity within 10% accuracy. Since this first paper, a number of publications emerged from this group.[38,42,56,58,62,64,66,68] An alternative route for (OM)CVD of $PbTiO_3$ was published by Yoon *et al.* in 1988 describing the use of metallic lead and $Ti(OC_2H_5)_4$.[39]

The first paper describing the use of low pressure in OMCVD of $PbTiO_3$ was by Kwak *et al.* in 1988 using $Pb(C_2H_5)_4$ and $Ti(^iOC_3H_7)_4$.[40] A

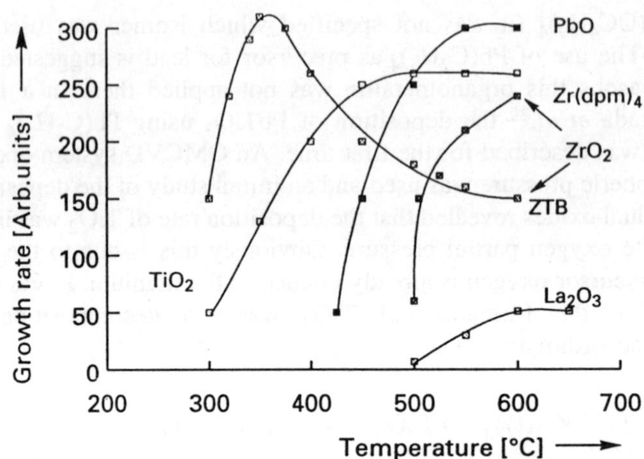

FIGURE 5. Deposition rates of oxides using OMCVD. The data were taken from Refs.62 and 64.

remarkable observation was that no additional oxygen was found to be necessary to obtain $PbTiO_3$. It was concluded that all the oxygen originates from the $Ti(^iOC_3H_7)_4$. This conclusion is not supported by data that have been published by others.[42,43,50] The use of low pressure OMCVD was also described in 1989 by Brierley *et al.*[35], Swartz *et al.*[41] and Okada *et al.*[42]

Hirai *et al.* report on the deposition of $PbTiO_3$ by the alternate introduction of $Pb(C_2H_5)_4$ and $Ti(^iOC_3H_7)_4$ into the deposition chamber.[52] In combination with the use of ozone as an oxidant a low deposition temperature of 510°C was possible.

For device applications $PbZr_xTi_{1-x}O_3$ is important and a number of studies on the OMCVD of this material are reported. The first paper on the OMCVD of $PbZr_xTi_{1-x}O_3$ using low pressure deposition conditions was by Okada *et al.*[56] In combination with $Pb(C_2H_5)_4$ and $Ti(^iOC_3H_7)_4$ the use of two different precursors for the zirconium was evaluated: $Zr(^iOC_3H_7)_4$ and $Zr(dpm)_4$, see Figure 5. With both precursors, deposition of single-phase perovskite type $PbZr_xTi_{1-x}O_3$ was obtained. Using $Zr(dpm)_4$, a higher preferential orientation on (001)MgO was obtained. In comparison with the deposition of $PbTiO_3$, a smoother surface topography was observed. This was attributed to a lowering of the strain with increasing ZrO_2 content. The high zirconium content resulted in a denser film. It was demonstrated by Kashihara *et al.*[59] that with OMCVD very good step coverage

can be obtained. This indicates that OMCVD may be very suitable for deposition on already structured wafers. The formation of $PbZr_xTi_{1-x}O_3$ thin films over 6 inch wafers was reported in the same paper. An inverted vertical reactor geometry was used with the gasses entering at the bottom of the reactor impinging directly on the wafer. An alternative lead precursor was studied, $Pb(dpm)_2$, which was evaporated at 135°C, in combination with $Zr(^iOC_4H_9)_4$ and $Ti(^iOC_3H_7)_4$. Additional O_2 was used in this process. Polycrystalline $PbZr_xTi_{1-x}O_3$ films were obtained on platinum coated silicon wafers at 600-700°C. Uniform deposition of $PbZr_xTi_{1-x}O_3$ on 4 and 6 inch platinum coated SiO_2/Si wafers has also been reported by other groups.[9,60,63] Variations in thickness and composition were below 5%. Also the integration of OMCVD grown $PbZr_xTi_{1-x}O_3$ films with CMOS is reported.[69]

The preparation of lanthanum substituted $PbTiO_3$ using OMCVD was reported for the first time by Tominaga *et al.*[64] The synthesis of $(Pb,La)TiO_3$ thin films was described using the precursors $Pb(C_2H_5)_4$, $Ti(^iOC_3H_7)_4$ and $La(dpm)_3$. Also in this work, the deposition of the individual components was studied similar to earlier work from this group, see Figure 5. It was found that the formation of La_2O_3 involved oxidation of the $La(dpm)_3$, comparable with PbO formation from $Pb(C_2H_5)_4$. The deposition rate of La_2O_3 levelled off at about 20% oxygen in the gas phase. Decomposition of the $La(dpm)_3$ started at approximately 500°C. This is a rather high temperature in comparison with $Pb(C_2H_5)_4$ (400°C) and $Ti(^iOC_3H_7)_4$ (250°C) illustrating the relative stability of this precursor. When the deposition rates of the individual components were matched, single-phase perovskite-type $(Pb,La)TiO_3$ thin films could be obtained. At 600°C, thin films of $(Pb,La)TiO_3$ on (001)MgO substrates were obtained with Pb/La-ratios up to 83/17. The films had a high preferential orientation with the *c* axis perpendicular to the (001)MgO substrate surface.

$(Pb,La)(Zr,Ti)O_3$ films deposited by OMCVD were first reported by Okada *et al.* in 1992 using the precursors $Pb(C_2H_5)_4$, $La(dpm)_3$, $Zr(dpm)_4$, and $Ti(^iOC_3H_7)_4$.[66] Single-phase perovskite type $(Pb,La)(Zr,Ti)O_3$ was deposited on (001)MgO, Pt/(001)MgO and $Pt/SiO_2/Si$ again by adjusting the deposition rates of the individual oxides. The film properties were examined as a function of the lanthanum content for $Zr/(Zr+Ti)<0.5$. The values for the dielectric permittivity depend on the $Zr/(Zr+Ti)$ and increased with increasing La content up to about 5 atomic %. The deposition of epitaxial $(Pb,La)(Zr,Ti)O_3$ films on (001)MgO substrates was reported by Funakubo *et al.*[67]

Most authors report deposition temperatures in the range of 600-700°C.

It may be necessary for future devices to lower the deposition temperatures as much as possible in order not to damage underlying structures. A way to decrease the deposition temperature is to make use of additional external energy sources like plasmas or photons. A second point of interest is that the use of such sources is expected to have an influence on crystal quality and, therefore, on the ferroelectric properties of the oxide thin film. As far as the OMCVD of $PbZr_xTi_{1-x}O_3$ is concerned, a few papers have appeared in the literature describing the use of such external power sources.

The photo-enhanced deposition of $PbTiO_3$ was reported using the precursors $Pb(C_2H_5)_4$, $Ti(^iOC_3H_7)_4$ and O_2 in combination with a Xe-Hg lamp of 500 Watt.[70-72] For the deposition rate of $PbTiO_3$, *a decrease* in deposition rate was observed when photons were used. $PbTiO_3$ with a (111) preferential orientation could be deposited on Al_2O_3 only with the photo-irradiation at 600°C. Using this photo-enhancement, an improved crystallinity was achieved and the deposition temperature of $PbTiO_3$ could be lowered to 530°C. When the O_2 was replaced by the more reactive NO_2, improvements of the dielectric properties of the photo-irradiated $PbTiO_3$ films deposited at temperatures as low as 550°C were observed.[71]

The use of a 13.56 MHz RF-plasma in the OMCVD of $PbZr_xTi_{1-x}O_3$ was also reported.[73-75] The precursors used were $Pb(C_2H_5)_4$, $Zr(^iOC_4H_9)_4$, $Ti(^iOC_4H_9)_4$, and O_2. Films were deposited at substrate temperatures as low as 425-500°C. At a pressure of 0.3 Torr and low RF-power, the pyrochlore type $PbZr_xTi_{1-x}O_3$ formed. Increase of the RF-power and lowering of the reactor pressure to 0.16 Torr yielded up to 92% of the perovskite-type $PbZr_xTi_{1-x}O_3$ in the thin films deposited on a platinized substrate. Typical deposition rates of 0.5-0.6 μm/hr were obtained, which are in the same range as for thermal OMCVD.

3.3 Control of the composition

The control of the composition is a matter of concern in the case of lead based (ferroelectric) thin films. This is due to the relatively high vapour pressure of both lead and lead oxide resulting in a relatively fast evaporation of these compounds from the growing surface. Many thin films techniques yield $(Pb,La)(Zr,Ti)O_3$ thin films only if compensation for this lead-loss is supplied. In the case of OMCVD the lead-loss can be compensated for by adjusting the lead precursor partial pressure. The high vapour pressures of lead and lead oxide turn out to be a benefit as the stoichiometry of the film is not very sensitive to variations in both the deposition temperature and the precursor partial pressures (within a certain range).[50,76] This is exemplified in Figure 6 for $PbTiO_3$ films grown on platinum coated SiO_2/Si substrates. In this figure the Pb/Ti of $PbTiO_3$ films grown at three temperatures is plotted against the gas-phase ratio of the precursors, $(Pb/Ti)_g$. For the films deposited at 500°C the gas-phase composition is reflected in the composition of the thin film. At this temperature, the gas-phase composition must be accurately chosen and controlled in order to obtain stoichiometric $PbZr_xTi_{1-x}O_3$ i.e. Pb/Ti=1. At 670°C, the composition of the film turns out to be stoichiometric and independent of the gas-phase composition within an interval of $(Pb/Ti)_g$. These observations indicate that a process window exists, with respect to both the gas-phase composition and the temperature, in which stoichiometric $PbTiO_3$ forms. Hence, the composition of the films is independent of small temperature variations over the wafer and on fluctuations in the precursor partial pressures. This is important for a reproducible and uniform deposition of such films over large wafers.

The occurrence of this process window may be explained by a mechanism involving a fast surface reaction of PbO with TiO_2 (and ZrO_2) to form $PbZr_xTi_{1-x}O_3$ in competition with the evaporation of the relatively volatile PbO. This mechanism is expected to be independent of the type of

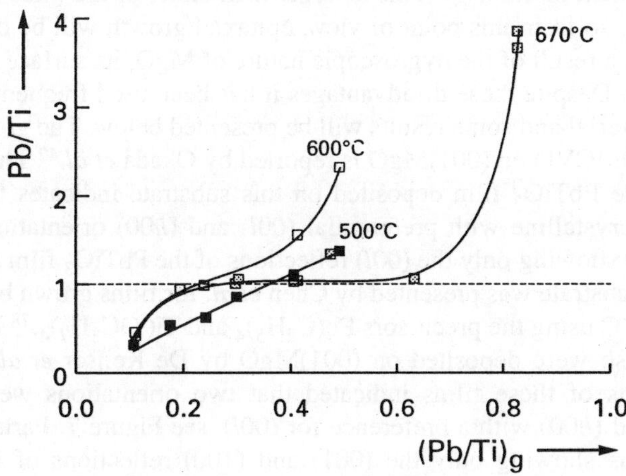

FIGURE 6. Pb/Ti of the solid as measured with X-ray fluorescence as a function of the gas-phase ratio $(Pb/Ti)_g$ for three deposition temperatures. Data taken from Ref.76.

substrate. The size of the process window depends on e.g. reactor geometry, substrate temperature and growth rate.[50,77] The growth rate, the ratio of the precursor partial pressures and the substrate temperature are the important parameters. This mechanism regulating the composition is not specific to OMCVD of lead-based films and comparable observations are expected for other in-situ deposition techniques and other materials. An example is $Bi_4Ti_3O_{12}$ where the presence of volatile Bi_2O_3 may induce a comparable mechanism.

3.4 Hetero epitaxial films

Although epitaxial ferroelectric films have a limited application range due to the small substrates, the study of such films may help the general understanding of growth mechanisms and the properties of ferroelectric capacitors. Some results will be presented of the growth of $PbTiO_3$ and $PbZr_xTi_{1-x}O_3$ on (001)MgO and (001)$SrTiO_3$. It should be noted that the description 'heteroepitaxial' is also used for films which could be described better as having some preferential orientation.

Growth on (001)MgO MgO has a cubic NaCl-type crystal structure with a lattice constant of 4.203 Å. This is larger than either of the three axes of $PbZr_xTi_{1-x}O_3$ so from this point of view, epitaxial growth will be difficult to obtain. As a result of the hygroscopic nature of MgO, its surface quality may be poor. Despite these disadvantages it has been used frequently as a substrate material and some results will be presented below. The growth of $PbTiO_3$ by OMCVD on (001)MgO is reported by Okada *et al.*[42] The XRD pattern of the $PbTiO_3$ film deposited on this substrate indicates that the film is polycrystalline with preferential (00*l*) and (*h*00) orientations. An XRD pattern showing only the (00*l*) reflections of the $PbTiO_3$ film and the (001)MgO substrate was presented by Chen *et al.* for films grown between 400 and 700°C using the precursors $Pb(C_2H_5)_4$ and $Ti(^iOC_3H_7)_4$.[78] $PbTiO_3$ thin films also were deposited on (001)MgO by De Keijser *et al.*[44] The XRD patterns of these films indicated that two orientations were present:(00*l*) and (*h*00) with a preference for (00*l*), see Figure 7. Parts of the XRD patterns showing only the (001) and (100) reflections of $PbTiO_3$ films with varying thickness are presented in Figure 8. If the film thickness decreases, a coalescence of the two reflections is observed indicating that the *c/a* ratio of the tetragonal structure decreases. Comparable effects were observed before for fine grained $BaTiO_3$ and $PbTiO_3$ powders.[79,80]

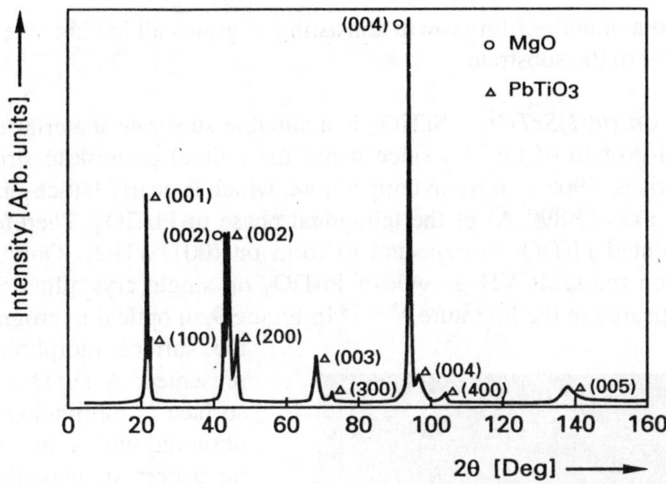

FIGURE 7. X-ray diffraction pattern of a $PbTiO_3$ film on (001)MgO. Ref.44.

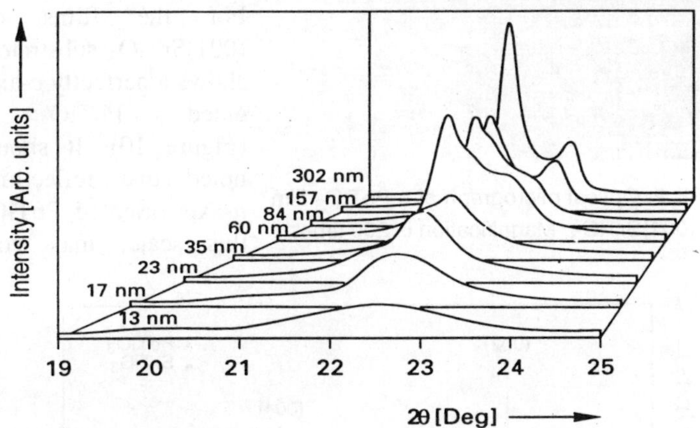

FIGURE 8. Part of the XRD pattern showing the (001) and (100) reflections of the $PbTiO_3$ as a function of thickness. The reflections tend to coalesce when the thickness decreases. Ref.44.

Funakubo reported the epitaxial growth of $PbZr_xTi_{1-x}O_3$ on (001)MgO.[57] The XRD patterns of the obtained films revealed the presence of strong (00*l*) and very weak (*h*00) reflections of the $PbZr_{0.5}Ti_{0.5}O_3$. The in-plane alignment of the films with the (010) and (100) substrate axes was confirmed by pole-figure analysis. Scanning electron microscopy of the films indicated a granular film growth consisting of grains all having an epitaxial relation to the substrate.

Growth on (001)SrTiO₃ $SrTiO_3$ is a suitable substrate material for the epitaxial growth of $PbTiO_3$ since it has the (cubic) perovskite structure. The *a* axis is 3.905 Å at room temperature, which is nearly lattice-matched to the *a* axes (3.899 Å) of the tetragonal phase of $PbTiO_3$. Therefore, *c*-axis oriented $PbTiO_3$ is expected to form on (001)$SrTiO_3$. Only a few reports on the OMCVD growth of $PbTiO_3$ on single crystalline $SrTiO_3$ have appeared in the literature.[43,49,78] In Figure 9 an optical micrograph of the surface morphology is presented. A slightly rough surface morphology is observed with a characteristic pattern of perpendicular lines. This pattern is attributed to an *a-c* twinning.[43,81] For the films on a (001)$SrTiO_3$ substrate XRD shows a perfectly *c*-axis oriented $PbTiO_3$ film (Figure 10). It should be noted that reflections of *a*-axis oriented $PbTiO_3$, on this scale, may coincide

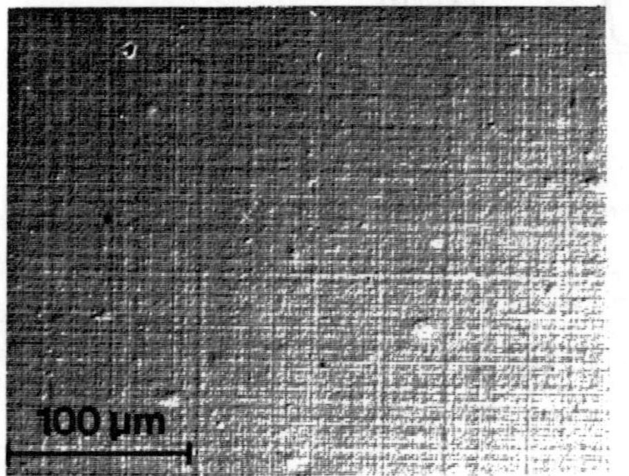

FIGURE 9. Optical micrograph of a $PbTiO_3$ film on (001)$SrTiO_3$. Magnification is 200 times.

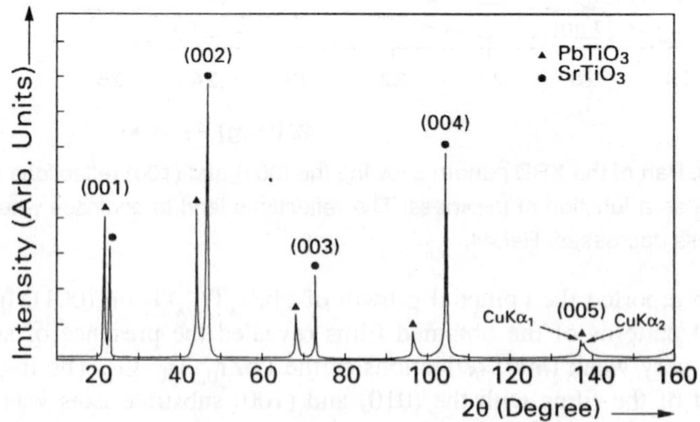

FIGURE 10. X-ray diffraction pattern of a $PbTiO_3$ film on (001)$SrTiO_3$.

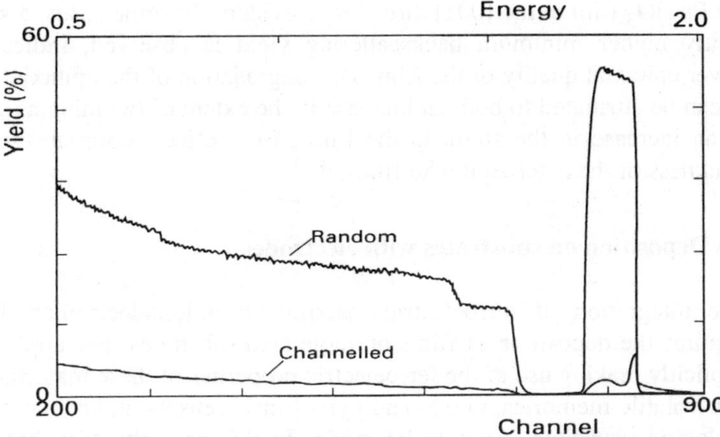

FIGURE 11. Random and channelled RBS spectra of an epitaxial $PbTiO_3$ film on (001)$SrTiO_3$. The minimum channelling yield in the (001) direction is ~3%, indicative of epitaxial growth.

with the substrate reflections because of the small difference in lattice constants of the $SrTiO_3$ (3.905 Å) and the *a* axes of the $PbTiO_3$ (3.899 Å). Pole-figure analysis proves that the principal lattice directions of the film coincide with those of the (001)$SrTiO_3$ substrate.[43] In Figure 11 the random and channelled Rutherford backscattering spectra of a $PbTiO_3$ thin film on (001)$SrTiO_3$ are presented. The minimum backscatter yield in the [001] direction is about 3%.[50] This low value for the minimum channelling yield is indicative of a heteroepitaxial film of high quality. The heteroepitaxial nature of these films, at least locally, was confirmed by high resolution electron microscopy (HREM), see Figure 12. The continuation of the atomic rows going from the $SrTiO_3$ substrate to

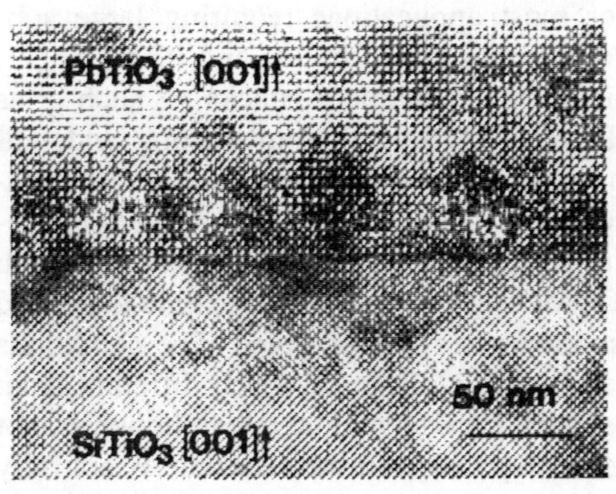

FIGURE 12. HREM image of a typical $SrTiO_3$/$PbTiO_3$ interface.

the $PbTiO_3$ film in the [001] direction is evident. In some cases, a significantly higher minimum backscattering yield is observed, indicating a lower epitaxial quality of the film. This degradation of the epitaxial quality can be attributed to both an increase in the extent of twinning as well as to an increase in the strain in the films. Both effects correlate with the thickness of the heteroepitaxial films.[81,82]

3.5 Deposition on substrates with electrodes

The integration of ferroelectric materials in microelectronics devices requires the deposition of films on large area substrates. For applications explicitly making use of the ferroelectric properties of these materials (e.g. non-volatile memories, piezo- and pyroelectric sensors and actuators) ferroelectric capacitors need to be made. In this case the film has to be deposited on an electrode. The most frequently used electrode material until now is platinum which is one of the few materials that can withstand high temperatures and oxidising conditions. Alternatively, oxidic conductors like ruthenium dioxide can be used. Such electrodes are compatible with $PbZr_xTi_{1-x}O_3$ processing and good ferroelectric properties can be obtained with these electrodes.[83]

Unlike the deposition of $PbZr_xTi_{1-x}O_3$ on lattice-matched substrates, the films on the electroded substrates are polycrystalline. Although it seems advantageous to use single crystalline (001)$PbZr_xTi_{1-x}O_3$ (since for tetragonal $PbZr_xTi_{1-x}O_3$ the polarisation is directed along the [001] crystal axis), polycrystalline films, for most applications requiring large substrates, have to be applied. Work on epitaxial $PbTiO_3$ and $PbZr_xTi_{1-x}O_3$ films on conducting single crystalline stacks, such as $La_{0.5}Sr_{0.5}CoO_3/SrTiO_3$ and $YBa_2Cu_3O_{7-\partial}/SrTiO_3$ is interesting and will help to understand the relation between the crystal structure and orientation on the one hand and switching mechanisms, fatigue problems, electrode effects, etc. on the other hand.[84,85] However, the substrates are small.

In this section some data of OMCVD of $PbZr_xTi_{1-x}O_3$ films deposited on silicon wafers with platinum electrodes will be given. The conditions for the deposition on electroded wafers will not be much different from those in the case of oxidic substrates. Care should be taken, however, not to damage the electrode during the deposition. If the electrode is not fully dense a reaction of the $PbZr_xTi_{1-x}O_3$, PbO or Pb with silicon or silicon dioxide deteriorating the electrode and forming undesirable lead silicates.

Figure 13 shows SEM micrographs of typical surface morphologies of

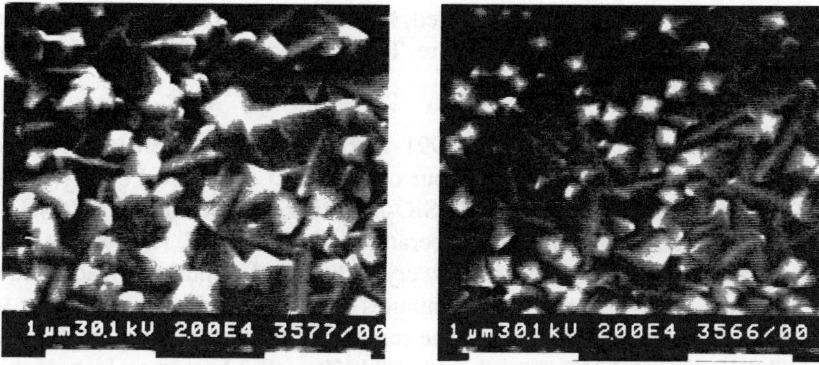

FIGURE 13. SEM micrographs of $PbZr_xTi_{1-x}O_3$ on platinized silicon wafers. x=0.45 (left) and x=0.75 (right). Ref.76.

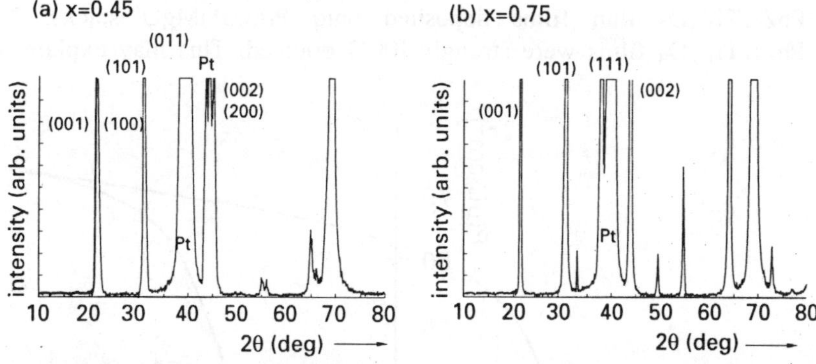

FIGURE 14. XRD patterns of $PbZr_xTi_{1-x}O_3$ on platinized silicon wafers. Ref. 76

polycrystalline $PbZr_xTi_{1-x}O_3$ films with x=0.45 and x=0.75 deposited at 700°C on platinized SiO_2/Si wafers. Rather rough morphologies are observed. The film consists of crystallites with an estimated average size of 0.5 μm for the film with x=0.45 and 0.2 μm for the film with x=0.75. These crystallites are faceted, pointing at a three-dimensional growth of the film with some crystal planes growing faster than others. The morphology of the films tends to improve somewhat with increasing zirconium content.

The XRD patterns of the $PbZr_xTi_{1-x}O_3$ thin films from Figure 12 are presented in Figure 14. As expected, for x=0.45 the diffraction pattern can be indexed by a tetragonal structure. The layer with x=0.75 can be indexed by a cubic structure.

Ferroelectric properties In 1991 Dormans *et al.*[45] were the first to demonstrate ferroelectric behaviour of OMCVD deposited $PbTiO_3$ films. The films were deposited on $Pt/SiO_2/Si$ using the precursors $Pb(C_2H_5)_4$ and $Ti(^iOC_3H_7)_4$ at substrate temperatures of only 400°C. At this low temperature the $PbTiO_3$ was poorly crystalline and contained lead oxide. Therefore, a post anneal for 30 minutes in air at 700°C was necessary to obtain ferroelectric properties. The remanent polarisation was 55 μC/cm² and the coercive field-strength was 100 kV/cm. The hysteresis loop of this film is presented in Figure 15. At higher field strengths of around 200 kV/cm a non-linear and time-dependent conduction through the layers was observed.

Good ferroelectric behaviour was observed by Sakashita *et al.*[58] for $PbZr_xTi_{1-x}O_3$ thin films deposited onto Pt/(001)MgO stacks. The $PbZr_xTi_{1-x}O_3$ films were strongly [001] oriented. This may explain the

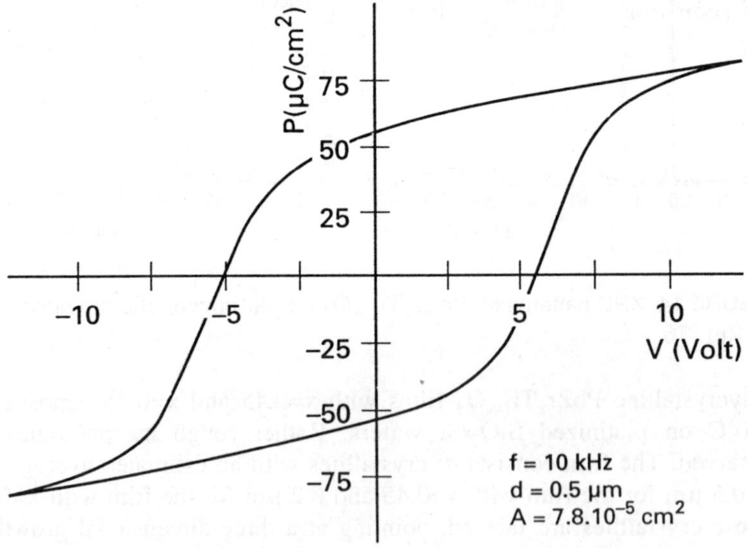

FIGURE 15. Hysteresis loop of a $PbTiO_3$ film deposited at 400°C and annealed for 30 minutes at 700°C in oxygen. Ref. 45

very square-shaped hysteresis loops that were presented by these authors. Values for the remanent polarisation were between 30-40 μC/cm² and coercive field-strengths were between 40-65 kV/cm. The latter decreased with increasing zirconium content in the film. Values for P_r were only slightly affected by changes in the composition of the layer, see Figure 16. The films showed a resistivity of 10^{12}-10^{14} Ωcm and had a breakdown field-strength: of 0.1-1 MV/cm The relative dielectric constant, ε_r, was found to be 200-350 almost independent of the composition for these highly *c*-axis oriented films. Randomly oriented films, deposited onto platinum coated oxidised silicon wafers, showed a higher value for ε_r which was dependent on the composition, comparable to what is found for bulk material.

An example of a hysteresis curve measured for a $PbZr_xTi_{1-x}O_3$ layer with x=0.75 deposited at 700°C onto a platinum coated SiO_2/Si substrate is presented in Figure 17.[76] A rather rectangular shaped hysteresis curve is observed for this film. This indicates a small distribution of domain orientations in these films. The values for P_r and E_c are high as compared to data reported for layers of similar composition but deposited by other techniques.

The high values for the remanent polarisations of OMCVD films, 40-60 μC/cm², are probably caused by the high degree of crystallisation and the preferential orientation of the crystallites. The unfavourably high values measured for the coercive field-strength as compared to values for

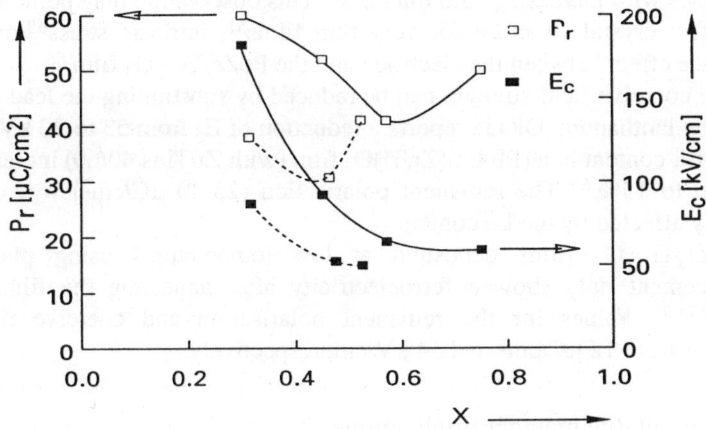

FIGURE 16. P_r and E_c versus x for $PbZr_xTi_{1-x}O_3$ films. Data taken from Ref.58 (solid) and Ref. 9 (dashed).

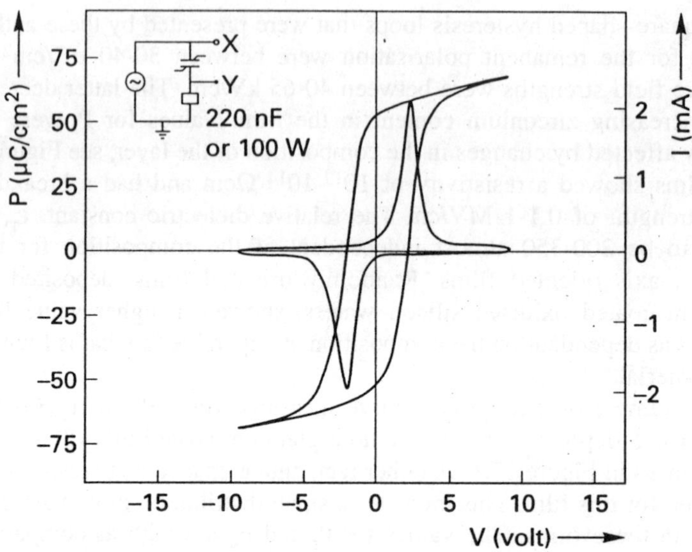

FIGURE 17. Hysteresis loop for x=0.75. Taken from Ref. 76.

bulk ceramic material, may have several causes.[58,60,65] Using optimal deposition conditions films were obtained having a coercive field strength of about 40 kV/cm. An important observation is that the apparent E_c decreases with increasing film thickness. This observation may point at an imperfect crystal structure for very thin films[51], intrinsic stress[62] or an interface effect between the electrode and the $PbZr_xTi_{1-x}O_3$ film.[86]

The coercive field-strength can be reduced by substituting the lead partially by lanthanum. Okada reports a reduction of E_c from 53 to 30 kV/cm as the La-content in a $(Pb,La)(Zr,Ti)O_3$ film (with Zr/Ti is 40/60) increases from 0 to 15%.[66] The remanent polarisation (25-30 μC/cm²) was only slightly affected by the La content.

$PbZr_xTi_{1-x}O_3$ films deposited at low temperatures using plasma enhancement only showed ferroelectricity after annealing the films at 650°C.[74,75] Values for the remanent polarisation and coercive field-strength were 12 μC/cm² and 54 kV/cm, respectively.

3.6 Non-volatile memory applications

For memory applications ferroelectric films should meet a number of

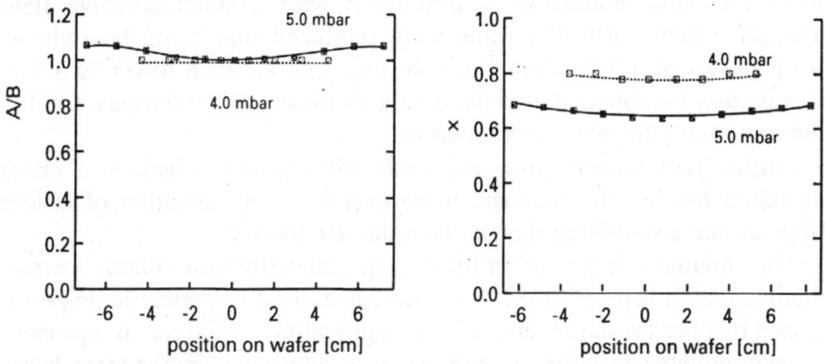

FIGURE 18. Uniformity in stoichiometry (left) and in x of $PbZr_xTi_{1-x}O_3$ over 4 and 6 inch wafers. Ref. 9.

device oriented requirements. These include demands on the deposition technique such as the capability of depositing uniform films over large areas with a good step coverage as well as demands on the capacitor properties. It will be shown that OMCVD is capable of producing films that meet the specified needs.

The uniformity of $PbZr_xTi_{1-x}O_3$ films over 4 and 6 inch wafers is presented in Figure 18. It demonstrates that both variations in Pb/(Zr+Ti) as well as Zr/(Zr+Ti) are below 5% .

The deposition of $PbZr_xTi_{1-x}O_3$ over a structured platinized silicon wafer is shown in Figure 19. It appears that OMCVD is capable of depositing layers conformally over such structures. Only in the trenches the film thickness is somewhat smaller but this is mainly due to a reduced thickness of the sputtered platinum electrode.

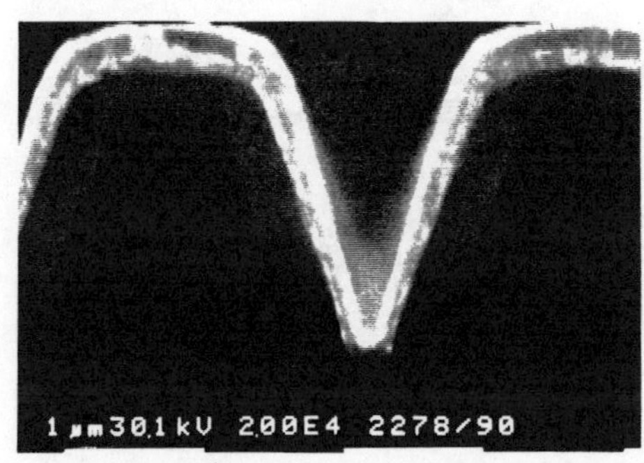

FIGURE 19. SEM micrograph of a $PbZr_xTi_{1-x}O_3$ film over a structure platinized substrate.

At present, for high density memories the operating voltage is nominally 3.3V. However the voltage available for switching of the ferroelectric capacitor will be about 2V. This requires very thin films with a small coercive field-strength. Using OMCVD, films were produced that could be switched using pulses of 1.5 V amplitude.[9] In principle an even lower switching voltage may be obtained by reducing the thickness of the ferroelectric film and improving the interface properties.

Further requirements to the ferroelectric capacitors include a virtual unlimited number of read- and write cycles, a data retention of at least 10 years and a switching time in the range 10-100 ns.

The endurance is a point of major importance for non-volatile memory applications. It is determined by the degradation of ferroelectric properties caused by polarisation reversals. This degradation, or fatigue, is a property of the complete capacitor structure, i.e. it is determined by the ferroelectric thin film properties, the electrodes and test conditions. It was demonstrated that for films deposited using OMCVD the difference in switched charge between a switching and a non-switching pulse still was sufficiently large after 10^{11} switching cycles.[9] In Figure 20 the influence of pulse amplitude on the endurance is presented. Clearly, the use of a lower

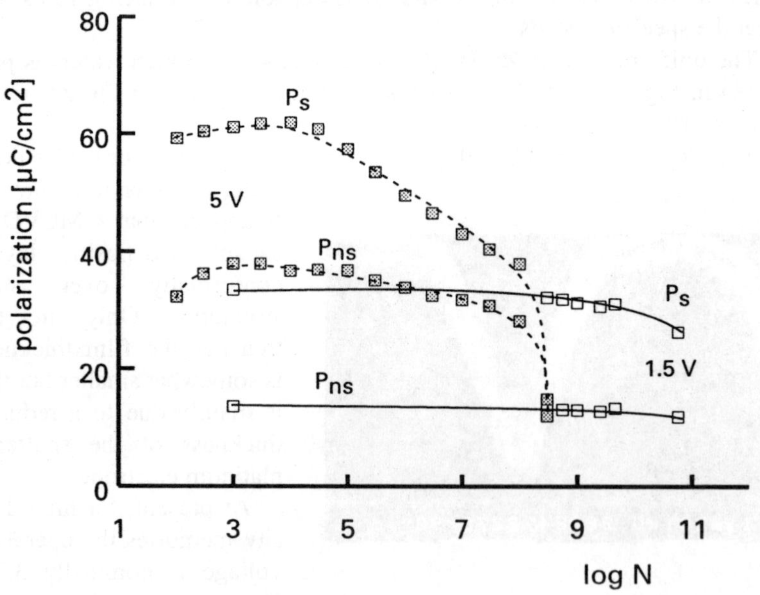

FIGURE 20. Endurance for a capacitor measured at two switching voltages. Thickness is 200nm and x=0.60. Ref. 9.

switching voltage leads to a better endurance. This effect has been observed before.[87] Obviously, the quality of the Pt/$PbZr_xTi_{1-x}O_3$/Pt capacitors with OMCVD ferroelectrics, approaches the regime where application of such films in ferroelectric random access memories is possible.

The retention of the ferroelectric capacitor is its capability of maintaining the polarisation with time. In some cases a considerable retention-loss of a ferroelectric capacitor is observed. This effect, caused by a partial back-switching of ferroelectric domains, is undesirable for memory operation. The result of such an experiment is presented in Figure 21. The delay time between the write- and read pulse was varied between 100 μs and 6.10^4 s. No large retention-loss is observed here. The ratio between the switched- and non-switched pulses is about 6 and is almost constant over this range of delay times.

In Figure 22 the switching and non-switching response of a ferroelectric capacitor is presented. The switching time is only a few nanoseconds. Switching times smaller than 1.8 ns have been reported.[88]

The successful integration of $PbZr_xTi_{1-x}O_3$ films grown by OMCVD with CMOS technology was reported by Dormans *et al.*[69] Both the CMOS part as well as the ferroelectric capacitors were only slightly affected by the processing. The proper functioning of ferroelectric memory cells was demonstrated. In Figure 23 the hysteresis loops measured at 1 kHz with amplitudes of 3, 5 and 8V for a fully integrated capacitor with a 300 nm

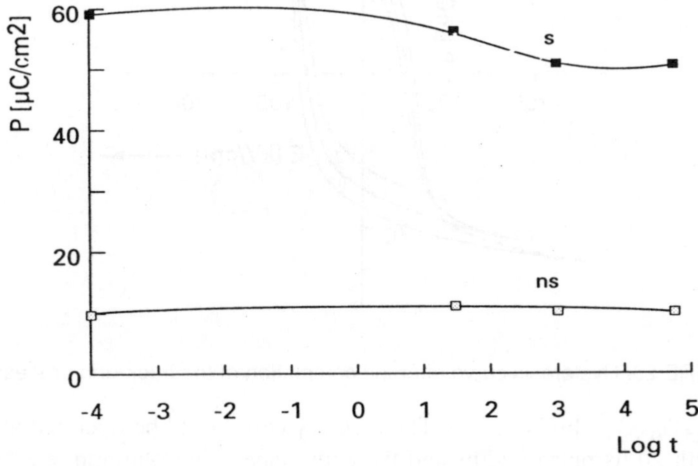

FIGURE 21. Retention measurement at room temperature. Only a small effect of back-switching is observed.

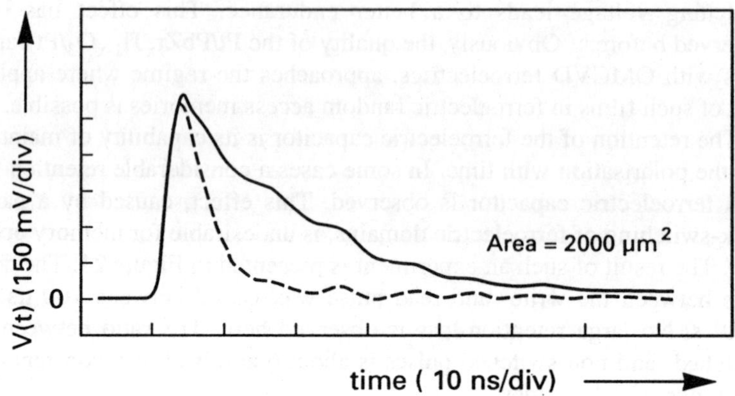

FIGURE 22. Pulse measurement using double bipolar pulses. Fast switching with a large difference between the switching and non-switching response is observed.

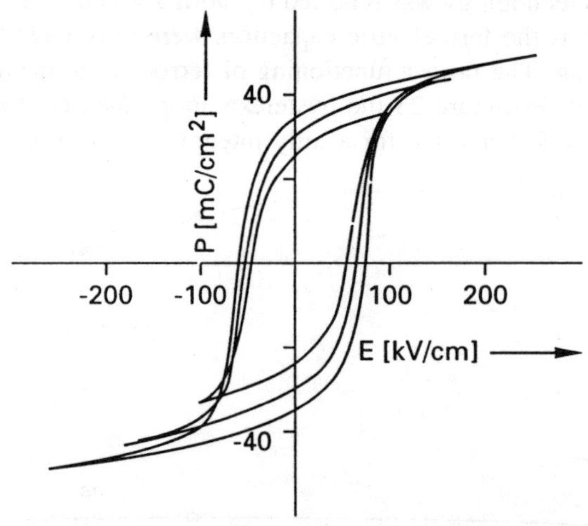

FIGURE 23. Hysteresis curve after the completion of the back-end processing.

$PbZr_{0.5}Ti_{0.5}O_3$ film is shown. The memory cells could be operated at only 3V with 20 ns pulse widths and the endurance using platinum electrodes exceeded 10^{13} read-write cycles.

4. SUMMARY

The process of organometallic chemical vapour deposition of ferroelectric thin films has been discussed with emphasis on the OMCVD of $PbZr_xTi_{1-x}O_3$. Both technical aspects as well as results and achievements were presented. Solutions for technical problems in connection with the handling of precursors with low vapour pressures were presented. Although suitable precursors are already available, the development of precursors with increased vapour pressures will simplify the OMCVD process.

Epitaxial films on lattice matched substrates have been grown. Deposition on electroded silicon wafers yields films having high values for the remanent polarisation. The values for coercive field-strength are still somewhat high as compared to values for bulk $PbZr_xTi_{1-x}O_3$. Endurance data as well as retention measurements indicate that these films are of sufficiently high quality for application in thin film devices such as ferroelectric memories.

Uniform deposition can be obtained over 4 and 6 inch substrates. It has been shown that with OMCVD ferroelectric capacitors can be made, using platinum electrodes, that meet the requirements for their application in high density memories. $PbZr_xTi_{1-x}O_3$ films grown by OMCVD were successfully integrated with CMOS devices without serious degradation of the CMOS part nor of the ferroelectric capacitors.

The acceptation of OMCVD as an industrial technique for large scale production of ferroelectric films will depend on the development and availability of IC-compatible OMCVD equipment. Such equipment should combine a high throughput with a high quality of the ferroelectric film. When using such dedicated equipment the cost per substrate will be competitive with other deposition techniques.

ACKNOWLEDGEMENT

The authors would like to thank H.W. Zandbergen, P.J. van Veldhoven and D.J. Taylor for assistance.

REFERENCES

1. H.M. MANASEVIT AND W.I. SIMPSON, J. Electrochem. Soc., **116,** 1725 (1969).
2. M.R. LEYS, C. VAN OPDORP, M.P.A. VIEGERS, AND H.J. TALEN-VAN DE MHEEN, J. Crystal Growth, **68,** 431 (1984).
3. See for instance: J. NISHIZAWA, T. KURABOYASHI, H. ABE, AND N. SAKURAI, J. Electrochem. Soc., **34,** 945 (1987).
4. R.A. ROY, K.F. ETZOLD, AND J.J. CUOMO, Mat. Res. Soc. Symp. Proc., **200,** 141 (1990).
5. G.B. STRINGFELLOW, Organometallic Chemical Vapour Phase Epitaxy: Theory and Practice (Academic Press, San Diego, CA) 1989.
6. D.M. FRIGO, Appl. Phys. Lett., **61,** 531 (1992).
7. J.P. STAGG, J. CHRISTER, E.J. TRUSH, AND J. CRAWLEY, J. of Crystal Growth, **120,** 98 (1992).
8. T. SUGIMOTO, M. YOSHIDA, K. YAMAGUCHI, K. SUGAWARA, Y. SHIOHARA, AND S. TANAKA, Appl. Phys. Lett., **7,** 928 (1990).
9. M. DE KEIJSER, G.J.M. DORMANS, AND P.J. VAN VELDHOVEN, Mat. Res. Soc. Symp. Proc., **310,** 223 (1993).
10. P.C. VAN BUSKIRK, R. GARDINER, P.S. KIRLIN, AND S.B. KRUPANIDHI, Proc. 8th Int. Symp. Appl. Ferroelectrics, Greenville, IEEE 92CH3080-9 p.340 (1992).
11. Z. LU, R.S. FEIGELSONN, R.K. ROUTE, S.A. DICAROLIS, R. HISKES, AND R.D. JACOWITZ, J. Crystal Growth, **128,** 788 (1993).
12. P.J. ROKSNOER, C. VAN OPDORP, J.W.F.M. MAES, M. DE KEIJSER, AND C. WEBER, J. Electrochem. Soc., **136,** 2427 (1989).
13. K.F. JENSEN, D.I. FOTIADIS, AND T.J. MOUNTZIARIS, J. Crystal Growth, **107,** 1 (1991).
14. T. SUGIMOTO, M. YOSHIDA, K. SUGAWARA, Y. SHIOHARA, AND S. TANAKA, Appl. Phys. Lett., **58,** 1103 (1991).
15. H.J. FRENCK, E. OESTERSCHULZE, R. BECKMANN, W. KULISCH, AND R. KASSING, Mat. Sci. Engin., **A139,** 394 (1991).
16. H. WENDEL, H. HOLZSCHUH, H. SUHR, G. ERKER, S. DEHNICKE, AND M. MENA, Mod. Phys. Lett., **B4,** 1215 (1990).
17. See for example: D.C. BRADLEY, Chem. Rev., **89,** 1317 (1989).
18. M. BALOG, M. SCHIEBER, M. MICHMAN, AND S. PATAI, Thin Solid Films, **47,** 109 (1977).
19. M. BALOG, M. SCHIEBER, M. MICHMAN, AND S. PATAI, J. Electrochem. Soc., **26,** 1203 (1979).
20. C. TRUNDLE, AND C.J. BRIERLEY, Appl. Surf. Science, **36,** 102 (1989).
21. T. NAKAGAWA, J. YAMAGUCHI, M. OKUYAMA, AND Y. HAMAKAWA,

Jpn. J. Appl. Phys., **21,** L655 (1982).

22. A.D. BERRY, D.K. GASKILL, R.T. HOLM, E.J. CUKAUSKAS, R. KAPLAN, AND R.L. HENRY, Appl. Phys. Lett., **52,** 1743 (1988).
23. T. NAKAMORI, H. ABE, T. KANAMORI, AND S. SHIBATA, Jpn. J. Appl. Phys., **27,** L1265 (1988).
24. Y. YAMANE, H. KUROSAWA, H. IWASAKI, H. MASUMOTO, T. HIRAI, N. KOBAYASHI, AND Y. MUTO, Jpn. J. Appl. Phys., **27,** L1275 (1988).
25. See e.g. Mat. Res. Soc. Symp. Proc., **169,** High-T_c Supercond.: Fundam. Prop. Novel. Mater. Process.
26. B.S. KWAK, K. ZHANG, E.P. BOYD, A. ERBIL, AND B.J. WILKENS, J. Appl. Phys., **69,** 767 (1991).
27. H. NAKAZAWA, H. YAMANE, AND T. HIRAI, Jpn. J. Appl. Phys., **30,** 2200 (1991).
28. L.A. WILLS AND B.W. WESSELS, Mat. Res. Soc. Symp. Proc., **243,** 217 (1992).
29. P.C. VAN BUSKIRK, P.S. KIRLIN AND R. GARDINER, Mat. Res. Soc. Symp. Proc., **243,** 223 (1992).
30. C.S. CHERN, Mat. Res. Soc. Symp. Proc., **243,** 573 (1992).
31. W.A. FEIL, B.W. WESSELS, L.M. TONGE, AND T.J. MARKS, J. Appl. Phys., **67,** 3858 (1990).
32. L.A. WILLS, W.A. FEIL, B.W. WESSELS, L.M. TONGE, AND T.J. MARKS, J. Crystal. Growth, **107,** 712 (1991).
33. H. WANG, L.W. FU, S.X. SHANG, S.Q. YU, X.L. WANG, Z.K. LU, AND M.H. JIANG, Mat. Res. Soc. Symp. Proc., **243,** 213 (1992).
34. A. GREENWALD, M. HORENSTEIN, M. RUANE, W. CLOUSER, AND J. FORESI, Mat. Res. Soc. Symp. Proc., **243,** 457 (1992).
35. C.J. BRIERLEY, C. TRUNDLE, L. CONSIDINE, R.W. WHATMORE, AND F.W. AINGER. Ferroelectrics, **91,** 181 (1989).
36. M. KOJIMA, M. OKUYAMA, T. NAKAGAWA, AND Y. HAMAKAWA, Jpn. J. Appl. Phys., **22,** *supplement* 22-2, 14 (1983).
37. M. OKADA, H. WATANABE, M. MURAKAMI, A. NISHIWAKI, AND K. TOMITA, J. Ceram. Soc. Jpn. Inter. Ed., **96,** 676 (1988).
38. M. OKADA, AND K. TOMITA, United States Patent no. 4,792,463, (1988).
39. S-G. YOON AND H-G. KIM, J. Electrochem. Soc., **135,** 3137 (1988).
40. B.S. KWAK, E.P. BOYD, AND A. ERBIL, Appl. Phys. Lett., **53,** 1702 (1988).
41. S.L. SWARTZ, D.A. SEIFERT, G.T. NOEL, AND T.R. SHROUT, Ferroelectrics, **93,** 37 (1989).
42. M. OKADA, S. TAKAI, M. AMEMIYA, AND K. TOMINAGA, Jpn. J. Appl. Phys., **28,** 1030 (1989).
43. M. DE KEIJSER, G.J.M. DORMANS, J.F.M. CILLESSEN, D.M. DE LEEUW, AND H.W. ZANDBERGEN, Appl. Phys. Lett., **58,** 2636 (1991).

44. M. DE KEIJSER, G.J.M. DORMANS, P.J. VAN VELDHOVEN, AND D.M. DE LEEUW, Appl. Phys. Lett., **59,** 3556 (1991).
45. G.J.M. DORMANS, M. DE KEIJSER, AND P.K. LARSEN, Paper presented at the 3th International Symposium on Integrated Ferroelectrics, Colorado Springs 1991. Paper published in: Integrated Ferroelectrics, **2,** 297 (1992).
46. B.S. KWAK, A. ERBIL, B.J. WILKENS, J.D. BUDAI, M.F. CHISHOLM, AND L.A. BOATNER, Ceramic Transactions, **25,** 203 (1992).
47. A. ERBIL, W. BRAUN, B.S. KWAK, B.J. WILKENS, L.A. BOATNER, AND J.D. BUDAI, J. Crystal Growth, **124,** 684 (1992).
48. G.R. BAI, H.L.M. CHANG, H.K. KIM, C.M. FOSTER, AND D.J. LAM, Appl. Phys. Lett., **61,** 408 (1992).
49. X-T. CHEN, H. YAMANE, AND K. KAYA, J. Phys. III France, **2,** 1439 (1992).
50. G.J.M. DORMANS, P.J. VAN VELDHOVEN, AND M. DE KEIJSER, J. of Crystal Growth,**123,** 537 (1992).
51. H. FUNAKUBO, T. HIOKI, M. OTSU, K. SHINOZAKI, AND N. MIZUTANI, Jpn. J. Appl. Phys., **32,** 4175 (1993).
52. T. HIRAI, T. GOTO, H. MATSUHASHI, S. TANIMOTO, AND Y. TARUI, Jpn. J. Appl. Phys., **32,** 4078 (1993).
53. Y. GAO, G. BAI, K.L. MERKLE, Y. SHI, H.L.M. CHANG, Z. SHEN, AND D.J. LAM, J. Mater. Res., **8,** 145 (1993).
54. Y. GAO, G. BAI, K.L. MERKLE, H.L.M. CHANG, AND D.J. LAM, Thin Solid Films, **235,** 86 (1993).
55. F.W. AINGER, C.J. BRIERLEY, M.D. HUDSON, C. TRUNDLE, AND R.W. WHATMORE, Mat. Res. Soc. Symp. Proc., **200,** 37 (1990).
56. M. OKADA, K. TOMINAGA, T. ARAKI, S. KATAYAMA, AND Y. SAKASHITA, Jpn. J. Appl. Phys., **29,** 718 (1990).
57. H. FUNAKUBO, K. IMASHITA, N. KIEDA, AND N. MIZUTANI, J. Ceram. Soc. Jpn. Int. Ed., **99,** 248 (1991).
58. Y. SAKASHITA, T. ONO, H. SEGAWA, K. TOMINAGA, AND M. OKADA, J. Appl. Phys., **69,** 8352 (1991).
59. K. KASHIHARA, H. ITOH, K. TSUKAMOTO, AND Y. AKASAKA, Extended Abstracts of the 1991 International Conference on Solid State Devices and Materials, Yokohama, 192 (1991).
60. G.J.M. DORMANS, M. DE KEIJSER, AND P.J. VAN VELDHOVEN, Mat. Res. Soc. Sym. Proc., **243,** 203 (1992).
61. J-H. CHOI AND H-G. KIM, J. Appl. Phys., **74,** 6413 (1993).
62. Y. SAKASHITA, H. SEGAWA, K. TOMINAGA, AND M. OKADA, J. Appl. Phys., **73,** 7857 (1993).
63. M. SHIMIZU, M. FUJIMOTO, T. KATAYAMA, T. SHIOSAKI, K. NAKAYA, M. FUKAGAWA, AND E. TANIKAWA, Mat. Res. Soc. Symp. Proc., **310,** 255

(1993).

64. K. TOMINAGA, M. MIYAJIMA, Y. SAKASHITA, H. SEGAWA, AND M. OKADA, Jpn. J. Appl. Phys., **29,** L1874 (1990).
65. S. WATANABE, T. TAKAMITSU, AND A. TANAKA, Appl. Phys. Lett., **62,** 3377 (1993).
66. M. OKADA AND K. TOMINAGA, J. Appl. Phys., **71,** 1955 (1992).
67. H. FUNAKUBO, K. IMASHITA, AND N. MIZUTANI, J. Ceram. Soc. Jpn. Int. Ed., **99,** 1130 (1991).
68. K. TOMINAGA, A. SHIRAYANAGI, T. TAKAGI, AND M. OKADA, Jpn. J. Appl. Phys., **32,** 4082 (1993).
69. G.J.M. DORMANS, P.K. LARSEN, G.A.C.M. SPIERINGS, J. DIKKEN, M.J.E. ULENAERS, R. CUPPENS, D.J. TAYLOR, AND R.D.J. VERHAAR, paper presented at the 5th International Symposium on Integrated Ferroelectrics, Monterey 1994.
70. M. SHIMIZU, T. KATAYAMA, T. SHIOSAKI, AND A. KAWABATA, Proc. 7th IEEE International Symposium on the Applications of Ferroelectrics, Urbana 1990, IEEE catalogue no. 90CH2800-1, 669. (1990)
71. T. KATAYAMA, M. FUJIMOTO, M. SHIMIZU, AND T. SHIOSAKI, Jpn. J. Appl. Phys., **30,** 2189 (1991).
72. T. KATAYAMA, M. FUJIMOTO, M. SHIMIZU, AND T. SHIOSAKI, J. Crystal Growth, **115,** 289 (1991).
73. W.T. PETUSKEY, D.A. RICHARDSON, AND S.K. DEY, paper presented at the 3th International Symposium on Integrated Ferroelectrics, Colorado Springs (1991).
74. D.A. RICHARDSON, C.J. KUO, W.T. PETUSKEY, AND S.K. DEY, paper presented at the Annual meeting of the American Ceramic Society, Cincinatti (1991).
75. W.T. PETUSKEY, Paper presented at the 4th International Symposium on Integrated Ferroelectrics, Monterey (1992).
76. M. DE KEIJSER, G.J.M. DORMANS, P.J. VAN VELDHOVEN, AND P.K. LARSEN, Paper presented at the 4t^{h} International Symposium on Integrated Ferroelectrics, Monterey (1992). Integrated Ferroelectrics, **3,** 131 (1993).
77. M. DE KEIJSER AND G.J.M. DORMANS, unpublished.
78. O. CHEN, K. KAYA, H. YAMANE, AND T. HIRAI, Extended abstracts (The 38th Spring Meeting, 1991); The Japan Society of Applied Physics and Related Societies, **2,** 28a-P-1 (1991).
79. M. ANLIKER, H.R. BRUGGER, AND W. KÄNZIG, Helv. Phys. Acta, **27,** 99 (1954).
80. K. ISHIKAWA, K. YOSHIKAWA, AND N. OKADA, Phys. Rev., **37,** 5852 (1988).
81. M. DE KEIJSER, D.M. DE LEEUW, P.J. VAN VELDHOVEN, A.E.M. DE VEIRMAN, D.G. NEERINCK, AND G.J.M. DORMANS, unpublished.

82. B.S. KWAK, A. ERBIL, B.J. WILKENS, J.D. BUDAI, M.F. CHISHOLM, AND L.A. BOATNER, Phys. Rev. Lett., **68,** 3733 (1992).

83. C.H. PENG AND S.B. DESU, Appl. Phys. Lett., **61,** 16 (1992).

84. R.M. WOLF AND J.C. CILLESSEN, paper presented at the MRS fall meeting, Boston (1991), unpublished.

85. R. RAMESH, W.K. CHAN, B. WILKENS, T. SANDS, J.M. TARASCON, AND V.G. KERAMIDAS, Integrated Ferroelectrics, **1,** 1 (1992).

86. P.K. LARSEN, D.J. TAYLOR, G.J.M. DORMANS, AND P.J. VAN VELDHOVEN, J. Appl. Phys. In press.

87. P.K. LARSEN, R. CUPPENS, AND G.A.C.M. SPIERINGS, Ferroelectrics, **128,** 265 (1992).

88. P.K. LARSEN, G.L.M. KAMPSCHÖER, M.J.E. ULENAERS, G.A.C.M. SPIERINGS, AND R. CUPPENS, Appl. Phys. Lett., **59,** 611 (1991).

CHAPTER 13

PULSED LASER ABLATION-SYNTHESIS AND CHARACTERIZATION OF FERROELECTRIC THIN FILMS AND HETEROSTRUCTURES

Orlando Auciello,[1,2] Rovindra Dat,[2] and Ramamoorthy Ramesh[3]

1. MCNC, Electronics Technology Division, 3021 Cornwallis Rd., Research Triangle Park, NC 27709-2889

2. N.C. State University, Department of Materials Science and Engineering, Raleigh, NC 27695-7919

3. Bellcore, Red Bank, NJ 07701

1. INTRODUCTION

This chapter is focused on reviewing the fundamentals and application of the pulsed laser ablation deposition (PLAD) technique for synthesizing ferroelectric and other oxide thin films for applications to non-volatile memories. The work discussed here relates mainly to: (a) recent advances in understanding ablation, transport of species, and deposition processes related to the synthesis of Pb $(Zr_{1-x}Ti_x)O_3$ (PZT) and conductive oxide films, and (b) the synthesis and characterization of ferroelectric thin films and heterostructures with the objective of developing materials integration strategies for producing structures with device-compatible properties.

Intensive research on the synthesis, characterization, and determination of processing-microstructure-property relationships of ferroelectric thin films and their integration into layered heterostructures has been performed during the last five years because of the potential application to the fabrication of electronic memory devices. The high dielectric permittivities of perovskite-type materials can be advantageously used in dynamic random access memories (DRAM),[1] while the large values of switchable remanent polarization of ferroelectric materials are suitable for non-volatile ferroelectric random access memories (FRAM).[2-7] The PLAD method has palyed a major role in advancing materials integration strategies, due to its versatility for rapid exploration of film synthesis.

The technological uses of oxide thin films are not limited to ferroelectric memories, but extend to a wide range of applications in discrete devices, utilizing the full range of bulk oxide properties, including dielectric, ferroelectric, piezoelectric, electrostrictive, pyroelectric, optical, electro-optic and magnetic properties, as well as electronic conduction, ionic conduction and superconduction. Applications include multilayer capacitors for memories, boundary layer capacitors, varistors, gas sensors, radiation detectors, temperature sensors, transducers, switches, shutters, MHD electrodes, fuel cell electrolytes, concentration cell electrolytes, and electrolytes for high energy density batteries. These ceramics represent an important world market and their use in thin film form is being explored to a large extent using the PLAD method.[8]

Various vapor phase deposition techniques [plasma and ion beam sputter-deposition (PSD and IBSD respectively), pulsed laser ablation deposition (PLAD), electron beam or oven-induced evaporation for molecular beam epitaxy (MBE), and chemical vapor deposition (CVD)] have been applied to produce ferroelectric films and layered heterostructures (see Refs. 9-14 for recent reviews). As pointed out in these reviews,[9-14] the deposition methods mentioned above have not yet been optimized, despite intensive efforts by various groups. However, remarkable progress has been achieved in understanding many of the basic processes involved in the different deposition techniques and in their use to develop materials integration strategies that resulted in heterostructures with microstructures and properties suitable for commercial devices.

In order to integrate ferroelectric/metal oxide heterostructures into a commercially viable memory technology several aspects have to be considered. The issue of interest in this chapter is thin film deposition, for which a scalable process (i.e., ability to grow on 6" wafers or larger) is needed. In this respect, techniques such as chemical vapor deposition (CVD) or sputter-deposition are currently more advanced than PLAD. For high density memories, where the aspect ratio of contact and via holes may be greater than unity, the film synthesis technique has to provide also the ability to grow thin films on non-planar surfaces with good conformal coverage. In this respect, CVD is presently better than sputter-deposition and presently it is not kown how these techniques compare with PLAD. The PLAD technique has been extensively investigated because of its versatility, and cost effective instrumentation.

PLAD has achieved popularity, primarily in the oxide thin film deposition community, through the demonstration of the growth of very high quality thin films and heterostructures. Thin films of the desired material and composition are deposited by the ablation of material from solid target surfaces by a short laser pulse. Under appropriate deposition conditions, the composition of the thin film is the same as that of the target within a small central area of the film, but is non-stoichiometric outside this central area. The ability of producing stoichiometric films is perhaps the main intrinsic advantage of PLAD, in comparison with other techniques such as sputtering when using multicomponent oxide PZT targets. For sputter-deposition of PZT films, enrichment of Pb in the PZT target is necessary to compensate for preferential sputtering. The compositional congruency in the central part of the ablated plume is translated to the film depending, among other factors, on the substrate temperature, the interface chemistry between the substrate surface and the film and, the oxygen partial pressure during deposition. Another advantage of PLAD, as implemented by various groups, is that multiple targets can be loaded inside the chamber, before deposition, on a rotating holder that allows sequential exposure of different targets to the laser beam, thereby enabling the in-situ growth of heterostructures and superlattices with relatively clean interfaces. The heterostructures can be cooled in a variety of background atmospheres after the deposition is completed. Other advantages include a rapid throughput of films that is useful in process optimization and the ability to include in-situ processing such as etching through the use of a load-lock and etching chamber in conjunction with the deposition chamber.

The PLAD technique has some deficiencies that need to be overcome. The first is that uniform coating of large area wafers (for example 4-6" diameter) has still not been routinely demonstrated. This will require not only the uniform heating of large area wafers but also the deposition of material with uniform thickness and composition over the entire surface. The second issue that remains to be demonstrated is the ability to carry out conformal growth. Finally, films with complete absence of particulates have yet to be demonstrated, although the particulate density and size distribution can be reduced considerably through the use of high density targets and/or the implementation of velocity selectors.[15]

2. BASIC LASER ABLATION, SPECIES TRANSPORT, AND DEPOSITION PROCESSES

The PLAD technique gives researchers the capability to fabricate ferroelectric capacitors with extremely good properties. However, there are aspects of the PLAD process that are not completely understood nor have been thoroughly studied. Understanding the deposition process for PZT ferroelectric film growth is important in order to minimize Pb loss during film growth, eliminate formation of pyrochlore (non-ferroelectric) phases, eliminate surface particulates, and control the film microstructure. Each of these factors will affect device properties to some degree. Knowledge and control of the ablated flux distribution is also important for scale-up to produce uniform films on large-area substrates.

Many studies of the pulsed laser ablation deposition process have been performed,[16-23] although most studies have concentrated on the ablation of YBCO.[17-22] These studies have demostrated that background gas pressure during deposition, substrate to target distance, laser energy and wavelength, and target-substrate relative geometric arrangement, among other parameters have a significant effect on YBCO film composition, microstructure and properties. The issues described above for YBCO films are also relevant to the synthesis of PZT films. In addition, Pb incorporation into the films is a very important issue in PZT film deposition, because of the high volatility of Pb and the easy formation of deleterious non-ferroelectric phases. For this reason, initial studies of PZT ablation processes have been focused on examining the Pb incorporation as a function of oxygen pressure,[23] and on the energy distributions of ablated species.[24] More recently, the North Carolina State University (NCSU)-MCNC group performed a systematic and detailed investigation of the ablation characteristics of PZT, $La_{0.5}Sr_{0.5}CoO_3$ (LSCO), and MgO targets to understand and control the various parameters mentioned above for the case of YBCO films, since they also play fundamental roles in controlling the composition, microstructure and properties of PZT, LSCO, and MgO films used for fabricating PZT-based capacitors for non-volatile memories. PZT was studied because of its potential as a ferroelectric material for memory device applications. LSCO was investigated because it is a conductive oxide material and was demonstrated to be an excellent electrode material for ferroelectric capacitors.[25-27] MgO ablation was studied because MgO thin films can serve as a good buffer layer for film growth on either GaAs or Si substrates.[28] In addition, the ablation characteristics of these materials were expected to be very different, because of their different absorption coefficients and melting points.

Various PLAD systems have been used to perform research on the ablation process as well as on thin film synthesis. The systems used by different groups are similar (as an example, the NCSU-MCNC system is shown in Fig. 1). Most systems, including those of the authors, involve: (a) an excimer laser beam (generally a KrF-248 nm laser beam - 1-2 J/cm^2 - 1-5 Hz is used because it results in films with better properties than those produced by other lasers such as CW's or Nd:YAG's); (b) a rotating target holder with capacity for various targets to sequentially position them under the laser beam in order to fabricate multilayered heterostructures without breaking vacuum; (c) a heated substrate holder compatible with relatively high oxygen background pressures needed to produce oxide films; and (d) desirably but not absolutely necessary, a load lock system to exchange targets and substrates without breaking vacuum. The last feature is recommendable since it dramatically increases the film production and research output.

A unique target holder that offers the capability for exchanging targets without breaking vacuum was designed and constructued by Auciello et al. (see Fig. 2),[29] and it has been used in part of the work discussed in this chapter. The ability of exchanging targets without breaking vacuum is important since work on film deposition using the PLAD method has demonstrated that generally it is necessary to polish the targets before each film deposition to obtain high quality films.

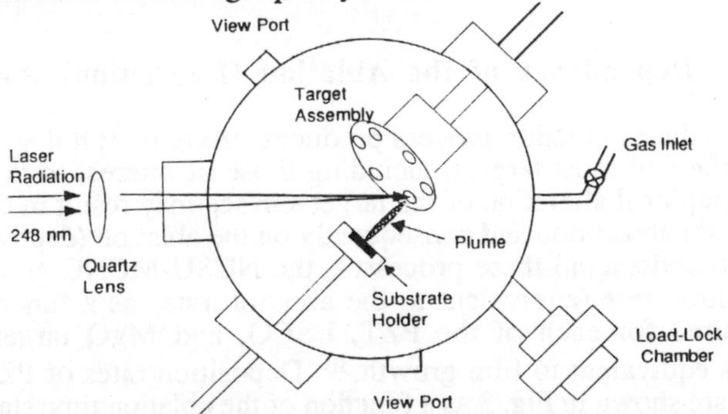

FIGURE 1. Schematic of a PLAD system as implemented by the NCSU-MCNC group.

FIGURE 2. NCSU - MCNC rotating target holder with capability for continuous rotation of targets during ablation (to minimize and uniformize target erosion) and for target exchange without breaking vacuum.[29]

In view of the discussion presented above, the NCSU-MCNC group investigated the effects of ablation time, laser energy, and processing gas pressure on the film deposition rate, as well as the effects of the oxygen pressure and laser energy on the angular distribution of deposited species.[30] Evidence that the oxygen pressure affects the deposition dynamics was clearly seen by visual inspection of the plume. The deposition flux characteristics were examined using a quartz crystal thickness monitor positioned at the substrate location. The angular distribution of the deposited species was examined using Rutherford Backscattering Spectrometry (RBS) to measure deposited film thickness variations across the substrate.[30]

2.1. Time Dependence of the Ablation (Deposition) Rate

The laser-induced ablation process produces strong morphologial changes on the surface of most targets, including those of interest to this chapter. The topographical alteration of the target surface may result in changes on the laser light absorption and consequently on the ablation (deposition) rate. In order to understand these processes, the NCSU-MCNC group studied the deposition rate (equivalent to the ablation rate) as a function of the ablation time for each of the PZT, LSCO, and MgO targets, using conditions equivalent to film growth.[30] Deposition rates of PZT, LSCO, and MgO are shown in Fig. 3 as a function of the ablation time starting from a virgin target surface. For all three curves, the measured data are indicated by points, and the lines represent fits to the data using Eq. (1) discussed below.

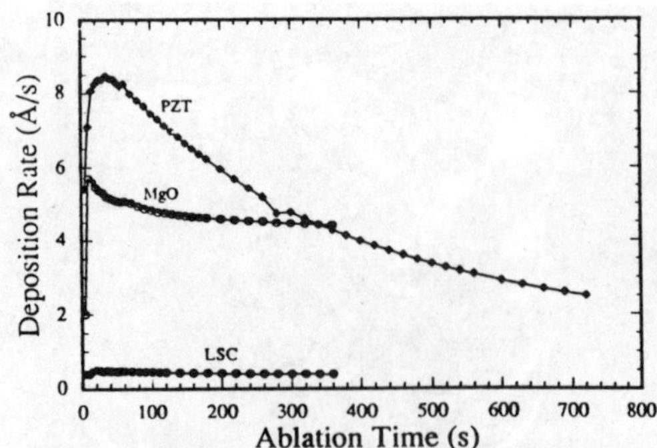

FIGURE 3. Deposition rate as a function of ablation time for PZT, LSCO, and MgO targets, which are initially in the "unablated", or virgin, state. The PZT and LSCO targets were ablated using a 400 mJ laser energy, 3 Hz repetition rate, in 300 mTorr oxygen. The MgO was ablated using a 900 mJ laser energy, 6 Hz repetition rate, in vacuum. The data are indicated by the points, while the lines are fits to the data using Eq. (1).[30]

The two characteristic features of all the curves shown in Fig. 3 are an initial transient period of decreasing deposition rate, followed by a steady-state, constant deposition rate regime. The initial (0 to 25 sec) low deposition rates observed for the ablation of the sintered MgO target (and the PZT to a lesser extent) can be attributed mainly to transient charging and heating effects on the quartz crystal rate monitor. It is also possible that the absorption coefficient of the MgO target increased after the surface became damaged, a phenomenon known to occur in materials with little absorption.[31] However, this effect should be relatively small in a polycrystalline ceramic target since grain boundaries will initially serve as absorption sites. The steady - state deposition rates observed in Fig. 3 implies that a steady-state target surface condition has been achieved. Both the target surface morphology and composition are expected to affect the ablation rate.[22] The appearance of the PZT and LSCO targets changes visibly after ablation. Both surfaces change color, with the PZT surface changing from yellowish to black. The black color results from a highly absorbing, rough surface characterized by a high density of protruding cones, as shown in the scanning electron microscope (SEM) image in Fig. 4(a). The LSCO target changes from tan to silver, with a relatively smooth surface morphology [Fig. 4(b)] with clear indications of surface melting and recrystallization. The MgO target, on the other hand, shows no visual change after ablation, although SEM analysis [Fig. 4(c)] reveals the presence of surface melting. The surface features observed by SEM are the result of liquid phases which solidified after ablation (between laser pulses). The morphology difference must be dependent on the time for the liquid to solidify (which will depend on the thermal conductivity and melting point of each material), and the magnitude of the shock wave generated at the target. All SEM micrographs were taken with the targets in the steady-state condition (> 10 min ablation time).

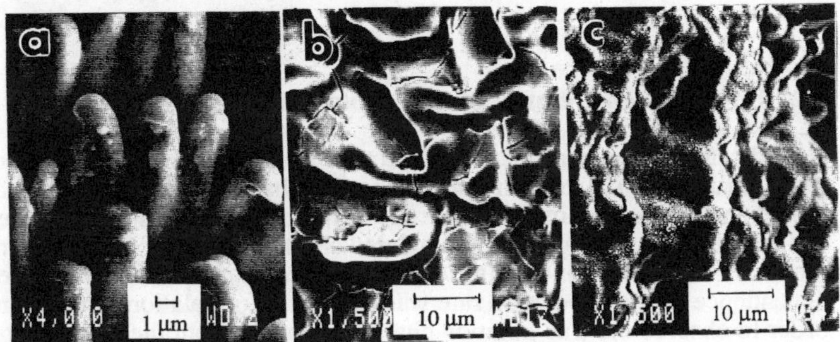

FIGURE 4. Scanning electron microscope images of the ablated surfaces of a) PZT, b) LSCO, and c) MgO polycrystalline ceramic targets, upon reaching steady-state deposition rates (> 10 min).[30]

A simple model has been developed, based on the deposition rate data, to describe the changing deposition rate (or changing surface conditions) over time. Taking into account the initial decay of the rate, followed by the attainment of steady-state, the deposition rate as a function of time can be expressed as:

$$R(t) = (R_o - R_\infty)\cdot \exp(-\tau/t) + R_\infty \quad (1)$$

where Ro is the rate at time zero, R_∞ is the rate at long (infinite) time, and τ is the decay time constant. The value of Ro depends on the ablation power density (which contains the energy, repetition rate, and area of the laser beam), the target material ablated (i.e., the melting point, thermal conductivity, and absorption coefficient of the target material), the gas pressure, and target-substrate distance during ablation. R_∞ is dependent on these same parameters, but with a modified "effective" absorption coefficient due to the changing surface of the target. Changes in the angular distribution or ablated composition could also occur as the target surface changes. The time constant t is a measure of the time needed to obtain a uniform, steady-state surface condition.

Equation 1 was fitted to the data in Fig. 3 using parameters described in Table I of Ref. 30. The drop in the ablation rate of PZT is believed to be mainly due to the dramatic change in surface morphology, as shown in Fig.4 (a). The change in the ablation rate of LSCO and MgO ablation is believed to be caused mainly by a phase change at the target surface, resulting from the rapid cooling of the liquid phase formed.

It is important to understand the time dependence of the ablation rate because the ablation rate affects the ratio of deposited species to reactive gas species at the substrate. As will be discussed later, the ablation rate may also affect the angular distribution, and indirectly the energy, of the ablated species.

2.2. Energy Dependence of the Ablation (Deposition) Rate

As indicated in the introduction of Section 2, the energy of the laser beam plays a fundamental role in controlling the characteristics of the ablated plume and consequently the properties of the films. The laser beam energy can affect the amount of material ablated and therefore the deposition rate, which in turn can control the characteristics of the films. The energy dependence of the deposition rate was examined for ablation in vacuum, in order to observe the true ablation characteristics without the complications of gas scattering effects. Figure 5 shows the energy dependence of the deposition rate for ablation of PZT, LSCO, and MgO in vacuum. Notice that the deposition rates from PZT and LSCO are linearly proportional to energy, while the MgO deposition rate shows two distinct regimes with different slopes. Also, note that doubling the laser pulse rate doubles the MgO deposition rate, without affecting the energy at which the two regimes intersect. Two distinct regimes in the energy dependence of the ablation rate are actually expected (at least for most inorganic materials); a low energy "evaporation" regime, and a higher energy "ablation" regime. The onset of the true ablation regime is termed the ablation threshold (E_{th}).

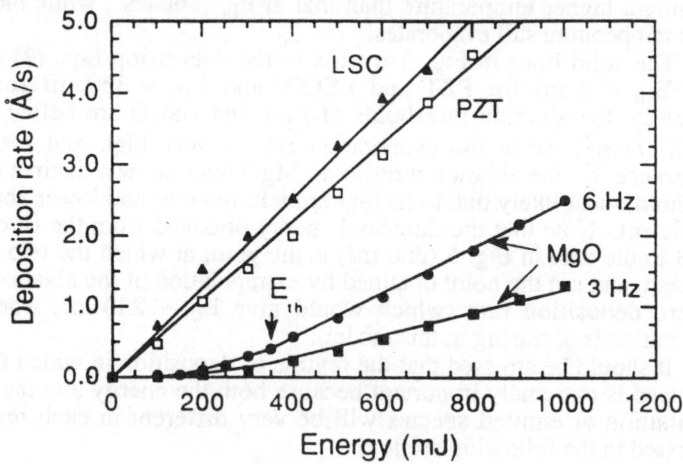

FIGURE 5. Deposition rate as a function of laser energy for PZT, LSCO, and MgO targets ablated in vacuum. The PZT and LSCO were ablated using a 400 mJ laser energy, 3 Hz repetition rate, at a 6 cm substrate-target distance. The MgO was ablated using a 900 mJ laser energy, and either a 3 or 6 Hz repetition rate, at a 6 cm distance. The data are indicated by the points, while the lines are fits to the data using Eqs. (2) and (3).[30]

Other groups have determined that the ablation rate is exponentially dependent on laser energy over the low energy, evaporation regime,[32] while at higher energies the ablation rate can increase linearly with energy.[33] Therefore the ablation rate as a function of laser energy R(E) can be modeled as:

$$R(E) = A \cdot \exp(-B/E) \quad \text{for } 0 \leq E \leq Eth; \qquad (2)$$

and

$$R(E) = A \cdot \exp(-B/Eth) + C \cdot (E - Eth) \quad \text{for } E \geq Eth; \qquad (3)$$

where A, B, and C are constants, E_{th} is the ablation threshold, and E is the laser energy. The values of A and C depend on virtually the same parameters as R_o and R_∞ in Eq. (1); i.e., the laser beam area, the pulse repetition rate, the target material properties, the gas pressure, and the target-substrate distance. The constant B is an effective "activation energy", related to the temperature rise of the target per pulse (or, more directly, a measure of the vapor pressure). The reason that the evaporation component is made to maximize at E_{th} is that it can be assumed that any material reaching a higher temperature than that at E_{th} "ablates", while the rest at lower temperature still evaporates.

The solid lines in Fig. 5 are fits to the data using Eqs. (2) and (3), with E_{th} = 0 mJ for PZT and LSCO, and E_{th} = 350 mJ for MgO. Evidently, the ablation thresholds of PZT and LSCO are below 100 mJ (0.275 J / cm^2), since the evaporation rate is very high and masks the appearance of the ablation threshold. MgO does show a distinct ablation threshold, most likely due to its higher melting point and lower absorption coefficient. Note that the threshold energy obtained from the fit of eqs. 2 and 3 to the data in Fig. 5 (350 mJ) is the point at which the two regimes intersect, and not the point obtained by extrapolation of the ablation region to zero deposition rate (which would give E_{th} ~ 250 mJ, where only evaporation is occurring in this model).

It should be stressed that the regime of deposition in which films are produced is extremely important because both the energy and the angular distribution of emitted species will be very different in each regime, as discussed in the following section.

2.3. Pressure Dependence of the Angular Distribution of Ablated Species

The angular distribution of the ablated flux plays a fundamental role on the resulting composition and thickness uniformity across multicomponent oxide films such as PZT, LSCO, and others of interest in this chapter. This angular spread depends on the backgroud gas pressure during deposition. The effects of increasing the oxygen pressure during ablation are clearly visible by observation of the plume. The photographs in Fig. 6 show the plume produced during ablation of a PZT target ($PbZr_{0.52}Ti_{0.48}O_3$ + 10%PbO), where each frame corresponds to one laser pulse, with the camera shutter open much longer than the plume lifetime. Narrowing of the plume is clearly visible as the oxygen pressure is increased. The color of the PZT plume is a strong white-blue in vacuum. When oxygen is added, the plume becomes yellow/orange/red at the plasma periphery, while remaining bluish inside.

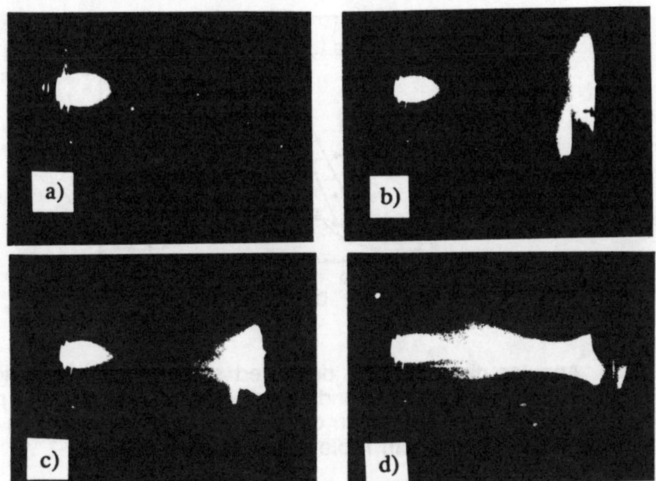

FIGURE 6. Photographs of the PZT ablation plume produced by a 900 mJ laser pulse in a) vacuum, b) 100 mTorr, c) 300 mTorr, and d) 900 mTorr oxygen. The actual target to substrate distance was 6 cm.[30]

Observations of the plume (visually and photographically) in argon gas show a similar effect on the plasma shape, but the plasma color remains totally blue. Therefore, the change in the plasma color can be attributed to chemical interactions between the oxygen and the ablated species. In addition, by ablating individual PbO, TiO_2, and ZrO_2 targets, the NCSU group was able to determine that the color change is due to the Ti and Zr interaction with oxygen. The blue in the center suggests that this region has not interacted substantially with oxygen. The shape of the LSCO plume is virtually identical to that of PZT, showing a narrowing in oxygen gas, although almost no color change is observed, as mentioned by Cheung et al.[25]

The film thickness uniformity was examined for PZT films deposited on unheated glass slides in O_2 pressures of 10, 100, 300, and 900 mTorr to determine if the change in plasma shape correlates with a change in the angular distribution of ablated species.[30] Films were deposited using a laser energy of 900 mJ (~2.2 J/cm^2), 3 Hz repetition rate, 6 cm target-substrate distance, and a 2 minutes deposition time. The film thicknesses (correlated with deposition rate), measured using RBS, increased at the center position as oxygen pressure was increased, followed by a decrease at very high pressures (Fig. 7). Most importantly, the angular distribution of ablated species was significantly affected by the pressure.

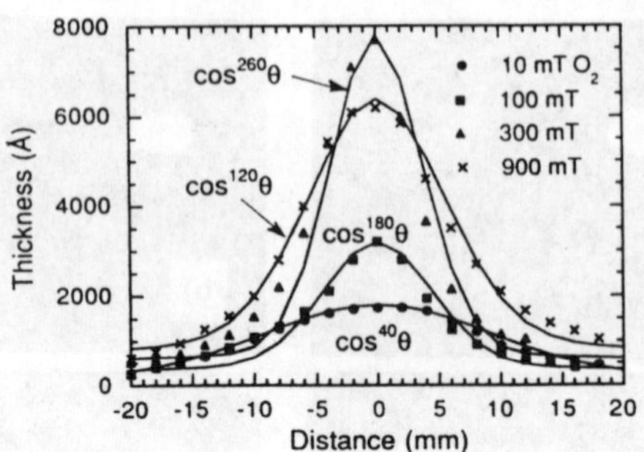

FIGURE 7. Angular distribution of deposited species during ablation of PZT. The laser energy is 900 mJ, and the oxygen pressures are 10 mTorr, 100 mTorr, 300 mTorr, and 900 mTorr oxygen. The lines are fits to the data using Eq. (4) and parameters from Table I of ref. 30.

The data of Fig. 7 was fitted using a sum of two cosine functions, taking into account the change in distance from the target as the angle θ is increased.[30] The complete expression for the thickness distribution t (θ) is:

$$t(\theta) = (M/r^2)\cos\theta + (N/r^2)\cos^n\theta, \tag{4}$$

where t is the film thickness, θ is the angle from the target normal, r is the target-to-substrate distance (which is a function of θ), and M and N are constants. The first term in Eq. (4) accounts for any evaporation - like component of the flux distribution, and the second term accounts for the ablated flux distribution.

Note from Fig. 7 that the angular distribution, characterized mainly by the exponent 'n', rises from a $\cos^{40}\theta$ in 10 mTorr oxygen, up to $\cos^{260}\theta$ for ablation in 300 mTorr oxygen, and decreases back to $\cos^{120}\theta$ for 900 mTorr. At this time, it is not clear whether the large exponents are realistic from the physics point of view or they just represent a mathematical parameter that happens to provide a good fit to the experimental curves shown in Fig. 7. In any case, the film thickness distribution measured by RBS is highly anisotropic. Also, the magnitude of the evaporation-like component (the constant M) is highest at the highest oxygen pressure. These findings reveal that the depositing species are actually tending to converge, or focus, as oxygen gas is added. This results in an increased deposition rate at the plume center. However, once the pressure is too high, gas scattering effects will tend to broaden the distribution out again. Gas scattering effects appear to be the origin of most of the cos θ component of the flux at high pressures, with an evaporation or sputtering component probably contributing the rest (due to the high fields generated in the plume, plasma ions repelled back to the target could possibly produce a sputtered component as well.) The angular distributions observed here for PZT ablation are much narrower than those observed by Foltyn et al.[34] and Venkatesan et al.[17] for ablation of $YBa_2Cu_3O_{7-\delta}$ (YBCO), probably due to differences in the ablation conditions.[17,34]

2.4. Combined Effects of Gas Pressure and Laser Energy on the Ablation (Deposition) Rate and Angular Distribution

In order to determine the range over which the above-mentioned plume focusing (or narrowing) occurs, the NCSU group examined the dependence of the deposition rate on both the gas pressure and the laser energy. Measurements of the PZT and LSCO deposition rates (after target equilibration), as a function of the laser energy and gas pressure (O_2, Ar, and Kr), revealed two distinct regimes. At low laser powers, the deposition rate drops as the pressure is increased, presumably due to a spreading of the ablated flux caused by gas scattering. At higher laser power, the deposition rate increases significantly with oxygen pressure (Figs. 6 and 7 of Ref. 28). Based on the plume photographs and the measured narrowing flux distribution, the NCSU group suggested that the deposition rate increase may be induced by the observed plume focusing effect (when the oxygen pressure is increased beyond 300 mTorr, the rate again begins to drop even at high power due to the random gas scattering effects.).

The combination of time and pressure dependence on the deposition rate reveals that when ablating from a fresh target surface, the plume will be in the focused regime because of the high deposition rate. Over time, the deposition rate drops to a much lower value, corresponding to the plume broadening regime.[29] Therefore, over time, the ablation rate and angular distribution should change dramatically when starting from a fresh polished (virgin) surface.

In order to ensure that the effects discussed above, when adding oxygen, are not chemical in nature, the NCSU group also investigated the effects of adding noble gases on the deposition rate. The use of Kr or Ar as the background gas resulted in a behavior similar to that observed when adding O_2; i.e., for low laser energy, the addition of gas resulted in a decrease in the deposition rate, while at higher laser energies, the addition of gas increased the deposition rate. When adding Helium gas, there is essentially no focusing observed, only a decrease in deposition rate due to random gas scattering.

The functional dependence of this change from gas scattering to focusing depends on the laser energy, the gas pressure, the target surface condition, and the measuring distance. The focusing is expected to occur when the ablated flux density and the background gas density are near the viscous flow level. The leading edge of the plume must push aside the background gas (which has less energy) from the front of the plume (the plume front has the highest flux density, and may travel as a shock front through the background gas[35]). This would create a brief transient period in which the central region of the plume is at a lower pressure than the surrounding area. The following ablated species (from the same laser pulse), which are slower or ejected at a later time,[35] would be scattered towards the plume center by collisions with the background gas "boundary layer", since the center of the plume is at a lower pressure during this brief transient period. The photographs of the plume in Fig. 6 indicate that no observable background oxygen interacts with the central portion of the plume (because of no observable color change), consistent with this explanation. We believe that this plume focusing is a general phenomena, and will always be observed if the ablation rate and gas pressure are high enough (the values will be dependent on the target material and gas type). The effect of the type of gas added is clearly seen when adding He gas. Helium gas has a much smaller scattering cross section than the other gases used, and a longer mean free path. Therefore, it is expected that a much higher pressure would be required to produce the plume focusing.

Finally, another set of experiments were performed to determine the dependence of the deposition rate of PZT as a function of the target-substrate distance. Figs. 8 (a) and 8 (b) show an example of this dependence for laser energies of 400 mJ and 800 mJ, respectively. The background O_2 pressure during ablation was either 300 mTorr, 100 mTorr, or vacuum.

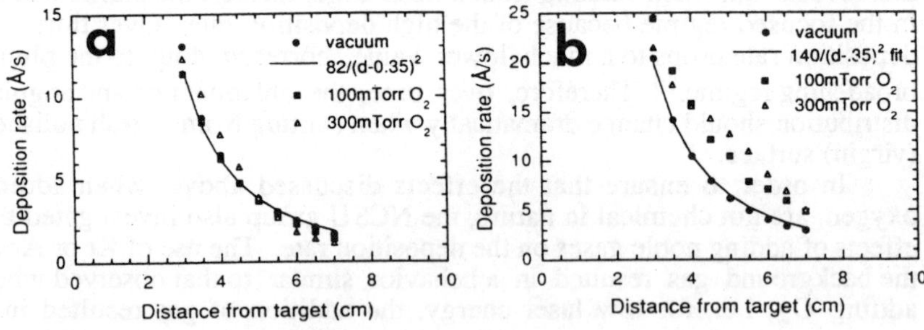

FIGURE 8. Deposition rate versus target-substrate distance for PZT ablated with a laser energy of 400 mJ (a) and 800 mJ (b), a 3 Hz repetition rate, in vacuum and pressures of 100 and 300 mTorr O_2. The line is a fit using Eq. (5) to the data obtained in vacuum.

In vacuum, the deposition rate is expected to decrease as the square of the distance, i.e.:

$$R(d) = R_{d=0}/(d+\delta)^2 ; \qquad (5)$$

where d is the substrate-target distance, $R_{d=0}$ corresponds to the deposition rate at the target surface (zero distance), and δ accounts for the fact that the laser spot is not a true point source. (Because the laser spot is of finite size, an extrapolated fictitious point source is located a distance d beneath the target.) Fits to the deposition rate data in vacuum, using Eq. (5), are shown by the solid lines in Figs. 8 (a) and 8 (b). Note that the inverse distance squared relation holds well. In Fig. 8, it is clear that the deposition rate drops faster than that expected from Eq. (5) when oxygen is added, revealing that gas scattering is broadening the ablation distribution. As shown in Fig. 8 (b), it is clear that at the ablation energy of 800 mJ, the deposition rate is higher than expected from Eq. (5) when oxygen is added, revealing that the added gas causes a relative narrowing of the ablation distribution. So, although the ablated species do continue to spread out over distance, the angular spread is slowed by the presence of background gas when ablating at high energy. This measurement of the deposition rate versus distance unequivocally proves that the gas pressure affects the angular distribution of the ablation plume, and doesn't merely alter the energy density or laser absorption.

The importance of these two deposition regimes (associated with focusing or broadening of the ablation plume) is that they have a significant effect on 1) the energy of ablated species, 2) their chemical state, and 3) the oxygen-to-ablated species flux ratio at the substrate. In the plume broadening regime, the gas scattering will result in a significant lowering of the energy of ablated species, an increase in gas phase reactions with oxygen, and an increase in the O_2 to ablated species ratio. In the plume focusing regime, the energies of ablated species should not be affected as much, because the ablated species are traveling through a rarefied region. There is also less interaction of ablated species with O_2 in the central portion of the plume, and relatively less O_2 available to interact with the substrate (during the deposition pulse).

Further work is necessary to correlate the effects of the ablated plume characteristics to the composition, microstructure, and dielectric properties of oxide films deposited under the various ablation conditions used for the synthesis of ferroelectric, conductive oxide films and layered heterostructures.

3. SYNTHESIS AND CHARACTERIZATION OF FERROELECTRIC/ METAL OXIDE HETEROSTRUCTURES

Realization of a commercially viable ferroelectric memory technology has been hampered by one or a combination of problems related to either the reliable performance of the PZT ferroelectric capacitor or to the growth and processing of capacitors that translate to high density memory elements. The electrode-ferroelectric interface is very crucial in determining the reliability characteristics of the ferroelectric capacitor, such as fatigue, which is the loss of switchable polarization when subjected to repeated bipolar pulses. Fatigue can be overcome, in PZT-based capacitors, for all practical purposes by replacing metallic Pt electrodes with metal oxide electrodes such as RuO_2[36-38] or any of the large number of perovskite metal oxides such as Y-Ba-Cu-O[39,40] and La-Sr-Co-O[25-27] among others. Similarly, internal interfaces in the ferroelectric, such as domain walls and grain boundaries and microstructural defects such as oxygen vacancies influence the ferroelectric properties. In another approach, a hithertofore undisclosed ferroelectric material, termed as "Y1",[41] has also yielded capacitor structures with desirable ferroelectric properties. Since the exact nature of this material still remains undisclosed and laser ablation deposition has not yet been used for producing Y1-based capacitors we shall not discuss or compare the properties PZT vs. Y1-based capacitors.

The synthesis of films and heterostructures described in this Section were performed in PLAD systems involving excimer laser beams ($\lambda = 248$ nm, 1-3 Hz, 1-2 J/cm^2) directed at a position inside the deposition chamber where PZT and other perovskite oxide targets (e.g., YBCO and LSCO) were sequentially positioned under the laser beam by means of rotating target holders with capability for continuous rotation of each target during laser ablation. PZT layers were synthesized using 10%Pb-rich or stoichiometric PZT targets, while YBCO and LSCO films were produced by ablation of stoichiometric $YBa_2Cu_3O_{7-x}$ and $La_{0.5}Sr_{0.5}CoO_3$ targets. In all cases, the layers were deposited in oxygen background atmospheres (300-900 mTorr), and at substrate temperatures in the range 550-640 °C.

3.1. Materials Integration Strategies

Work performed at NCSU-MCNC and Bellcore laboratories, during the last two years, has been focused on using perovskite metal-oxides as top and bottom electrodes instead of Pt. The perovskite oxide electrodes, processed under optimized conditions, yields capacitors that show negligible or no bipolar fatigue at least up to 10^{12} cycles.[26,27] Two different types of perovskite oxide electrodes have been discussed, namely, the layered superconducting Y - Ba - Cu - O (YBCO)[39,40] and the cubic metallic

La-Sr-Co-O (LSCO).[26,27] These perovskite oxides have room temperature resistivities of the order of a few hundred μΩ-cm (for example, epitaxial LSCO has a resistivity of 90 μΩ-cm at room temperature). In comparison, Pt has a resistivity in the range of 5-20 μΩ-cm at room temperature. In an attempt to lower the effective sheet resistance of the bottom electrode while maintaining the good fatigue characteristics of the LSCO-based capacitors, heterostructure LSCO/PLZT/LSCO capacitors were grown on epitaxial or highly [001] oriented Pt. The integration of these capacitors with metal oxide electrodes on Si can be accomplished using a novel approach that involves the growth of a thin layer of an anisotropic perovskite template layer such as bismuth titanate (BTO) on the thermally oxidized Si wafer prior to the deposition of a Pt layer and subsequent LSCO/PLZT/LSCO capacitor heterostructure.[26,27] The BTO template layer grows with a preferred c-axis texture, even on an amorphous surface such as SiO_2/Si. It not only provides a good chemical barrier between the overlayers and the Si wafer but also provides a perovskite-like surface (since it is c-axis oriented and hence the a-b plane is exposed) for the subsequent growth of the cubic perovskite layers. The perovskite structure and crystal chemistry of the BTO template layer is advantageously used to nucleate the [001] orientation of a perovskite phase both in LSCO and PZT layers preferentially over other competing orientations of the perovskite phase and other competing phases. Fig. 9 schematically illustrates both these concepts.

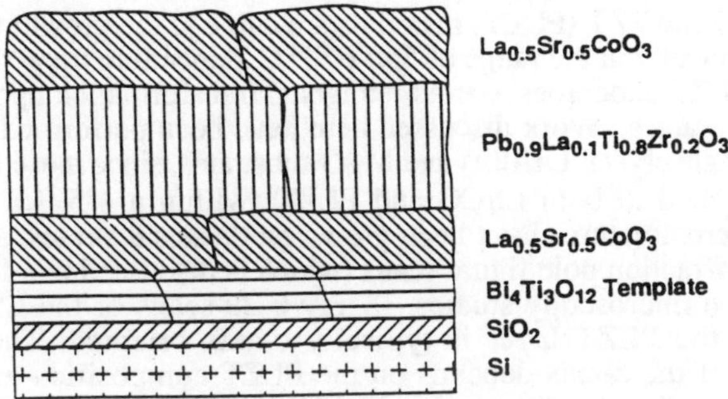

FIGURE 9. Schematic demonstrating the concept of template BTO layer-induced perovskite phase in electrode and PZT layers.

The materials integration strategies discussed in this Section will be illustrated with examples primarily from research currently in progress in our laboratories, although most of the conclusions are quite relevant to those obtained from other research laboratories. Wherever appropriate, results from other research groups relevant to this topic will also be discussed. Details of the laser deposition conditions used for the work discussed in this review can be found elsewhere.[10-15, 21, 22] The discussion in this Section is centered around the LSCO / PLZT / LSCO heterostructure capacitor, because this system is not only interesting from the fundamental thin film science point of view, but also has a very strong chance of being used in a commercial memory technology in the future. The compositions of the PZT and PLZT ferroelectric layers typically investigated in our programs include Pb ($Zr_{0.47}$ $Ti_{0.53}$) O_3,[27] $Pb_{0.9}$ $La_{0.1}$ $Zr_{0.2}$ $Ti_{0.8}$ O_3 (10 / 20 / 80),[26] $Pb_{1.0}Zr_{0.2}Ti_{0.8}O_3$ (0/20/80)[27] and $PbZr_{0.5}Ti_{0.5}O_3$(0/50/50)[27] although the whole range of compositions in the PLZT phase diagram can and have been grown by PLAD.

Following the general introduction presented above, a description is presented in subsequent sections, about the key advances made in the field of ferroelectric film growth by PLAD, through the use of the approaches discussed above, namely the use of anisotropic perovskite templates and metallic oxide electrodes to influence the phase stability and orientation distribution as well as the ferroelectric properties of the test capacitors.

3.2. Structure and Microstructure of PZT-Based Capacitors

3.2.1. Highly Oriented PZT-Based Capacitors on Oxide Substrates

Both LSCO and PZT (PLZT) materials are cubic at the growth temperature, which is typically in the range of 550-670 °C, depending upon the substrate. PZT and PLZT undergoes a cubic-tetragonal transformation upon cooling to room temperature. Work discussed here have been performed using [100] oriented single crystal $LaAlO_3$ and MgO substrates, since these materials are lattice matched to both LSCO and PLZT (within a 4-5 %). Very high quality heterostructures have been grown on these substrates, as evidenced by x-ray diffraction pole figure scans (which is discussed in a later section) and electron microscopy studies. X-ray θ-2θ scans of the LSCO/ PLZT show that the PLZT layer is typically c-axis oriented. The degree of alignment of the c-axis depends on the PLZT composition and the post-deposition cooling conditions. PLZT(10/20/80) compositions yield only c-axis oriented films while the 0/20/80 composition has a stronger propensity to yield a mixture of c-axis and a-axis (i.e., c-axis in the substrate plane) oriented films. The volume fraction of the film with the c-axis normal to the substrate increases as the cooling rate from deposition temperature increases.

Transmission electron microscopy studies of the films grown on $LaAlO_3$ substrates confirm the x-ray diffraction results with respect to their crystalline quality. Planar sections reveal the presence of only low angle grain boundaries and dislocations. The presence of 90° domain walls have been demonstrated in 0/20/80 films, both by x-ray diffraction (i.e., the presence of a-axis oriented regions) and by diffraction contrast electron microscopy. Figure 10 shows one such example of 90° domains observed in a planar section micrograph. These domains appear to nucleate at the interface between the electrode and the ferroelectric film, as illustrated in the cross-section micrograph in Fig. 11.

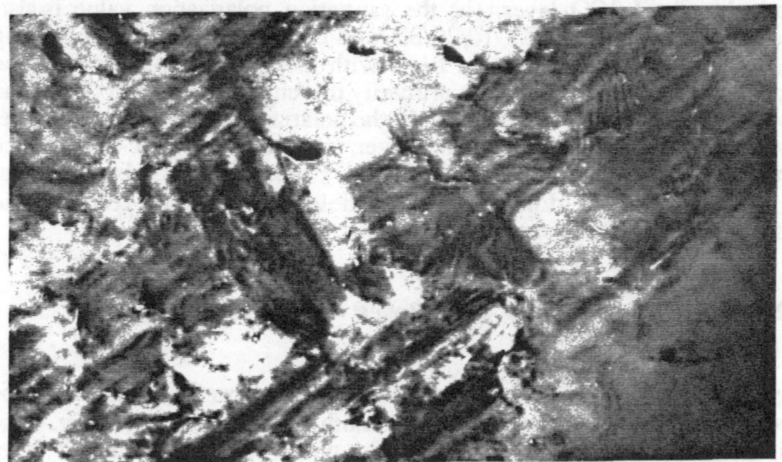

FIGURE 10. Planar section micrograph of a PLZT-0/20/80 film showing the presence of 90° domain walls.

FIGURE 11. Cross - section micrograph showing that domains appear to nucleate at the interface between the electrode and the ferroelectric film.

The growth crystallography of these films on $LaAlO_3$ is the cause for the 45° angle these domain walls make with the substrate normal. On the other hand, 10/20/80 films have very little 90° domains, if any. This observation is consistent with x-ray diffraction data which shows a completely c-axis oriented film. This difference can be understood in terms of two facts : (i) the c/a ratio for the PLZT 10/20/80 films is much smaller than for the PLZT 0/20/80, and the a-axis lattice parameter is much better matched to the LSCO layer; (ii) the saturation polarization value is also much smaller for the PLZT 10/20/80 (and consequently the depolarizing field is also much smaller). In films with the 0/20/80 composition, the tendency to form 90° domains is primarily driven by the desire to lower the depolarizing field, which can happen if the polarization direction is in-plane; however, this is opposed by the larger mismatch of the c-axis lattice parameter with that of LSCO or the $LaAlO_3$ substrate. It is also clear that the presence of 90° domain walls is deleterious to the ferroelectric performance, as will be shown later.

3.2.2. Highly Oriented PZT-Based Capacitors on Silicon Substrates

The growth of PZT layers on either YSZ-buffered Si or SiO_2/Si can also be accomplished through the use of a bismuth titanate template layer. The efficiency of this layer in controlling the phase stability and PLZT layer orientation can be established from x-ray diffraction patterns. Fig. 12 shows the effect of using the BTO template layer on the crystalline quality of PLZT-based heterostructure capacitors grown on [100]YSZ/Si substrates. In Fig. 12 (a), the x-ray diffraction pattern show the existence of predominant [110] PLZT and LSCO peaks along with the [002] YSZ peak, although one might expect to see the [001] peaks instead. The reason for this is the intrinsic mismatch between the LSCO (a=3.83Å) lattice parameter and that of YSZ ($a/\sqrt{2}$=3.64Å; mismatch of 4.7%). This, in conjunction with the inherent lack of structural anisotropy in the cubic perovskites, leads to a situation where the growth of the LSCO layer is predicated by the minimization of the growth surface energy. In general, this is accomplished by the stacking of the planes with the highest atomic density (i.e., the close-packed planes). In the case of the perovskites, (110) type planes exhibit the highest atomic density and consequently the films have a predominant [110] surface normal texture. It is relevant to notice that weak [001] and [002] diffraction peaks are also observed in Fig. 12 (a). The BTO template layer dramatically alters the orientation of the PLZT layer because of its intrinsic structural anisotropy. When the BTO layer is deposited at the appropriate substrate temperature, it grows with a strong preferred [001] direction normal to the substrate, i.e., the a-b plane is the lowest energy surface. A thin layer, sufficient to cover the YSZ substrate surface can subsequently provide a perovskite - like surface for the subsequent growth of the cubic layers. The distinct effect of such a template layer on the orientation of the LSCO/PLZT/LSCO ferroelectric heterostructure is demonstrated in the x-ray scan in Fig. 12 (b), where the main noticeable feature is the complete absence of the [110] diffraction peak. It should be noted that this is not completely unique to BTO; indeed, earlier research on high temperature superconductors (such as YBCO) had already demonstrated the growth of high quality c-axis oriented YBCO thin films on the same YSZ/Si substrate. While the difference in structural quality may itself be of fundamental interest from the growth mechanism point of view, the films with and without the template layer show a dramatic difference in ferroelectric properties, which is discussed in the next section.

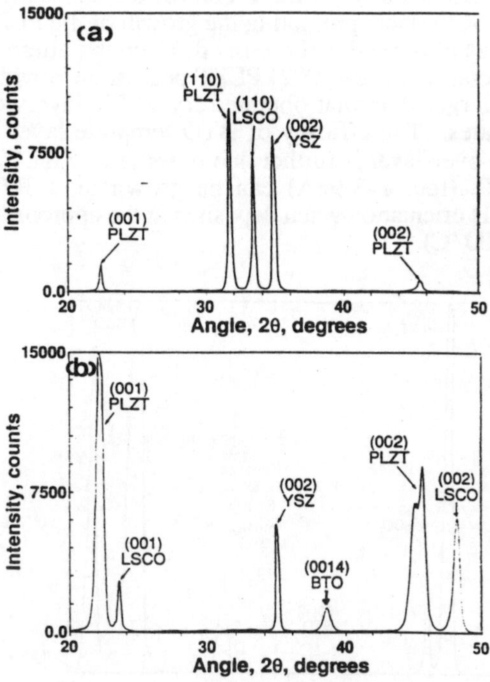

FIGURE 12. a) X - ray diffraction pattern of a PLZT/LSCO/YSZ heter-structure, showing the existence of predominant [110] PLZT and LSCO peaks along with the [002] YSZ peak. b) X- ray pattern of a PLZT / LSCO / BTO / YSZ heterostructure showing the highly preferred [001] PLZT orientation induced by the BTO template layer.

Transmission electron microscopy studies have shown that the LSCO and the PLZT layers in the PLZT/LSCO/BTO/YSZ heterostructure are very highly c-axis oriented with very little, if any, large angle grain boundaries. In addition, the same TEM studies have shown that the BTO layer is c-axis oriented and has a high density of stacking defects, which is well known in these layered perovskites.

High resolution TEM studies have also shown that there is no indication of interfacial reactions in LSCO/BTO heterostructures. However, the PLZT/LSCO interface is highly defective, primarily due to interface dislocations as a consequence of the lattice mismatch.

Probably the best demonstration of the effect of the BTO crystallographic anisotropy on its growth is the studies of growth on SiO_2/Si. The presence of the thermal oxide on Si is important since it will form the basis for fabricating the pass-gate transistors in the Si-CMOS wafer. At first glance, it should be expected that the BTO layer would be completely polycrystalline when grown on an amorphous surface such as SiO_2; however, this is not the case. At the appropriate growth temperature, the BTO layer grows completely c-axis oriented, although the film has very little crystallographic long-range correlation in the plane. The BTO layer provides a template with a perovskite structure to control the subsequent growth of the LSCO/PLZT/LSCO stack promoting the growth of the PLZT layer with [001[orientation, as illustrated in the x-ray diffraction pattern in Fig. 13. Rocking measurements about the [002] PLZT peak show a width of about 1.0-1.2°, which is larger than that obtained for PLZT layers on $LaAlO_3$ and YSZ/Si substrates. The efficacy of BTO template layer to control the orientation of the over-layer is further demonstrated by the fact that even metals, such as Pt (fcc; a=3.92Å) can be grown on a BTO template with a preferred [001] orientation when deposited at the appropriate substrate temperature (400-500 °C).

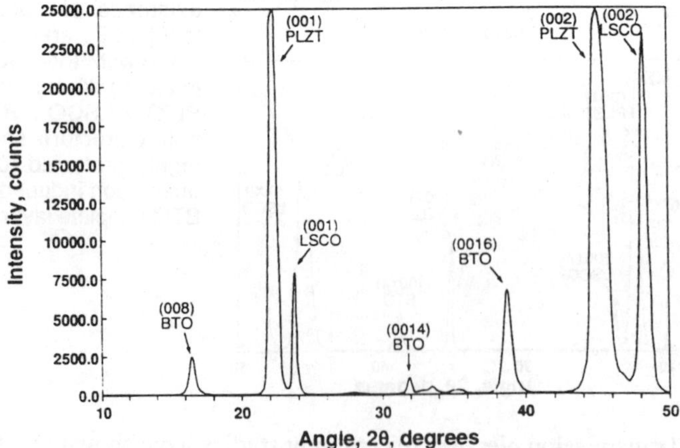

FIGURE 13. X - ray pattern of an LSCO/PLZT/LSCO/BTO heterostructure grown on a SiO_2/Si substrate, at 600 °C, showing the prefered [001] orientation of the PLZT layer induced by the BTO template.

Fig. 14 shows a θ-2θ X-ray scan from a heterostructure that is similar to that schematically illustrated in Fig. 8, but has a [001] Pt layer (~ 800 Å thick) grown on the BTO template layer before the growth of the LSCO/PLZT/LSCO heterostructure capacitor. As it can be seen in Fig. 14, only the [002] Pt peak is observed with a very strong intensity.

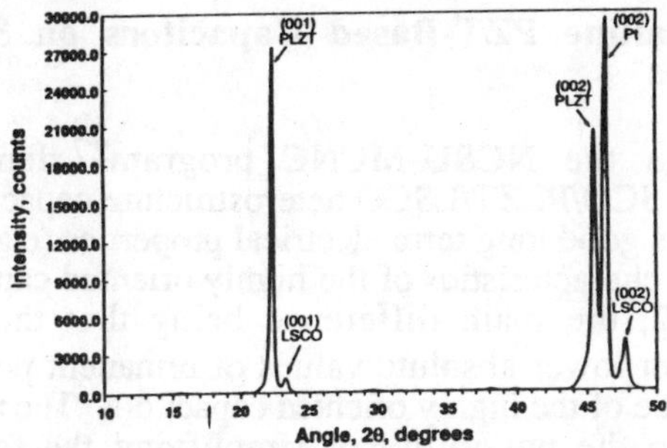

Figure 14. A θ-2θ X-ray scan from an LSCO/PLZT/LSCO/Pt/BTO heterostructure grown on a SiO_2 / Si substrate, at 600 °C, showing the preferred [001] orientation of the whole structure induced by the BTO template. The inset shows a similar X - ray scan from a PLZT/Pt heterostructure without the BTO template.

X-ray pole figure scans of LSCO/PLZT/LSCO heterostructures grown on the three types of substrates discussed here are shown in Fig. 15. In this experiment, the pole figure is obtained about the [101] PLZT diffraction peak.

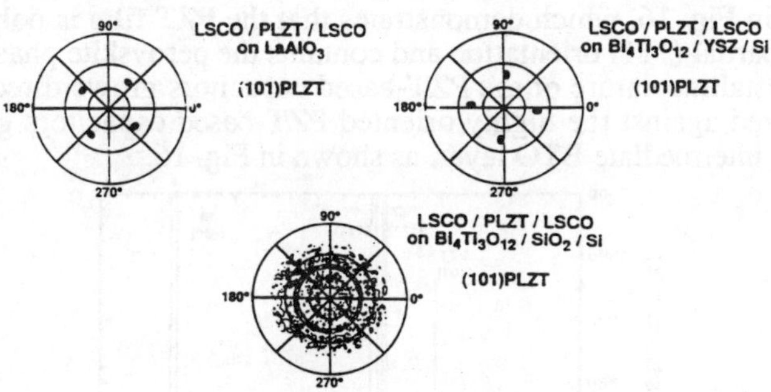

FIGURE 15. X - ray pole figure scans of LSCO/PLZT/LSCO heterostructures grown on $LaAlO_3$, YSZ/Si and SiO_2/Si.

In the case of films grown on $LaAlO_3$ and YSZ/Si substrates, a strong four-fold intensity distribution is observed, indicating that there is a significant degree of orientational locking of the in-plane [010] and [100] directions on these substrates. On the other hand, the films grown on SiO_2/Si appear to be completely random in the plane of the film. The effect of these microstructural differences on the ferroelectric properties are discussed in Section 4.

3.2.3. Polycrystalline PZT-Based Capacitors on Silicon Substrates

Recent work in the NCSU-MCNC program[27] demonstrated that polycrystalline LSCO/PLZT/LSCO heterostructure capacitors grown on Si substrates have as good long term electrical properties (e.g. negligible or no fatigue) as those characteristics of the highly oriented capacitors discussed in Section 3.2.2, the main difference being that the polycrystalline capacitors present lower absolute values of remanent polarization[27] than those characteristic of the highly oriented capacitors. The relevance of these results relates to the possibility of simplifying the fabrication of the capacitors for non-volatile memories.

For example, heterostructure LSCO/PZT/LSCO capacitors were produced on Pt/Ti/SiO_2/(100)Si substrates using the PLAD technique in such a way that all layers of the capacitor were deposited in-situ without breaking vacuum,[27] at 600 °C, in an oxygen background pressure of 300-900 mTorr. The Pt and Ti layers were deposited in a separate system using e-beam evaporation, which produced (111) Pt layers with a columnar structure perpendicular to the substrate surface.

An XRD pattern of an LSCO/PZT/LSCO/Pt/Ti/SiO_2/Si structure is shown in Fig. 16, which demonstrates that the PZT film is polycrystalline with a partial (111) orientation and contains the perovskite phase only. The polycrystalline nature of the PZT-based capacitors grown directly on Pt is contrasted against the highly oriented PZT-based capacitors grown on Si with an intermediate BTO layer, as shown in Fig. 16.

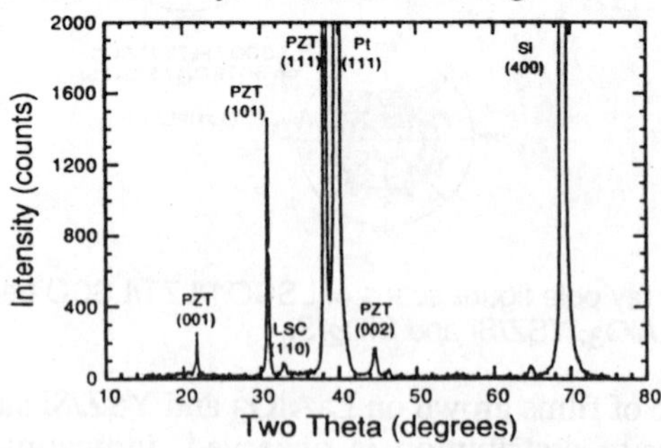

FIGURE 16. XRD spectrum of an LSCO / PZT / LSCO / Pt / Ti / SiO_2 / Si heterostructure, showing the polycrystalline nature of the PZT layer.[27]

4. SHORT AND LONG TERM ELECTRICAL PROPERTIES OF PZT - BASED HETEROSTRUCTURE CAPACITORS

4.1. Highly Oriented PZT-Based Capacitors

The main reason for the considerable interest in growing PZT-based heterostructure capacitors with oxide or hybrid metal-metallic oxide electrodes is the recent demonstration of very desirable ferroelectric properties using such structures. Many of the details of the ferroelectric properties of LSCO/PLZT/LSCO capacitors produced by the PLAD method can be found in the recently published literature.[26,27] In this Section, only the salient features and some particular aspects of films prepared by the PLAD technique are discussed. The PZT-based capacitors used for the electrical tests discussed in this section were fabricated using blanket bottom electrodes and a combination of photolithographic lift-off in conjunction with either wet etching or dry ion beam milling to pattern the top electrodes. Capacitors with 600 to 2 μm diameter were fabricated using the methods mentioned above. The details of device fabrication are described in earlier papers. The electrical tests were performed mainly using the ferroelectric tester RT66A from Radiant Technologies, primarily in the pulsed testing mode to simulate as close as possible the memory operation. For example, the differential pulsed remanent polarization, ΔP, defined as switched polarization, P* - nonswitched polarization, P^ is in general twice the conventionally measured remanent polarization, P_r, which can be measured from the hysteresis loop.

Fig. 17 shows a family of pulsed hysteresis loops of an LSCO/PLZT/LSCO capacitor grown on $LaAlO_3$. Typically, pulsed remanent polarization values in the range of 25-35 $\mu C/cm^2$ were obtained at room temperature, with an applied field of 5V. The coercive field of the PZT-based capacitors with oxide electrodes, which is also an important memory element design variable, is typically in the range of 0.7-1.5V. The field dependence of the hysteresis properties is illustrated in Fig. 17. The film resistivity at 5V is typically in the range of 5×10^9 - 5×10^{10} Ω-cm, a value which is considerably higher compared to those obtained in the early days of the use of these metal oxide electrodes. This result can be attributed to better control over the deposition parameters, including the film thickness.

Fig. 18 shows the pulsed remanent polarization and film resistivity as a function of temperature, for the same capacitor for which the hysteresis loops are shown in Fig. 17, measured with an applied voltage of 5V. This temperature dependence is expected, since we are approaching the Curie temperature. Concurrently there is more thermal activation for point defects such as oxygen and lead vacancies to move, thus contributing to transport in the film. Ferroelectric hysteresis properties for films grown on YSZ/Si are very similar to that on $LaAlO_3$.

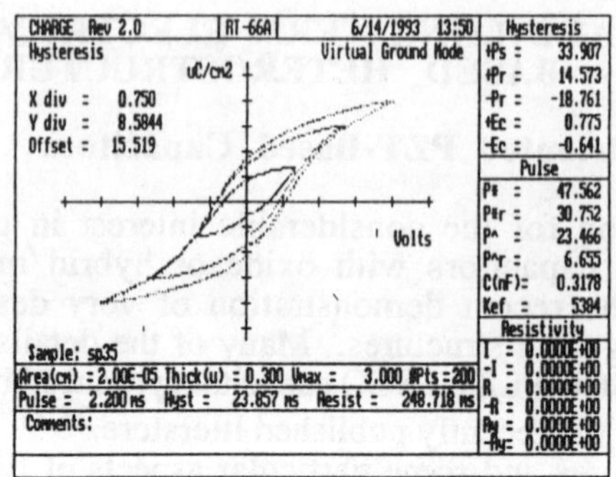

FIGURE 17. Family of pulsed hysteresis loops obtained from a highly oriented LSCO/PLZT/LSCO capacitor grown on $LaAlO_3$.

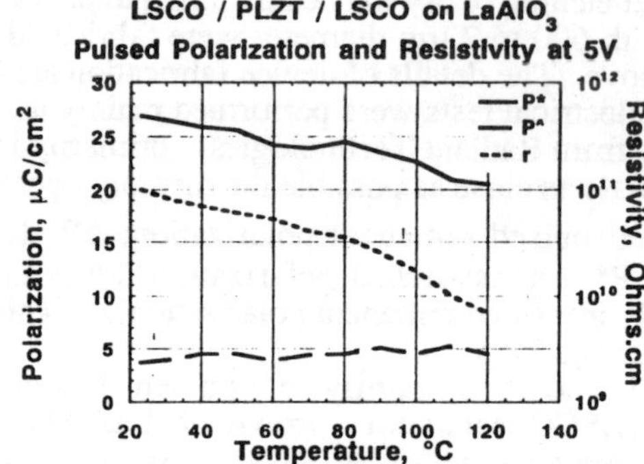

FIGURE 18. Pulsed remanent polarization and resistivity vs. temperature curves for the same LSCO/PLZT/LSCO capacitor for which hysteresis loops are shown in Fig. 17.

Fig. 19 shows pulsed hysteresis loops for a LSCO/PLZT/LSCO heterostructure grown on SiO_2/Si with the BTO template and a Pt underlayer. The overall trend is very similar to that of the capacitors on $LaAlO_3$, with the difference being that the coercive fields are slightly larger and the pulsed remnant polarization is smaller (typically in the range of 15-20 $\mu C/cm^2$ at 5V). It should be noted, however, that these values are more than sufficient for nonvolatile memory operation, if they satisfy the reliability requirements. Consequently, very detailed pulsed testing of the capacitors described above is being performed to understand the effect of process variables and microstructure on the key reliability parameters. The initial work was focused on finding solutions to the problem of bipolar fatigue in PZT-based capacitors. As outlined in the introduction, it has been found that using metallic oxide electrodes instead of the conventional Pt electrodes makes a dramatic difference in the fatigue characteristics. This is illustrated in Fig. 20, for the case of a LSCO/PLZT/LSCO capacitor grown on $LaAlO_3$. In capacitors grown under the optimum conditions, we do not observe any significant fatigue even after 10^{12} cycles. These results are very promising, considering that fact that capacitors with Pt electrodes typically fatigue and lose more than 50% of their initial polarization after just 10^8 cycles, as shown in Fig. 20. This marked difference in fatigue characteristics with the type of electrode (i.e., Pt or metal oxide) is observed irrespective of the crystalline quality of the PLZT thin film. It is important to note that although many research groups have demonstrated similar fatigue results using metallic oxide electrodes, a clear and rigorous understanding of the cause for this difference in fatigue behavior is still not available.

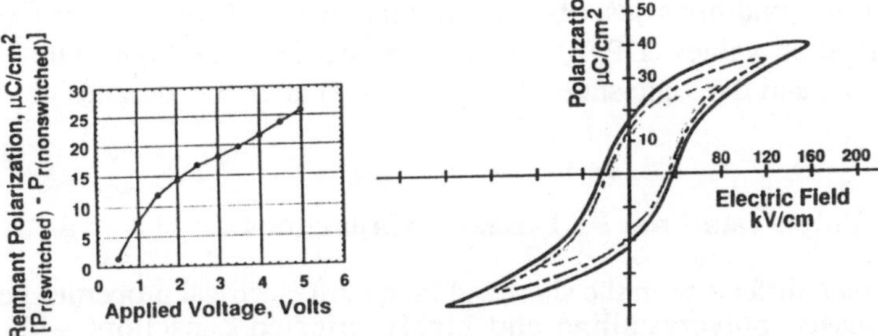

FIGURE 19. Pulsed hysteresis loops for an LSCO / PLZT / LSCO heterostructure capacitor grown on a SiO_2 / Si substrate with the BTO template and a Pt under-layer .

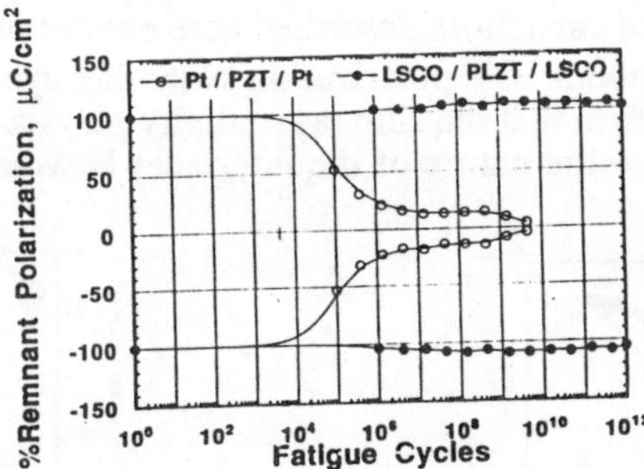

FIGURE 20. Results of a fatigue test on an LSCO / PLZT /LSCO capacitor grown on a $LaAlO_3$ substrate and compared with that of a Pt/PLZT/LSCO capacitor on $LaAlO_3$.

Capacitors grown on YSZ/Si and SiO_2/Si show essentially identical fatigue behavior as for the systems shown in Figs. 19 and 20. These heterostructures also show similar fatigue characteristics when tested at 100 °C. The almost complete absence of any loss in pulsed remanent polarization under these testing conditions is quite desirable for the ultimate use of these capacitor structures in nonvolatile memories.

Another important parameter related to the reliable electrical behavior of a capacitor for non-volatile memories is the polarization retention as a function of time after a capacitor has been polarized in one logic state. Retention tests have been performed using a sequence of pulses [-5V, 4V, 4V] at room temperature and 100 °C. Under these test conditions, the first read pulse probes a switched polarization (P*), i.e. the logic state "1", indicating the ability to maintain a polarization charge in the logic state "1". The second read pulse probes the non-switched polarization (P^), which corresponds to the logic state "0". The switched polarization ($\Delta P = P^* - P^\wedge$) followed a log-linear relationship with retention time. The capacitors tested under the conditions described above presented sufficient difference in the polarization values (DP 10 - 12 $\mu C/cm^2$ between the logic state "1" (10 $\mu C/cm^2$) and the logic state "2" (12 $\mu C/cm^2$) after five decades of waiting time.

4.2. Polycrystalline PZT-Based Capacitors

The main difference in the short and long term electrical properties between PZT-based polycrystalline and highly oriented capacitors with oxide electrodes, discussed in this chapter, are that the polycrystalline capacitors exhibit slimmer hysteresis loops with lower remanent polarization (compare Figs. 21 (a) and 21 (b)). However, both the polycrystalline and highly oriented PZT-based capacitors described here exhibit practically no fatigue when tested up to about 10^{10} polarization switching cycles (Fig. 22). These results further confirm that the film crystallinity plays a minor role vis-a-vis fatigue compared to the nature of the interfaces between the electrode and the ferroelectric.

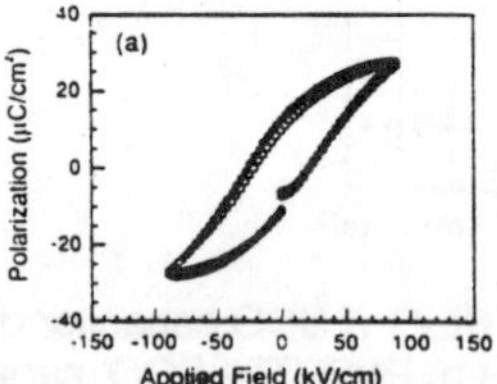

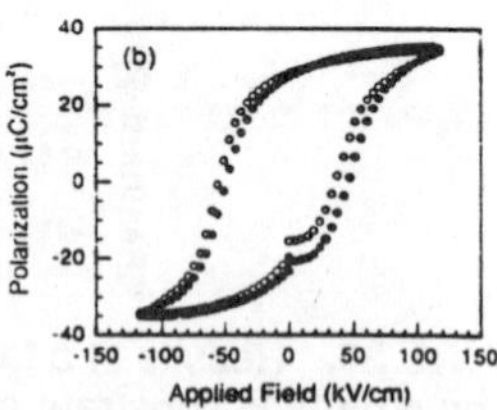

FIGURE 21. (a) hysteresis loop [(•) before and (o) after fatigue] of a polycrystalline LSCO/PZT/LSCO capacitor deposited on a $Pt/Ti/SiO_2/Si$ substrate; (b) hysteresis loop [(•) before and (o) after fatigue] of a highly oriented LSCO/PZT/LSCO capacitor deposited on a (100) MgO substrate presented for comparison .[27]

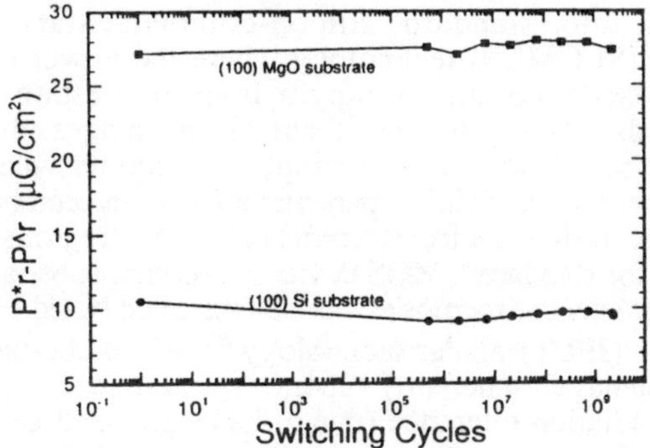

FIGURE 22. Fatigue data for polycrystalline and highly oriented LSCO/PZT/LSCO heterostructure capacitors produced by the PLAD technique.[27]

Polarization retention measurements for the PLAD polycrystalline PZT capacitors discussed here revealed that they will maintain a minimum of 2 $\mu C/cm^2$ of switched polarization after a cumulative aging period of $1x10^{10}$ seconds ($3x10^2$ years!). Therefore, retention is a parameter that is currently under control for both highly oriented and polycrystalline PZT-based capacitors with oxide electrodes.

In summary, PLAD polycrystalline PZT-based capacitors with hybrid LSCO-Pt electrodes fabricated on Si exhibit excellent fatigue and retention characteristics at room temperature, demonstrating that it may not be absolutely necessary to have oriented PZT in order to achieve fatigue-free ferroelectric capacitors. This result, coupled with the fact that Pt/Ti adheres well to SiO_2 and is stable in an oxygen ambient, makes the integration of ferroelectric PZT capacitors with Si much simpler for fabricating PZT-based non-volatile memories. In addition, resistivity measurements indicate that the hybrid LSC/Pt electrode have much lower resistivity than the pure LSCO electrode, which would improve the device performance. However, further testing under a variety of operating conditions are required to verify the device performance described above.

4.3. PZT-based Heterostructure Capacitors on Alternative Oxide and Semiconductor Substrates

Besides Si and the oxide substrates discussed in prior Sections, gallium arsenide and sapphire are two other important substrate for ferroelectric capacitors because of their potential application to fabricating devices with specific properties. On silicon, the ferroelectric capacitors can be integrated into standard silicon-complementary metal oxide semiconductor (Si-CMOS) technology where the power consumption is low, but the CMOS circuits occupy a large real estate and have low switching speeds. To circumvent these disadvantages, the ferroelectric capacitors can be fabricated on a sapphire substrate where the CMOS circuits are faster (due to reduced parasitic effects) and can be laid out more densely (due to reduced circuit coupling across regions of insulating sapphire) than for standard CMOS devices. Another substrate of interest is GaAs where ferroelectric memories can be fabricated based on junction field effect transistor (JFET) planar technology.[42] When compared to silicon, gallium arsenide has an inherently superior speed (due to its higher electron mobility) and radiation tolerance (due to its larger band gap). Because of these properties, it is advantageous to integrate ferroelectric capacitors with gallium arsenide integrated circuit technology for fabricating non-volatile ferroelectric memories.

Heterostructure LSCO/PZT/LSCO PLAD capacitors have been fabricated by the NCSU-MCNC group on sapphire and on a pre-metallized structure of Pt / Ti / Pt / GaAs in the case of the gallium arsenide substrate.[43] The Pt/Ti/Pt metallization scheme is generally used to form Schottky barrier contacts to n-type GaAs. A previous study has successfully demonstrated that ferroelectric capacitors with good long term properties can be fabricated on Pt/Ti/SiO_2/Si substrates.[27]

The LSCO/PZT/LSCO heterostructure PLAD capacitors grown on R-cut sapphire presented polycrystalline PZT layers with the highest intensity peak along the (100) direction. In the case of the Pt/Ti/Pt GaAs substrate, the PZT layer is also polycrystalline with high intensity (101) and (111) peaks. In both cases, there is no evidence of impurity phases. Other studies have shown that without an intermediary LSCO layer, PZT grown on Pt or RuO_2 using PLAD will contain mixtures of pyrochlore (non-ferroelectric) and perovskite (ferroelectric) phases.[27]

A series of hysteresis loops taken at different bias voltages are shown in Figs. 23 (a) and 23 (b) for PZT films deposited on sapphire and GaAs substrates, respectively. In Fig. 23 (a), the loops show some enhanced conductivity which can be observed as a roll-off in polarization as the applied field is increased. The absence of any roll-off in Fig. 23 (b) may be indicative of a higher quality interface between the PZT and LSC layers for the capacitors produced on GaAs substrates. The resistivity of the PZT films on both substrates was in excess of 1 x 10^9 ohm-cm.

Fatigue tests using peak-to-peak voltage pulses of 10 volts and 500 kHz excitation frequency revealed that both PZT-based capacitors on sapphire and GaAs practically fatigue free up to about 3 x 10^{10} switching cycles (Fig. 24). These results are similar to those obtained with PZT-based capacitors on other substrates, as discussed in Sections 3.1 and 3.2.

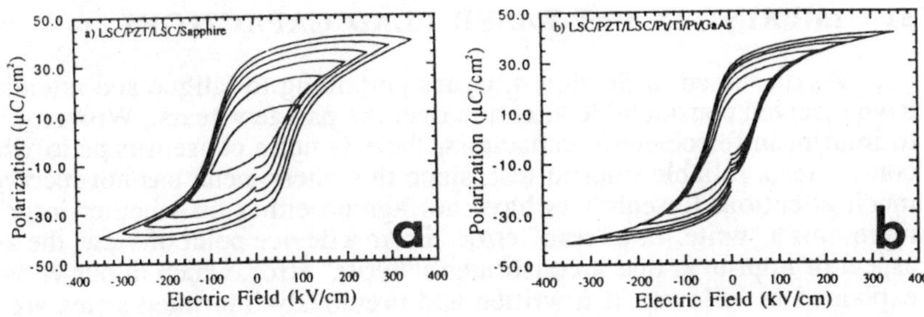

FIGURE 23. Hysteresis loop of LSCO/PZT/LSCO capacitors fabricated by PLAD on R-cut sapphire (a) and on Pt/Ti/Pt/GaAs (b) substrates. For both cases, loops were taken at 4, 6, 8, 10,12,and 14 volts.[43]

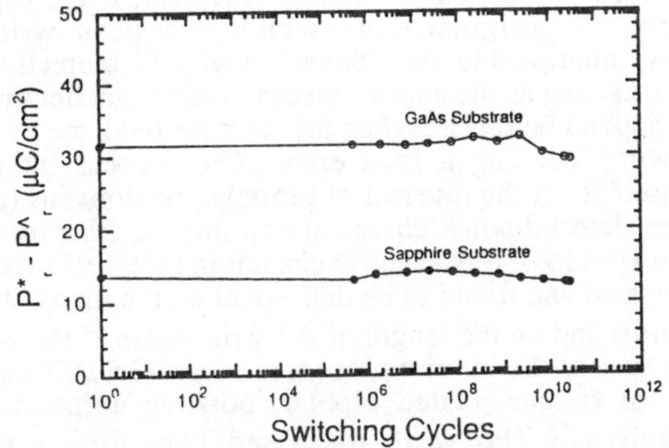

FIGURE 24. Fatigue characteristics of LSCO/PZT/LSCO heterostructure capacitors on R-cut sapphire and Pt / Ti / Pt / GaAs substrates.[43]

Polarization retention measurements for the PZT-based PLAD capacitors on sapphire and GaAs revealed that a threshold polarization of about 2 $\mu C/cm^2$ (minimum value needed by the sense amplifier to discriminate between the two logic states) will remain in the capacitors after 10^{12} seconds.

The results discussed above indicate that heterostructure LSCO/PZT/LSCO capacitors fabricated on sapphire and GaAs also have properties suitable for their application to non-volatile ferroelectric memories.

5. IMPRINT IN PZT-BASED PLAD CAPACITORS

As discussed in Section 4, issues pertaining to fatigue and retention have received considerable attention over the past few years. With regards to imprint in ferroelectric capacitors, there is not a consensus as to what constitutes a reliable imprint test, since this phenomena has not received much attention. Researchers have not agreed either on whether imprint represents a "write" or a "read" error. From a device point of view, the key aspect of imprint is that it constitutes a "write" error, which in our view is explained as follows: If a written and previously imprinted states are in opposite directions (e.g., up and down, respectively), the written (up) state will, as a function of time, show a tendency to revert to the imprinted (down) state. Hence, the written state will, after some time, assume the complementary (down) orientation of the intended (up) logic state causing a "write" error. If it is assumed that the capacitor is read non-destructively, the write error will be easily verified. However, if the capacitor is read destructively, the original state which had been written "up" and subsequently imprinted to the "down" state will correctly be read as a "down" state as long as the applied electric field is greater than the coercive field. If the applied field is less than the coercive field, the imprinted dipoles will not switch causing a read error. The process by which imprint manifests itself is via the reversal of ferroelectric domains (possibly due to space charge, defect dipoles, charge at trap sites, or other mechanisms).

In thin film PZT ferroelectric capacitors, the self reversal effect has been investigated and found to be dependent on the time between the write and read pulses and on the length of the write pulse.[44] Recently, work was discussed related to the imprint properties of Sol-Gel PZT capacitors, using a sequence of fatigue-related bipolar, positive unipolar, and negative unipolar pulses.[45] The work discussed here differs from previous investigation in the following ways: (a) the imprint resuits are not affected by fatigue effects, since fatigue-free capacitors were used for the tests; (b) the effects of retention loss and imprint are separately identified; (c) the test methodology and instrumentation are much simpler; (d) alternative test sequences are provided; and (e) the tendency to imprint is accelerated by a DC bias rather than by temperature, where pyroelectric effects may influence the measurement.

The imprint test methodology recently used by the NCSU and Bellcore groups involves testing the imprint characteristics of ferroelectric capacitors using either a DC bias or an asymmetric AC signal to accelerate the process and determine the "rate of imprint". This term is used to describe the rate of approach towards imprint failure.[46] In analyzing the experimental data, polarization retention effects were identified and separated from the rate of imprint. It is relevant to try to understand these effects separately to better interpret their combined effects which may determine the time it takes for device failure to occur.

5.1. Imprint Criteria and Tests

The criteria for determining the occurrence of imprinting in a ferroelectric capacitor is discussed in context with the hysteresis loop of Fig. 25, where the polarization parameters are obtained from measurements involving a set of triangular pulses as discussed in the operating manual of the RT66A ferroelectric tester[47] and are defined the same way as in the RT66A fatigue and retention test programs (i.e., P* and P^*_r are the switched polarizations, while P^ and $P^\wedge_r$ are the non-switched values). There is a 1 second wait period between the switched logic states, causing the relaxation which is shown as $P^\wedge_r$. According to the literature, it is assumed that failure occurs in a memory cell when the difference between the switched and non-switched polarizations is less than about 2 μC/cm^2. However, this value can vary depending on the memory architecture and type of sense amplifier.

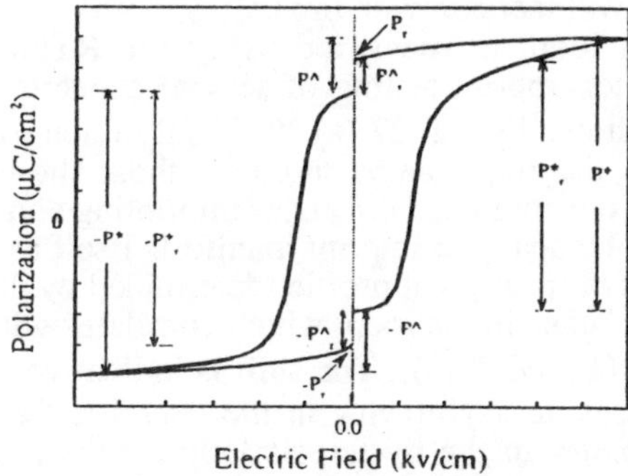

Figure 25. Hysteresis loop showing the various pulse-polarization parameters taken into account to determine the conditions for imprint.

Assuming that the voltage pulse train of Fig. 26 is applied to a ferroelectric capacitor, causing some dipoles to be imprinted in the "up" state, when the first read signal is applied, the dipoles will switch "down" and -P* will be measured. During the interval between the two read pulses, some dipoles will return back to the "up" (imprinted, preferred) state. The second read pulse will switch the "imprinted" dipoles back to the "down" state and -P^ will be recorded. If $|-P^*-(-P^\wedge)| < 2$ μC/cm^2, then an imprint failure has occurred. It should be noted that the above criterion is the same as that which has been established for fatigue and retention failures. Thus, a distinction is required in order to be able to identify a particular failure mode. In the case of fatigue and retention tests, the switched (P* and P^*_r) and non-switched (P^ and $P^\wedge_r$) polarizations will decrease with time (including the possibility that P^ and $P^\wedge_r$ may remain constant). The distinction between these failure modes is determined by the device history, i.e., whether failure is due to switching cycle effects or simply time-related effects. The evaluation of retention using the pulse sequence of Fig. 26 without the DC bias, will show loss of switchable polarization.

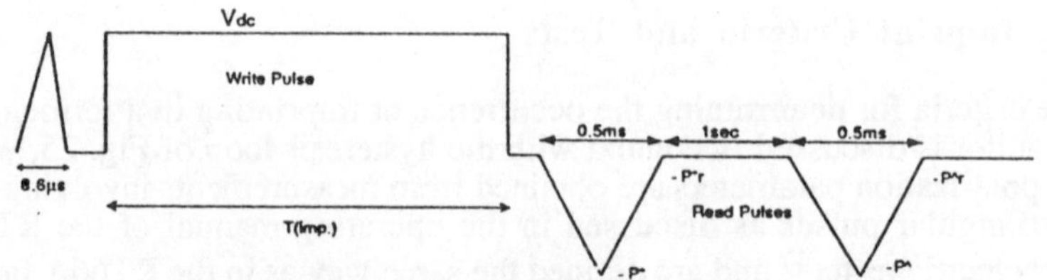

FIGURE 26 . Voltage wave form used to test for an imprinted state that is in the opposite direction of the read voltage.[46]

A failure by imprint (without loss of retained polarization) as measured using a pulse sequence such as that of Fig. 26 will show an increase in P^ and $P^\wedge_r$, while the switched components (P* and P^*_r) will stay constant. When the loss of retained polarization is simultaneously occurring during an imprint failure, the switched polarizations will decrease and the non-switched components will increase with time.

A representative hysteresis loop, as measured using the RT66A ferroelectric tester, before and after imprint testing of several capacitors (with minimal retention loss) is shown in Fig. 27 (a).[46] If only retention loss occurs during the imprint test, the loops can be similar to those shown in Fig. 27 (b). Figure 27 (c) shows the combined effects of imprinting and a loss of switched polarization. The tendency to imprint manifests itself as a displacement of the loop to the left of its original profile (determined by the imprinting direction), and a larger "gap" in the loop, which correlates with the increased $P^\wedge$ and $P^\wedge_r$ (Figs.27 (a) and 27(c)). The shift is indicative of an internal bias field, which causes an asymmetry in the coercive field values. This time-evolved asymmetry in the hysteresis loop results in a tendency for the capacitor to exhibit a preferred logic state.

In addition to the general discussion based on Fig. 27 above, it is relevant to present the hysteresis loops shown in Fig. 28 to illustrate an important point that needs to be considered when performing imprint tests. In general, the NCSU-MCNC group has observed that hysteresis loops of PZT - based capacitors, which have not been annealed after top electrode patterning (by photolithography and ion beam etching) and prior to the electrical test, are shifted along the voltage (electric field) axis (Fig. 28).[46] This shift is strongly reduced or eliminated by subjecting the entire capacitor structure to an annealing step (550-600 °C for 5-10 minutes). It has been observed also that if the annealed capacitor is subjected to another ion beam etching step, the hysteresis loop is shifted again along the voltage (electric field) axis but the shift is much smaller than the original shift observed prior to the first annealing of the whole capacitor structure. These results suggest that the as-deposited capacitors present an imprint-like behavior (manifested in a hysteresis loop shift along the voltage axis) prior to any imprint test. The initial loop shift observed in non-annealed capacitors may be due to a combined effect of a poor as - deposited top electrode/ferroelectric interface quality (due to composition, microstructure, charged or uncharged defects) and ion beam-induced radiation effects such as damage and/or ion charge incorporation into the PZT layer adjacent to the patterned electrodes. The experiment involving the second ion beam etching after the first annealing suggests that the ion beam etching contribution to the initial hysteresis loop shift is much smaller than the contribution of the as-deposited top electrode/ferroelectric layer interface quality. The annealing of the whole capacitor structure strongly reduces or eliminates defects/structural damage at the top electrode/ferroelectric layer interface, which results in the reduction or elimination of the loop shift. Further work is currently underway in the NCSU-MCNC and Bellcore laboratories to investigate the effect illustrated in Fig. 27. Based on the discussion presented above, it is recommended that capacitors to be tested for imprint be annealed prior to the test.

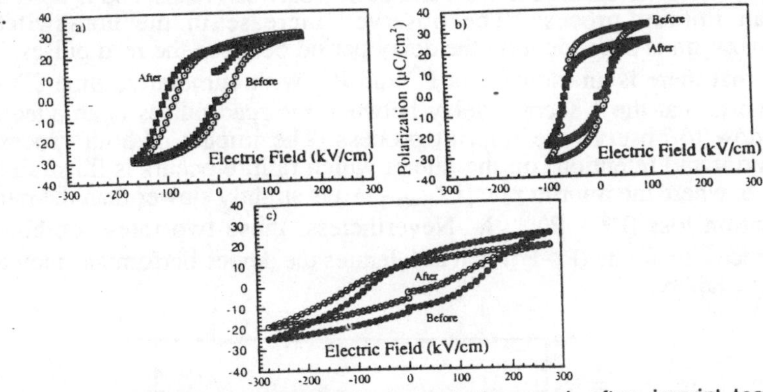

FIGURE 27. Comparison of hysteresis loops before and after imprint testing. (a) loop shift reveals tendency towards imprint with minimal loss of retained polarization. (b) loop shift shows loss of retained polarization with minimal tendency to imprint. (c) loop shift in the coercive field is due to an internal bias field. There is also a loss of retained polarization.[46]

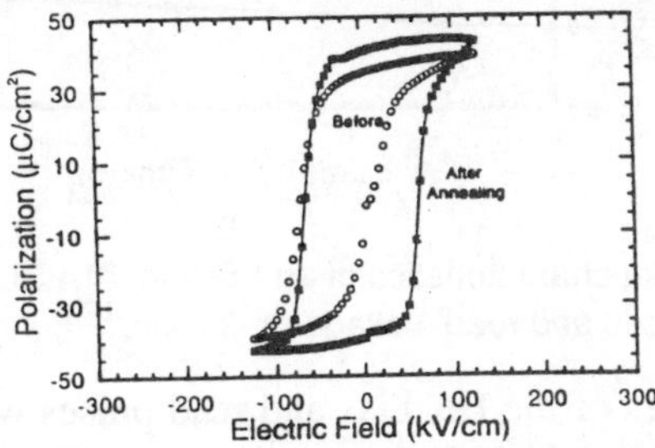

FIGURE 28. Hysteresis loops before and after annealing an LSCO/PZT/LSCO capacitor with top electrode patterned by photolithography and ion beam etching.

Figure 29 shows the result of an imprint test performed on a (100) oriented PLAD ferroelectric PZT capacitor. The retention program was used to generate the voltage sequence of Fig. 26. The values of the DC bias and read signals were +5 and -3 volts, respectively. Based on the imprint criteria and the pulse sequence described above, the absolute value of the non-switched polarizations ($P^\wedge$ and $P^\wedge_r$) should increase as a function of time. This is demonstrated in Fig. 29 (notice sign of polarization), although the $P^\wedge_r$ curve shows a very slow rate of decrease. Also, the switched polarization (P^* and P^*_r) decreases with time and is described as a retention loss process that was accelerated by the application of the +5 volt DC bias. On the other hand, the increase in $P^\wedge$ and $P^\wedge_r$(caused by the cumulative T_{imp} periods of the DC bias) can be interpreted as a manifestation of domain reversal during the 1 second period between reads, and is described as an imprint process. The observed increase in the non-switched polarization is dependent on the delay period between the read pulses. The fact that there is an increase in $P^\wedge$ and $P^\wedge_r$ with cumulative time (T_{imp}) suggests that the 1 second delay between the read pulses is an adequate window to observe the imprint process. The impact of both processes (imprint and retention) on the rate at which failure occurs is illustrated in Fig. 6, where the imprint rate [$P^*_{t=o}$ - $P^\wedge$] is slightly slower than the rate of retention loss [P^* - $P^\wedge_{t=o}$]. Nevertheless, these two rates combine to produce a total rate (P^*-$P^\wedge$) that accelerates the device performance towards a logic failure.

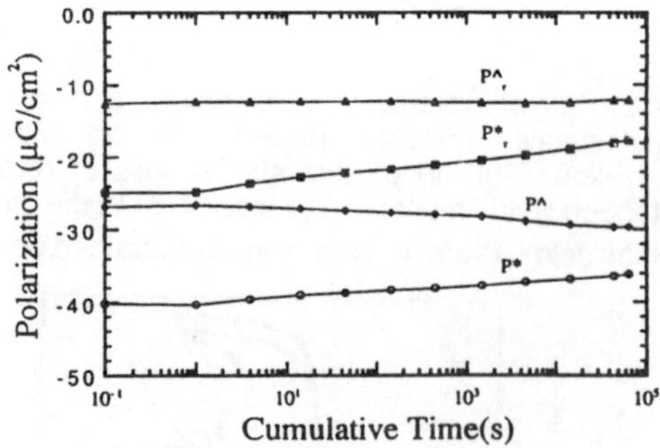

FIGURE 29. Imprint characteristics of an LSCO/PZT/LSCO/MgO capacitor using a DC bias = 5 volts and read voltage of -3 volts.[46]

When the polarities of the DC bias and read pulses were changed from +5 volts and -3 volts to -5 volts and +3 volts, the combined imprint + retention loss rate did not change. However, a reduction in the total rate was observed as the magnitude of the read voltage was increased from +3 volts to +4 volts (see Fig. 30). The main contributor to this reduction was a decrease in the apparent imprint rate. The rate of loss of retained polarization was independent of the magnitude of the positive read pulse. From these observations, it is clear that as the magnitude of the read voltage pulse is increased, in the opposite direction of the DC bias, the total non-switched polarization measured during the second read is reduced. Essentially, the read pulse magnitude tends to remove the imprint effects if the device is weakly imprinted. For a strongly imprinted device, a higher read voltage may not affect the imprinted state, although a reduction in retention loss may be observed.

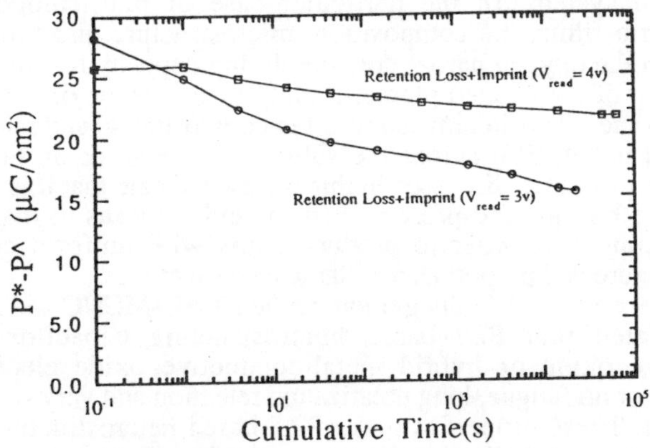

FIGURE 30. Effect of the magnitude of the read voltage on the rate of approach towards imprinting. DC bias voltage = -5 volts.[46]

Imprint tests performed for heterostructure LSCO/PZT/LSCO capacitors on sapphire and GaAs also revealed,[43] as for the capacitors on MgO and Si discussed above, that failure due to imprinting would not occur until 10^{13}-10^{14} seconds of operating time, under the present test conditions.

Based on the imprint test results discussed above, it appears that failures by imprint/retention loss mechanisms are not an issue for the type of capacitors that were evaluated (including size), at least under the test conditions used in the work described here.[43,46] However, because the failure mechanism is not understood, extrapolations of this type have limited applicability. In addition, it is very important to perform imprint tests on capacitors much smaller than those used in the tests reported here, i.e., tests should be performed with micron and submicron-size capacitors to fully evaluate possible imprint effects on capacitors that will be used in high density non-volatile memories. Further work is obviously necessary to better understand the implications of imprint on the operation of ferroelectric non-volatile memories.

6. CONCLUSIONS

Pulsed laser ablation deposition (PLAD) is a very versatile technique to perform fundamental and applied research related to developing materials integration strategies for thin film-based devices.

In order to use the full capabilities of this method, it is important to understand the ablation, species transport, and deposition processes relevant to PLAD. Work discussed in this review and in the literature have demonstrated that the ablation process, transport of species, and film growth depends strongly on the background pressure conditions in the film deposition system. In the particular case of multicomponent oxide ferroelectric films, the composition, microstructure and properties may depend on the oxygen partial pressure during deposition, which can result in focusing of the ablated plume, leading to an uneven distribution of the ablated species. This in turn can result in compositional and microstructural variations in the film across the substrate. The work on laser ablation deposition processes discussed in this review indicate that it is necessary to control the background pressure and to device means to manipulate the ablation plume in order to produce films with uniform composition, microstructure and properties over large area substrates.

Work on materials integration by the NCSU-MCNC and Bellcore has demonstrated that PZT-based heterostructure capacitors involving conductive oxide or hybrid metal-conductive oxide electrodes have negligible or no fatigue, long polarization retention and very small tendency to imprint. These properties make PZT-based heterostructure capacitors suitable for non-volatile ferroelectric memories. However, further work is necessary, particularly in relation to producing small size capacitors (≤ 1 μm), to determine if scaling down to dimensions compatible with high density memories will not introduce unexpected undesirable effects.

In conclusion, PLAD will remain a very useful technique for fundamental research, but extensive work is necessary to make PLAD a viable method for device fabrication on large area substrates.

7. ACKNOWLEDGEMENTS

The authors gratefully acknowledge the support of Dr. Jane Alexander (ARPA) through contracts at Bellcore (N00014-93-C-0127) and at NCSU and MCNC (N00014-93-1-0591). The authors have benefited from many discussions with numerous colleagues around the world and would like to acknowledge each and every one of them. The authors acknowledge specific on-going interactions with J.T.Evans, Jr. (Radiant Technologies), A.I. Kingon (NCSU), A.R. Krauss (Argonne National Laboratory), W.E.Warren and G.Pike (Sandia National Laboratory), J.Lee (Bellcore) and E.Goo (University of Southern California).

8. REFERENCES

1. J. CARRANO, C. SUDHAMA, J. LEE, A. TASCH, W.H. SHEPHERD, AND N. ABT, IEEE Trans. Ultrasonics, Ferroelectrics and Freq. Control, **38**, 690 (1991).
2. J.T. EVANS AND R. WOMACK, IEEE J. Solid State Circuits, **23**, 1171 (1988).
3. J.F. SCOTT AND C.A. PAZ DE ARAUJO, Science, **246**, 1400 (1989).
4. S. DEY AND R. ZULEEG, Ferroelectrics, **108**, 37 (1990).
5. Proceedings of the 3th International Symposium on Integrated Ferroelectrics, edited by C.A. Paz de Araujo (University of Colorado Press, Colorado Springs, 1991). Also, Proceedings of the 4th International Symposium on Integrated Ferroelectrics, edited by C.A. Paz de Araujo (University of Colorado Press, Colorado Springs, 1992).
6. Ferroelectric Thin Films I : MRS Symposia Proc. **200**, (Materials, Research Society, Pittsburgh, 1990), edited by E.R. Myers and A.I. Kingon . Also, Ferroelectric Thin Films II: MRS Symposia Proc. **243**, (MRS, Pittsburgh, 1991) edited by A.I. Kingon, E.R. Myers, and B. Tuttle
7. Multicomponent and Multilayered Thin Films for Advanced Microtechnologies: Techniques, Fundamentals and Devices, NATO/ASI Book Series E, **234**, edited by O. Auciello and J. Engemann (Kluwer Academic Publishers, The Netherlands, 1993).
8. D.B. CHRISEY AND G.K. HUBLER (EDS.), Pulsed Laser Deposition of Thin Films, John Wiley and Sons, Inc., New York, (1994).
9. O. AUCIELLO, A.I. KINGON, D.J. LICHTENWLANER, and A.R. KRAUSS, in Multicomponent and Multilayered Thin Films for Advanced Microtechnologies: Techniques, Fundamentals and Devices, NATO/ASI Book Series E, **234**, edited by O. Auciello and J. Engemann, (Kluwer Academic Publishers, The Netherlands, 1993) 151.
10. G.K. HUBLER (Ed.), various papers in Pulsed Laser Deposition: Special Issue of the MRS Bulletin (USA), **17**, (1992).
11. O. AUCIELLO, A.R. KRAUSS, AND K.D. GIFFORD, a chapter in this book.
12. R.E. LEUCHTNER AND K.S. GRABOWSKI, in Pulsed Laser Deposition of Thin Films, Chapter 20, edited by D.B. Chrisey and G.K. Hubler, pp.473 (John Wiley and Sons, Inc., New York, 1994).
13. G.J.M. DORMANS, M. DE KEIJSER, AND P.J. VAN VELDHOVEN, Ferroelectric Thin Films II: MRS Symposia Proc. **243**, (MRS, Pittsburgh, 1991) edited by A.I. Kingon, E.R. Myers, and B. Tuttle, PP. 203.

14. J.T. CHEUNG AND H. SANKUR, CRC Critical Rev. in Sol. State and Mat. Sci., **15**, 63 (1988).
15. L.C. CHEN, in Pulsed Laser Deposition of Thin Films, D.B. Chrisey and G.K. Hubler (eds.), John Wiley & Sons, Inc., New York, pp. 167 (1994).
16. C.L. CHAN AND J. MAZUMDER, J. Appl. Phys. **62**, 4579 (1987).
17. T. VENKATESAN, X.D. WU, A. INAM, AND J.B.WACHTMAN, Appl. Phys. Lett. **52**, 1193 (1988).
18. O. AUCIELLO, S. ATHAVALE, O.E. HANKINS, M. SITO, A.F. SCHREINER, AND N. BIUNNO, Appl. Phys. Lett. **53**, 72 (1988).
19. D.B. GEOHEGAN, Appl. Phys. Lett. **60**, 2732 (1992).
20. D. FRIED, G.P. RECK, T. KUSHIDA, AND E.W. ROTHE, J. Phys. D: Appl. Phys. **24**, 1065 (1991).
21. H.S. KIM AND H.S. KWOK, Appl. Phys. Lett. **61**, 2234 (1992).
22. S.R. FOLTYN, R.C. DYE, K.C. OTT, E. PETERSON, K.M. HUBBARD, W. HUTCHINSON, R.E. MEUNCHAUSEN, R.C. ESTLER, AND X.D. WU, Appl. Phys. Lett. **59**, 594 (1991).
23. J.S. HORWITZ, K.S. GRABOWSKI, D.B. CHRISEY, AND R.E. LEUCHTNER, Appl. Phys. Lett. **59**, 1565 (1991).
24. R.E. LEUCHTNER, J.S. HORWITZ, AND D.B. CHRISEY, in Ferroelectric Thin Films II, edited by A.I. Kingon, E.R. Myers, and B. Tuttle, Mater. Res. Soc. Proc. **243**, pp 525 (Pittsburgh PA 1992).
25. J.T. CHEUNG, P.E.D. MORGAN, AND R. NEURGAONKAR, in Proc. of the 4th Inter. Symp. on Integrated Ferroelectrics, p. 158, Monterey, CA (1992).
26. R. RAMESH, J. LEE, T. SANDS, V.G. KERAMIDAS, AND O. AUCIELLO, Appl. Phys. Lett. **64**, 2511 (1994).
27. R. DAT, D.J. LICHTENWALNER, O. AUCIELLO, AND A.I. KINGON, Appl. Phys. Lett. **64**, 2673 (1994).
28. D.K. FORK, F.A. PONCE, J.C. TRAMONTANA, AND T.H. GEBALLE, Appl. Phys. Lett. **58**, 2294 (1991).
29. O.AUCIELLO, J. EMERICK, J. DUARTE, AND A. ILLINGWORTH, J. Vac. Sci. Technol. **A11**, 267 (1993).
30. D.J. LICHTENWALNER, O. AUCIELLO, R. DAT, AND A.I.KINGON, J. Appl. Phys. **74**, 7497 (1993).
31. J.P. DICKINSON,S.C. LANGFORD, L.C. JENSEN, P.A. ESCBACH, L.R. PEDERSON, AND D.R. BAER, J. Appl. Phys. **68**, 1831 (1990).
32. S. KÜPER, J. BRANNON, AND K. BRANNON, Appl. Phys. A **56**, 43 (1993).
33. G.H. PETTIT AND R. SAUERBREY, Appl. Phys. A **56**, 51 (1993).

34. S.R. FOLTYN, R.E. MEUNCHAUSEN, R.C. DYE, X.D. WU, L. LUO, D.W. COOKE, AND R.C. TABER, Appl. Phys. Lett. **59**, 1374 (1991).
35. D.B. GEOHEGAN, Appl. Phys. Lett. **60**, 2732 (1992).
36. C.K. KWOK, D.P. VIJAY, S. B. DESU, N.R. PARIKH, AND E.A. HILL, Proceedings of the 4th Intern. Symp. on Integrated Ferroelectrics, C.A. Paz de Araujo (ed.), University of Colorado Press, Colorado Springs, pp. 412 (1992).
37. H.N. AL-SHAREEF, K.R. BELLUR, O. AUCIELLO, AND A.I. KINGON, Proc. 5th Intern. Symp. on Integrated Ferroelectrics, C.A. Paz de Araujo (ed.),University of Colorado Press, (in press, 1993)
38. O. AUCIELLO , K.D. GIFFORD, AND A.I. KINGON, Appl. Phys. Lett. **64**, 2873 (1994).
39. R.RAMESH, A. INAM, B. WILKENS, W.K. CHAN, D.L. HART, K. LUTHER, AND J.M. TARASCON, Science **252**, 944 (1991)
40. R.RAMESH, W.K. CHAN, B. WILKENS, H. GILCHRIST, T. SANDS, J.M. TARASCON, V.G. KERAMIDAS, D.K. FORK, J. LEE, AND A. SAFARI, Appl. Phys. Lett. **61**, 1537 (1992).
41. T. MIHARA, H. WATANABE, C.A. PAZ DE ARAUJO, J. CUCHIARO, M. SCOTT, L.D. MCMILLAN, Proc. 4th Intern. Symp. on Integrated Ferroelectrics, C.A. Paz de Araujo (ed.), University of Colorado Press, Colorado Springs, pp. 137 (1992).
42. S.K.DEY AND R. ZULEEG, Ferroelectrics **112**, 309 (1990). Ferroelectric Tester Operating Manual, v **2.0** (1990).
43. R. DAT, D.J. LICHTENWALNER, O. AUCIELLO, AND A.I. KINGON, Appl. Phys. Lett. (submitted, 1994).
44. J.M. BENEDETTO, R.A. MOORE, AND F.B. MCLEAN, J. Appl. Phys.**75**, 460 (1994).
45. T. MIHARA, H. WATANABE, AND C.A. PAZ DE ARAUJO, Jpn. J. Appl. Phys., **32**, 4168 (1993).
46. R. DAT, D.J. LICHTENWALNER, O. AUCIELLO, AND A.I. KINGON, Integrated Ferroelectric(in press, 1994).
47. J. BULLINGTON AND J. EVANS, RT-66A Standardized

AUTHOR INDEX

Al-Shareef, H. N.	193	(Chapter 7)
Araujo, C., Paz de	1	(Chapter 1)
Auciello, O.	393	(Chapter 10)
	525	(Chapter 13)
Bell, J. M.	93	(Chapter 4)
Dat, R.	525	(Chapter 13)
Dey, S. K.	329	(Chapter 9)
Dormans, G. J. M.	485	(Chapter 12)
Fox, G. R.	427	(Chapter 11)
Gifford, K. D.	393	(Chapter 10)
Hu, H.	427	(Chapter 11)
Ishibashi, Y.	135	(Chapter 5)
Johnston, G. R.	93	(Chapter 4)
Keijser, M., de	485	(Chapter 12)
Kingon, A. I.	193	(Chapter 7)
Knight, P. C.	93	(Chapter 4)
Krauss, A. R.	393	(Chapter 10
Krupanidhi, S. B.	427	(Chapter 11)
Larsen, P. K.	485	(Chapter 12)
Ramesh, R.	525	(Chapter 13)
Scott, J. F.	1	(Chapter 1
Shur, V. Ya	153	(Chapter 6)
Smyth, D. M.	47	(Chapter 3)
Taylor, G. W.	vii	(Foreword)
	1	(Chapter 1)
Tilley, D. R.	11	(Chapter 2
Turova, N. Ya	233	(Chapter 8)
Yanovskaya, M. I.	233	(Chapter 8)
Waser, R.	47	(Chapter 3)

SUBJECT INDEX

A

ablation plume, 535
ablation rate, 530, 532
  angular distribution, 534, 536, 537
  energy dependence, 532, 537
  magnesium oxide, 531
  lanthanum strontiuim cobalt oxide, 531
  lead zirconium titanate, 532
  pressure dependence, 534, 537
acceptor, 51
activation, 157
activation energy, 488
activation field, 144
adaptive transitor, 1
adatom mobility, 441, 463
adsorption, 487
$Al(OEt)_3$, 244, 280
$Al(OR)_3$, 236, 243, 257, 263, 274, 275, 284
aliovalent impurities, 50
alkoholysis, 246, 247-250, 252, 288, 292
alkoxide, 233-311, 495-499
alkyl, 495-499
allyl, 495-498
analog memory, 1
anion, 47
applied field, 150
Arrhenius, 488
Auger Electron Spectroscopy (AES), 120, 122, 347

B

$Ba(OC_2H_4OMe)_2$, 259
$Ba(OEt)_2$, 242, 284, 286, 296, 298, 301, 302, 308

backscattered ions, 463
Barium Strontium Titanate
(BST), 3, 5
Barium Titanate ($BaTiO_3$), 1,
99, 102, 111, 113, 119, 122,
144, 158, 240, 286, 296, 304,
496, 499, 500, 506
base metal electrodes, 59
Beer's Law, 13
Bessel functions, 32
$Bi(OC_2,H_4OMe)_3$, 251, 260,
272
$Bi(OEt)_3$, 260
bimetallic alkoxides, 236, 238,
282, 284, 291, 296
binding energies, 432
Bismuth Titanate ($Bi_4Ti_3O_{12}$),
2, 496, 499, 500, 506
breakdown field, 79
breakdown mechanism, 6
bubbler, 490-493, 497, 499
bulk ferroelectric, 1

C

Cadmium Selenate (CdSe), 1
Cadmium Sulphide (CdS), 1
capacitance, 98, 111
capacitance-voltage
measurements, 116, 117,
127
depletion capacitance, 97,
107, 118
space-charge capacitance,
96, 111, 116
casting, 2
catastrophes (see also
geometrical transformations),
154, 181-184
cation comosition, 458
characteristic length, 141
characteristic time, 141, 156,
184, 187
characterization of growth
processes, 413
in situ, real-time, 413
ion scattering spectroscopy,
413, 415, 416, 418
time-of-flight, 413
chelates, 495, 498
Chemical vapor deposition
(CVD), 330, 362, 380, 396,
485
plasma-enhanced, 362, 370
radio frequency (rf) plasma,
370-374
electron cyclotron resonance
(ECR) plasma, 374-378
simulation, 380-383
Clausius-Clapeyron, 495
cluster boundaries, 451
CMOS silicon chip, 2
coherence length, 20
cold spot, 492, 493
computer simulation, 154, 183,
185, 187
concentration polarization, 84
conductivity
ionic, 61
field enhanced, 76
electronic, 52
cone formation, 436
critical thickness, 15, 25f, 39,
41f

crystallization, 438, 441, 442, 450, 464
$Cu(OC_2H_4OMe)_2$, 245, 260
$Cu(OEt)_2$, 242, 244, 258, 284, 292
$Cu(OMe)_2$, 245, 260
$Cu(OR)_2$, 244, 245, 260, 261
Curie-von Schweidler law, 74
CVD, 7
cylinders, ferroelectric, 32

D

DC resistivisty, 459
Debye relaxation, 67
defect chemistry, 6
defect complex, 64
delay time, 159, 171, 176
depletion layer, 65, 97, 116
depolarization, 154, 156, 172, 173, 176-178, 180
depolarization effects, 12, 14, 32ff, 39ff
deposition
- parameter, 344, 372, 378-379
- rate, 374

deposition processes, 527
deposition rate, 431, 432, 433, 435, 438, 486, 488, 489, 491, 499, 501, 503, 504, 530, 538
- energy dependence, 533
- magnesium oxide, 533
- lanthanum strontium cobalt oxide, 533
- lead zirconium titanate, 533, 537

desorption, 487
dielectric, 2
dielectric constant, 110, 111, 112, 114, 115, 117, 127, 442, 461, 468
- dispersion, 108, 111
- thickness dependence of, 109, 110, 111, 112
- two-layer capacitor model for, 108, 109, 110, 112, 113

dielectric loss, 468, 477
diffusion, 94, 98, 100, 103, 104, 108, 115, 120, 125, 126, 128, 487, 489, 494
- barrier, 120
- cation diffusion, 99, 106, 107
- constant, coefficient, 99, 100, 103, 104, 106
- equation, 99, 100, 108
- grain boundary, 99, 102, 106, 115
- kinetics, 126
- oxygen, 99, 101, 106, 107

dipole field, 33
DOBAMBC, 142
domain, 136
- backswitching, 156, 158, 173, 177
- dynamics, 155, 158, 159, 160, 470
- forward growth, 160, 170, 173
- nonthrough, 166, 170, 171, 173, 179, 180

residual, 176
residual/remnant, 156, 160, 164, 166, 170, 171, 176, 187
shape, 156, 161, 164, 166, 173, 182, 185
charged, 173
sideways motion, 155, 15ˎ 162, 164, 165, 166, 171, 183
domain reversal, 469
domain structure, 21, 43
domain walls, 6, 146
donor, 51
dopant compensation
acceptors, 358-360
donors, 358-360
effects on leakage current, 361
DRAM, 5, 195, 196, 393
driving force, 154, 168
DRT model, 74

E

EEPROMs, 4
effective-field theory, 21
electric field
weak, 162, 165-170, 172, 177
local, 154-156, 188
strong, 156, 162, 164, 166, 168, 171, 172, 179
superstrong, 165-167, 176, 179, 180
electrical properties, 549
polarization hysteresis, 550-552, 555, 557, 559
fatigue, 551, 553, 555
imprint, 556-558, 560, 561
of ferroelectric device, 94
of ferroelectric film, 93, 101, 107
electrocolouration, 83
electrode interfaces, 70
electrode materials, 7
aluminum, 98, 107, 111
gallium arsenide, 98, 109, 121
gold, 95, 98, 110, 117, 119, 120, 121
nickel, electroless nickel, 95, 98, 99, 101, 102
platinised silicon, 110, 117, 118, 120, 121
platinum, 94, 95, 98, 102, 110, 120, 121, 127
RuO_2, 95, 98, 107
silicon, 95, 98, 100, 101, 110, 118, 125
SnO_2, $Sn:In_2O_3$, 95, 106, 107, 108, 111, 121
stainless steel, 95
titanium, 95, 98, 100, 102, 107, 118, 119, 123, 124, 126, 127
transition metal silicides, 95
$YBa_2Cu_3O_7$, 95, 106, 107, 121
electrodes, 32, 34ff
electrodes, perfectly conducting, 35ff

electron beam evaporation, 395
electron cyclotron resonance, 479
electrophoresis, 2
electrostatic energy, 33
endurance, 2, 4
epitaxy, 485, 486, 500, 503, 506, 507, 509, 510, 519
equivalent network, 66
erase-write cycles, 4
etching, 373, 381
Euler-Lagrange equation, 12, 17, 20, 28, 30, 36, 145
evaporation, 2, 450
excimer laser, 528
experimental technique, 154, 157, 158, 163
  electron microscopy, 153, 158
  nematic liquid crystal, 158
  optical methods, 158-163
  pyroelectric probe, 158
explosive crystallization, 443
exponential law, 144

F

fatigue, 218, 411
Fermi Level, 96, 98
Ferroelectric Memories (FeRAM, see also FRAM), 4, 5
Ferroelectric Microwave Monolithics (Fe-MMICS), 4, 7
ferroelectric
  capacitor, 156, 172, 173, 174, 177
Ferroelectric - Semiconductor, 5
ferroelectric films, 393, 397, 407, 409
ferroelectric liquid crystal, 142
Ferroelectric Thin Films, 1, 2, 5, 6, 427
ferroelectrics, 486, 487, 496, 498-500, 504, 506, 510, 512, 514 - 519
film deposition
  laser ablation, 121
  sol-gel, 95, 102, 106, 109, 123
  spin-casting, 121
  sputter deposition, 95, 106, 121, 123
FRAM (see also FeRAM), 195, 196, 393

G

Gadolinium Molybdate [$Gd_2(MoO_4)_2$], 162-166, 177, 178
Gallium Arsenate (GaAs), 3, 7
gas system, 490, 492-495, 499
geometrical transformation, 154
grain boundaries, 65
growth processes, 402, 409-413

H

halides, 485, 496
heater tape, 493
heterojunction, 94, 98

heterostructure capacitors, 542, 544, 545, 547, 548, 550-555, 560
heterostructures, 397, 409, 415, 418
$Hf(OC_2H_4OMe)_4$, 260
high density, 2
Hillocks, 197
hole trapping, 58
holes, 57
homogeneous case, 139
hot spot, 492, 493
hydrolysis, 2, 238-241, 245, 247, 257, 258, 268, 282, 284, 288-290, 296-299, 303, 305, 307, 310, 311
hysteresis loop (curve), 145, 150

I

I-V curves, 224
incident ion, 433, 466
inhomogeneous case, 140
injection model, 74
integral data, 154
integration, 2
integrated ferroelectric devices, 3
interfacial
  buffer layer, 345-346
  reactions, 346-351
interfacial interactions, 6
internal electrical field, 156, 170, 172, 175, 178
Internet, 5
interstitial, 62
interstitial defects, 49
ion assisted deposition, 460
ion beam, 7
ion beam sputter-deposition, 402
  electrical characterization, 410-412
  lead zirconium titanate (PZT), 409-412
  microstructure, 406
  phenomena, 402-405
  PZT capacitors, 407
  systems, 402-404
  TEM studies, 409
  XRD studies, 408, 409
ion channeling, 465
ion fraction, 448
ion/atom ratio, 430, 474, 476
Ising model in a transverse field (IMTF), 12, 16, 17ff, 29

K

KDP, 2, 16
kinetics, 149, 488, 489, 501
KNO_3, 13
Kolmogorov-Avrami model (theory), 7, 135, 181
Kröger-Vink, 56

L

$La(OPr\text{-}i)_3$, 259, 313
Landau free energy expansion, 17, 25f, 29f, 35f, 40
laser ablation, 427

laser ablation-deposition, 394, 395
latent nucleus, 142
lattice model, 135, 136
layered perovskites, 3, 4, 6
Lead Germanate, 161, 163
Lead Lanthanum Zirconate Titanate (PLZT), 427, 496, 500, 503, 504, 514
Lead Scandium Tantalate, 499, 500
Lead Titanate (PT), 310, 311, 499-506, 508-510, 512
Lead Zirconate Titanate (PZT), 2, 3, 298, 310, 311, 487, 494, 500, 502-507, 510-515, 517-519
leakage, 472
leakage current, 6, 221, 223, 224, 412
LiOR, 242, 251, 253, 261, 262, 284, 305, 307
liquid-source CVD, 493
Lithium Niobate ($LiNbO_3$), 95, 101, 284, 305, 307
lithium tantalate, 307
$Ln(OPr-i)_3$, 259
load resistance, 471
local potential, 145
localized plasma, 460
low energy ion bombardment, 428
LSCO, 214, 215, 218, 219

M

magnetron sputtering, 427
markets, 2, 3, 5
mass flow controller, 491
materials integration, 541
  bismuth titanate, 541
  lead lanthanum zirconium titanate, 541
  lanthanum strontium cobalt oxide, 541
mathematical, 187
mathematical treatment, 153, 154, 180, 185
Maxwell-Wagner polarization, 65
mean free path, 431
mechanical properties, 2
memory, 499, 510, 516-519
metal precursors
  alkoxides, 334, 364, 367
  alkyls, 7, 364, 366
  B-diketonates, 334, 367
  carboxylates, 334
  combination of ligands, 368-369
  cyclopentadienyls, 364, 366
  liquid delivery, 390
  organic salts, 334, 364
$Mg(OC_2H_4OMe)_2$, 259
$Mg(OMe)_2$, 243, 244, 258, 259
MIBERS, 427, 438, 441, 445, 446, 461, 478
microcontrollers, 5
microstructure, 206, 208, 212, 213, 428, 449, 450, 451, 458, 542

lead lanthanum zirconiuim titanate, 543
microwave circuits, 3, 4
MOCVD, 5, 7
MOD, 7
model objects, 154, 161, 185
molecular beam epitaxi, 396
$MoO(OEt)_4$, 260
$MoO_2(OC_2H_4OMe)_2$, 260, 284
multi ion beam reactive sputtering, 428
multicomponent oxides, 427

N

$NaNO_2$ microcrystals, 14
nanocrystals, 455, 456
NaOR, 243, 250-252, 254, 288
$Nb(OBu\text{-}n)_5$, 260, 309
$Nb(OC_2H_4OMe)_5$, 260
$Nb(OPr\text{-}i)_5$, 260
$(NH_2CH_2COOH)_3.H_2SO_4$ (TGS), 141
negative ions, 427
neodymium-YAG laser, 528
$Ni(OR)_2$, 245, 251, 271
non volatile memory, 1, 2
non-ohmic contact, 93, 97, 116
nucleation, 6, 7, 155, 184, 456, 463
  correlated, 156, 180
  process, 168
nucleation probability, 138
nucleation-growth mechanism, 135
nucleus, 136

O

ohmic contact, 93, 95, 97, 99, 116, 118
ohmic regime, 77
order-prameter profiles, 21f, 27
organometallic, 485, 486, 595, 501
oxoalkoxides, 237, 257, 263, 280, 282, 290-294
oxygen, 486, 496, 497, 500-503
oxygen partial pressure, 52
oxygen sensors, 72
oxygen species, 430, 479
oxygen vacancies, 54
ozone, 502

P

p^4-model, 147
parasitic reactions, 489
partial pressure, 491, 492
$Pb(C_2H_5)_4$, 496, 501-504, 506, 512
$Pb(^iOC_2H_5)_2$, 500, 503
$Pb(OC_2H_4OMe)_2$, 259, 292
$Pb(OPr\text{-}i)_2$, 259, 292
PbO, 452, 456
$PbTiO_3$ particles, 14, 42
PCS - band phones, 3, 5
perovskite, 440, 442, 447
perovskites - layered, 2, 3, 6
phase diagram, 14, 23
phase transition
  first order, 155
physical vapor deposition, 427
plasma, 500

plasma sputter deposition, 397-401
- barium titanate, 397
- deposition conditions, 400
- large area films, 401
- lead titanate, 397
- lead lanthanum zirconium titanate, 399
- lead zirconium titanate, 397, 398

PLT, 427, 445, 447, 450, 455, 456, 478
point defects, 358-361
Poissou's equation, 33
polarization, 95, 99, 111, 444, 466
- reversal, 7, 135, 153
- spontaneous, 154

polarizing microscope, 143
Poole-Frenkel effect, 78
power law, 144
precursor synthesis, 336, 339
- alcoholation, 340
- complexation, 342
- composition, 341-342
- condensation, 343
- dehydration, 340-341
- esterification, 341

presputtering, 430
primary ion source, 429
process,
- fast reversal, 154, 155

Pt/Ti, 197, 207, 217, 219, 221, 223, 224, 225
Pt/TiO_2, 202
pulsed laser ablation, 7, 8
pulsed laser ablation deposition, 525, 527
- lead titanate, 529
- lead lanthanium zirconium titanate, 531
- lead zirconium titanate, 529, 530

pyrochlore, 440, 447
pyrochlore vs. perovskite formation, 345, 350, 371
pyrolytic behavior, 337
PZT, 93-95, 101, 106, 107, 117-121, 124, 427, 444, 461, 463
- crystallization, 207, 209, 211
- microstructure, 208, 212, 213
- properties, 217, 218, 221, 222, 223, 224, 225

R

rapid thermal annealing (RTA), 339, 345
rare-earth molybdates, 157
rate equation, 149
reactor cell, 490, 492, 494
real-time, 153
reduction, 53
reentrant transition, 13
resistance degradation, 80
resputtering, 462
retention, 444
RF diode sputtering, 427
Rochelle salt, 160
RTA, 441

Ruddlesdon-Popper-phases, 71
RuO_2, 209, 210, 218, 222, 223, 224, 225
Rutherford Backscattering Spectrometry, 122, 123, 125, 126, 127, 128

S

Sc(OPr-i)$_3$, 259
Scanning Auger Microscopy, 122, 123, 124, 125, 126, 127, 128
Scanning electron microscopy (SEM), 351, 352
Schottky barrier, 69
Schottky disorder, 49
Schottky equilibrium, 63
screening, 173
    bulk, 156
    external, 156
    process, 156
Secondary ion bombardment, 430
Secondary ion mass spectroscopy (SIMS), 346
SEM images, 531
    magnesium oxide target, 531
    lanthanum strontium cobalt oxide target, 531
    lead zirconium titanate, 531
    lead zirconium titanate capacitors, 542
semiconductor integrated circuits, 1
semiconductor-thin film, 1, 2
sideways velocity (see domain: walls; sideways motion), 158
sideways domain growth, 145
sol gel, 7, 234, 235, 238, 239, 241, 282, 298, 304, 311, 329, 332, 333, 379
    acid/base hyrolysis, 334
    densification with excess PbO, 357
    gel structure evolution, 337-339
    microstructure evolution, 351, 354, 356, 380
    sol synthesis, 335, 339, 342
solid phase epitaxy, 358
solid state chemistry, 358, 369, 374
solid-source CVD, 493
space charge limiting currents, 78
speciation, 448
Spectroscopy and analytical chemistry, 335, 337, 341, 343
speed, 4
spontaneous polarization, 139
sputtered flux, 439
sputtering, 2, 7
sputtering yield, 433
$Sr(OC_2H_4OMe)_2$, 259
$SrRuO_3$, 209, 214, 218
step coverage, 362, 378, 380, 486, 589, 502, 515
stress, 198, 199

Strontium Titanate (STO), 2, 3, 13, 101, 104, 121, 240, 297, 300, 301-303, 496, 499, 500, 506, 508-510
Strontium Barium Niobate, 499, 500
superconductor - ferroelectric, 5
superconductors, 486, 498
superlattices, 11, 28ff
surface extrapolation length, 20
surface morphology, 436
surface ordered state, 18, 20, 28
susceptor, 494
switching, 6, 153, 470
    fast, 157
    process, 154
    superfast, 159
    time, 157
switching current, 139
switching time, 144, 150
symmetry group invariants, 17
SZM, 446

T

$Ta(^{i}OC_2H_5)_4$, 500
$Ta(OBu\text{-}n)_5$, 260
$Ta(OPr\text{-}i)_5$, 260
Tellerium (Te), 1
texture, 457
TGS, 13, 158
thermionic emission, 78
Thermodynamic phase stability diagram, 363, 365
thin film, 72
thin film semiconductor, 1, 2
thin plate, 136
Thomas-Fermi dielectric function, 41
Thomas-Fermi screening, 34
threshold electric field, 157, 158
Ti interdiffusion, 201, 202, 203, 204
$Ti(^{i}OC_2H_5)_4$, 501
$Ti(^{i}OC_3H_7)_4$, 499-504, 506, 512
$Ti(OBu\text{-}n)_4$, 259, 309, 313
$Ti(OC_2H_4OMe)_4$, 259
$Ti(OPr\text{-}i)_4$, 259, 302, 303, 313
$Ti(OR)_4$, 250, 252, 256, 257, 263, 364, 269, 271, 284, 290-303, 312, 313
$TiCl_4$, 488
time dependent dielectric breakdown, 472, 473
tin oxide, 1
TIOEt, 246, 270
titanates
    acceptor doped, 51
    donor doped, 62
    undoped, 52
transient current, 153
transistors adaptive, 1
Transmission electron microscopy (TEM), 348, 353
transport theory, 84
tunneling, 78

V

vacuum deposition, 2
vapor pressure, 491, 492, 495, 497, 499
varistor regime, 77
viscosity, 146
$VO(OR)_3$, 246
Vollmann, 92

W

$WO(OC_2H_4OMe)_4$, 260
$WO(OEt)_4$, 260, 284, 290
write cycles, 4

X

X-ray difraction, 120, 122
XRD patterns, 545-548

Y

$Y(OPr\text{-}i)_3$, 259
YBCO, 214, 215, 218, 219

Z

$Zr(dpm)_4$, 502, 503
$Zr(OC_2H_4OMe)_4$, 254
$Zr(^iOC_3H_7)_4$, 502
$Zr(OPr\text{-}i)_4$, 256, 259, 275, 276
$Zr(OR)_4$, 233, 250, 252, 253, 259, 276, 284, 296, 312, 313

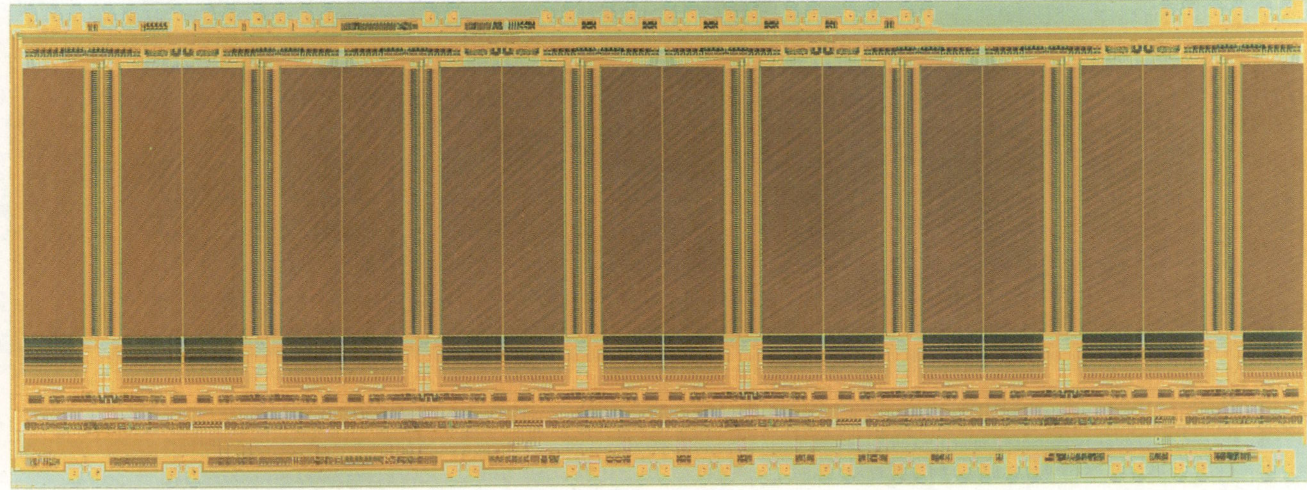

Figure 1. One megabit ferroelectric memory by NEC[9]
(See Chapter 1, page 4, C. Paz de Araujo, J. F. Scott and G. W. Taylor)